PERSPECTIVES IN INFORMATION SCIENCE

NATO ADVANCED STUDY INSTITUTES SERIES

Proceedings of the Advanced Study Institute Programme, which aims at the dissemination of advanced knowledge and the formation of contacts among scientists from different countries.

The series is published by an international board of publishers in conjunction with NATO Scientific Affairs Division

A	Life Sciences	Plenum Publishing Corporation
B	Physics	London and New York
C	Mathematical and Physical Sciences	D. Reidel Publishing Company Dordrecht and Boston
D	Behavioural and Social Sciences	Sijthoff International Publishing Company Leyden, The Neth. and Reading, Mass., USA
E	Applied Sciences	Noordhoff International Publishing Leyden, The Neth. and Reading, Mass., USA

Series E: Applied Science - No. 10

PERSPECTIVES IN INFORMATION SCIENCE

edited by

ANTHONY DEBONS

Professor of Information Science
University of Pittsburgh

and

WILLIAM J. CAMERON

Dean, School of Library and
Information Science
University of Western Ontario

SPRINGER SCIENCE+BUSINESS MEDIA, B.V. 1975

Proceedings of the NATO Advanced Study Institute
on Perspectives in Information Science
held in Aberystwyth, Wales, UK, August 13 - 24, 1973

ISBN 978-94-011-7761-0 ISBN 978-94-011-7759-7 (eBook)
DOI 10.1007/978-94-011-7759-7

Originally published by Noordhoff International Publishing in 1975
Softcover reprint of the hardcover 1st edition 1975

In a cloud of unknowing
but with the hope to know
what it means to know

PREFACE

What is Information Science? What does a person do as an Information Scientist? The yearning for replies to these queries so far is unabated. Meanwhile those who have dedicated their interest to information ponder the possible answers and issues. It is not surprising to realize that man has always been interested in information - in producing it, in using it, in preserving it. But the past two or three decades have added new dimensions and new emphases in information studies. With the increase in the production and availability of knowledge, the need to understand more fully how to use it effectively has gathered greater momentum. During the same period the nature and incidence of crisis situations - whether public or private - has demanded greater dependence on timely delivery of data to make decisions and undertake action. Moreover, there has been a harvest of vast and significant technological developments to accompany these changes in social and individual states. Information has gathered and continues to gather growing importance in the lives of individuals as well as nations.

Despite the importance of information to human welfare, information as a concept remains tangled in loosely defined terminology - in jargon and speculative propositions about its nature and purpose. Information environments, or systems, as they are usually called, are being designed and created without any understanding of how to estimate their worth. These systems, creating a web of human and technological links, underlie a current debate on privacy and challenge the focality of the human decision process. Yet these very systems are creations without a soul - without a clear definition of goals or reasons, except that they presumably save time and effort and augment the human intellect.

Furthermore, the commitment to understanding is far more serious if correlated to vocation. During the past decade the number of U.S. educational institutions providing educational programs on Information Science of one variety or another has almost doubled. It is safe to postulate that this number will continue to grow in the coming years. If indeed we are to educate for information science, in what areas are we to educate? Again, what is the domain of Information Science and what competencies do we need to develop to ensure that Information Scientists do what they are required to do?

The Advanced Study Institute supported by NATO and held in August 1972 at Seven Springs, Champion, Pa., was an initial attempt to establish the boundaries of what may be defined as a Science of Information. As was recognized then and remains clearly in perspective now, a science essentially consists of three basic elements - a body of law, theories, and constructs, a statement of the focus of interest in the human condition, and a corpus of knowledge which provides the basis of an educational philosophy. The present Advanced Study Institute held in August 1973 in Aberystwyth, Wales, had the purpose of reviewing the insights gained in Pennsylvania and extending these insights to further formalize the three germane interests of an information scientist - an understanding of that experience we call information, the practical aspects that center around this understanding, and establishment of the requirements for educating a corps of competent individuals to pursue the related issues and problems.

The organization of the present Institute is similar to that established in Seven Springs. The issues remain, and the lectures/ discussion focus on these issues, namely, the nature and function of information, the state of technology and whether or not it is meeting human needs, the impact of such technology on society and culture as well as the individual, and last but not least, the professional responsibilities for education and research. Down deep we are hoping that this organization will provide the first approximation to a useful statement about what constitutes the technical substance of a science of information, the kinds of questions that information scientists attempt to answer, and what such professionals are concerned with as well as what they do.

The main objective of the Institute was tutorial. The interests of the Scientific Affairs Division of NATO are directed towards education. More specifically, the mission of the Institute was instructional and this mission was initiated by the tutorials given at the beginning of the Institute. The purpose of the tutorials was to bring the audience up to date as to our state of understanding regarding the issues and problems facing Information Science. In this connection, the individuals presenting the tutorials were asked to review carefully the proceedings of the

1972 NATO Institute on Information Science and to further the findings by summarizing whatever important work had been done by scientists and scholars that were directly applicable to the mission of this Institute.

The hope that we could proceed just one step beyond the point at which we ended the '72 Institute was part of the philosophy underlying the organization of the present program. One device for furthering the development of concepts was the inclusion of four working groups (one group for each of the specific areas of the program discussed previously). Each working group chairman was assigned a colleague as well as a graduate assistant prior to the convening of the Institute so that material could be prepared that was relevant to the group's deliberations. Interaction among the chairman and his associates continued over several months prior to the convening of the Institute. The working groups were charged with extending the arguments and discourses presented by the distinguished lecturers and to formulate a statement which would reflect the current status of the issues and problems entertained by that working group.

The Institute was particularly fortunate in its recruitment of distinguished scientists. Of great significance, perhaps, is the high level of commitment that these distinguished scientists applied to the Institute. Without exception, their involvement was outstanding, particularly when one realizes the overwhelming demands our culture places on individuals of this stature. As the result of this commitment, the Institute reaped an unusual privilege in having a number of our leading scientists summarize their impressions of the conclusions at the terminal session of the Institute. These have been included in a separate section of the proceedings.

Indeed, these proceedings contain far-ranging discourses, for, although the organization of the Institute was directed to fall within specific boundaries, no limits were imposed and in many cases there were diversions. These diversions proved highly exciting and beneficial, although they at times did violence to program structure. In this connection, the organization of the presentations as they are contained in the present volume do not reflect the exact sequence of the presentations but rather the logic applied by the editors. No apologies are expressed for the violence that such a logic may present to the perceptions of readers who may in fact regard the structure and process of argument included in the summaries, or in the ordering of the papers, to be an exercise in propaganda. The editors are willing to "stick their neck out", as it were, in the hope that the material and its ordering would receive the critical attention we feel the amount of human involvement it represents deserves.

The collection of the volume in its present form is indebted to many: in particular, to Ishbel MacGregor of the University of Western Ontario for the preparation of the final manuscript; to the Office of Information Science Dissemination, National Science Foundation, for a grant in aid; to Howard Vandersluis and Klaus Otten for assistance in writing the summaries; and to Charlene Robertson for editing work on them. The editors are particularly grateful to those individuals who aided in the organization and administration of the Institute, namely, Joan Montgomery of Pittsburgh, Frank Gibbons, University of Liverpool, Frank Hogg, Principal, College of Librarianship, Aberystwyth, Wales, as well as many others of the same institution who helped in the administration of the Institute. Acknowledgment is also due to the U.S. Army Research Division, Durham, North Carolina for providing funds to enable some of the speakers to attend and participate in the Institute.

TABLE OF CONTENTS

OPENING REMARKS

OPENING REMARKS

Anthony Debons

University of Pittsburgh, Pittsburgh, U.S.A.

Ladies and gentlemen, it gives me great pleasure, on behalf of the North Atlantic Treaty Organization to welcome you to the second NATO Advanced Study Institute in Information Science.

Before I introduce our distinguished guests, I would like to bring the context of the institute into perspective and indicate the nature of our objectives.

The mission of this institute is instructional. The task therefore is to bring forth to your attention the state of affairs concerning information science, both from the scientific and technological points of view.

Because Information Science is a new science in terms of what Kuhn would call "paradigmic science", the institute has an additional task of attempting to define, delineate issues in a manner that could provide advances in the present ways of thinking, to suggest hypotheses. All of these efforts would be in the direction of development in theory formulation perhaps, but certainly in hypothesis generation.

It is my strong belief that the present Institute can move us directly in what I consider the greatest need in Information Science, that is to find interrelationship between concepts, between theories - bridges, if you will, that will tie formulations given birth in the various formal disciplines, to the current needs of a science of information. May I add that some measure of success was achieved in this direction during the first NATO Advanced Study Institute in Information Science that was held in Seven Springs, Pennsylvania last year. I am sure that we can achieve greater things in this direction in the present institute.

A word or two about the organization of the Institute. The instructional aspects of the Institute are centered in four major tutorials that will be presented. The tutorials will attempt to summarize the context and the conclusions arrived at the Seven Springs institute. The tutorials will deal with the four areas of concern:

The nature of information.
Information technology and its ability to meet human needs.
The impact of information technology on man.
Professional aspects.

The main lectures of the institute will concern these areas and will attempt to bring us all up to date as well as present new food for thought.

The major developmental aspects of the institute are contained in the working groups. All of the participants of this institute have been assigned to working groups, headed by a group leader. The group leaders have been working behind scenes for several months. Early in the formulation of the institute the working group leaders were given technical guidelines concerning the issues facing the respective areas. During this institute working group participants are asked to develop cohesive positions regarding the respective technical and theoretical issues. A written report is expected to emerge that will be presented during the second week, possibly the last day of the Institute.

At Seven Springs some attempt was made to evaluate the Institute along several dimensions. This evaluative effort not only provided feedback on the conduct of the institute but also contributed to generating much room for thought about the science and the field itself. With this in mind, we will attempt again to evaluate the activities that we will undertake here in Aberystwyth and I would like to request your cooperation in this endeavor.

A word or two about the main lecture sessions and the discussions that follow the lectures. Learning from last year's evaluation, I believe it would be to the advantage of all of us if the question and answer period following the presentation be carefully managed. The discussions following the presentation should be directed towards clarification of points presented by the speaker, not the elucidation of new points. The latter should be entertained during the working group sessions. I have asked the speakers not to read their papers, rather to abstract the major points. This should provide opportunity for discussion centered around the issues advanced by the speaker. If we restrict ourselves to these issues, much can be achieved.

A comment about the program is in order. I am sure that all of you will share with Dr. Cameron and I the desire to achieve the maximum output from the program that has been established. Now Bill and I favor a degree of flexibility on this score consistent with some semblance of order. Consequently, we may experience some changes in the scheduling, in the order of presentations, etc. We ask your indulgence in this respect.

Because every program has some theoretical perspective that provides the basis for its structure, I believe that some mention should be made concerning the main thoughts that underlie our present efforts here.

In Seven Springs, I pleaded for clarification of boundaries of concepts. This plea was essentially based on what I presently perceived and experienced as confusion concerning terminology among those of us who were attempting to develop or establish a science of information. This plea is extended to the present institute but perhaps with less fervor. I am pleased to realize that the distinctions of information as a human process, information as a commodity have received some consideration and acceptance among us. This will enable greater clarity when we discuss our present state of understanding concerning these two aspects of our field.

It has been proposed, and I reaffirm the proposal that was presented in Seven Springs that Information Science must deal with three aspects of Information, namely: *generation, utilization* and *transfer*. I believe the task of a science of information is not primarily concerned with principles underlying the *separate* aspects (for these can certainly be dealt with by individual disciplines) but rather, the concern of a science of information is to discover the *relationships* between these aspects. As I have indicated previously, it is my belief that theories and laws now exist that could explicate part of the problems that are of concern to us. We need to go a step further and determine the interrelationship between these theories. My colleagues advise me that bridges also exist. I am pleased to learn this and perhaps we should explicate these bridges in support of the instructional role of this institute and build new ones in support of our developmental tasks.

I do not wish to clutter your thinking further on matters that can be best handled in the program, through the lectures or working group sessions.

Let us proceed with the program. But one more word - and this a word of appreciation. I want to thank all of you for your interest in this institute. To our sponsor, the North Atlantic Treaty Organization, to our co-sponsors the U.S. Army Research Divisions, Liverpool Polytechnic, University of Western Ontario and of course,

College of Librarianship here in Aberystwyth.

At this time, I would like to introduce Mr. Morgan, Member of Parliament for the County of Cardiganshire.

OPENING REMARKS

Dafydd E. Morgan

Member of Parliament for Cardiganshire, Wales.

Mr. Moderator, members of the NATO Advanced Institute for Information Science, I am extremely grateful indeed for this opportunity of giving you a warm and hearty welcome to the County of Cardigan and to the country and nation of Wales. This is rather an early hour for politicians to operate, although I have experienced more early hours than this in the House of Commons in perhaps more turbulant circumstances. Some of you may have heard the tale of that London police officer who was on points duty in Leicester Square some months ago, in the early hours of the morning, who saw a shadowy figure flit from doorway to doorway, and gave chase and eventually brought his quarry to ground to find he was a person who had escaped from a totalitarian country, and all that this person would say to the police officer was "political asylum, political asylum", at which the bobby turned to him and said "There you are, mate, behind you is the finest political asylum in the world".

Now, as far as the County of Cardigan is concerned, it is an ancient community, about 15 centuries old, shaped on the anvil of the centuries, a place which still has a deep sense of separate identity and community. And no doubt you have heard a great deal about the characteristics of the strange people who live here, our wariness, our caginess, our appreciation of our own material advantage brought about by the conditions of life imposed upon us for so many centuries. In part, no doubt, that is true, but you will also find us people with a warm and friendly welcome at all times.

I am very glad that you are coming to this particular part of Wales. To many people outside, Wales is a land of narrow, cold valleys and the dark satanic mills of steel and tinplate, and few people realise the delicate beauty and peace of this countryside here in mid-Wales. We are a resourceful people and no doubt you have come across Welshmen in your different walks of life all over the world. The nation of Wales is very much an enigma; we are an ancient nation with the sharp outline of nationhood long before most European nations and states were formed, and yet we still exist in almost a half light. Introspective self-pity, with a thousand years of robust disunity behind us in almost every field of our national life. At any rate, we are a distinctive people and we are glad to welcome you here to Wales.

May I say one word about this College of Librarianship. Over the few years of its existence, of the 10 or 11 years of development, it has achieved so much that must have been beyond the wildest dreams of its founders at that time. It did not come into being by a deliberate national decision to found an institution of this nature, much needed though it was, but its existence stems from the vigour and vision of a small dedicated band of people. The most dynamic and able and energetic very clearly is Principal Hogg, and were it not for his great assiduities and the great foresight which he has brought to bear on the development of this college, I doubt very much whether it could be the institution of national, and indeed international, repute that it is at the present moment. And it is on account of the breadth of his vision and the willingness at all times to be progressive and adventurous in development that you find yourselves here to-day. There is no other school of learning in Wales that caters in any way for the field of study in which you are involved and I appreciate very much, in mentioning your field of study, that I am coming to the boundaries of a very dark continent as one who is about as untechnical as any creature that treads this earth. But even as a layman, Mr. Moderator, I do appreciate the tremendous significance of your subject - the study of communication, the appreciation of the impact that such can make upon human relationships, and the charting of the various theories and principles which govern this all-powerful and extremely important subject, a subject whose significance to human life in general is enormous and whose scope must be regarded as practically illimitable. It stretches from higher calculus, it seems to me, to psychology; on the other hand, from cryptology in one direction to almost the mathematical basis of languages on the other.

Now, we are very glad indeed that Aberystwyth has had the opportunity of being the venue for your course of study. This is not the first time that Aberystwyth has played host to NATO itself. I appreciate, of course, that you are a NATO-sponsored body and are not really part of the military organisation at all, but some five or six years ago this place was stiff with senior brass when NATO held a Strategic Planning Conference in conjunction with the International Politics Department here. It was of a very high level;

one had to be a Major General before one had the chance of getting to the bar at all in the evenings. We appreciate that you are a body sponsored by NATO, but wholly separate and independent from it. We appreciate, even those of us who are innocent laymen like myself, the tremendous significance of your subject and we appreciate, too, that NATO must be very much more than a mere system of defence and has to be a body for development as well. And it is for that reason that we hope that your stay at Aberystwyth will be a happy and fruitful one, and that there will be every bounty of success on your deliberations.

OPENING REMARKS

Tilo Kester

NATO

Dr. Debons, Mr. Morgan, Ladies and Gentlemen, it is my pleasure to welcome you to the second Advanced Study Institute on "Information Science" on behalf of the NATO Science Committee and the NATO Scientific Affairs Division. Amongst the distinguished participants of this meeting I should particularly like to welcome Professor Alves Martins, the Portuguese representative at the NATO Science Committee since 1961. His presence demonstrates the interest of the Science Committee and the Advisory Panel on the Advanced Study Institutes Programme in the topic to be treated at your meeting. These NATO bodies since their very existence have always tried to give substantial support to scientific activities in fields which need this and where breakthroughs can be expected.

The first Advanced Study Institute on "Information Science" held in Seven Springs in the U.S.A. at about the same period of the year 1972 was mainly a North American meeting. It is for this reason that Dr. Debons decided to propose another Advanced Study Institute on the same subject to take place in Europe and to organize it such that the knowledge developed previously will be carried over to Europe to the benefit of scientists involved with research in Information Science and related fields. In this effort, Dr. Debons was assisted by Dr. Cameron, his co-director, his local organizer, Dr. Gibbons, and a number of devoted people. - We also gratefully acknowledge the co-sponsorship of the "College of Librarianship Wales", "The University of Western Ontario", the "Liverpool Polytechnic" and the "United States Army".

Last but not least, a tutorial institute depends on the design of the scientific programme and the willingness of experts to teach chapters of this programme, and having had the opportunity to attend the last year's institute - regrettably for a few days only - and to meet personally most of the lecturers present here, I can say that both conditions are fulfilled. Participants from many countries are looking forward to an exciting Advanced Study Institute.

As I expect to be with you for the next two weeks and there will be many opportunities for all of you to discuss any questions you might have in relation to this institute, the NATO Advanced Study Institutes Programme and the scientific programmes of the NATO Science Committee in general, I should like to restrict myself to saying that the ASI Programme continues to be the backbone of the Science Committee activities. This fact is demonstrated through the increasing interest of scientists in participating in the institutes and the increase in the number of proposals for Advanced Study Institutes discussed with our office. We have to do a severe preselection in order to prevent a further increase of the rejection percentage of formal applications for support.

Most of the proceedings of previous Advanced Study Institutes have been published in book form, but as many of the directors of the institutes found it very difficult to find adequate publishers, publishers who could guarantee fast publication, efficient dissemination of the knowledge and publication at their own risk, we have now come to an agreement with four well-known publishers to publish a uniform ASI Series. This series consists of five sub-series: Life Science, Mathematics, Physics, Social Sciences and Applied Sciences. The first 25 volumes of the 1973 Advanced Study Institutes are expected to be available by the end of this year. Whether the proceedings of your ASI on Information Science will start the sub-series of Applied Sciences will almost entirely depend on the collaboration of the lecturers of this meeting. I am not sure whether the way of dissemination of knowledge chosen by us is considered by yourself as a good one, but we shall continue printing "black on white" as long as more sophisticated methods are not made available to us by the information scientists.

Heaps of scientific journals and books, documents and correspondence on the desks of most of us are the witnesses of the existing information problem. If we are not willing to be the victims of "information pollution" in future, we have to find ways to reduce the data produced and to up-date those data continuously which are subject to frequent changes. This should be done before they become "coded facts". Your research and the availability of modern technology, pooled together, should eventually result in satisfactory systems. How ever such systems would look like, the designers should bear in mind that *information* is probably more than "matter" or "energy". Information means *power*, which can be used to the benefit of our society or, given to individuals, groups of people, countries, nations or even continents could be detrimental to the peaceful development of our world. The point which I want to make here is that those who through their contributions help to get a breakthrough in the field of "Information Science" should be aware of their social responsibility.

OPENING REMARKS

Herbert S. White

President-Elect, American Society for Information Science

I deeply appreciate this honor of being asked to speak to you as the representative of the American Society for Information Science. Perhaps I should be hesitant to speak to you as the representative of ASIS at all. The group assembled here this morning contains many prominent members of ASIS, including a past president. However, that very fact is perhaps more indicative than anything I can say of the close rapport between our Society and the theme and objectives of this Institute.

For those few of you who may not know much about the American Society for Information Science, it is an organization of about 3500 members. The name is perhaps somewhat misleading. We are not exclusively an American organization, although the majority of our present membership resides in the United States. We have an active Western Canadian chapter, with further indication of potential chapter activity in Eastern Canada, Mexico and Central America, Europe and Israel. In addition, many of our members, perhaps a majority, would be more properly called technologists than scientists.

My own personal fascination with ASIS is in the cross-section of interest and outlooks it represents. A simple reading of the names of the active Special Interest Groups will illustrate this point. They include groups on Arts and Humanities; Automated Language Processing; Biological and Chemical Information Systems; Behavioral and Social Sciences; Costs, Budgeting and Economics; Classification Research; Education for Information Science; Foundations of Information Science; Information Analysis Centers; Library Automation and Networks; Non-Print Media; Reprographic Technology; Selective Dissemination of Information; Technology, Information and Society; and User On-Line Interaction. Although as incoming president of the society I am not permitted any display of partiality, I must confess a particular pride in the Special Interest Group on Costs, Budgeting, and Economics, and on the fact that both our 1972 regional conference in Dayton, Ohio, and our upcoming national conference in October of this year in Los Angeles deal heavily with this subject. There has, I think, been a historical and unfortunate confusion between the laudable principle that information access should be as unrestricted as possible by the economic constraints of inability to pay - and the, in my judgement, totally fallacious position that information is therefore free. It is certainly not cost-free to produce, to analyze, or to disseminate, and it would indeed be unfortunate and even incompetent of us as information scientists if we allowed our feeling that access to information should be as open as possible to lead us to ignore the cost-benefit alternatives which all of us must consider in any environment which operates with limited resources - and that is, realistically, every environment. I mention this only because it is sometimes fashionable for administrators of large information collections to decry the need to analyze and justify their activities, and in an administrative responsibility I consider such a position to be indefensible. As an operating head of a working information system my approach is perhaps unusual in this conference setting, and since that system is in fact operated on a return-on-investment premise, it is a probably unique viewpoint here.

In my judgement, the most significant contribution ASIS makes is that it provides the arena for a cross-fertilization of ideas and an exchange of viewpoints from a membership which includes a concern for information theory and information research, linguistics, manual as well as computer systems, reprographic and non-print techniques and media, on-line as well as off-line systems, information networks as well as small specialized information collections, a subject interest in fields as diverse as arts and humanities on the one hand and organic chemical substructure analysis on the other. It includes a range from basic research through applied research to information technology, and from there to the librarians, chemists, and information systems managers who use the output of information science in a strictly pragmatic sense to get their specific jobs done.

Our membership resides in the academic community, in government, and in private industry. It includes information users and information suppliers, and among information suppliers it subdivides further into the "for profit" and "not for profit" sectors, although that distinction is sometimes difficult to discern by paying customers. I can think of no clearer example of the diversity and strength of ASIS than with regard to the current discussion about revisions to and interpretations of the U.S. copyright law, particularly with regard to fair use copying, and specifically embodied in the landmark Williams and Wilkins case. The ASIS membership, and, for that matter, the ASIS Council, includes representatives of every point of view of this complex and thorny question, the equitable disposition of which is so important, and ASIS ability to consider this question knowledgeably but without an entrenched position to defend is probably unique among professional groups in this field.

I need not go on. I think it is clear why ASIS is so very pleased to participate in this Advanced Study Institute in Information Science. The development and exchanging of ideas must precede or parallel research, research must precede development, development must precede application, and this in turn precedes evaluation and measurement in the operating environment. ASIS is concerned with all of these steps. The institute promises to be, from the program and the list of distinguished participants, a landmark conference, and I personally look forward with great eagerness and anticipation to the next two weeks.

TUTORIALS

TUTORIAL, GROUP 1: NATURE OF INFORMATION

Eugene J. Rathswohl

University of Pittsburgh, Pittsburgh, Pennsylvania, U.S.A.

The proceedings of the 1972 NATO Advanced Study Institute in Information Science reported that there is a great need to create, develop and strengthen a theoretical foundation for information science. This theoretical foundation was seen to focus on a "solid, consistent, theoretical explication of the notion of information". Thus, the task at hand is to study the fundamental principles underlying the structure and use of information. The first pay-off in studying the nature of information, as seen in the 1972 proceedings, is in terms of developing a "true" science of information. A second, and equally important, pay-off is in the practical domain of information system design, implementation, management, and usage. Thus, the task of studying the nature of information is to facilitate information science's becoming a *science* and to facilitate information science's contributing to *society* in terms of better designed information systems.

WHAT IS THE NATURE OF INFORMATION?

The 1972 NATO ASI proceedings pointed out that our "intuitive notion of information is deeply inconsistent ... we see information in apparently incompatible ways". Indeed, Wellisch[1] examined something like 39 published definitions of information and information science, all of which were evaluated as inadequate or inconsistent in a number of ways. For the sake of brevity and organization, a matrix is proposed with which to systematize the many seemingly inconsistent "fundamental principles of information".

The argument is this: that the nature of information has been talked about in terms of four concepts - commodity, process, state of knowing, and environment - and that two approaches have been taken in each case - definitional and criterial.

Four concepts of information

Commodity. This concept is intended to capture empiricist notions of information. It refers to the empirical content of communications and knowledge. Information as a commodity represents something, e.g., some event, some state of the world.

Process. This concept is intended to capture formalistic or rationalist notions of information. It refers to the structural and organizational aspects of communications and knowledge. Information as a process is represented by something, e.g., genetic structure.

State of Knowing. This concept is intended to capture pragmatic notions of information. It relates directly to the effect of information in terms of changes in a human knower. Information as a state of knowing relates directly to the structural-functional aspects of human knowing activity.

Environment. This concept is intended to capture systems notions of information. It refers to man-machine and man-man systems, e.g., command and control systems, within which certain data-information transformations occur.

Two approaches to information

Rescher has suggested that two different but related motivations characterize our approach to understanding the concept *truth*.[2] I suggest the same be applied to understanding the concept *information*.

Definitional approach. The definitional approach is an attempt to offer a definition of information; the approach is oriented toward explicating the meaning of information.

Criterial approach. The criterial approach is an attempt to specify the test-conditions for determining whether or not certain commodities, processes, states, or environments could be characterized as informational, e.g., under what conditions would one be able to recognize this commodity as information commodity? In other words, the criterial approach is oriented toward implementing the concept of information and thus has practical consequences for information system design.

Discussion

Given the above, there are eight categories with which to identify and classify our efforts to understand the nature of information, each of the two approaches applying to each of the four concepts of information.

Commodity. (1) Definitional. Examples here are instances where information is defined as a commodity. For example: "information *is* data of value",[3] "information *is* coded fact".[4] Information *is* something to be used in decision-making, planning, problem-solving, etc. Information *is* the empirical content of a decision; it *is* relevant facts.

(2) Criterial. Examples here are instances where information is conceptualized as a commodity and where the emphasis is on the necessary conditions that make information as a commodity possible. The Conference Board Report is a good source for identifying various principles of information technologies,[5] information technologies in general being criterial approaches to information. Other examples include classification theories, coding theorems, and statistical signal detection theories.

Process. (1) Definitional. Examples here are instances where information is defined as a process. Last year's NATO ASI proceedings contained a number of examples. Debons[6] wrote that "information *is* the transformation of data", transformation being used as a verb. Both Auerbach[7] and Samuelson[8] regard information as a flow of signals and both equate information with the notion of a processing structure.

(2) Criterial. Examples here fall mainly within the area of the general problem of computability. Examples include concepts of Turing machines, automata theories, cybernetics, data and information structures, and related problems of software design and management.

State of Knowing. (1) Definitional. Examples here are instances where information is defined as a state of knowing. In the 1972 NATO ASI proceedings, Furth[9] wrote that information *is* a process of knowing. MacKay[10] writes that "information *is* that which promotes or validates representational activity ... this activity amounts to the gradual construction of a mental model of the external world. Information is that which adds to our model".

(2) Criterial. Examples here are instances of conceptualizing information in the context of other concepts such as "weltanschauung",[11] "paradigm",[12] and "world view".

Environment. (1) Definitional. One excellent example here is McLuhan's work.[13] His now famous phrase "the medium is the message" defines information as man's technical environment.

(2) Criterial. Most systems concepts in inform-

ation science refer to the conditions under which information can be generated and utilized. The command and control tradition is most relevant here.[14] Man-machine system experimentation is also an example.[15] Other examples are found in sociology and especially in the analysis of knowledge production systems.[16]

Conclusions

The above discussion was intended to juxtapose certain concepts and ideas concerning information in some coherent manner. Still, a response to our original question (i.e., what is the nature of information), is lacking. My proposal is that, given the above discussion, the nature of information can be said to be primarily *technical*. More correctly, the research-guiding interest in the present literature of information science seems to be mainly a technical interest. By a technical interest I refer to an interest in the manipulation of the physical and social environment, including the manipulation of ourselves, when we attempt to adjust to our immediate environment.[17] In this view, the nature of information is seen as being inextricably intertwined with a motivation for keeping certain processes, certain structural-functional relationships, under control. Namely, information is seen to be a technique for maintaining a certain status-quo. The nature of information is a certain power to manipulate and control our lives.

There are a number of manifestations of the technical interest in information science. First, it shows up when one reflects about information in terms of commodity, process, state of knowing, and environment. Information as a commodity relates to the empirical content of knowledge and, as such, to the regularities in nature, an understanding of which is assumed to be instrumental in improving decision-making. Information as a process refers to the logical or rational quality of representation in the context of efficient and effective computation. Information as a state of knowing refers to forming, revising, and using a representation internal to a human learner. Information as an environment relates to the structure of man-machine and man-man relationships so as to aid man in making the best use of the data available to him for decision-making.

The technical interest is clearly formulated by Kochen:

> Information science has a central concern with the growth of knowledge, understanding and wisdom. Information is useful for its role in coping with problems. It is the raw material for knowledge, which is information processed with a point of view, through a representation. Information is transferred from the environment to a "learner" where it is used to revise a representation and to help utilize such a representation for selecting actions ...[18]

Still another manifestation of the technical interest is reflected in the final report of the working group concerned with the nature of information at the 1972 Institute. The group asked the question - what might a science of information look like? Their summary response was:

> It is likely to look quite different from the other sciences - more like what Weinberg call 'trans-science' - because of its concern with real-world decision-making and problem-solving. It will probably be concerned with the relations between various levels and forms of aggregation and organization to a greater extent than is economics and physics. Its central focus seems likely to be the impact of an incoming message on a decision-maker operating in some context, increasing (or decreasing) his capability to so operate ...[19]

LEGITIMACY OF THE NATURE OF INFORMATION

The question I want to raise at this point is that of the legitimacy of our concept of information. The argument here is that our understanding of and our ability to understand the "nature of information" depends on our making presuppositions in at least three areas. We must make, and indeed have made, presuppositions concerning science and technology, society, and fundamental principles, per se.

Science and Technology

The opinion is becoming more and more widely discussed that contemporary science and technology have become a new form of legitimate power in society. Indeed, science and science-based technology have become basic to the production processes of advanced industrial societies.[20] What is the contemporary self-image of science and technology? At the risk of oversimplifying, the images are at least three in number: the ideal of objectivity, the ideal of unified science, and the ideal of naturalistic approach to gaining knowledge. "Objectivity" refers to the idea that knowledge is historically and inherently neutral. It refers to the techniques of science, that they are neutral as to the motives and values of particular researchers. "Unified science" refers to the idea that there is a unitary scientific method or technology possible for generating information. Specifically, unified science refers to a concept of reductionism, a concept that posits a certain hierarchy of the sciences and proposes that explanation can be "reduced" to terms, concepts, and theories of the "basic" sciences in the hierarchy. The "naturalistic approach" refers to the idea that the standard of certainty and exactness in the physical sciences is the *only* explanatory model for scientific knowledge. Indeed, it is commonly said that one of the primary goals of science is to explain, explanation itself being defined in terms of pure theory, value-free facts, and historically neutral method.

Our task here is to pose certain questions and problems for information systems. I suggest that it is important that information science begin to understand how these "fundamental principles" or "ideals" of contemporary science are reflected and indeed are implemented in the design of information systems. This is important because there are good reasons for questioning the legitimacy of these ideals of science and, thus, to question their validity as presuppositions or assumptions for the design of information systems and for information science, per se. What are some of the reasons? First it is reasonable to consider what Husserl has called the fallacy of objectivism. When knowledge is conceived as a neutral picturing of fact, what is overlooked is how the societal framework within which research activity takes place affects the processing of theory and data. The fallacy of objectivism is that the ideal of objectivity ignores the existence of certain psychological and sociological processes that are necessary for knowledge generation and utilization. Thus, even though we are able to control the systematization of knowledge, contemporary science is criticized on the basis that it cannot reflect about its own presuppositions and is left without any way of legitimately objectifying its structuring framework. A second reason relates to the ideals of both reductionistic unified science and of naturalistic approach to knowledge. The criticism is that an exclusively or predominately naturalistic approach leads to an impoverishment in the understanding of man himself - that is, leads to a certain dehumanization. Furthermore, such an approach, e.g., behaviorism, leads to the view that human beings may be manipulated and exploited by social and human sciences in the same way nature has been manipulated and exploited with the help of the natural sciences. Our understanding of man becomes clouded by a "nothing but" syndrome. To use Marcuse's expression, the understanding of man becomes more and more "one-dimensional". Or to use Rozak's phrase, our understanding becomes a "waste-land".

In summary, the criticism of the ideals of contemporary science is that we have an unreflective science as well as an impoverished science with respect to the understanding and emancipation of man. The major suggestion to be made here is that these ideals *are* active forces in contemporary society and *do* transform the human world. I pose it as a problem that information science must begin to understand how it incorporates these ideals into its own work. The greatest problem information science faces is not whether theories of information generation and usage (e.g., behaviorism, systems analysis) are valid but whether they might not become valid through a self-fulfilling prophecy instrumented by contemporary information systems justified by a technocratic ideology of a "true" science of information.

Society

The concern here is to pose this question: are our fundamental principles concerning the nature of information appropriate to our practical concerns in information science, i.e., our conceptualizing of and dealing with real-world problems. Recently we have been hearing a number of characterizations of real-world problems. Beer[21] has characterized them as problems of managing complexity. Sackman[22] has emphasized the evolutionary aspects of society, especially the purposeful control and management of societal evolution and change. Kochen[23] has identified the pressing real-world problems as ill-structured rather than well-structured problems and argues that our need is to learn to cope with ill-structured problems rather than to attempt to solve just our well-structured objectified problems. There is a growing concern that how we think about societal problems is inappropriate - inappropriate for handling the complex, evolutionary, ill-structured aspects of society. The argument is that modern systems analysis is guided merely by an interest that is technical, that is merely data-collecting. We must begin to try to understand society as a whole, or as Churchman has written "... as a manifold of interconnected problems". In short, a growing part of our socially conscious literature in information science is arguing for fundamental principles of information that facilitate holistic, interdisciplinary thinking about society. I pose it as a problem for information science to begin to understand how our present concepts about the nature of information both facilitate and preclude the development of holistic, interdisciplinary attitudes.

Fundamental Principles

Recall the suggestion made previously that the fundamental principles concerning the nature of information seem to be motivated by a research-guiding interest which is technical. This meant that the nature of information, as we presently conceive it, is to give us increased power over our environment, to manipulate and control it. There is a growing literature, however, in the area of metascience[24, 25] which is arguing that the technical interest is not the only legitimate research-guiding interest in science and society. This second interest could be referred to as emancipatory in nature. It does not take the existing organization and structure of science and society for granted, as does the technical interest. It knows that the "laws" of science and society are not "given" in the same sense as are the laws of nature. It sees that there is no 'fundamental principle' of science and society which could not be changed. It understands that science and society is created by man and can and should be changed by man. It is an interest that believes man can become aware both of how he is limited by the "laws" of nature and communication, and how and in what direction he can change or redesign those laws.

Our 'fundamental principles', unlike (for example) Kant's categories of understanding which were posited as inherent and fixed, can be considered neither as fixed, nor as ahistorical, items of information. We can in fact talk about the appropriateness of our concepts of the nature of information and then posit and study alternative conceptualizations. McKeon[26] has pointed this theme out very clearly when he writes:

> We have ceased to seek a
> fundamental system of things
> or of thoughts on which
> to base order but we
> have gained confidence in
> the discovery and
> institution of a system
> of communication and a
> system of communities
> by which to order what
> we say and do and
> even what we think
> and what we know
> about the order of
> things.

This theme is an extension and, in many ways, a revision of many pleas in information science to "synthesize the various formalistic descriptions into one unified set of theories which is equally applicable to all contributor sciences and disciplines".[27] The revision is that any unified set of theories concerning the nature of information must be synthesized with an eye toward facilitating human development and understanding. Rossman[28] writes in the context of education that the problem is not so much to learn how to change culture but rather that we must change culture in order to learn. Applying this attitude to information science, the practical problem is not only to learn how to design information systems but also to design information systems so as to enable man to learn. We must build information structures within which fundamental principles of the nature of information can be detected and used in the advancement of human freedom and holistic, interdisciplinary thinking.

References

1. Wellisch, H., From Information Science to Informatics: a terminological investigation, *Journal of Librarianship*, 4, 157, 1972.

2. Rescher, N., *The Coherence Theory of Truth*, Oxford University Press, Oxford, 1973.

3. Yovits, M., Information Science: Toward the Development of a true Scientific Discipline, *American Documentation*, 20, 369, 1969.

4. Furth, H., Paper to appear in *Information Science: Search for Unity*, Marcel Dekker, New York, 1971.

5. The Conference Board, *Information technology: initiatives for today - decisions that cannot wait*, Report No. 537, 577, Conference Board, New York, 1972.

6. Debons, A., ed., *Information Science: Search for Unity*, Marcel Dekker, New York, 1974.

7. Auerbach, *ibid.*

8. Samuelson, *ibid.*

9. Furth, *ibid.*

10. MacKay, D.M., *Information, Mechanism, and Meaning*, M.I.T. Press, Cambridge, 1969.

11. Churchman, C.W., *The Design of Inquiring Systems*, Basic Books, New York, 1971.

12. Kuhn, T., *The Structure of Scientific Revolutions*, 2nd ed., International Encyclopedia of Unified Science: University of Chicago Press, 1970.

13. McLuhan, M., *Understanding Media: the extensions of man*, Bantam Books, New York, 1964.

14. Debons, A., Command and Command and Control: technology and social impact, in *Advances in Computers*, 11, 319, 1971.

15. Parsons, M., *Man-Machine System Experiments*, Johns Hopkins Press, Baltimore, 1972.

16. Etzioni, A., *The Active Society: a theory of societal and political processes*, Free Press, New York, 1968.

17. Radnitzky, G., *Contemporary Schools of Metascience*, Akademiforlaget Gumpert, Göteborg, 1968.

18. Kochen, M., Views on the foundations of information science, *Information Science: Search for Unity*, Marcel Dekker, New York, 1974.

19. Final Report, Proceedings of working group 1. The nature of

information. *Information Science: Search for Unity*, Marcel Dekker, New York, 1974.

20. Shroyer, T., Toward a critical theory for advanced industrial society, *Recent Sociology*, 2, 210.

21. Beer, S., Managing modern complexity, *The Management of Information and Knowledge*, U.S. House of Representatives, 1970.

22. Sackman, H., *Mass Information Utilities and Social Excellence*, Auerbach, Princeton, N.J., 1971.

23. Kochen, M., Views of the foundation of Information Science, *Information Science: Search for Unity*, Marcel Dekker, New York, 1974.

24. Radnitzky, G., *Contemporary Theories of Metascience*, Akademiforlaget Cumpert, Göteborg, 1968.

25. Habermas, J., *Knowledge and Human Interest.*

26. McKeon, R.P., *The Future of Metaphysics.*

27. Otten, K. and Debons, A., Towards a metascience of information: Informatology, *Journal of the American Society for Information Science*, 21, 93, 1970.

28. Rossman, M., *On Learning and Social Change*, Random House, New York, 1972.

TUTORIAL, GROUP 2: INFORMATION TECHNOLOGY AND HUMAN NEEDS

Esther E. Horne

Westinghouse Electric Corp., Electro-Mech. Division,
Cheswick, Pennsylvania, U.S.A.

PURPOSE

The purpose of this tutorial is to help you recall and recognize current issues, trends, and problems in Information Technology; to develop a point of departure for further discussion and study, to raise questions and stimulate inquiry. This presentation deals with the transformation and transfer of information and human needs. Is Information Technology now adequate in meeting human needs? Not quite, just yet. On the one hand, more efficient utilization of Information Technology must be realized and on the other, a fuller, deeper understanding of the human component - the User of Information Technology - is needed. (1) The price is still too high for most individual, institutional and corporate users. (2) The trend to networking and resource-sharing must accelerate. (3) The results are not that good. There is still too much irrelevancy and too long a time lag. (4) The interactive mode is frustrating. Present day query languages restrict the user. (5) On-line, real time often means off-line and/or down-time. BUT costs are declining on hardware and designs are getting better. Management Information Systems are integrating decisions at the operating levels of many manufacturing divisions of companies. There are four major components in Information Technology: (1) Human (info. proc.); (2) Information (data base); (3) Transmission (com. tech.); (4) Computer (data proc.). One and two are necessary but not sufficient to effect information transfer. Some transmission channel or carrier is needed, that is three. Four is optional in some cases. (Four does facilitate accessing and processing with speed, ease and efficiency.) The *human* is the ultimate information processor and user, but has a limited mental capacity being able to handle only 7 ± 2 items at a time. The *carrier* is the medium, a transmission device or channel. The *computer* is an artifact, a tool, a huge slide rule for some, a great way to sort for others. A logic machine which works well if the user uses his well. *Information* is a usable answer to a user's question derived by a processor from a data base.

It might be argued that the human will some day be a non-essential in the system when computers equipped with sensors process collected data, solve problems, make decisions, and generate new data, all the while keeping each other informed and in good repair. But - until that day the human must be reckoned with.

INFORMATION TECHNOLOGY IS INFORMATION SYSTEMS AND COMMUNICATION MEDIA

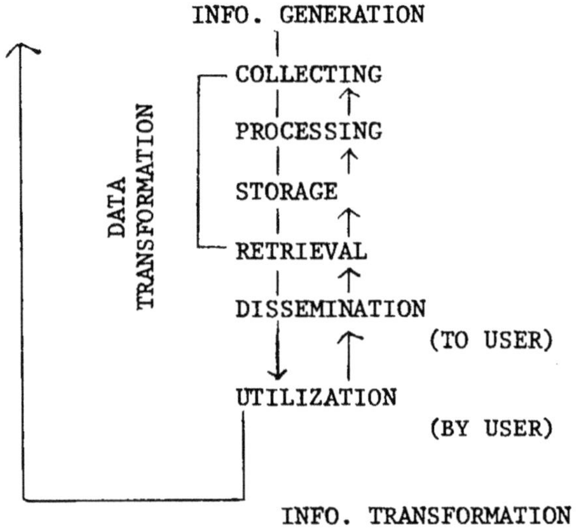

Communication Technology creates an environment through the use of various media of transmission enabling information flow as it proceeds from generation to dissemination for utilization. This flow is maintained by further generation and re-use of previously generated information. The flow from utilization to generation certainly exists though it is not clearly understood. The flow from generation to utilization is the transmission we are explicitly dealing with. The flow backward through the system or feedback is known to be necessary but seldom implemented. In this model the external *data* transformer is the machine (computer) and the internal *information* transformer is the human. There are different levels of Information Technology.

Here we have a model of Information Technology without the computer but with a natural transmission channel, bi-directional

SPOKEN LANGUAGE

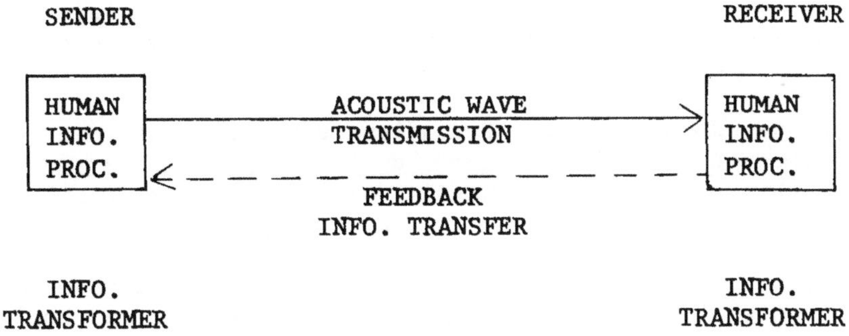

flow, and an internal memory. Spoken language is an old information technology and still our most important and most used. It facilitates transformation and effects transfer, between sender and receiver. The next level of technology, written and printed, extended and amplified this capability. This had to suffice and still does to a very large degree today. Concurrent with the development of the written and printed media came the development of a type of information system to fulfill the need for an external memory from which to retrieve on demand. This was called a *library*. Today because of electronic media and devices, Information Technology has become more complex and better organized; at the same time society has become more complex. Change has accelerated. The human information processor needs more external assistance than a book, however useful it still is.

There is a basic human need and that need is for information. But the spoken language is constrained by many factors *some* of which are: (1) Acoustics and codes; (2) Differing human experience; (3) Perceptual difficulties; (4) Emotional and motivational factors; (5) Human processing capability. The last four are human factors. All of these limit the effectiveness of Information Technology. (2) is also called the "Frame of reference" of sender and receiver; (4) is the behavioral factors; (3) and (5) are cognitive factors.

The written and printed language fares no better. It too has limited effectiveness and in addition, it expects a more sophisticated User. Written and printed language capability is constrained by: differing codes; differing human experience; format and presentation of material; perceptual difficulties; emotional and motivational factors; human processing capability.

Then there are policies affecting its distribution and use. Very early policy determinations limited access and availability and still do. Also many would-be Users simply cannot afford information. Here we have the same four human factors but in the last factor the limited human capacity to process, store and retrieve also includes a limited capability to design truly effective information systems. According to Allen Kent there are two unsolvable problems: one in selection and one in relevance.

In what way does the new Information Technology ameliorate the transformation/transfer problem? Can the computer offer more than the capability to speedily handle volume? Can communication channels do more than efficiently and speedily transmit volume? The new information technology capability is constrained by: same limiting factors as before;but amplifies and extends them over a much wider range of users. These capabilities do expand and amplify the human powers of computation, and file management, effect collection and transmission of tremendous amounts of data, perform control functions as well as collapse time and distance.

In order to solve problems and make decisions, the user must be able to correlate, integrate, analyze, synthesize information. Without the computer's computational and processing capacity and capability many problems would be unsolvable because of the sheer number of man-years required for their solution. And the human predilection to error. Computer processing would seem to relieve the constraint of human limited capability to process, store and retrieve. Englehart thinks it is more than that. There is a synergistic effect. As the possibilities for utilization of Information Technology and Information increase it becomes very important to ascertain whether or not information so derived is utilized and utilizable. And how, or why not, as the case may be. In the case of ERIC, the potential users, the practitioners and administrators never became users. If we are indeed developing a man-machine symbiosis, then more bi-directional flow should be evident between the system and the User, in the areas of design, use, and policy as it affects the former two. The literature would indicate otherwise, although according to Brittain there is a movement away from source-controlled and toward receiver-controlled systems. SMART System is one example. There the User feedback affects the structure of the system. The conclusion to be reached is that without feedback, systems may become inflexible and thus inefficient in satisfying User needs.

NEED

Now that the technology has been examined, what about NEED? When the user needs information: it must be accessible and available to him; it must be relevant, reliable and timely; it must be in a readily usable form; it must be at a cost he is able to afford or better, free.

The User receives information (or documents which are presumed to contain it) and goes his way. We know very little about his intentions toward or actual utilization of this information. Furthermore why did he think he needed it in the first place? There are many questions about the User that point to a need for more studies on User behavior: that is studies on the individual. Brittain points out that we only study demands or requirements for meeting need, not need itself. The User may be unaware of his need; or unable to correctly perceive it. In Vol. 7 of ARIST,[1] information needs and demands are equated. In order to meet the criteria of "relevant and timely" it seems sensible to inquire after the User's perception of his problem and his intended, versus actual, use of the information.

What can be defined as "readily usable" form in light of individual differences and needs? How can we know when the User has too little or too much at any given time and what effect does that have on problem-solving? More easily answered in the following question - does the potential User know what is available to him in the way of information services? SUPARS showed that knowledge of service increases use and the INFROSS study[2] of social scientists in England showed that such was the case. But in the real world the majority will remain non-users unless made aware. Brittain states that we don't understand the behavioral effects of information on the user and the user on information. How do these effects impinge on utilization of information and Information Technology? ARIST concludes that the important area of research is in studying the interaction between information and the User.

Going back again to "reliable" information, insuring reliability is easier said than done. Reliability suggests quality and quantity control. This is feasible for the system hardware and software but what of information itself? Physical science and technology information appears to be self-correcting in time but in the social sciences and other areas it may not be so - at least not now. Then there are everyday, common fact problems like consumer information (some call it "truth in selling"). One could confound the total situation further by considering Kent's unsolvable problems of selection and relevancy.

"Timely" usually means current but timely may be the 18th century for a historian.

"Costs" speak for themselves. The affluent sophisticated User is provided for with the best the technology can offer. The not-so-well off must make do with less information or none at all.

Further from Vol. 7 of ARIST - studies have been made of scientists and technologists as to *factors* that generate differential needs and to *types* of information needs. Some factors that generate differential needs are: type of work - whether it is basic or applied; discipline - work in the social, physical or biological sciences; availability - of services, the effect of demand on use (SUPARS Study).[3]

As for types of information needs: most of the emphasis is on media or channel not content (the INFROSS project clearly demonstrates this). Much work has been done in the area of exchange of information and patterns plotted. There are Price's and Crane's "invisible college" studies. Allen's communication 'stars' and other studies on the social communication network. These are informal exchange patterns. INFROSS studies the uses of primary publications and informal channels. At meetings and conferences formal papers and information discussions were considered equally useful. However, Van Cott's study indicates that in their present form international meetings need vast improvement to become effective media for information exchange. Lack of User feedback in the design of systems was explored in ARIST. This built-in feedback component should provide information about actual uses and responses of potential users. Many innovations in design falter for lack of it. For instance in the system ERIC microfilm was not accepted as anticipated and few practitioners or administrators were users. The User must be integrated as a necessary component affecting the system before, during, and after design.

In Vol. 7 of ARIST there were no utilization studies which would fit Brittain's criteria - those stressing individual differences. Or Bourne's aspects of user psychology of which he lists nine. Or Line's individual variables in information gathering and seeking behaviors. In Brittain's opinion the majority of studies being done are only observational.

Past volumes of ARIST have raised the subjects of utilization of information and the effect on human behavior - on the human as an information processor - and also of the effect of varying information flow situations on the user.

MANAGEMENT INFORMATION SYSTEMS

Management Information Systems for decision-making presents a crucial problem for Information Technology. This information situation includes all of the previously-mentioned problems. In the area of decision-making, most MIS are planned for the use and operation by middle and lower-level managers isolated from policy-making. There is voluminous literature on the subject of management but little on the implications and uses of Information Technology for management.

The Conference Board recommended that top management needs its own set of special information hardware, software and systems, but the initiative to generate the design and production must come from top management. Decision-makers need a variety of information which may be difficult and impossible to find, further complicating the situation. Grayson in HBR[4] states that the manager needs convenience, speed and accessibility from the system and reliable results. Grayson, who served as Chairman of the Price Commission, Phase II of the Economic Stabilization Program, established with 600 people a program to control prices and make decisions involving costs volume, prices, productivity, resource allocations, elasticities, multiple goals, trade-offs, predictions, politics, and risk value. Because of a 16-day time constrain he rejected using management science tools and mathematical and statistical models. In most design tasks:

(1) There is a shortage of time - managers work in "crisis mode". The Manager doesn't want to wait - can't wait on data-gathering and model-building.

(2) Management scientists call for inaccessible, non-existent or uncompiled data to suit "theoretically current models".

(3) Explicit recognition of financial and emotional cost of organizational effort to educate and change is lacking. (*Power, politics* and *change-resistance* factors are called the imperfect human variables.)

(4) Time constraints, data availability, questioning people problems, power structures and political pressures, are left out.

(5) The design methodology is heavy on math and statistics - weak on psychology, sociology and political science.

At least the design methodology should put people, time, power, data accessibility and response times into models thereby creating crude, workable solutions.

Causes of user dissatisfaction with management information systems: inaccessibility of input data; long response time; invalidating simplifications; resistance to change by the manager. Inaccessibility of input data, resistance by the user, long response time and invalidating simplifications lead to bad output and user dissatisfaction. Other bad results as far as the User is concerned are solutions by the system of non-existent problems and insignificant ones. Real variables are omitted from messy problems leading to unreliable output; therefore managers become confused and impatient.

Basically there exists a lack of understanding about the decision-making process and the decision-maker. Managers are inclined toward intuition and action. They see the situation as a mosaic. Their value systems often outweigh facts where the whole picture seems to justify it. Management Information Scientists dissect and analyse, are less action-oriented and averse to risk-taking. Management Information Scientists must think in terms of the manager's perceptions, needs, expectations and pressures. On the other hand Top Management should not isolate Management Information Scientists but give them line responsibility. Put a Management Information Scientist on an operating term. In Case Institute, Ackoff, Churchman, Arnoff and students worked on real problems in real time in real business settings. But this teaching approach has not caught on, because real problems are not neat and clean but very messy, exactly like the real world.

The general trend, as noted in ARIST, is toward greater use of computerized systems even though the User continues to complain of the lack of and irrelevancy of the information obtained. A strong movement is seen toward rapid interactive access to major public and commercial data bases and another toward cooperative system development and operations. As seen in both OECD and Conference Board studies there is an urgent need to extend the utilization of Information Technology via networking at national and international levels. Policy formulation on utilization of Information Technology is almost non-existent, except for Japan. Problems involving the individual user, interactive systems and networking remain to be solved.

INTERACTIVE SYSTEMS INTERFACE

It is important to provide the user with: ease of access and good response time; browsing capability; terminal and display must accommodate, not frustrate. Browsing capability should be provided for the User through natural language input; insuring flexibility to individual demand. All in all the User must be able to concentrate on understanding the problem he wishes to solve not the inner workings of the computer. Furthermore, networks should sometime in the future allow the User to access other databases, other computers, other Users, easily and speedily, maintain bi-directional flow, avoid breakdown, and prevent misuse and misdirection of information at a not prohibitive cost.

SUMMARY

Despite the outstanding accomplishments of Information Technology, actual performance does *not* meet expected performance.

Information Technology's capability is constrained by human limitations; that is by the human as an User of, designer of, and policy-maker for Information Technology.

In addition, cost considerations are not to be ignored in the design, implementation, and operation of systems.

The most difficult factor in the utilization of Information Technology lies in determining User need.

References

1. Cuadra, C.A., Ed., *Annual Review of Information Science and Technology*.

2. INFROSS, Information Requirements of the Social Sciences.

3. SUPARS, Syracuse University. Psychological Abstract Retrieval Service.

4. C.J. Grayson, *Harvard Business Review*.

TUTORIAL GROUP 3: SOCIAL IMPACTS OF INFORMATION

James L. Finnerty

Sogang University, Seoul, Korea.

An attempt to evaluate information impacts calls forth value judgments. Hence, in this paper the reader will find some value judgments congenial to his own, some perhaps not. The point at issue is not the value judgments, however. The points are that there are serious impacts of information flow in our society, that typical effects are easily identifiable, and that an attempt to measure their importance is worthwhile.

Information impacts are listed in three categories, not to suggest that they are so neatly displayed in real life, but to stress that familiar information phenomena have recognizable impacts on modern society, and that people have a role in moderating these effects through control of their information environment. What are labeled the two-edged, the adverse, and the creative impacts of information are offered as fruits of one man's reflection, and hopefully, as stimulus for another's, lest through neglect of information impacts we suffer, with MacBeth, the social consequences of "bloody instructions, which being taught, return to plague the inventor".

TWO-EDGED IMPACTS

Life Style Biasing is one of the two-edged impacts of information. We say 'two-edged' because these impacts are able to be productive or counterproductive, often depending upon one's viewpoint. We may be biased in regard to the life styles of others. Consider the following two passages that speak of Japan:

> JAPAN Experience the spectacular contrasts of Japan ... its busy cities, its quiet ancient temples and palaces, its scenic beauty in

mountains, lakes and countryside. (A travel brochure)

JAPAN A report on the pressures in a tight-packed society of 100 million people, for whom the next earthquake could be doomsday.[1]

Were we ignorant of Japan and the Japanese people, would we not envision quite different life styles from these two information sources? And what does one make of a fitting life style for a lady librarian from the following ad: "I was the mainstay of the public library until I discovered _________ vodka. The effect was shattering." (Smirnoff)

What we like about these kinds of persuasion we are inclined to call cultivation, especially if we are doing the cultivating, and what we don't like we call seduction or propaganda. For example, Americans who read such attractively formated news monthlies as *Soviet Life* or *China Pictorial* are apt to interpret them as propaganda, imagining they are being cultivated by suspect or enemy sources. On the other hand they accept publications of their own news media as normative; i.e., unbiased. In the meanwhile the portrayal of life styles in China, through publications like *China Pictorial*, goes to sixteen different language groups in the world and its impact is not negligible. The impact is two-edged because while in one country *China Pictorial* will be regarded as propaganda, in another population it will have a cultivating effect: "That is the life for me."

As for more visible effects upon life styles, the information impacts stemming from counterculture movements in America in the late 1960's may still be seen. Changes in the aims of the young, changes in their styles of dress, working habits, recreations, and political outlook have all been duly noted. Practically, one result of this life-style biasing of the young is to make doubtful the successful recruiting of a volunteer army in the near future, in the United States. In Britain, on the other hand, where the benefits of a military career are impressed upon the very young, and are heralded continuously in the *Sunday Times*, the recruitment of a volunteer army seems well on the way to success. Life-style biasing is a component in these two outcomes.

Life-style biasing, through information flow, is in fact a component of many outcomes. The average citizen might well be unaware of how much he has been cultivated toward such projects as missile construction, weapons development, and wars, to mention but a few. He might wonder, for instance, if a man or a nation has to be cultivated by someone in order to identify his enemies. How do nations select their enemies? What hand do information impacts have in making such identifications possible?

Consumer and voter biasing are further information impacts that are two-edged. Some of the biasing is helpful, some is not; but it ought to be recognized as operative and given due respect. Valuable information impacts here are consumer biasing towards quality products, budgeted expenditures, and responsible use of credit, and voter biasing towards thorough study, investigation, and attention to long-range implications of the ballot. Unfortunately, little advertising urges responsibility in buying, little campaigning urges study and reflection, and the information industry itself is a center of conflict over who has rights to the consumer (cf. Control Data and Telex suits in 1973 against IBM).

Information impacts upon consumer and voter are often from the less noble edge. Just as Miss America must be changed, the consumer is urged, through a yearly parade of fashions and models, to keep changing everything from homes and appliances to clothing and cars. Voters are also biased, sometimes towards change for change's sake, and sometimes towards adamancy. An example of the latter is the attempt in 1973, through bulk mailing from local congressmen, to bias laborers in the steel industry against the building of Japanese steel mills in Pennsylvania: "How would you like getting your check from the Bank of Tokyo?" A more subtle example of voter biasing towards change (from the Philippine voter's view) is the following costly paid advertising:

> PHILIPPINE PROSPECTS After 26 years of democratic stalemate, dynamic leadership directs the Republic toward the New Society.[2]

Revolution is another information impact that is two-edged. It is, of course, the extreme result of persuasion that is called, in the information stage, either "the truth" or sedition, depending upon whose side you are on. Revolution is an information impact - no information, no revolution. And this is as true of social and cultural revolutions as of political and military ones. The contemporary scene in America suggests that a number of revolutions have taken place with hardly a notice. Nudity is in at the beach, the laity are in in the Catholic Church, and Mayor Yorty is out in Los Angeles. In international politics, recent revolutions have occurred - Taiwan is out of the U.N., Americans are in Peking, and two Koreas have agreed to talk with one another. What further revolutions are brewing are brewing in an information flow. Of this we can be sure.

ADVERSE IMPACTS

For want of a better term, *adverse* has been chosen for a series of information impacts that seem to have but one edge. This is not to deny that outstanding social benefits are derived through the uses of adversity, but simply to label certain information impacts as unprofitable in themselves. In the following listing, the levels of adversity are seen as increasing. We will go through them quickly, employing a variety of witnesses.

The first of these adverse impacts is boredom. *Boredom* is the awful feeling we sense when faced with bad advertising, bad campaigning, lackluster entertainment, dull teaching, and filibusters. It is an adverse impact, but one that is capable of being challenged, if people care, or exploited if they do not. "No need to improve T.V. They will watch it anyhow."

The second of adverse information impacts is confusion. *Confusion* results from, among other things, a plethora of up-to-date news. We are in danger of letting up-to-the-minute reporting substitute for in-depth evaluation of what is reported.

> ... undigested up to the minute information is not enough. We have to know what to do with it and how to utilize it ...
>
> The very mechanism of the dissemination of news leads necessarily to a gap between the political leader's judgment and the impressions of the average citizen and, by the same token, between the whole political apparatus and the body of the nation.[3]

The gap, spoken of above, is a confusion gap. Our political representatives, barely able to get the pulse of society, and not having time to do all of the homework on all of the problems, have to borrow problem-solving skills from experts. The experts are often self-appointed, and if society does not question them, why should they question whether or not they themselves really see through the confusion that they are supposed to be seeing through? Unless society does its information homework, confusion will eventually reign.

Beyond confusion is *Error*, the result of the big lie in politics, fakery in advertising, slanting by news media, and repetitions of ingrained prejudices from generation to generation. But why ingrained? Is this not an information impact; i.e., ingrained prejudice? Is it subject to correction? How should society set about to correct its information errors? In this case "information errors" can be read as "the continual processing of bad data".

Disillusionment is the next adverse information impact on society. It is a strong impact because of its emotional component, and indeed what is adverse about disillusionment is not that people's eyes are opened, but that their emotions are down-trodden. But for the latter, disillusionment should be placed in the category of creative impacts of information. The consequences of Watergate on American politics are described in the following - the reader may judge for himself how to categorize the impact.

> The 181 hours of televised Watergate hearings made a major contribution to opening people's eyes to the misdeeds committed by this Administration and the result, in the aftermath of the hearings, is bound

> to be an even deeper mistrust of politicians and of government than there was already. This is one of the most dangerous consequences of this crisis.
>
> It was the young who in the face of the Vietnam war lost their belief in American political institutions and protested, rebelled against them, even thought they could overturn them: now it is the adults, who used to be very critical of the disillusioned young, who are disillusioned.
>
> There is a limit to the amount of disillusionment that any nation can take without losing confidence and optimism, of which Americans used to have ample reserves ...[4]

Without the 181 hours of information impact, through television, it is doubtful that the degree of disillusionment reflected in U.S. political opinion polls of summer 1973 would have been reached.

Chagrin is felt when things are going on around us that we might want to object to, but that we have no control over at the moment. The vastly-expanded construction of computer-accessed data bases in modern society is one source of chagrin, and how little control we have had over this cause may be estimated from the following passages:

> ... nongovernmental organizations have generally been free from judicial or legislative supervision of their record-keeping procedures. This has included the questions of whether an individual who is the subject of a record must be notified about its maintenance, has a right to inspect its contents, and is given any procedure for challenging its accuracy, completeness, or propriety.[5]
>
> In private law, the majority of American states, either by statute or judicial decision, recognize what is known as the common law right to privacy ... For the most part, however, the common law right to privacy did not play a role in the record-keeping area. The courts had not held that individuals could invoke the common law right to prevent the recording of personal information about themselves by a government agency or private organization.[6]

The problem is that we don't know the data, nor where it is being kept, when and to whom it will be revealed, or for what use. Here is an information impact that is unique in that it exists independently of whether or not any information actually flows once it has been obtained.

Closely related to chagrin is *Fear*. This too is an information impact for some and it is an adverse one where unjustified. People fear misuse of computer accessed data bases, through malice or mistake, even their own mistake. Credit bureau information, the Internal Revenue Service income tax files, and Federal Bureau of Investigation fingerprinting practices lead to some of this fear. There are now 83 million fingerprint records and an equal or greater number of credit bureau records. That there is some cause for fear may be seen in the following:

> Computerization of arrest records is advancing swiftly, for the F.B.I.-run National Crime Information Center is creating a network that will ultimately tie together the files of over 6,000 law-enforcement agencies. ... computerization had eliminated some errors in these records and updated others to include the ultimate dispositions. But computerization has clearly increased the storage and circulation of such arrest data and made them far more accessible to many state agencies, employers and other law-enforcement agencies ...
>
> Congressional studies have found that many in the credit-checking industry engage in inadequate and indiscriminate investigations, often exerting a pressure on their investigators to compile derogatory and often gossipy information, including multiple hearsay; 2,000 of these credit bureaus are being tied together in a computerized network with files on 100 million Americans.[7]

Impotence is the information impact on society seen by Weizenbaum. He is alarmed that the self-fulfillment of an entire population may be curtailed by computer technology, not so much by what the computer can do, but by what people imagine the computer can do. He suggests anonymous, irresponsible forces are the cause of impotence.

> The potentially tragic impact on society that may ensue from the use of systems such as I have just discussed ... the psychological impact on individuals living in a society in which anonymous, hence irresponsible, forces formulate the large questions of the day and circumscribe the range of possible answers ... large numbers of perceptive individuals living in such a society experience a kind of impotence and fall victim to the mindless rage that often accompanies such experiences.[8]

We will return to Weizenbaum shortly on the subject of responsibility in society.

Injury is another adverse impact of information. The U.S. spy ship *Pueblo* suffered injury when, in January of 1969, she was captured off the shores of North Korea. Information-seeking was the mission of the vessel, but it was not the information sought that brought about the injury. It was the reliance on navy codes of communication, misinformation about available rescue teams, and lack of information at the time about how many planes could respond, where they were, and how they could be contacted. Deeper injury was sustained at Pearl Harbor, 28 years earlier, for lack of well-coordinated military information systems at a time when abundant data were available. Reliance on an unworkable system caused this and other significant adverse impacts in recent U.S. political and military history, but none perhaps as dramatic as the Pueblo incident. Here was a ship, with the world's most sophisticated communications and surveillance gear, unable to save itself in a crisis, due to information problems.

The point of *Ruin*, as an information impact, is made when we review the story "If Mao Had Come to Washington in 1945", by Tuchman.[9] This anecdote, taken from recent foreign affairs and political life, contains an information and communication drama whose impacts have been truly ruinous. (The anecdote given here could be matched by others of equal magnitude, depending upon the reader's choice of time frame and historical arena.) The story goes like this.

Mao and Chou En Lae wanted to visit Franklin D. Roosevelt in January of 1945 to explain their situation and the real-life situation of China. Their message to this effect was channelled through the American Embassy in Chungking. Though it was directed to General Wedemeyer, whom they trusted, in his absence it went through the hands of American Ambassador Hurley, whom they did not. The message was eventually forwarded to FDR, but only after being interpreted by Hurley in a highly unfavorable light. This portion of the drama is the communication part. Tuchman would have us suppose it had gone otherwise:

> ... let us imagine instead that, following a more normal process, the message had been duly forwarded to the "highest officials", and had received an affirmative response which is 99 and 44/100 percent unlikely but not absolutely impossible. If Mao and Chou had then gone to Washington, if they had succeeded in persuading Roosevelt of the real and growing strength of their sub-government relative to that of the decadent Central Government, and if they had gained what they came for - some supply of arms, a cessation of America's unqualified commitment to Chiang Kai-shek, and firm American pressure on Chiang to admit the Communists on acceptable terms to a coalition government (a base from which they expected to expand) - what then would have been the consequences?[10]

A "more normal process" might have been through the U.S. Diplomatic Corps in Chungking, some of whose members, knowledgable in Chinese affairs, saw the cause of Chinese unification in a different light than Hurley. Herein lies the information part of the drama. Roosevelt, Hurley, and the American public were persuaded that Chiang Kai-shek was the only one who could unite China and rally her in support of the U.S. war effort still going on against Japan. "Persuaded" seems too mild a term for the information fixation that directed U.S. foreign policy in favor of Chiang throughout World War II, the subsequent Chinese Revolution, and to the shores of Taiwan. In addition to persuasion, mistrust was another one of the information channel blocks.

According to Tuchman, Roosevelt trusted only Hurley, not his diplomatic corps in Chungking, and would listen only to his Ambassador. Hurley in turn believed it his mission to achieve unity in China - his way, which was Chiang Kai-shek's way. Again we can say that maybe it would not have made any difference if the message from Mao to FDR did get through, and if Roosevelt was not governing in isolation from this problem, but could it be this very isolation in government that is the trouble? George Reedy states that it is increasing,[11] and that

> Power breeds isolation. Isolation leads to the capricious use of power. In turn the capricious use of power breaks down the normal channels between the leader and the people whom he leads. This ultimately means the deterioration of power and with it the capacity to sustain unity in our society. This is the problem we face today.[12]

"Today" in the preceding passage is 1973, and perhaps things are not so different in government now from 1945. Even though Nixon has visited the same Mao and Chou En Lae, and information flow has greatly expanded between Asia and America, what cannot be erased is the cost of the past. There was the revolution in China, the Korean war, war in Vietnam, 27 years of distrust and standoff between the U.S. and China, the largest representative of the Third World. Even today there is continuing information output that stability in Asia is somehow an American problem. It is, of course, an Asian problem, just as was the problem of unity in China in the 1940's a Chinese problem. Information impacts here are crucial, for they may be ruinous.

CREATIVE IMPACTS

Creative impacts of information are the less obvious as they are taken the more for granted. Who can estimate the "widening of horizons" that takes place in the imaginations of young and old through productions such as *Sesame Street*, TV coverage of moon walks and space exploration, time laps and microphotographic presentations of our oceans, fauna, flora, and even the human blood stream? Through communication satellites we are led through a cultural open door to the winter Olympics in Japan and, upon occasion of the visits of national leaders, to the Forbidden City, Peking. The sky is truly the limit for creative impacts of information. Speaking of computer services, Sackman writes:

> Computer services, if we so wish, over the long-range evolution of the human species, can enhance effective human intelligence toward unlimited and undiscovered heights. If this value and vision are accepted, then it follows that mass computer services be dedicated to lifelong education and timely information for every individual.[13]

The impact of *Stimulus to Responsibility* is a potentially great information benefit to society. In the view of some writers (like Weizenbaum) there is danger here. The danger is that society will relinquish its powers of mind and judgment to the machine. At the moment that society, believing that technology can do the job better, stops asking questions, there is trouble. For, if society's problems are computer-generated, the possible answers are very much circumscribed. If bombing targets are to be selected, there is no possibility of debate on bombing; if war games are to be played, war games they must be; and if missiles are to be guided and controlled, no machine will come up with the question - why missiles? And if society's questions and answers are handled by computer, responsibility is also relinquished. Nobody calls the computer responsible. Society, having abandoned its specifically human role of using mind and judgment to the computer, then functions as a machine without a mind. Already we hear around us "the computer says", and we do. Decisions are made without reference to the human components of the decision.

> Modern technological rationalizations of war, diplomacy, politics, and commerce such as computer games have an even more insidious effect on the making of policy. Not only have policy makers abdicated their decision-making responsibility to a technology they don't understand, all the while maintaining the illusion that they, the policy makers, are formulating policy questions and answering them, but responsibility has altogether evaporated. No human is any longer responsible for "what the machine says". Thus there can be neither right nor wrong, no question of justice, no theory with which one can agree or disagree, and finally no basis on which one can challenge "what the machine says".[14]

On the other hand, as Sackman says, society may grasp technology to serve its inquiry ends. In this view, society will save itself by using its mind and judgment to place information technology at the service of its problem solvers. Society will frame the questions, structure the answers, and be continually responsible for use of its powers of mind and judgment. Lifelong education at one's fingertips could be a leading result of the employment of information technology, and a genuine participatory democracy might become possible. It will depend largely upon the people's choice.

> The crucial problem with competing values and priorities is that the early reinforcement of some will tend to suppress the emergence of others. With unrestricted laissez faire, mass computer services would follow the path of least resistance and greatest profits, probably along the lines of pay-TV entertainment on demand. With well-formulated and widely supported public policy and adequate government support, universal lifelong education and home voting systems could be available at the outset and could continually dominate the long term evolution of computer services.[15]

In conclusion, it would seem that information goals are what society is greatly in need of. Although we have measures of successful information systems, without goals for information systems, measurement of their success will be meaningless. Information "goal tending" may well be the challenge of our times.

References

1. *Observer Review*, August 19, 1973, 25.

2. *New York Times*, June 10, 1973, Section 11 (Advertisement), 1.

3. Ellul, J., *The New York Times*, July 1, 1973, Section E, 13.

4. Brandon, H., *The Sunday Times*, August 12, 1973, News Section, 8.

5. Westin, A.F. and Baker, M.A., *Databanks in a Free Society; Computers, Record-keeping, and Privacy*, National Academy of Sciences, Washington. Project on Computer Databanks. Quadrangle Books, N.Y., 1972.

6. Westin, A.F. and Baker, M.A., *Op. cit.*, 19.

7. Schwartz, H., *New York Times Book Review*, July 8, 1973, 19.

8. Weizenbaum, J., On the impact of the computer on society, *Science*, 176, 613, 1972.

9. Tuchman, B.W., If Mao had come to Washington in 1945, *Foreign Affairs*, October 1972, 51, 44 (Reprinted in *Notes from China*, Collier Books, New York, 1972).

10. Tuchman, B.W., *Op. cit.*, 51, 45.

11. Reedy, G.E., *The Presidency in Flux*, Columbia University Press, N.Y., 1973, 72.

12. Reedy, G.E., *Op. cit.*, 16.

13. Sackman, H. and Borko, H., Eds., *Computers and the Problems of Society*, AFIPS Press, Montvale, N.J., 1972, 25.

14. Weizenbaum, J., *Op. cit.*, 613.

15. Sackman, H. and Borko, H., *Op. cit.*, 25.

TUTORIAL GROUP 4: PROFESSIONAL ASPECTS

Frank Gibbons

Liverpool Polytechnic, Liverpool, England.

In the 1972 NATO Advanced Study Institute at Seven Springs, Pennsylvania, working group 9 considered the professional and educational aspects of information science. Two stages of the deliberations must be distinguished. Because several members of the working party had to leave the Institute at the end of the first week, the group agreed to issue a statement of "points of agreement".

POINTS OF AGREEMENT

The members of the group were able to formulate nine points of agreement which, in effect, focus the content of their deliberations. It is important to recall from the introduction to this report that the group members did not perceive it as their responsibility or their position to phrase the points that follow as recommendations for the information field. It was their belief, however, that other groups, which had been configured to be representative of the field, might find the statements of this group, which were representative of only its members, interesting and useful.

(1) Specific and concentrated attention should be devoted soon to the differences, both actual and potential, between information academics and information professionals.

(2) On all levels of education in the information field, increasing attention should be given to inter-institutional and international development and use of resources and comparison of curricula.

(3) There should be active discussion on all levels of educ-

ation in the information field of two major educational objectives, namely the preparation of confident and adaptable graduates.

(4) On all levels of education in the information field increasing attention should be given to continuing education.

(5) Appropriate measures should be taken by educational institutions to insure more complete integration of programs in the information field including basic disciplines and various fields of application. This aim could be achieved through many means, some of the most important being:

(a) At the administrative level, co-ordinating courses taught in various disciplines and fields.
(b) At the faculty level, encouraging co-operation among individuals of varied subject origins.
(c) At the student level, providing for the tailoring of courses according to the requirements of the student's future careers.

(6) It should be stressed that an adequate preparation at "higher educational" levels should include acquaintance with the methodologies of the following basic disciplines (among others): logic, mathematics, statistics, languages and linguistics, behavioral and social sciences, computer science, and cybernetics. NOTE: This statement did not achieve total agreement in the group because of the ambiguity of the phrase "higher educational" which means radically different things in the educational systems of different nations.

(7) Efforts should be undertaken to evaluate, co-ordinate, and extend the few studies of manpower needs in the information field.

(8) Efforts should be undertaken to develop profiles of people actually and potentially in the information field and to develop profiles of actual and potential employment opportunities in the field for the purpose of enabling active mutual interaction to provide a better match.

(9) There is a great need for better integration of research and teaching in the information field than is currently the case. A comprehensive evaluative study of the present state of research in teaching institutions in the field across various countries should be encouraged. Such a study should cover methodology, interdisciplinary aspects, results, and practical implementation.

The second stage consisted of the deliberations of the remaining members of the group during week two of the Institute. It was decided to concentrate attention on a limited number of topics arising from the discussions of the first week. The group agreed on the following statements.

CLARIFICATION OF POINTS OF AGREEMENT

As noted in the main body of this report, point of agreement No. 6 did not receive total agreement in the full working group because of the differences in inter-national interpretations of the phrase "higher educational levels". The members of the residual working group concluded that the following wording provides for the difficulties:

> It should be stressed that an adequate preparation at some level of education should include an acquaintance with the methodologies of the following basic disciplines (among others): logic, mathematics, statistics, languages and linguistics, behavioral and social sciences, computer science, and cybernetics.

In discussion of point of agreement No. 8 the members of the residual group considered a possible study to approach the question of the types of jobs which employers associate with the terms "information" and "data". It was concluded that such a study could be approached through a systematic, retrospective examination of employment advertising (perhaps over a period of as long as ten years). As well, the study could be segmented by nation for interesting opportunities for comparison. A comment was made that the major shortcoming of the study as discussed would be in the area of verifying its results. Nevertheless, the study was regarded as immediately "do-able" and capable of producing thought-provoking data.

During the period from the Seven Springs Institute to the present day, Dr. Debons, Director of the Institute, and the joint-group leaders of Working Group 4 in the 1973 Institute, Professor Jack Belzer of the University of Pittsburgh, U.S.A., and Mr. Frank Gibbons, of the Liverpool Polytechnic, United Kingdom, have continued discussions with a view to developing an appropriate programme for the deliberations of the group in the 1973 Institute. Our work in this respect has been materially aided by the work of Mr. A.B. Gupta, our Graduate Assistant.

As the summary of the 1972 Institute indicates, we have considered that three points emerge as vital topics for further discussion at the 1973 Institute.

(A) What are the professional identity problems of information scientists?

(B) Is interdisciplinary orientation versus speciality orientation a problem?

(C) Can the diversity in information science education be justified in the light of current needs for information professionals?

These topics have been elaborated for your deliberations as indicated below:

Working Group 4 Agenda

The program will follow on directly from the discussions at the 1972 Institute and will centre on the following areas:

(1) (a) Identification of job descriptions and job levels in the professional practice aspects of information science.

(b) How can we overcome present difficulties in obtaining manpower estimates for the future?

(2) Discussion of the social image, professional status, of information science practitioners.

(3) Identification of personnel and institutions engaged in research and development which is relevant to information science.

(4) Is there a need for an Institute or Learned Society for the development of information science as a discipline rather than a purely professional activity?

(5) What steps are necessary to establish a synthesis of various national viewpoints?

(1)(a) Identification of job descriptions and job levels in the professional practice aspects of information science. The essential problem lies in the identification of groups of practitioners who can be styled information sciences professionals. The questions below delineate the problem in more detail:

(i) The professions based on library science, computer science, scientific information work, etc., all include practitioners who can be described as information scientists. Can we distinguish the information scientists in these areas from the rest of the professionals in those sectors?

(ii) How can we identify information science practitioners outside those fields?

(iii) Can we detail the professional operations which we would accept as being within the domain of information science?

(iv) Can we designate different *levels* of professional operation related to various types of information science practitioner?

(v) There will also be sub-professional levels of information science operation. Can we define these job levels?

(vi) What educational and training programmes are required to meet the needs of the various categories of information science personnel listed above?

(vii) Continuing education for information professionals is an accepted need.

(a) What types of educational programme are required to meet this need?

(b) Information scientists from one area are likely to have to switch to other areas of information work. What types of "conversion course" are needed to enable them to make the transition?

(viii) New developments in information technology are likely to change drastically present job descriptions and job levels of information workers. Can we anticipate some of these changes?

(ix) Increasingly, the information scientist will have a vital role to play in management. What management training is required and at what stage in the information scientist's career?

(1)(b) How can we overcome present difficulties in obtaining manpower estimates for the future?

(i) Qualitative as well as quantitative studies are vital.

(ii) Problems encountered in obtaining manpower assessments.

(iii) How do we deal with the problem of the changing nature of information work requirements?

(iv) Can we design a technique for more effective estimating and monitoring of manpower needs?

(v) Questionnaires, Delphi methods, interviewing are methods used in the past. Can we produce a model scheme for future use?

(2) Discussion of the social image, professional status, of information science practitioners.

(i) What is the social status of the information scientist?

(ii) What status does the information scientist have in the hierarchy of his organisation?

(iii) What is the status of information science as a profession or discipline?

(iv) Why has information science failed to make the same impact on society and achieve the same growth rate as, for example, the management or computer sciences?

(v) Within the profession what avenues of career progress are open to the newly-entered information scientist?

(3) Identification of personnel and institutions engaged in research and development work related to information science.

(i) Can we define the areas of research or development activity which can be encompassed under the heading information science?

(ii) What facilities already exist to identify and record such research workers and research projects?

(iii) What procedures are needed to extend and co-ordinate existing facilities?

(iv) What part can the various educational institutions engaged in information science education play in the expansion and development of the research which are vital to the emergence and recognition of information science as a discipline?

(v) How can research workers in this area best be put into contact with each other and how can such contacts best be maintained?

(4) Is there a need for an Institute or Learned Society for the development of information science as a discipline rather than a purely professional activity?

(i) OECD have already proposed one (Anderla Report).

(ii) An international centre might draw support (including finance) which national centres may not attract.

(iii) Several international organisations are already at work in the field, e.g., OECD, UNISIST, EEC, NATO. What measures are needed to ensure co-ordination and co-operation?

(iv) If the need for an Institute is established what is the next step to be taken? By whom?

(5) What steps are necessary to establish a synthesis of various national viewpoints?

(i) What areas of general agreement already exist?

(ii) Which are the major areas of conflict of opinion?

(iii) Can, *and should*, the conflicts be resolved?

(iv) If yes, by what process?

(v) In what ways can international co-operation be encouraged and extended during a period when differences have still to be resolved?

Chapter 1

NATURE AND FUNCTION OF INFORMATION

Nature and Function of Information

Editorial Summary

One of the foremost problems facing the development of a science of information is the uncertainty as to what constitutes information. Because the term derives essentially from non-scientific application, the reference to the terms in the scientific, technical, and engineering literature as well as non-scientific sources is unclear. Implied in these sources is the concept that information is some sort of thing that one acquires and uses to achieve ends. Consequently, reference is made to storing the thing, giving it to others, or making it accessible to others. Some discuss the "thing" (information) as having value (e.g. "information is power", "information is data of value to decision making").

Fairthorne maintains that what information is depends upon one's own views, and he regards this as essentially desirable because it provides flexibility. Such flexibility seems especially desirable in the early stages of the development of a science. Nevertheless, Fairthorne maintains that types of information must be identified. The different types would lead to a classification. This is a limited view if we assume that whatever information is, it indeed changes the state of the receiver (he is "informed"). It would appear that the process of generating or producing information needs to be linked to the concept of information as a commodity to be used. The confusion in terminology is given emphasis by what Marzocco alludes to as "Synecdochism" - the substitution of words out of context with reality. In a kalaedescopic treatment of the practices of information scientists with words, Marzocco provides an inclusive description of the various interests of different segments of the field in information and how they regard information as a phenomenon.

The most direct statement as to what the information is, is expressed by Harmon, who considers information as meta-energy - "that which occurs between, with, or after energy". He sees information as that which is instrumental in the transformation and control of energy. Otten also stresses the role of energy or matter as an information carrier; with energy, no information can exist. Nicolis likens information to energy and develops his arguments along engineering lines to show that system adaptation is the result of changes in information. Information in this sense expresses established rhythms or oscillations of energy.

By far the most popular approach to the determination of the nature of information is the expression of models which attempt to describe the functions underlying information as an experience. The papers by Barnes, Otten, Watanabe, Smith, and Conway fell within this category. Barnes and Watanabe view information as a socio-physiological experience, with varying underlying connotations. Both recognize the importance of the role of sensory mechanisms in information. Watanabe applies current understanding of the visual processes and couples such understanding with formal logic in dealing with the use of information. Barnes links sensory input with values that man brings to bear in applying these inputs to decision making. Barnes subscribes to the Yovits camp, that information is "data of value in decision making". Barnes further takes the position that information is generated as the result of decisions. Accordingly, the event world, to Barnes, is the event world of decisions. Decisions generate the need for information. This position contrasts with that of Debons, who maintains that the event world continuously triggers the process that transforms data to information. Barnes makes no such distinction in the detailed model which shows the application of his concepts. By claiming that once one makes a decision, the choices for action have been reduced, and thus one has gained information, Barnes shows the proximity of his development to the Shannon school. Barnes in this sense refers to the changed state of the organism and does not provide, from our understanding, a position as to the processes that lead to that changed state. It should be emphasized that it is not necessary that a science of information makes these distinctions and provide such theoretical linkages.

Otten presents a model of information which is intended to be valid for all processes and phenomena referred to as information. Information, according to the model, is relative and functional (systems specific). According to his claims, any reference to information is meaningful only if it specifies the level within a system, the relation of the system to its environment, and the context in which the stimulus is interpreted at the specific level, considering all stimulus inputs. In addition, Otten maintains that the formation of information is system-dependent; the processing complexity of the system sets absolute limits to the interpretation of the stimulus.

Despite the predilection of information scientists for establishing models to describe the functions underlying information, or the information experience, Hugh Smith claims that almost all will come to naught if the complexity of the human dimension is not taken into account. Smith is concerned with the experimental verification of the Barnes model and other models which attempt to describe the phenomena in question. Apparently, Smith finds it difficult to justify the enthusiasm for model building in this area, particularly when little is known about the underlying human processes. The crucial issue remains; that is, given a particular model, what can be said about the distribution of decision making and problem solving functions between man and machine. The issue is expressed in Smith's terms not for the development of more sophisticated models, but rather for the integration of existing theories that address the complexity of the human dimension.

Hugh Smith's position is given emphasis by Monasterio. Monasterio claims that information cannot be regarded as homogeneous phenomena and attempts to formalize the underlying processes. In all of this, Monasterio asserts that a reductionist view of information can only lead to false hopes in our attempt to unravel the complexity underlying the experience of information. This approach thus seems fruitless for our further understanding of information.

INFORMATION: ONE LABEL, SEVERAL BOTTLES

Robert A. Fairthorne

This is to be a simpleminded talk, perhaps in more ways than I intended. The sense which I did intend is that of simplifying, that is, to try to uncover the barest essentials without which we cannot carry on whatever it is or they are that information scientists and technologists do carry on with. Such things, assuming that they do exist, are necessary but not sufficient. In no social situations, which include medical ones, can one ever formulate the situation explicitly and at the same time exhaust all possibilities. In our profession we must face the fact that we are going to be lucky if we can deal with things that really are necessary, because there always will be things we have not thought of or could not have thought of. The problem is that of Tristram Shandy who took one year to write the first day of his life. The problem is whether he could ever finish his autobiography.

To alleviate the situation I am going to run over some of the uses and, briefly, the abuses of the word 'information'. This word is used, at least it is uttered or written, in all branches and applications of our fields of studies and activities. Presumably it has some significance or significances. It need not, of course, necessarily 'stand for' any independent entity in the external world. Nor, if it should, need it be the simplest entity involved.

I will try to show that underlying all reputable uses of this word is the notion of variety, diversity, or, in a naive sense, complication. If this be so, then variety, diversity, or complication of what? One can talk about diversity without having to mention what in fact is diverse. This I think is one of the causes of confusion. Diversity may be of material objects, locations, noise, interpretations, types of subject interests, and so on. It can be of patterns as such, and of the realizations of patterns, two rather different things. The difference is illustrated by the remark of C.S. Pierce that there is only one word 'the' in the English language but that a page of printed English contains on the average around thirty 'the's'.

I will identify, roughly, some of the things that are diversified and which I consider to be the elements of discussion and contemplation and application in our fields though not usually one at a time. Then I will argue plausibly, I hope, that what we normally mean by 'information' can be any of several relations that involve the diversity of these ingredients.

We are not concerned here with the naming of either these diversities or of the field of our activities. Enjoyable and possibly profitable debate can take place on whether 'documentation' has too many and too narrow connotations, and whether 'informatics' is an appropriate and acceptable replacement. What matters here is not the common name of the varieties of information but their separate natures.

Every use of the word 'information' that is not downright inane or vacuous presumably refers to something useful to someone. When sorted out these uses therefore should yield some elementary and independent types of diversity that are necessary - but not sufficient - ingredients of Information Sciences and Technologies. One method for sorting out and identifying these will emerge in due course, or even earlier. The bulk of the task I will leave to you. Leaving work to be done by other people, is I believe, the mark of a good administrator.

A few words about the Word itself. The first, or zeroth, comment is that one does not have to use it at all, certainly not in ordinary discourse. Before the July and September 1948 issues of the Bell System Technical Journal it was not used much by other than communication engineers, and even then more as a linguistic convenience. For instance 'He gave me information about so and so'. When most people who were not trying to be impressive would say 'He told me about so and so' or even just 'He told me so and so'. The last means merely that the speaker made certain noises with his mouth, and this is the communication engineer's idea of information. A useful one, but one that does not go far enough.

One disadvantage of saying 'He gave me information' is that it gives the impression that information is handed around in plastic cartons, an impression that I fear is rather deeply imprinted these days. We will return to this later. It is worth noting that Ranganathan's Five Laws, more exactly Five Commandments, of Library Science, published in 1931, do not use the Word. The five are:

Books are for Use.

To every Reader his Book.
To every Book its Reader.
Save the Reader's time.
A Library is a growing Organism.

Most people would agree that Library Science in the narrow sense is not the whole of our activities, but few would disagree that it is wholly within them. Probably it is at the heart of them. Nevertheless it can be discussed without using That Word.

But to point out that one can do without a word does not mean that it is desirable or sensible to do without it. One can, I suppose, discuss analytical mechanics solely in terms of mass, length, and time, without using terms like force, energy, acceleration, moment, power, and the rest. But anyone doing that would soon go raving mad, without the already heavy inducements to lunacy inherent in that study. (I was once a sort of aeronautical engineer.)

The Oxford English Dictionary gives one whole page on 'inform' as a verb. It gives us also 'inform' as an adjective meaning 'without form' now, fortunately, as obsolete as 'inflammible' meaning 'flammible' should be. The substantive 'information' gets about two-thirds of a page of subdivisions of seven major senses of the word. None of these are the technical ones. Indeed, the most recent citation is dated 1890. This need not bother us. The Oxford English Dictionary is an authority on the history of English words. It is not an authority on their proper use. For our purposes we can regard the word in the first instance just as an orthographical unit of ten letters.

As a word of ten letters it is used, first, as a blanket term; e.g., 'The Information Sciences'. This is just as legitimate as the use of, say, 'dimension' by mathematicians. There is no precise general sense, but there are several very precise particular senses in particular branches of mathematics. None of these are inconsistent with the others in connotation.

Second, as a labour-saving device. It is after all a nuisance to make up one's mind as to whether one is talking about signals, the marks on paper, the marked paper, changes in the state of one's knowledge caused by the meanings of the marks on the paper, and so on. It is easier to let the reader do this for you.

Very soon this labour-saving device becomes a substitute for thought: 'Don't think, write information'. This can be valid; most mathematical notation is thoughtless thinking, but only because we or our reader can translate it unambiguously if necessary.

Third, 'information' as the name of some 'stuff'. This substance is squeezed from books like water from a sponge, and can be stored and pumped around. This notion of an 'informational essence' has less excuse than the notion of Phlogiston or of the 'luminiferous aether'. It is meaningless unless applied to something that has some permanent identity and can be conserved. In this company I do not have to warn against the grosser abuses of this notion, only to warn that it has a hasty habit of operating under subtle disguises.

Fourth, 'information' in communication engineering. This has been around quite a while, but really dates from Shannon's published papers in 1948, so far as other disciplines are concerned.

It is a *measure* based on the reduction of a given finite number of alternatives caused by the reception, and interpretation, of a signal. Hence the complete name 'selective information'. So far as it goes, this measure has properties in agreement with common notions about information; and the signalling model in which it appears is the simplest model in which it can agree. This model has also the great merit of connecting the idea of information, however narrowly, to the facts of life of discourse. With given physical resources and conditions, you have to take at least so much time or expend so much power, on the average, to transmit so many distinguishable signals. Therefore this type of information must contain some of the essential ingredients of the Information Sciences and Technologies.

Let us look at the elements involved in Shannon's 'selective information'.

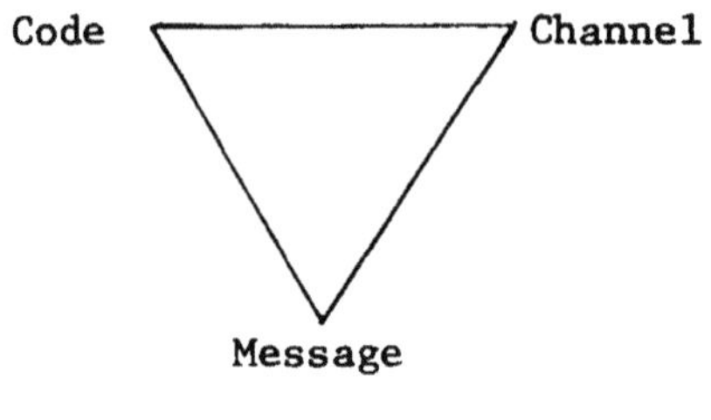

Roughly - very roughly - we have

Messages. i.e. what someone wants to say (not what may be going on in his mind in other respects).

Code. i.e. the *agreed* patterns (not their physical representations) that signify individual members of an *agreed* set of messages.

Channel. i.e. the physical environment in which and out of which the code patterns are realized.

The selective information is here basically a measure of the maximum diversity of distinct signals that can be achieved under the given conditions. In this restricted model, and only in this model, this can be taken as a reasonable measure of selective information. There is no guarantee that wider or deeper or higher notions of 'information' have any metric at all, let alone a numerical measure.

This measure of selective information is of the same mathematical form as the measure of 'entropy', more precisely of negative entropy. This does not mean that it *is* entropy. Many measures have the same mathematical form but do not measure the same thing.

Under certain rather bizarre conditions entropy and selective information may be identified, maybe. Fortunately this is not relevant to this discussion, but it may be as well to recall that both selective information and entropy are dimensionless; that is, they are pure numbers. They do not involve length, time, mass, or electric charge - only a relation between these, or none of these, that is unaffected by units of their measurement. Hence, even if entropy and any kind of information be the same thing - more properly relation between same things - they are not like power, energy, and the like. They are not entities or relations that can exist when there is no-one around to make decisions; e.g., as to what is to be regarded as the standard order to be disordered.

The only relevance the measure of selective information has fòr us is that it is a measure of diversity of any proper set of entities that has the essential property that it can never increase, only decrease (with respect to a defined standard). The signalling triad is the simplest set of entities that can connect the physical with the social aspects of discourse.

Obviously this triad is inadequate to cover more than just signalling. For one thing, we are not only interested in which message has been transmitted, (for instance, 'message no. 8', 'This letter is E.') but also in what it *is*, either as a linguistic expression or utterance and in what it *designates*, if anything other than itself.

Let us look now at the elements necessarily involved in Discourse; that is, in the activity "A tells B about C".

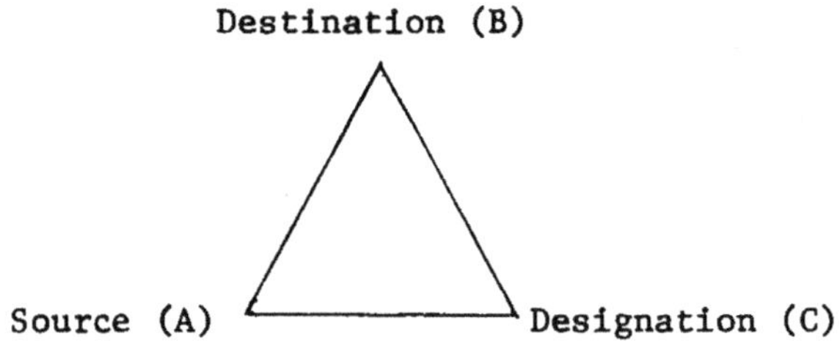

I use the term 'designation' because there is no reason that what A tells B should denote anything. A is at liberty to talk about anything whether it exists or not, and often takes that liberty.

The elements involved in Utterance, "A tells B something" are:

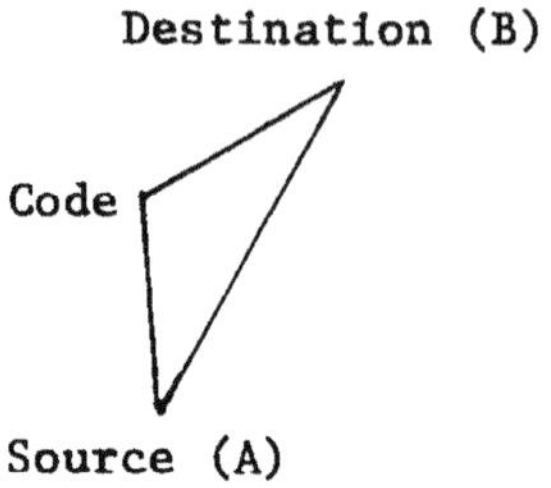

Here the Code is the something that A tells B, the linguistic expression in the sense of a linguistic pattern, not the physical representation of that pattern.

Communication in the engineers' sense aims to give instructions as to how to perform, for instance how to reproduce an utterance whether of words, sounds, plays, musical compositions, or any of the events of discourse. This may be done in more than one stage. One may signal instructions to reproduce the instructions for performance, so this task impinges on a part of the task of theoretical directors and human performers. So Communication Engineering involves four elements thus:

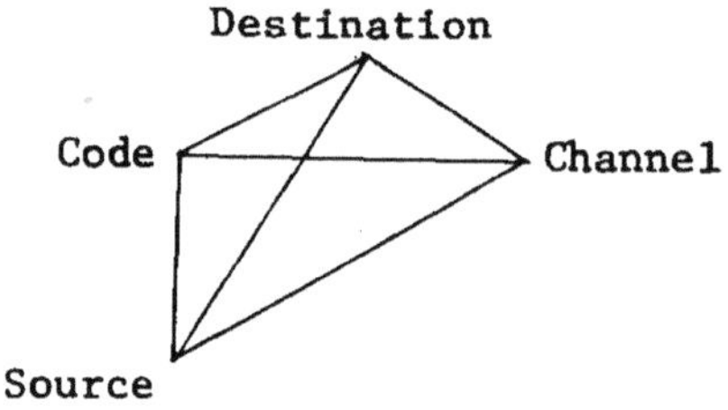

Here are embedded the sub-activity Reception;

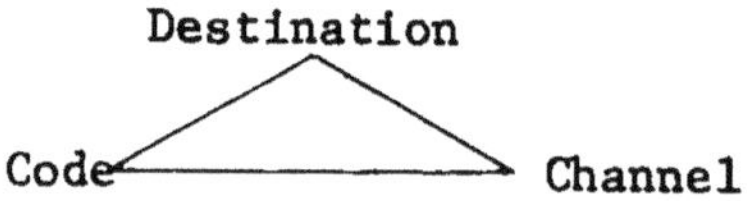

and Transmission;

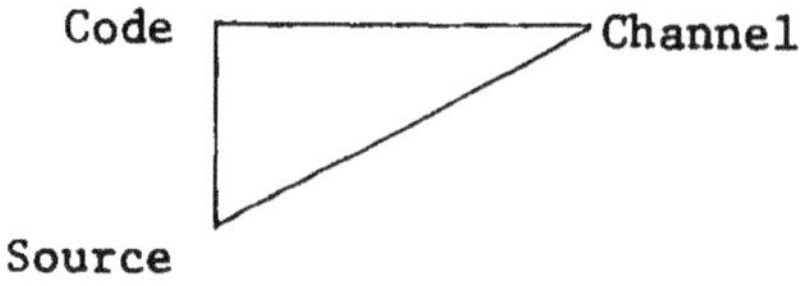

and Utterance;

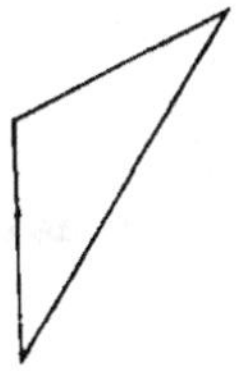

and Routing (Transport)

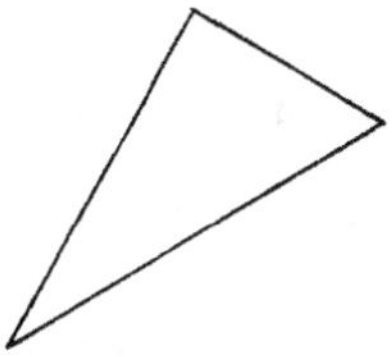

These are the basic activities of Communication Engineering in the narrow sense. They cannot stand on their own, but they can be studied separately in the first instance, and profitably so.

Also they have their own types of diversity. For instance, Shannon's measure applied to the Routing triad, interpreted as material transport in the sense of traffic, gives the average diversity of traffic density, types, and so forth, according to the property you examine. If you wish, and find it useful, you can also interpret this as the average unpredictability of this property of the traffic, and therefore a measure of the information needed for a statistical model of it. But you are not compelled to do so.

These four elements taken together, (Destination, Code, Channel, Source), are the elements of a self-subsistent function. It can exhibit 'flows' in the ordinary reasonable sense if, and only if, two triads have two elements in common. For instance, if the Source and the Destination are in accord about the Code and the Channel, there can be flow from Source to Destination. There certainly will not be if they are not.

But though this triad is the basis for a viable function, and can permit flows in the accepted sense, it is incomplete. It does not include Discourse and other essential sub-activities. For these we must add Message and Designation. Then we have all twenty triads from Signalling to Discourse:

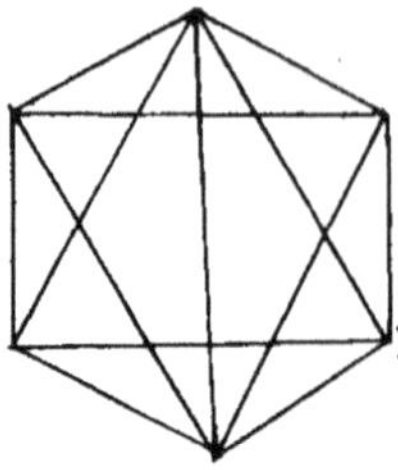

I must emphasize that this is not a geometric figure, it has no angles, lengths, or areas. It is a diagram showing which elements coexist with which, and no more.

Identification of the various triads is an instructive exercise, but listing them here would waste time and space. Each of the twenty can legitimately use the Shannon measure of diversity in different ways, some of which can be called information, though what they inform about differs in at least twenty ways. Also one has three choices as to which of the three elements is to vary and which pair are to be taken as fixed. For instance, the two well known problems associated with the Signalling triad: optimum coding for a given set of messages in a noisy channel of given characteristics, and detection of signals in the presence of noise.

Calvin Mooers has shown me, but has not yet published, how the six elements and their triads can be treated formally like thermodynamical variables. That is they respond interestingly to partial differentiation and such manipulations. This however would be out of place here. It seems fruitful to take the lowest task of the Information Sciences and Technologies, to wit "Notification" (bringing together Message and Destination, or conversely) to involve the six elements of our complete hexagram, with the connexion between Message and Destination missing. Since then there is no direct route, we can create 'flow' only by using one pair out of Source, Designation, Code, and Channel to create a tetrad of which Destination and Message are the opposed pair. This in fact is what we do do in practice, with author indexes, subject catalogues, shelf arrangement, 'profiles', and the lot. There are six basically distinct tetrads. Each has at least two types of diversity that can be identified with a distinct type of information; i.e., the information is about different things. So even for Notification, the lowest level of our activities there are twelve types of diversity or 'information', all of which have to share the one label.

Whatever the label, and whatever the level, it seems to me that the underlying task of the Information Sciences and Technologies is to manage, maintain, and make available diversity of discourse, diversity of its users, diversity of its uses, and diversity of its sources. In this we cannot fall back on the usual expedient of management, abolition of diversity by imposition of uniformity. A library consisting of one book, one author, and one reader, would not be diverse in one sense of 'information' however varied the bindings, type faces, and pseudonyms. These of course would involve a great deal of 'information' in relation to other types of diversity.

We have to clarify and distinguish the types of 'information' not only for our own intellectual satisfaction, our professional satisfaction, or our practical satisfaction. We have to sort things out because the task of our profession is one form of the task of all professions; how to compromise between multitude and diversity for the benefit of all.

HUMAN DIMENSIONS OF INFORMATION NEED

Frank N. Marzocco

The Cleveland State University, Cleveland, Ohio, U.S.A.

Synecdochism is an action commonly performed in rhetoric ritual magic, and information science. In each case it involves a substitution of part for whole or whole for part. The nature of the substitution is somewhat different from one field to another, but there are strong similarities. The substitutions that occur in rhetoric and magic are straightforward, while the more complex substitutions in information science tend to include elements of both other kinds as well as some of its own.

In rhetorical usage synecdochism is practiced by mentioning the part to designate the whole or vice versa. A temporary distortion is introduced in the interest of effective speech or writing by using an inaccurate word that nevertheless emphasizes some selected aspect of whatever is being discussed. The distortion is obvious in Antony's request: "Friends, Romans, countrymen, lend me your ears". Generations of schoolchildren have laughed at the picture of a crowd of Romans proffering their ears in outstretched hands. The distortion is subtle as Antony continues his speech, painting Brutus and the other assassins of Caesar over and over as honourable men, convincing the crowd of its falsity by the very repetition of that one-sided characterization. Heightening or changing the awareness of the listener is brought about by the choice of words, not by changing the meaning of whatever words are chosen. Synecdochism in rhetoric is like other rhetorical devices in attempting to influence the listener while leaving the rest of the world alone.

As a magical device synecdochism is very definitely an attempt to influence the world. Belief in sympathetic magic is practically universal among primitive peoples and fairly common among the civilized. Almost all practitioners of sympathetic magic accept that charms worked on a part are much more effective than those worked on a model or something only loosely associated with a person or object in the past. Hair clippings and nail parings are especially potent parts according to a belief that goes back at least to Zoroaster, who flourished two and a half millenia ago. The seventeenth chapter of the *Vendidad*, the sacred Zoroastrian book, prescribes a complex set of rites to assure the harmless disposal of hair and nails.

Burying hair is synecdochistically equivalent to burying its former owner. An attempt to procure hair for such use was responsible for a notorious killing that took place in 1928 in York, Pennsylvania, U.S.A. As John Blymire felt progressively less able to sleep well, eat well, or talk well, he visited a succession of what in that time and place were known as powwow doctors, folk healers who professed to achieve cures by magical means. All decided he was bewitched, but only one was able to identify the origin and suggest a remedy. She worked a spell around a dollar bill to determine that a known sorcerer, Nelson Rehmeyer, had laid the hex that caused the illness. To lift the hex the victim would first have to take from the magician a particular grimoire and a lock of hair. After that the victim would bury the hair deep in the ground behind a barn, presumably with an appropriate incantation read from the book that had been taken along with the hair. Symbolic death of the sorcerer would bring relief from the sorcerer's evil spell, just as a real death would.

Blymire enlisted two young fellow-believers to help him carry out the plan. Confronted by the trio at his isolated farmhouse, Rehmeyer did not object to parting with the book of spells, but he did object to giving up any of his hair. He preferred to fight. During the course of the battle that resulted, one or another intruder struck the ill-starred sorcerer, kicked him, knocked him down, beat him with a stick of firewood, battered him with a chair, tied a rope around his neck, dumped kerosene from a lamp all over him, and finally set fire to what by then must have become a lifeless body. The intended dispossessors departed without either book or hair. Of course, neither was any longer needed to remove the hex. Any connection with information science ends at this point of the story, but it may be of interest that Blymire and both his companions were found guilty of murder a few months later. Blymire and one youth were sentenced to life imprisonment, the other youth to twenty years. By a quirk of fate, Pennsylvania law required leniency because no robbery had been committed in connection with the murder.

Magical synecdochism is not always so exciting, despite a general disposition to employ magic only in situations having considerable emotional content. The state of feeling has nothing to do with the theory or practice of any magical rite. If the magician's art cannot be applied directly to the intended object of enchantment, then some surrogate is found so that sympathetic magic can be used. The principle of synecdochism is that an actual part of the object of interest is usually the best surrogate. Unable to achieve mastery over what is really of concern, the magician devotes attention to some fragment instead. Lo! By magic, the same purpose is accomplished as if the magician's power had not been limited in the first place.

When a word like *information*, rich in meaning and full of human significance, is restricted to refer only to a measure of some particular physical quantity, that is a kind of synecdochism. When a word of fairly limited meaning like *decision* is generalized so that it refers to the totality of observable human behavior, that too is a kind of synecdochism. These examples typify the use of the mechanism in information science, where elements of rhetorical and magical synecdochism both occur, but where some additional, unique features are also found. By analyzing similarities and differences, it is possible to reach some conclusions about information science that might not emerge without the comparisons.

Synecdochism in information science is similar to that in rhetoric on several counts. Both involve words in whole-for-part or part-for-whole substitutions, and both emphasize selective aspects of whatever the words are used to name. The two usages are different in that information science substitutions are not purely one word for another but include words and meanings simultaneously. A word that normally refers to a whole is given the restricted meaning of the part; a word that normally refers to a part is given the generalized meaning of the whole. There is another difference, too, in that substitutions in information science are intended to be permanent. Instead of a temporary emphasis, every use of the synecdochistic term is an attempt to affirm some unusual feature of the concept or thing described.

Magical synecdochism is done with things, not words, which makes it different from synecdochism in information science, but the purposes appear to be identical. In both fields, if knowledge or control cannot be gained over the real object of interest, it is established over a surrogate instead. In forming the surrogate, the magician must always substitute a part for the whole, while the information scientist has the additional freedom of taking a whole for the part in some applications. In either case, for maximum effectiveness at least a part of the real object of interest is always included. The presence of that piece of whatever can be identified with the real problem seems to be enough to convince either the magician or the information scientist that the substitute problem is the same as the original one. Perhaps the strongest similarity in the two fields is that both practitioners end up satisfied that such an approach is perfectly valid.

Meanings inevitably undergo some sharpening and shifting as they are taken from everyday discourse and put to scientific use. That effect doubtless accounts for many substitutions that can be noted in information science. Something else seems to be at work too, though. An unnecessary and possibly willful urge to do violence to accepted meanings seems to characterize at least some attempts at what is usually presented as terminological reform. Some efforts of information scientists to shape words to ill-fitting employment are reminiscent of the efforts of Procrustes to shape his guests to ill-fitting beds.

Any perceptive reader of the information science literature will note many examples of misuse of words. Fairthorne[1] remarks that terminological corruption in the field has gone so far as to constitute scandal. He observes that confused terminology confuses both aim and effort and relates a number of instances supporting that contention. Shera[2] gives an example that is specifically synecdochistic in the description of past efforts to substitute the more limited term *documentation* for the more inclusive one *librarianship*. His discussion of the reasons makes it quite clear that the attempt was undertaken more to satisfy personal needs of members of the documentation faction than to respond to any need of science.

Information, the very word from which information science takes its name, is for that reason probably the most stretched and chopped term in the whole field. Balked in the desire to understand and control information, either the concept or the thing, workers typically respond by focussing attention on some related aspect that is easier to deal with. Then, having achieved mastery over something else, they try to convince themselves and others that the fundamental problem of information has been solved thereby.

Violence committed on the term *information* seems to be rooted in a gentle attempt to improve telegraphy. In a piece written about a half century ago, Nyquist[3] considers the problem of choosing codes that will maximize the amount of what he calls intelligence that can be transmitted over telegraph wires in a given time. His use of the term *intelligence* for whatever he is talking about seems not to reflect any philosophical position. He simply responds to the need for an English language word in the prose describing the meaning of his equations and appears content to use any word that happens to come to mind.

Not so when Hartley[4] takes up the flag. In extending Nyquist's mathematical theory, Hartley also points out that it applies to a great deal more than transmission of telegraphic codes. He discusses even television. Hartley chooses the word *information* to name the measure defined as the logarithm of the number of symbols available to be chosen by the sender and received by the listener. He is perfectly certain that this limited, partial conception accounts completely for any problem that may come up in understanding information. He insists vehemently that information is a quantitative concept based solely on physical considerations and absolutely independent of any psychological consideration whatever. While not convincing, the argument concerning the independence of information from psychology is so passionate that one is reminded of a Pygmalion in reverse, desperately trying to caress every trace of life out of his creation.

Passion abates in the work of Claude Shannon[5] who manages to remain equally calm whether discussing entropy of an information source, capacity of a noisy channel, or the level of redundancy in *Finnegans Wake*. His concern is the rate at which information can be produced and transmitted under a wide variety of physical conditions. He does not really care about any uniquely human aspect of communication. Emphasizing the sender's freedom to choose messages rather than symbols, even though the message may consist only of a string of symbols, is a kind of humanizing of the concept of information. Limiting information to correspond only to the initial freedom of the sender's choice is a kind of dehumanizing. Shannon is interested in neither, discusses only the mathematical implications, and does it clearly and coolly. Even the mathematics is treated very deliberately. In connection with the expression for the amount of information produced by a Markoff process representing a discrete information source

$$H = -K \sum_{i=1}^{n} p_i \log p_i$$

Shannon remarks that neither the theorem nor the assumptions required for its proof are necessary for development of the rest of the theory, that they are there only to lend plausibility to some later definitions. Shannon's low key approach is apparent again when he observes that meaning has no relevance to the engineering problem in communication and whenever he uses the term *information*, which he treats as an ordinary word rather than as a technical term.

It is often remarked that Shannon's work in information theory has little bearing on information science. The remark is probably true of the mathematical theory, which is elegant, deep, and extremely limited in scope. It is not true of Shannon's lucidly explained conceptual model of the communication process. The communication model is simple, clear, and so general that its traces are easily discerned in works on neural functioning, person-to-person communication, spread of rumors, diffusion of innovations, libraries and information centers, and others. The model divides a communication system into five parts; an information source, a transmitter, a channel, a receiver, and a destination. The information source produces a message, the transmitter converts the message into a signal, the channel carries the signal across space, the receiver turns the signal back into a message, and the destination takes the message. Description of the model, while clear, is not exhaustive. The last step in the process, the destination, receives so little attention that it appears to have been added only for reasons of symmetry. Other steps are described more fully, but the interpretation is literal throughout. No alternate interpretations outside of electronic communication are drawn. Words signify their everyday meanings, neither more nor less.

The rhetorician and magician of the Shannon-Weaver team is Weaver. His interest is in meaning, and nothing is irrelevant to that interest. The problem of communication Weaver sees as existing at three levels. There are technical problems of the accuracy with which the symbols of communication can be transmitted. There are semantic problems of the precision with which the transmitted symbols convey their desired meaning. There are effectiveness problems of the success with which the meaning of a message leads to the conduct desired in the recipient. The latter two levels, consisting of the semantic and effectiveness problems, are obviously related to meaning, if in no clear fashion. The first level, consisting of technical problems, is not. But the first level is the one Shannon's formulas fit. Ergo, the first level too must relate to meaning. Weaver shows how.

Adopting the tactic of rhetorical synecdochism, Weaver begins by using the word *information* to denote a miniscule part of its usual meaning. With a fervor that would do credit to a gospel singer, he develops the thesis that information is nothing but a measure of the freedom of choice available to someone selecting a message from an ensemble of messages, any one of which might be chosen to be sent. It is no matter that in thus revising the definition he takes no account of the content or meaning of the message. Weaver argues that the information is not in the message anyway, but in the process of choosing it.

In everyday discourse, to speak of information is to speak about the communication of knowledge from someone who knows something to someone who does not. This procedure is initiated whenever a communicator decides what is to be said from all the things that could be said. In this sense it is true that the everyday notion of information involves a sender's choice. The everyday notion also involves a great deal more, most of it having to do with the content or meaning of the messages passed. In answer to any possible disagreement as to the completeness of a concept of information that makes no reference to the content or meaning of messages, Weaver argues that the definition is only intended to hold at the technical level. At that level, he assures us, it is necessary to characterize the statistical nature of the information source rather than any property of particular messages the source might produce. Only so can an engineering system handle all the possible messages that might be produced. Why that characterization has to be called *information* is another matter.

Having established mastery over the part, Weaver proceeds to invoke the magical synechdochism that gives mastery over the whole. In this case it is done by taking advantage of the very emptiness forced on the concept of information by dealing with it as no more than a term at the technical level. Weaver calls attention to the great generality of the concept so considered. Since information defined as a measure of the sender's freedom of choice in selecting a message does not depend in any way upon the meaning of the messages available to be selected, it applies to all such messages regardless of possible meaningfulness. Hence also it applies to meaningful messages. Thus it is established that the limited, technical, first level analysis of information also relates to meaning. Weaver reads something profound into this observation. He observes that a real theory of meaning may now be possible for the first time. He also observes that the concept of information developed in his discussion may seem bizarre. It does. It is.

When a tree falls in a remote forest and no one is around to hear the crash, is there any sound? The answer depends on the way sound is defined. If sound is taken to be a physical event, vibration in some transmitting medium, then the answer is "yes", but if sound is taken to be a psychological phenomenon, a sensation, then the answer is "no". The question is valuable, not to elicit one answer or the other, but to emphasize that there are grounds for either answer. Once it is accepted that there are two quite different aspects to sound, it becomes easy to see that study of the physical characteristics, frequency and amplitude, is not sufficient to gain knowledge of the psychological attributes, pitch, loudness, volume, and density. Indeed, recognition of two aspects brings awareness of a third, the relationship between the physical and the psychological. A complete knowledge of sound requires study of physical characteristics and psychological attributes and their psychophysical interactions as well.

When a message travels through a communication system and no one is around to interpret the output of the receiver, is there any information? The answer depends on the way information is defined, whether as physical event or psychological phenomenon. Each observation made above about sound can be restated in a parallel form for information, although what comprises the psychological attributes of information and their interactions with the physical aspects is not at all clear.

All the studies considered so far have tried to deal with

information as a set of physical events; the psychological aspect has either been disregarded or denied. That an approach limited to the physical aspects of information is at best one-sided has been recognized. Thus for Ackoff[6] the meaningfulness and value of information are the central issues to be resolved.

Like a Chinese cook, who cuts and chops all the ingredients in advance of the actual cooking, Ackoff gets all his preliminary definitions ready before he uses any of them. Some of the principal ones are value, individual, purposeful state, course of action, efficiency, and possible outcome. Each of these and many more are given mathematical expression plus an explanation in natural language. Once that is done, Ackoff is able to assert that the value an individual puts on a purposeful state is a joint function of three multivariate arguments. The assertion is made as an elegant mathematical statement,

$$V(S) = f\ (\ P_i,\ E_{ij},\ V_j\ ).$$

In this statement, P_i is a set, each member of which is a probability that the individual will choose a particular course of action. Since each course of action has some probability, the total number of members of P_i is the same as the number of courses of action. E_{ij} is another set. Each of its members is the efficiency of some course of action in achieving some possible outcome. Since every course of action is assumed to have some efficiency to achieve each possible outcome, the total number of values in E_{ij} is obtained by multiplying the number of courses of action by the number of possible outcomes. V_j is still another set, with each member the value the individual places on some possible outcome. The total number of values in V_j is the number of possible outcomes.

Of course, the mathematical statement about V(S) only describes V(S). It says nothing about a way to calculate the value an individual places on a state. Not only would all the components have to be measured (Ackoff talks about that problem, but he doesn't really say anything), it would also be necessary to specify the exact functional relationship. One example of a possible relationship is given, but as an illustration, not for real use. The reader who decides at this point that this is all a lot of mathematical puffery and gives up trying to understand the theory makes the right decision but takes the wrong action. The mathematics is banal, concealing meaning where some good prose would explicate it, but the meaning is there. The V(S) turns out to be another stage of preliminary preparation - like marinating the meat after slicing but before cooking. Its part in the final production becomes apparent when communication is discussed, for receipt of a communication is taken to bring about a change in the purposeful state of the recipient, and thus a change in the value of the state. But the change in value of the purposeful state is necessarily accompanied by a change in one or more components.

When the value an individual puts on a purposeful state is changed because a communication changes the P_i, the probabilities of choice associated with available courses of action, the individual has been informed of something. When the value is changed because the communication changes the E_{ij}, the efficiencies of various courses of action to achieve various possible outcomes, the individual has been instructed. When the communication changes the V_j, the values of the possible outcomes, the individual has undergone a change in motivation. Ackoff goes on to derive some equations that purport to quantify the amount of information, instruction, and motivation communicated by a message, but this is more mathematical garnishment, useless because there is no way to assign meaningful measures to any of the variables. What is not useless, at least potentially, is the observation that there are several important ways communication can affect the recipient of a message. The agent of communication, the message, causes a change to take place in the recipient. The message acts, and because it acts it has to be studied as a dynamic force, not ignored or treated as a mere passive carrier of information. More than that a person responds to a message in no simple fashion. The response depends on the current state of several distinct personal process variables, so that the person has to be considered as a manifold responder to the message rather than a simple receiver of it.

Enlarging the area of usual concern about information in apparently fruitful fashion, as Ackoff does in forcing attention to the message and its recipient, leads to the question whether further enlargement might also be worthwhile. Yovits and several colleagues report an affirmative answer in several publications, most recent of which is Whittemore and Yovits[7], although a fuller discussion of points discussed here is found in Yovits.[8] The fundamental notion is that by following what happens to information after it has been presented to the recipient, a closed loop can be traced in several stages back to the point of presentation. Yovits identifies four essential functions as occurring during the circuit and describes them in terms of a model called the generalized information system.

Of the four, the information acquisition and dissemination function of a generalized information system most nearly corresponds to the source in a conventionally represented communication system. Inputs to this function consist of data, some gathered from the immediate environment of the system, some from a data base internally stored as a result of past interactions with the environment, and the rest from the system itself as output from the preceding function. Processing of data may be done by storing or operating on them. If the latter, the information acquisition and dissemination function may restructure, filter, weight, select, reject, analyze, sequence, order, predict, display, or otherwise treat the data. Processed outputs of the function are also data, but they become information if they are used in the next stage.

The decision making function of a generalized information system takes all the output of the information acquisition and dissemination function as its input and processes it initially in much the same way as the preceding function processes its data. The decision making function goes on, however, to make decisions. It does so basically by choosing an appropriate course of action with the aid of a predictive model developed with whatever information is available. The choice is transmitted to the next stage to be transformed into observable actions. As data pertaining to any past decision is received, the decision making function may choose to continue or modify either the predictive model or the related course of action.

Two more limited stages follow. Output of the decision making stage goes to an execution function, where the selected course of action is transformed into an actual observable action. The execution function is influenced by the outside world as well as the decision function, for which reason its outputs are not completely determined in advance. Since they are physical events, the observable actions may be observed and measured. In addition to acting on the environment, then, observable actions serve as input to the last stage. A transformation function takes observable system actions as its input, turns them into data, and provides the data to the information acquisition and dissemination function to begin another iteration of the information cycle.

Virtually all situations involving a flow of information are said to behave as generalized information systems. Applications as varied as the use of information by an individual research scientist, command and control of a military engagement, and switching on or off of a furnace-thermostat system are suggested. All these are closed-loop systems, and describing information flow in closed-loop terms seems to offer some intriguing new possibilities of analysis.

Closed-loop systems with a human component have been studied extensively by human factors engineers. Typical systems are the tracking tasks in which a set of continuous system processes are directed toward a changing or changeable goal and in which data on system operation are presented to a human operator who manipulates a set of control devices so as to correct for discrepancies between actual and desired system outputs. For an example, consider how an automobile driver regulates the path or speed of the vehicle on the road. Other examples are obviously provided by many of the systems proposed as generalized information systems, for the two definitions parallel one another fairly exactly. Of course, not all generalized information systems are tracking systems, but many are, and it appears that all tracking systems are generalized information systems.

A considerable body of literature on the design and description of tracking systems exists. Unfortunately, very little of what is known can be applied to generalized information systems because very little of what is known can be applied even to all tracking tasks. One problem is the tremendous variation among classes of tracking systems. The behavior of any one system is strongly influenced by its particular physical characteristics, and the physical characteristics of, say, an automobile, a commercial airliner, and an anesthetized patient are too different to allow for much common theory. Another problem in describing systems with a human component is that people are not machines and do not behave like machines. Many serious attempts have been made to describe human tracking performance by means of transfer functions, assuming that a human operator behaves to some degree of approximation like a linearly responding servomechanism. All have foundered on the shoals of human complexity even when restricted to single systems. Mathematical equations that purport to describe human performance across a variety of systems are either a sham or express some trivial factor at best. What is true of a subset cannot be denied for a set that includes it; hence the claims for all-embracing applicability of the generalized information system notion can be taken to apply only at the level of analogies and metaphors, the level of synecdochism.

Looking back over these many attempts to explain the nature of information, it is apparent that none of them says much about the reason a person may pick up a telephone or enter a library. Nyquist, Hartley, Shannon, Weaver, Ackoff, and Yovits differ in rhetorical stridency and rhetorical effectiveness, but all are like Blymire battling for that lock of hair that will give control. Each grabs a part that he knows what to do with; when people must be considered, they are treated as mathematical or symbolic abstractions. These may be invaluable pieces, but the whole picture of information can only be constructed with some regard for the real people who create, manipulate, use, and otherwise affect information, and who are affected by it.

Fortunately the practice of information-related activities is not constrained by the limitations of theory. Just as farmers managed to grow their crops for many years before they got much help from science, so librarians and information center personnel still manage to do their jobs without an adequate theory of information. The decoupling of practice from theory is quite evident when considering the relationship of information systems to people. Libraries, information centers, information systems of all kinds exist to serve people. The principle that an information system is completely the tool of its users and potential users, that their purposes are its purposes, that their needs are its opportunities is nearly universal in its acceptance. The principle is not always honored, but it is never abandoned for long. The theorist is free in portraying a long chain of information activities to put an abstract consumer at the very end, but experience forces the practitioner to put the real patron or client first.

In addition to users, at least four other sets of people can be identified as playing distinct, important roles in relation to information systems. The functions of some of these groups can be seen nicely in a library or information center interpretation of the Shannon communication system model discussed previously. The Shannon model includes a source, transmitter, channel, receiver, and destination. Looking at a library as a communication system, the part of the source is taken by the authors represented in the library, while that of the transmitter is taken by publishers and other document-producing organizations. The channel, which connects production functions to consumption functions, becomes for the library the various means of document distribution on the outside as well as acquisition, cataloging, and storage functions carried out inside the library. The receiver in the library interpretation consists of the equipment, procedures, and people comprising the retrieval capability, since these are what it takes to get the "message", the document, to its destination. The destination, of course, is the reader or other user.

Information systems have a definite communication purpose, a feature that lends legitimacy to their identification with communication systems. Correlating the two conceptions shows how complex an information system is when looked at as a whole, when information is traced, as the engineer might put it, from source to sink. Still, for all the complexity, human contributions to the total process can be summarized under the three headings of beginning, middle, and end. At the boundaries are people who originate information in the beginning and people who use it in the end. Inside the boundaries are all the people between, performing system functions to meet system goals of channeling information appropriately from originator to user.

Two other groups do not contribute directly to the information transmittal process, but they are vitally important to any information system. One consists of the system designers and developers. It is rarely forgotten that information systems are created for people and that they consist largely of people, but it is seldom remembered that they are also put together by people. The last group also is seldom considered, despite its tremendous size. Inside, outside, and all around every information system are people who make up a social nexus connecting the system with all the entities and events and people of everyday life. The environment of a system consists not only of things and what things do, but also of people - all the people.

Least likely to be overlooked in any discussion of people who relate to information systems are the users. Every volume to date of the *Annual Review of Information Science and Technology*[9] has carried a chapter on *Information Needs and Uses*. The fact indicates a considerable interest in the consuming end of the communication chain if not necessarily in the consumer.

Studies of the user tend to deal with the particular requirements of a given field and often are limited even further to user needs that can be fulfilled by a specific information system. Current state of the art in user studies, with special attention to those in the social sciences, is discussed by Brittain.[10] He has little good to report. Most studies employ survey methods to determine user demands for documents. They are weak in conception, often depending on such strange assumptions as that the behavior of scientists is rational and orderly. They are weak in execution, often using methods borrowed uncritically from other disciplines and then applied inappropriately. They are weak in interpretation, taking expressed need, for example, as the equivalent of actual need.

Very few user studies deal with the actual needs created by a scientist's work, and very little is known about the use of information once it has been acquired by the scientist. As for studies of the needs created by the scientist's being a person:

> It would be pertinent to enquire about the stimulus value of information, user assimilation capacity and overload, redundancy in information systems, optimisation of information intake, preferred signal/noise ratio, and relationship between information and action ... Other aspects of use about which little is known include the effect of informal contacts on use of information, the function of seemingly casual conversations, the nature of scientific writing, the way in which information is abstracted and used in the creation of new material, and the sociological aspects of users and patterns of use. There are also ... the relationship between motivation and information seeking and information usage, and the relationship between personality factors and creativity and productivity.[11]

Other commentators share Brittain's conclusions that the typical user study is a bad piece of work and the typical user has still to be studied. Allen[12] takes a generally optimistic view except when he discusses the trivia that result from assuming that questionnaire design and administration can be mastered in a single attempt without training. Paisley[13] takes methodological exception to shallow conceptualization. He suggests that the work and social context of the scientific information user has been neglected and cites many studies that need to be done. He constructs a framework of concentric systems, ten of them, to help organize the effort. Rees and Schultz[14] also suggest that the study of psychological variables might be of some help to information science, but they suggest that not much of that help is likely to bear fruit in the near future. Where they see a great lack of suitable methodology in information science, they also see great opportunity, for they suggest wholesale adoption of the experimental methodology of psychology.

Some of what Rees and Schultz suggest no doubt makes good sense. Every scientific field needs its paradigms, methods accepted without question by all the workers in a common arena. A discipline without paradigms must get them from somewhere, but to take them from psychology while leaving behind the concern with people that the methods represent - that is perverse. Information science no less than psychology is a science of human behavior, although it is also more than that. It is true that the concern of the information scientist is not with all behavior, but only with the information-handling part of it. Yet information-handling behavior is like love-making behavior or bicycle-riding behavior - something done by people. Learning, motivation, personality, and other kinds of psychological variables affect all behavior, not just that tagged as psychological. Information science may be able to use some of the psychological methods, but what it needs is some of the psychology.

To introduce a chapter on human factors within information systems, Saracevic[15] makes some telling comments about the typical treatment of the human component. People inside the system are often considered nuisances, stubbornly defying orderly manipulation. People are treated as adjuncts to the system, not the other way around. Such attitudes exist in spite of knowledge that it is not a hypothesis but a tautological statement that systems and people within systems affect one another strongly. There does not appear to be any great interest in the problem, either, since Saracevic was able to find only four good papers demonstrating good human factors practice with people inside information systems.

There are really two problems in dealing with people inside the information system. The first is to find ways to make the people truly effective as parts of the total system. People perform a vast number of functions in helping information move between its creator and consumer. There are many publication and distribution functions[16] that precede acquisition of a document by an information center, and there are plenty of functions after that too. All these involve some human effort, and practically all could be improved with better knowledge of related human capabilities and limitations.

Improving the system is not the only reason to pay attention to the people inside the information system. The second problem in dealing with this group of people is to be sure to treat them as people and not as system components. The human factors approach is a way to assure systems that are highly efficient and effective when evaluated against limited, short-range, system-oriented criteria. It can be destructive in the long range, though, if people fight the system because their needs and wants take second place.

In addition to users and people inside the information system, another group of people important for information systems and active in the communication process consists of the creators of information. Rarely are authors considered part of a formal information system, but they are certainly part of the total communication process. Both Klare[17] and Orr[18] emphasize the importance of including the author in any account of information transfer. Both are concerned with scientists as authors, of course, and they might not feel so strongly about poets or novelists. Whatever the class of writing, full understanding of information transfer processes probably requires as good a knowledge of the writer as of the reader. Little labor has gone into scientific study of the factors influencing the writer. Perhaps the greatest effort has been directed at the productivity of scientists and engineers, work reported by Pelz and Andrews.[19]

A fourth group of people is almost never mentioned as having psychological needs and wants satisfied in some way by an information system. That is the group engaged in design, development, production, and installation of such systems. The existence of the group has not gone totally unremarked, for Kent[20] suggests four possible objectives underlying design of a system. These are to maximize profits, fulfill an obligation, enhance the designer's reputation, or to provide novel or attractive services in a competitive situation. Kent remarks on the lack of altruism in these objectives, presumably to warn his students about the facts of information system life.

As with most descriptions of human purposes, the objectives discussed by Kent form too narrow a set to apply accurately to real people, but the more usual conception, that the sole purpose of the system designer is to do good for someone else, does not apply accurately either. The forces that power human behavior are part of the human estate. Forces considered shameful do not retire from action just because some more acceptable forces also help impel the system designer toward socially welcome goals. Every person in the world represents a coming together of a lifetime of influences, all contributing to what the person is and does. Since the system designer is also a person, it is no occasion for surprise or dismay that the designer behaves like a person.

Effects produced on people by an information system touch many more than those with a direct relationship to the system. Likewise, effects produced on the system come from many more than the system's particular users, workers, authors, and designers. The system is like a traffic officer in the middle of a busy intersection. Only the cars that actually enter the intersection arouse and respond to the officer's signals, but any event in the intersection may be the cause or effect of happenings all over the city.

Public libraries, perhaps because their constituencies tend to be the largest and most varied, seem most ready of all information systems to accept the concept of wide-ranging influence. A 1969 report on the Chicago Public Library projected only half a dozen years ahead to the time when: "The library within four walls will have given way to the concept of the library as an immediate and integral part of the lives of people - on the block where they live, in the groups with which they associate, in the jobs they hold, in the cultural and recreational activities which give life and color to their existence."[21] It is true in a way that this is a statement about users, but these are users so broadly considered that they coincide with the public at large.

A fearsome aspect of failure to consider that a library or information center relates to society as well as its users is the tendency to fragment knowledge into ever finer pieces. The problem has been described as a cognate tendency of new-wave thinking and pictured as especially disturbing for a universal library,[22] but it should be equally disturbing for a small technical library or information center. With any user population whose principal information needs are limited and well defined, strong forces operate to see that the principal needs are satisfied quickly and completely. There is a tendency to make all that and only that material known to be relevant easily available to the user. Everything else is relegated to the background or not made available at all. The information system becomes a mere procurer of predetermined data, engendering intellectual starvation, when it should be carrying out the vital social and personal function of encouraging intellectual development.

User, worker, author, designer, person - these are five dimensions in which to locate any discussion or investigation of the relationship of people and information systems. They are human, not physical dimensions; they are not orthogonal; they have no measure; but they do label different concerns for the way information systems provide satisfaction of basic human needs. They relate to the idealistic concept of library strategy,[23] but they indicate a realistic way to implement the strategy. Each of these human dimensions identifies a large set of common human experiences. At least a basic understanding of each dimension can be gained by adopting the corresponding point of view long enough to get to feel like a user, worker, or what have you.

The proposal seems almost too simple, but it works. Collison[24] uses essentially this method. He points out what a difference point of view makes by an anecdote of a librarian's sketch to be included in a users' guide. A group of art students observed that the sketch would not do because it had been drawn literally from the librarian's point of view. It showed the library as it appeared when looking toward the street, whereas a reader needed it the other way around, as it appeared on entering from the street. Collison takes the story seriously and writes a good book on reader services by putting himself constantly in the place of the reader. The information scientist confronting any of the human dimensions would do well to go and do likewise.

References

1. Fairthorne, R.A., "Use" and "mention" in the information sciences, in *Education for Information Science*, Heilprin, L.B., Markuson, B.E., and Goodman, F.L., Eds., Spartan Books, Washington, 1965, 9.

2. Shera, J.H., Of librarianship, documentation, and information science, *UNESCO Bulletin for Libraries*, 22, 58, 1968.

3. Nyquist, H., Certain factors affecting telegraph speed, *Bell System Technical Journal*, 3, 324, 1924.

4. Hartley, R.V.L., Transmission of information, *Bell System Technical Journal*, 7, 535, 1928.

5. Shannon, C.E. and Weaver, W., *The Mathematical Theory of Communication*, University of Illinois Press, Urbana, 1949.

6. Ackoff, R.L., Towards a behavioral theory of communication, *Management Science*, 4, 218, 1958.

7. Whittemore, B.J. and Yovits, M.C., A generalized conceptual development for the analysis and flow of information, *Journal of the American Society for Information Science*, 24, 221, 1973.

8. Yovits, M.C., Information science: toward the development of a true scientific discipline, *American Documentation*, 20, 369, 1969.

9. Cuadra, C.A., Ed., *Annual Review of Information Science and Technology*, 1-2, Interscience Publishers, New York, 1966-1967; 3-7, Encyclopaedia Britannica, Chicago, 1968-1972.

10. Brittain, J.M., *Information and its users. A review with special reference to the Social Sciences*, Wiley-Interscience, New York, 1970.

11. *Ibid*, 148.

12. Allen, T.J., Information needs and uses, in *Annual Review of Information Science and Technology*, 4, Cuadra, C.A., Ed., Encyclopaedia Britannica, Chicago, 1969, 3.

13. Paisley, W.J., Information needs and uses, in *Annual Review of Information Science and Technology*, 3, Cuadra, C.A., Ed., Encyclopaedia Britannica, Chicago, 1968, 1.

14. Rees, A.M. and Schultz, D.G., Psychology and information retrieval, in *Information Retrieval; a critical view*, Schecter, G., Ed., Thompson Book Company, Washington, 1967, 143.

15. Saracevic, T., *Introduction to Information Science*, R.R. Bowker, New York, 1970.

16. Doebler, P.E., Publication and distribution of information, in *Annual Review of Information Science and Technology*, 5, Cuadra, C.A., Ed., Encyclopaedia Britannica, Chicago, 223, 1970.

17. Klare, G.R., Writing with a helping hand, in *People and Information*, Pepinsky, H.B., Ed., Pergamon, New York, 1970, 131.

18. Orr, R.H., The scientist as an information processor, in *Communication Among Scientists and Engineers*, Nelson, C.E. and Pollock, D.K., Eds., Heath Lexington Books, Lexington, Mass., 1970, 143.

19. Pelz, D.C. and Andrews, F.M., *Scientists in Organizations: productive climates for research and development*, Wiley, New York, 1966.

20. Kent, A., *Information Analysis and Retrieval*, Becker and Hayes, New York, 1971.

21. Martin, L.A., *Library Response to Urban Change*, American Library Association, Chicago, 1969.

22. Library of Congress, The Library of Congress as the national library: potentialities for service, in *Libraries at Large*, Knight, D.M., and Norse, E.S., Eds., R.R. Bowker, New York, 1969, 435.

23. Churchman, C.W., Operations research prospects for libraries: the realities and the ideals, in *Operations Research: Implications for Libraries*, Swanson, D.R. and Bookstein, A., Eds., University of Chicago Press, Chicago, 1972, 6.

24. Collison, R.L., *Library Assistance to Readers*, 4th ed., Crosby Lockwood and Son, London, 1963.

INFORMATION AND METAENERGY

Glynn Harmon

University of Texas at Austin, Texas, U.S.A.

The prefix *meta* is used to denote or imply *between, with,* or *after*. *Meta* is used to specify that which transcends, goes beyond, is more highly organized, and yet is associated with something else. *Meta* also implies change or transformation as in metamorphosis.[1] Energy, from the Greek *energeia* (*en* in; *ergon* work) was used in Aristotelian philosophy to denote "existence in actuality". In general, energy means activity, power of action or the capacity for doing work.[2] The term *metaenergy* is used in this paper as that which occurs between, with or after energy. Metaenergy is also treated as that which is instrumental in the change, transformation and control of energy.

Definitions of *information* are numerous and dynamic. It is now popular to correlate or equate information and energy: "Information is a commodity no less tangible than energy; if anything, it is more pervasive in human affairs."[3] Going further, "information, like electricity, is a form of energy ... information is energy for mind work".[4] Additional definitions and conceptions of information and energy are treated herein to clarify relationships between them. The hypothetical position of this paper is that information can be viewed as metaenergy or one of its forms.

EVOLUTION OF THE ENERGY CONCEPT

The energy paradigm has evolved from ancient and specialized notions of energy as a source of force, to modern and generalized notions of energy as the universal currency of nature and science. Modern notions of energy are recent results of a long and intricate process of conceptual transformation and refinement. Evolution of the energy concept may conveniently be divided into five consecutive stages.[5]

(1) Aristotelian conceptions of energy as a source of *force*, wherein force signified the immediate cause of motion and also its cumulative determination or energy content. Distinctions between force and energy were blurred and there were few, if any, real differentiations between different forms of energy.

(2) *Mechanical* conceptions of energy, popular during the Renaissance and the times of Galileo and Newton, viewed energy as that which can be transformed into work or that produced by *work*. Energy was measured in terms of force times velocity or distance, but natural philosophers were still preoccupied with mechanics - namely with force and work. Mechanical notions "gained priority and were regarded as more fundamental than the concept of energy, which thus appeared as a derived conception".[6] Even today most definitions of energy are generalizations of the notion of work in mechanics.

(3) Three notions were established by the middle of the nineteenth century: (a) that different *forms* of energy exist; (b) that one form could be *converted* to another; (c) that their sum total could be *conserved*. In particular, clear distinctions had been made between potential and kinetic energy, between heat and motion, and between food and organismic energy. The forces of nature, previously regarded as disparate and incommensurable, came to be viewed as one "great association" in which magnetic, electrical, chemical and organic phenomena were interrelated and interconvertible. Thus it was recognized that energy could be neither created nor destroyed, but only consolidated, dissipated or transformed.

(4) The *universality* of energy and its relationship to *space* and *time* was acknowledged in the last half of the nineteenth century. Ostwald, for example, argued that all phenomena of nature were merely manifestations or transformations of energy. He also related knowledge to energy in that sound waves, light waves and chemical changes are requisites of sensing and knowing. "The totality of nature appears as a series of spatially and temporally changing energies, of which we obtain knowledge in proportion as they impinge on the ... sense organs fashioned for the reception of the appropriate energies."[7]

(5) The mathematization of energy as an integral invariant came with the twentieth century recognition of mass as a form of energy, Einstein's generalization of the Maxwell-Hertz equations into $E = mc^2$, and atomic energy.

These five stages indicate how the energy concept gradually became more general and abstract in order to accommodate newly-discovered energy phenomena, and to reveal the universality of energy.

PHYLOGENY OF THE NERVOUS SYSTEM

Relationships between information and energy can be explored by an analysis of organic evolution, particularly by studying the phylogeny of the nervous system. From protoplasmic irritability (the capacity to respond to stimuli) to active cognition is a development which has required upwards of a billion years. The following quotation summarizes a book on neurological evolution and suggests relationships between information and energy:

> The change from a unicellular plan of construction to a multicellular one permitted animals to become larger. With multicellularity came a greater need for coordination; with the increase in size came a greater need for the conduction of information within the organism. To meet this need, certain cells of the multicellular organism specialized in enhancing both its irritability and the rapid conduction of irritability changes through the animal. With further specialization, these cells became so organised that there grew up a division of labor to detect environmental change (both inside and outside the organism), to conduct information, to integrate information, and to initiate a response. These cells became the nervous system, the matrix of behavior ...
>
> The main function of the nervous system is to control the release of energy through the contraction of muscles and the secretions of glands; it also regulates such other forms of energy as light (the firefly) and electromotive force (the electric eel). Of these effects, muscular movement is surely the most widespread and important. Until complex muscular systems evolved, intricate behavior was impossible even though the nervous system may have possessed the potentiality for such behavior. The control of muscular energy involves more than mere efficiency of contractile mechanisms. The contractions of the various muscles must be synchronized so that the work provided by them is directed to serve the requirements of the animal. Otherwise there is a useless waste of energy.[8]

In highly-evolved nervous systems such as the human brain, the energy required to fire a neuron is approximately one billionth of a watt. If all ten billion neurons in the human brain were active simultaneously (unlikely), the whole brain would require only 10 watts.[9] This is remarkably low energy requirement for such a complex mechanism, particularly when one considers the hierarchy of neurological, muscular and chemical systems which it controls. Relative to the brain, these systems expend enormous amounts of energy. Thus there is a tremendous step-up or multiplication of energy from the brain to hierarchies of systems under its control. This energy multiplication is even larger in instances where muscular systems manipulate the control mechanisms of large machines and weapon systems. Thus, brain-power, or the information processed in the brain, appears to possess an energy leverage of extremely high magnitude. In this sense information might be viewed not only as metaenergy, but as superenergy.

The precise mechanisms for the conversion of energy into information and vice-versa are not known at this time. For one thing, we lack adequate definitions of information. Each specialized discipline or occupation proffers its own definition of information, but none apparently has viewed the whole information elephant. For another thing, even when we invoke such specialized notions of information as organismic or behavioral views, we are still at a loss to explain the "black box" innerworkings of the brain. We are forced to make hypothetical constructs about possible internal mechanisms and intervening variables which act on stimuli and produce responses. Or we might be forced into armchair contemplation of the historically confounding mind-body or mind-reality problems.

Empirical investigations have, nevertheless, provided a clearer notion of energy-information conversion. Sensory receptors can be regarded as transducers which transform energy from the environment into chemical processes that produce electrochemical nervous impulses.[10] In the case of the eye, at least, nervous impulses are filtered out as they pass from simple receptor cells to increasingly complex neural cells. Through the selective destruction of sensory inputs at a cellular level near the eye, those inputs that are not filtered out can be regarded as summarized data. These summarized data are processed by cells of greater complexity and are either selectively destroyed or abstracted into information. There is further selective destruction of the information, but it is not clear how far this process of abstraction by convergence of communication channels does go. Sensory data become meaningful only after a series of such abstractions have transformed the data into a pattern that matches a pre-existing mental structure in the organism. The outer core of the brain, the cerebral cortex, appears to specialize in producing structures necessary for discrimination, choice, learning, and thinking. Certain association areas appear to provide an "executive process" by regulating and monitoring the flow of information in the total nervous system.[11] Nerve fibers in the central nervous system lead out to connect with effectors, which in turn, lead to responding organs, muscles and glands.

In summary, evolution of the nervous system provided the means by which organisms could graduate from innate, stereotyped, and stimulus-triggered behavior to the accumulation of past experience and symbolic behavior. On the lower part of the phylogenetic scale, such organisms as protozoa, metazoa, worms, and insects are dominated by taxes, reflexes, and instincts. On the higher part of the phylogenetic scale, animals possess the neural equipment to depend also on learning and reasoning. The phylogeny of learning further reveals such phenomena as imprinting, habituation, and conditioned behavior. With reasoning, organisms were freed from slavish dependence on sensory stimuli and habit, and could adapt to the environment with insight, the use of symbols, and ultimately language. Information processing appears to be a characteristic of higher organisms who possess complex nervous systems capable of converting energy inputs into data, and data into information.

EVOLUTION OF THE INFORMATION CONCEPT

The evolution of the information concept and its emergent relation to energy is traced in a recent article by Tribus and McIrvine.[12] These authors discuss early work in thermodynamics, particularly the notions of entropy and negentropy. The thrust of their argument is that the flow of energy in human societies is controlled by the minute amount of energy that is used for the flow of information. They also summarize the work of Shannon, Weaver and other information theorists, including that of Robert Evans. In 1969 Evans derived a measure of a quantity called "essergy", which means the essential aspect of energy. Evans' equation states that when a given "thermal-noise energy per degree of freedom" ratio is below a threshold of 100, one is dealing with information; when it is above 100, one is dealing with energy. Thus, the distinction between energy and information is considerably refined if not considerably weakened. Tribus and McIrvine conclude:

> ... it is certain that the conceptual connection between information and the second law of thermodynamics is now firmly established ... The use of energy, based on considerations of thermodynamic availability, of necessity involves information. Thus information and energy are inextricably interwoven.[13]

In the context of general systems theory, J.G. Miller points out that "matter-energy and information always flow together. Information is always borne on a marker."[14] The term marker applies to those observable bundles, units or changes of matter-energy which bear information symbols. Markers, then, can refer to such things as a door key with notches on it, smoke-signals, punched cards, a difference in electrical potential in a magnetic field, the structural arrangement of a DNA molecule, the molecular structure of a hormone, or radio waves. Any form of communication, however, requires that markers move in space, and this movement follows the same physical laws as the movement of any other kind of matter-energy.

The concept of information was also basic to Wiener's formulation of cybernetic concepts. Wiener also discussed relationships between information and energy:

> Information is a name for the content of what is exchanged with the outer world as we adjust to it, and make our adjustment felt upon it ... The transfer of information cannot take place without a certain expenditure of energy, so that there is no sharp boundary between energetic coupling and information coupling.[15]

To Wiener, information was more a matter of process than of storage. Information had real validity and meaning in the context of *action* - that is in the interactions of a system with its internal or external energy environments. Wiener thus stressed the concepts of feedback and control.

CONCLUSION

In summary, metaenergy has been defined as that which occurs prior to, with, or subsequent to energy transactions and is instrumental in the transformation of energy. In turn, the energy paradigm has evolved through successively higher levels of abstraction to embrace new discoveries of specialized forms of energy. Organismic nervous systems have evolved to enable the abstraction of information from energy and to control the release of various forms of energy. The information concept has evolved to the point where energy and information are recognized as inextricably interwoven phenomena. This paper has developed the notion that information can be viewed as metaenergy.

Otten and Debons were among the first in information science to pose questions about relationships between energy and information and to propose that information science be regarded as a metascience.[16] As relationships between energy, information, systems, control, and feedback become clearer, perhaps information will acquire the locus and stature that energy did in the history of science. If it does become increasingly profitable to view information as metaenergy and to engage in the metascience of information, then one locus of that metascience would be energetics. The contemporary "energy crisis" could then be viewed from more enlightened perspectives.

References

1. *Webster's Third New International Dictionary of the English Language*, 1971.

2. Jammer, M., Energy, in *The Encyclopedia of Philosophy*, 2, Edwards, P., Ed., 511, 1967.

3. McCarthy, J., Information, in *Information: A Scientific American Book*, W.H. Freeman and Company, San Francisco, 1966, 1.

4. Drucker, P.F., *The Age of Discontinuity: guidelines to our changing society*, Harper and Row, New York, 1969, 27.

5. Jammer, M., Energy, *The Encyclopedia of Philosophy*, 2, Edwards, P., Ed., 511, 1967.

6. *Ibid*, 512.

7. *Ibid*, 517. Actually, the cognitive aspects are not proportional to physical stimuli. Doubling the intensity of a stimulus does not double its perceived intensity. Stimuli are subjected to complex analysis by the nervous system. See Lindsay, P.H. and Norman, D.A., *Human Information Processing: An Introduction to Psychology*, Academic Press, New York, 1972, 641.

8. Dethier, V.G. and Stellar, E., *Animal Behavior: Its Evolutionary and Neurological Basis*, Prentice-Hall, Englewood Cliffs, New Jersey, 1961, 2.

9. Hilgard, E., *et al.*, *Introduction to Psychology*, Harcourt Brace Jovanovich, New York, 1971, 32.

10. Lowenstein, W.R., Biological Transducers, *Scientific American*, 203, No. 2, 98, 1960.

11. Hilgard, E., *et al.*, *Introduction to Psychology*, Harcourt Brace Jovanovich, New York, 1971, 43.

12. Tribus, M. and McIrvine, E.C., Energy and Information, *Scientific American*, 225, No. 3, 179, 1971.

13. *Ibid*, 188.

14. Miller, J.G., The Nature of Living Systems, *Behavioral Science*, 16, 281, 1971.

15. Wiener, N., *The Human Use of Human Beings: cybernetics and society*, 2d ed., Doubleday & Company, Garden City, New York, 1954, 17, 39.

16. Otten, K. and Debons, A., Towards a metascience of information: Informatology, *Journal of the American Society for Information Science*, 21, 89, 1970.

AN OPINION ON AN UNDERSTANDING OF INFORMATION SYSTEM DYNAMICS*

John Nicolis

University of Patras, Patras, Greece.

Information Science is certainly a science of systems with interacting parts. What we are trying to do here is to find ways and conditions under which the interacting parts can become coherent under the action of in-flowing information.

Not all systems, of course, are susceptible to receiving and processing information in a way which is purposeful - that is, increasing their coherence towards the achievement of a goal. For instance, let us consider two systems A and B: Information flows from A to B, with the ultimate objective to achieve its goal. The goal is achieved by increasing B's coherence to mutually interacting parts of the system. But I consider this to be a hopelessly difficult task if large social systems are involved. The mutually interacting parts of B are so tied together and so non-linear, that a non-wholistic approach is senseless. Consequently, one should ask what minimal criteria can be established in the form of algorithms so as to insure the ordering of information from one component to the other can be realized?

In the first place one has to admit that when we refer to the interacting parts of system B, we have reference to physical systems where the parts interact in accordance to pre-given dynamic laws to which a statistical fluctuating process is inevitably superimposed. In such a case there are interplays within B, way before the action begins - or before the curtain is raised as it were. You have an incoherent interplay between a dynamic law and a statistical process. This makes the system strongly non-linear but near equilibrium nevertheless. As such, it is completely incoherent at noise level. Eventually if the information flowing is to be exercised in B in a meaningful, all-organizing purposeful way, B must conform with a minimum set of pre-requirements. These pre-requirements are well known from statistical mechanics of non-linear processes. Briefly, the system B must be an open system which can exchange freely energy and information with involvement and must have the ability to be applied easily, that is, with minimal energy within a confined flow of information. In this case, an increment of x with additional increments of x must be able to achieve a state of equilibrium - within what is called a limit cycle. In other words, the system B must be structurally stable. If this dynamical ability is not achievable by system B - if it is not able to go through a modeling cycle, then the situation is hopeless. If the modeling cycle is not attainable, we cannot expect incoherent interacting parts to act coherently. This conceptualization of the flow of information from A to B is important because one cannot expect the flow to possess minimal qualifications to allow for the interactions. This is similar, if you will, to electromagnetic propagation such as that found in pressures involved in acoustical processes.

The issues of the dynamic properties of interaction are fundamental to the position that A and B interact coherently. The issues represent formidable problems to physicists and engineers. The underlying concerns relate to the identification of the *process* of organization rather than the concern for the prediction of the steady states. We know for example, that communications represent transient phenomena and cannot be described as falling within the classification of a steady state. It does assume that there is a choice on the part of A and that there is a corresponding expectedness on the part of B. There is always a process of communication between A and B which is transient. Let me say that we are in similar constraints in Lazer theory, for example. In the latter case we satisfy ourselves in accepting the fact that there is an underlying steady state regime. We place all first and second order dimensions to zero and assume that there are solutions which the parameters of the system satisfy in a steady state fashion. But the problem in the present case is not the steady state situation, we are not even in a position to know if such a state exists. It is the study of the transient conditions that we are not clear about and that we do not understand.

In the final analysis the need is to determine the dynamical laws that govern the behavior of system B. In essence there is an interplay between statistical and dynamic laws. We often refer to these as representing a stochastic process. By the application of mathematics, a stochastic equation is derived - the so-called "focal-plane" equation. This will describe the probability of the distribution of the parameter under study with time, and from this the Monte Carlo method can be applied and so on.

This is my concept as to how we can approach the problem of

the transiency that exists between system A and system B - so as to give us a notion of the coherent behavior of information systems.

*NOTE

This presentation is a transcription of informal comments from tape and is included in the proceedings in lieu of a paper "A frequency entrainment model with relevance to systems displaying adaptive behaviour" published elsewhere. (Eds.)

INFORMATION AND DECISION

Robert F. Barnes, Jr.

Lehigh University, Bethlehem, Pennsylvania, U.S.A.

The relation between decision and information is clearly an intimate one - in making decisions, we often need to seek information, and information we may acquire is generally put to use in some decision-making context. This close relationship has led some to argue that information can not (or should not) be treated in abstraction from decision-making, that a decision component should be a key feature in any conceptual framework for a science of information. I want first to argue against this position, though I'll shortly qualify my argument in an important way.

Science, we should note, not only concerns itself with the actual, but also with the possible. The laws of motion, the laws of thermodynamics, etc., govern not only real bodies and processes, but possible ones. Indeed, they can be read in ways so as to assert the very impossibility of certain kinds of events, not merely their non-occurrence. Moreover, certain kinds of properties of objects, even when they are real and not merely possible, are inextricably linked to possibilities, whether they occur or not. Consider an object placed in water, for example. It may dissolve to any of various degrees, or it may not. To the extent that it does or does not, we say that it is *soluble* or *insoluble*. But what of a substance not placed in water? Does it still make sense to speak of it as soluble or insoluble? Of course, for even though the substance is not in water, it would be physically or logically possible to put it in water - and we ascribe solubility or insolubility on the grounds of what would happen if the substance were in water. In contrast, imagine a substance whose molecular structure were to preclude its being placed in water - say, because it decomposes water into hydrogen and oxygen. Because such a substance could not be put in water, the question begins to assume a different character, and now it's no longer clear that the notion of solubility is even appropriate.

Let's see what this says about information. Suppose I need to make a decision - say, as to how to respond to a request I've received - but can't because I don't understand some of the vocabulary involved. If I can get information about the terms I don't understand, well and good. But suppose my information sources simply don't help at all in this decision-making context? Suppose for example, not the crucial unknown terms, but instead some entirely new ones, are defined for me? Does this constitute real *information*? In analogy to the substance which is not in water but is judged soluble or not in terms of what would happen if it were in water, I think we must judge this input to be informative or not in terms of what *would* happen if it *were* applicable to a decision context. In fact, we can argue the same point on other than analogical grounds, for although this input gives me no actual *assistance* in decision-making, we can see that it does give me an increased decision-making *capability*. If, for example, I now obtain a second set of definitions of the original unknown words in terms of my newly learned words, I'm now able to proceed with my decision. Without the earlier definitions, however, this new input would do me little, if any, good. On the other hand, had the earlier definitions not been in terms of words I know already, they'd have given me considerably less (decision-making) capability, for I'd still need more definitions even after obtaining the new set. This example and the differences it reveals force us to conclude, I believe, that we must distinguish input *information*, which is connected with the entire range of possible decision-making situations, from the *value* that the input has for any actual (or hypothetical) decision-making situation; only if the input data can assist us in no possible decision context should we deny its information-containing status altogether.

This point having been made, however, another point should also be brought out - that for a complete theoretical understanding of certain kinds of information systems, a decision component in the theory *is* necessary, but for a reason which is exactly the opposite of the one we considered earlier. That is, the feature which necessitates the tie between information and decision is *not* that information is used in making decisions, but the reverse. In systems of the kind I'm considering, one makes decisions in order to acquire information, not vice versa. What occurs in such a system is a cyclic sort of process - information-decision-information -decision, etc.; at the end, the information *is* put to service to assist some ultimate decision, but this is not a part of the functioning of the system. I suspect, moreover, that such systems are far more common and more important than the "one-shot" sort of information system that has been more thoroughly studied. (Human beings seem to function in just such a way in most of the things they do, for example.) This doesn't deny the earlier point about the separability of the two phenomena, however, since the behavior of such an interactive information system can be seen as the repeated alternation of a simple, decision-free information system and a decision-making component.

Let us, then, try to specify the essentials of such a combined information-decision system, to see what they are and how they fit together. In doing so, we can try to see, incidentally, what technical developments will be necessary before we can achieve any general theory of such systems, and also, perhaps, what directions in research may prove useful in devising systems of this sort which are adequate to model real, concrete phenomena in a fruitful way. As we have pointed out earlier, in such an over-all structure, we shall need both an information component and a decision component, with connections between them showing how each facilitates the functioning of the other.

For the development of the information-structure component of the system, I want to draw heavily on the work of Manfred Kochen and his concept of a "learning automaton".[1] I believe, however, that we shall need to make parts of our structure somewhat more complex than his; this additional complexity is not without its own advantage, though, as it apparently allows certain simplifications in other areas. Kochen begins by specifying an *environment*, consisting basically of two parts. One is a set of possible *environmental states* with appropriate relations defined among them. Of these states, one is designated as the *actual* (or real, or starting) state; the second is a finite set of *actions* which produce transitions from one environmental state to another. Now, for a number of reasons which will gradually become apparent, I believe we shall need to specify the nature of these states in considerably greater detail than this. In our structure, a state will correspond to everything (at least with respect to the specified subject matter) that is the case. Accordingly, we shall represent a possible state of the environment by a (logically consistent) *complete theory* (over some specified formal grammar). By "theory" I mean the logician's use of the term - that is, the set of logical consequences of some set of sentences; no particular axiomatization or deductive apparatus is assumed. By "complete", I mean that for any grammatical sentence S, either S or its negation is true (i.e., is in the theory). (This condition is imposed to express our presupposition that, in any possible state of the environment, either S or its negation must in fact be the case.) In terms of such a representation of environmental states, an action (e.g., shutting the window) may then be seen as a process which replaces certain sentences in the environmental state-theory (say, the sentence "the window is open") by others ("the window is shut"). This convention allows us to specify actions in as much detail as is desired and also to relate actions to each other as well as to state-changes. Our environment thus consists of: (1) a collection of complete theories (the possible states), with one designated as "real"; and (2) a finite number of rules (the actions) for replacing certain sentences in a state-theory with other sentences. An action, then, causes a transition from any state to whatever state is obtained by replacing sentences according to the specification.

A question is perhaps in order here. What on earth is the purpose of replacing such relatively straightforward things as environmental states by formal theories, and actions by replacement rules? There are several purposes, I think:

(1) It enables us to specify precisely what constitutes any given environmental state and in what ways states are related to one another.

(2) It prevents us from confusing a particular environmental state with certain features or aspects of that state (which it may well share with other states).

(3) A purely philosophical point, to be sure - the nature of formal theories and replacement rules is considerably clearer and better understood than that of environmental states and actions.

(4) In any actual application (say, in computer modelling), environmental states and actions will almost certainly have to be given some sort of linguistic storage representation anyway; our convention simply specifies ahead of time what seems to be a particularly clear-cut and useful kind of representation.

Corresponding more or less to the *real* state of the environment, we shall also need within the system some *representation* of that state to indicate what is known about the environment. Here, too, our decision to use formal theories as these states gives us an additional advantage. As the *internal representation* of the environment, we shall also use a formal theory, but not one of precisely the same sort. Certainly, not everything that is true in the external world is known by an information user, nor, unfortunately, need everything that is known be true. Therefore, as the system's internal representation, we shall also use a formal theory, but it will not in general be complete (i.e., there will be many facts which are unknown), nor need it be consistent with that formal theory which represents the real state of the environment (i.e., we may simply be mistaken in what we take the environment to be like). We may also, in order to ensure that this knowledge is subject to real sorts of limitations, wish to stipulate that the non-logical part of this internal theory contain only finitely many logically independent statements (or, alternatively, be finitely axiomatizable). This, however, we shall not make into a system requirement, but simply leave it as an option.

Defining the internal representation of the external world in the way we have has two convenient features. For the first feature, notice that, just as an action produces a transition from one environmental state-theory to another by replacing sentences in it, it may in the same fashion produce a change from one internal theory to another. (Of course, if the sentences to be replaced do not occur in the internal theory, the action has no effect on it.) This allows us to distinguish what we shall call *empirically significant* actions from those which are not; empirically significant actions are those which *do* produce a change in the internal theory. Our definition provides, as a second feature, partial grounds for a possible distinction between those environmental state-changes which merely occur and those which are brought on by the actions of the information recipient - we might require, for instance, that these latter actions must all be empirically significant ones.

This second feature leads us rather naturally to a formal measure of information. Notice that for the internal theory to be consistent with an environmental state (a complete theory, remember) is simply for the internal theory to be included as a subtheory within the state-theory. In general, since the internal theory is not complete, there will be more than one such state-theory with which the internal theory is consistent. These states are the *observationally possible* states; they are those states which, for all we know, may be the real state. If the designated real state is not among them, then something we "know" is mistaken. As more information is added to our internal theory, we know more about the state of the world; this lays more constraints on the ways that things might possibly be, so the number of observationally possible states decreases. That is to say, our uncertainty as to which is the *real* state is lessened; this is in accord with the classical information-theoretic notion of information as a measure of the reduction of uncertainty. Note that, if new data contradicts the old, adding it to the internal theory would produce a logically inconsistent theory - there would be no observationally possible states. If this occurs, we must choose between the new data and the old; in terms of our model, an obvious criterion for choice (which accords with our intuitive behavior in many cases) is to select the "largest" consistent subtheory of this inconsistent theory - that is to retain those data which, in sum, minimize the set of observationally possible states, offering us the least uncertainty.

We now need to represent the impact of an *information input* into the structure. First, of course, is the question as to how we shall represent such inputs. Though information is acquired in many non-linguistic ways, we shall assume our information inputs to consist of written material. In addition to the reasons previously mentioned for using linguistic representations for environmental states, which apply here as well, there is also the fact that even non-linguistic information, insofar as it is explicit and consciously received, can be expressed linguistically; hence, though we may appear to be dealing with only one specific type of information, in fact we lose no significant generality. Therefore, as our information inputs, we take *messages* - sets (or ordered sequences) of sentences. The receipt of a message by the system, then, adds to the internal theory those sentences contained in the message, together with those which are consequences of the thus-enlarged theory, to obtain a generally larger theory. This larger theory admits no more (and, as a rule, fewer) empirically possible states of the environment - so the *information content* of the message can be defined in terms of the decrease in the set of empirically possible states. (This is clearly related to the semantic information measures of Carnap and Bar-Hillel.)[2]

As an extreme example, suppose the internal theory consists of but a single sentence A. The set of empirically possible states consists of all those complete theories containing A. Now, if the message "if A, then B" is input into the system, the resulting internal theory consists of the sentences A, "if A, then B", and their consequences, which include the sentence B. The set of empirically possible states is correspondingly reduced to those states containing these sentences; the information content of the message "if A, then B" is a measure of the extent of this reduction. Consider now the following three possibilities. If the message B is now input, it contains no further information, since it does not further reduce the set of empirically possible states. If, on the other hand, the message "not-B" is input, the internal theory would become inconsistent (i.e., if "not-B" were incorporated into it) - the set of empirically possible states would be empty (i.e., there are no such states); in this case, we may say that the message contains an infinite amount of information - too much to handle. Finally, if instead the message "if B, then C" were input, the theory would consist of A, "if A, then B", "if B, then C" and all their consequences among which are both B and C. The set of empirically possible states is reduced even further, and the information content of "if B, then C" measures this reduction. Note that, had the message "if B, then C" been received *before* the message "if A, then B", the internal theory would at that point have contained only A and "if B, then C" (and, again, consequences), but neither B nor C. Under those circumstances, the set of empirically possible states would be reduced less, so that the message "if B, then C" would have a *smaller* information content.

Note that, according to this definition, information (or more properly the information content of a message) is neither a process nor a commodity; it is a purely quantitative measure of the change in the set of empirically significant states. One may look at information as a process which reduces this set of states by eliminating some of its elements, and for some purposes such a view may be productive. Alternatively, in terms of a more commodity-oriented viewpoint, one may look at information as either the message itself or as the propositional content of the message, which may be bought, sold, requested, provided, etc., as needed (recognizing, of course, that the quantity of information in the message will not be independent of the condition of the internal theory). This attitude, too, may be fruitful for other purposes. But we can see that these two views as to the nature of information are essentially two different but complementary ways of looking at the same global phenomenon. (This may be seen as a situation somewhat similar to that in physics, in which photons, for example, are neither exclusively waves nor particles, but entities with aspects of each.)

Before moving to the decision component of our structure, perhaps we should observe that within this framework we may also incorporate features representing question-asking (and answering). We have previously characterized actions as rules for replacing specified sentences in a state (and also, perhaps, in an internal theory) with others. Typically, such a rule will explicitly specify both the sentences to be replaced and their replacements, as well. We might, however, provide a rule with an undetermined specification for the replacement sentences, in which just *which* sentences serve as replacements may depend on certain features of the real state. Thus, suppose that a particular parameter a can take either the value 0 or the value 1. In some states, the sentence "a=0" will be in the state-theory, and in others, "a=1". To say that the real value of a is unknown is to say that neither of these sentences is in the internal theory. Assuming at this point that the grammar also admits statements of the form "It is known that S" for any object-sentence S, it follows that neither "It is known that a=0" nor "It is known that a=1" is in either the real state-theory or the internal theory, but the negations of both these are in the real state-theory and, we may assume, in the internal theory, as well - that is to say, we know what it is that we do not know. Now consider the replacement rule which goes as follows: Replace the two sentences "It is not known that a=0" and "It is not known that a=1" with (1) "It is known that a=0", if the sentence "a=0" is in the real state; but with (2) "It is known that a=1", if "a=1" is in the real state. This rule also makes the same replacement in the internal theory, bringing in as a consequence either "a=0" or "a=1". This resulting change in the internal theory is thus the way our framework models the inquiry "Does a=0 or a=1?" and the discovery (i.e., the environment's answer) of the correct value. It may be remarked that this seems a rather complicated way to model a rather simple phenomenon, and indeed it is. The point, however, is that the phenomenon can be represented in terms of the conceptual machinery already provided, so that we need not provide any additional machinery to handle question-asking.

Clearly, if we wish to add additional features to allow a simpler representation, it is possible, as long as we also connect them properly with those features we have just discussed.

To sum up before we move on to consider the decision component, the basic conceptual machinery of the information component consists of the complete theories which serve as environmental states, the sentence-replacement rules which serve as the actions, and a generally incomplete theory which serves as the internal representation of the environment. In terms of these, we can characterize state-theories as observationally possible or not (as they are or are not consistent with the internal theory). We can furthermore measure the information content of an incoming message in terms of the reduction of the set of empirically consistent states, and show how and why the information content of a message is not absolute, but is relative to the internal theory. Finally, we can also characterize certain actions as information-seeking (i.e., representing questions) and show how questions and answers provide information to the system. What we can not yet do is to show how and when we may *need* to ask a question. Also we are as yet unable to explain how a fixed quantity of information in a message may have differing values for different decision-making activities. To show these aspects of an information-decision system, we must now consider the decision component.

In this component, the structure will be much like that envisioned by Bruce Whittemore and Marshall Yovits,[3] in their notion of a "generalized decision-maker". Their decision structure consists of several parts: (1) a set of *states* of nature, with probabilities for the occurrence of each; (2) a set of *alternatives*; (3) a set of *outcomes*, with a *value* for each outcome; (4) an *outcome probability* for each state, action, and outcome - i.e., the probability that given the specified state, the specified action will result in the specified outcome.

Much of this we already have. For instance, the states of nature are exactly the environmental states we have discussed; we need only specify their respective probabilities. Here, it should be noted that these probabilities will really be conditional probabilities from the decision-maker's point of view - i.e., the probabilities of the various environment states being the real state given the internal theory (the "knowledge") of the system. A state-theory, therefore, has a non-zero probability of being real, given the internal theory, if and only if it is observationally possible; the particular values of the probability distribution can be obtained from the set of observationally possible states according to standard "state description" approaches. It should be recalled that, as the internal theory changes, the set of empirically possible states will change, some empirically possible states becoming impossible or some impossible ones becoming possible (as parts of the internal theory may be abandoned); as this occurs, the individual probabilities will also change. It is in this way that one of the "uncertainty" conditions posed by Whittemore and Yovits arises - that even the set of states of nature "may vary according to the decision-maker's current understanding of the decision situation".[4]

The various courses of action open to the decision-maker are represented by the alternatives in the Whittemore-Yovits model. In our structure, they will be those actions which we attribute to deliberate efforts of the decision-maker. That means that the class of actions must be divided into two subclasses: *decision choices*, the deliberate actions of the decision-maker (which, as we suggested previously, must all be actions which are empirically significant), and mere *environmental changes*. Note that it is necessary to specify both classes, since our equivalent of shutting the window (to recall our earlier example) is merely replacing sentences by sentences - e.g., replacing "the window is open" by "the window is shut", etc. This may be either an action deliberately performed by the decision-maker or an environmental change occurring independently of the decision-maker (as, for example, when a support breaks and the window slams shut). We should also note that the set of decision choices is not constant, as decision choices must be empirically significant actions, and the set of these actions depends on the internal theory, which itself changes. Furthermore, in order for any action to occur, certain appropriate conditions must obtain (e.g., one cannot shut the window unless it is open) and these depend on what the real state or the internal theory is. Hence, the set of decision choices which are available *and* the set of environmental changes which can occur both change as the internal theory and the real state change.

We must now consider the set of outcomes and the outcome probabilities. There is little difficulty with the outcomes themselves; they, too, can be seen as empirical sentences expressing outcome events, and an outcome occurs if and only if that outcome sentence is contained in the real environmental state-theory. The difficulty lies in dealing with the outcome probabilities. To see why this is so, let us first observe that to get an adequate picture of iterative decision-making, we must consider both objective and subjective outcome probabilities. The objective probabilities are relatively straightforward, but to see how they arise, it is nevertheless useful to consider actions again. Some outcomes are the direct results of actions - shutting the window has, as a direct outcome, the fact that the window is shut. Such outcome sentences are brought into the theory by the replacement itself. Others, however, are different; the room's growing warm, for example, is not a direct outcome of the window's being shut - it is an indirect result. How? The answer is that the window's being shut interacts with certain features of the environment to bring about this output -

either with complete certainty or with some degree of probability. For instance, if the environmental state contains the two sentences "the window is open" and "if the window is shut, the room is warm", then the action of closing the window replaces the first of these sentences by "the window is shut", and the sentence "the room is warm" is then brought in as a consequence. In a similar way, when the environmental state-theory contains probabilistic sentences (e.g., the probability of the room's being warm given the window's being shut is 70%) we get probabilities for outcomes. The point so far as these are concerned is that outcomes may "chain" into arbitrarily long sequences, so that a single action may have arbitrarily many outcome results, each with its own probability.

This is an unexpected complication, but it poses no great problem for us at this point, since the completeness of the environmental state-theories ensures that each one does contain an appropriate probability statement to provide an outcome probability for all these outcomes. The fact that all the objective outcome probabilities appear as conditional probability statements in the environmental state-theories, emphasizes that these probabilities are not absolute, but relative to the states of nature. It also explains why the set of outcomes may seem to change as the state of nature changes. Given a state, a *possible outcome* is one which in that state has a non-zero probability of occurring in response to a possible action; as the state of nature changes, possible outcomes may become impossible, so that the set of possible outcomes is not fixed.

The same sorts of conditions, moreover, relate actions and outcomes with respect to the internal theory (i.e., what is known about the world), except that since this theory is not complete, many probabilities will be unspecified, so such unknown outcomes will not appear in the set which we call *perceived outcomes* - that is, perceived (correctly or not) as possible outcomes. The distinction between these two kinds of outcomes emphasizes the difference between the real possibilities available to the decision-maker and the ones he perceives as real. It also allows the model to serve as a framework for exploring either the *process* of decision-making (which is carried out in terms of the internal theory) or the *effects* of decision-making (which occur in terms of the real environmental state), as well as for systematically relating the two.

Let us sum up briefly the basics of the decision component thus far; again we have the *environmental states*, the *actions*, and the *internal theory*. Now, however, we distinguish actions as *decision choices* or *environmental changes*, or both; decision choices must be empirically significant. *Outcomes* are merely empirical sentences; *possible* outcomes are those which have non-zero probabilities in the real state-theory, and *perceived* outcomes are those with non-zero probabilities in the internal theory. The set of choices, the set of perceived outcomes, and the perceived outcome probabilities will depend on the internal theory; the set of possible outcomes and the possible outcome probabilities will depend on the real state-theory. Finally, we have outcome values, which do not arise from within the model, but must be supplied externally.

We must also include some criterion for the decision-maker's choice - a *decision rule*. This may be of various sorts, but in Whittemore and Yovits it typically depends on: (1) the outcome values; (2) the outcome probabilities (which, in general, depend on the environmental state); and (3) the probabilities of the environmental states. Note that, if we consider the outcome probabilities in the *real state*, the decision rule tells us the best choice for the decision-maker, and if we consider the outcome probabilities in the *internal theory*, the decision rule tells us what the decision-maker perceives as the best choice. I believe that, as a factor in the decision rule, it will also be necessary to introduce a *cost* function for choices in addition to the value function for outcomes, and to have the decision rule take such costs into account; a choice with a low cost and a given probability of producing a certain high-value outcome is (other things being equal) certainly preferable to a choice with a high cost and the same probability for the same outcome. Be that as it may, the decision rule typically produces an expected value for each choice and then, generally, directs the decision-maker to take that choice whose expected value is greatest.

We can now begin to specify the interaction of information and decision in greater detail. Whittemore and Yovits discuss the effect on the decision-maker of several types of uncertainty. First they distinguish structural uncertainty (uncertainty as to what are the possible choices, states, outcomes, etc.) from relational uncertainty (uncertainty as to how such factors are related). Then, independently, they also distinguish environmental, executional, and goal uncertainty; this gives six possible kinds of uncertainty. We can now pinpoint each of these uncertainties in terms of our structure and say exactly what kind of information is needed to resolve each kind of uncertainty. For example, structural executional uncertainty is uncertainty as to what the possible choices and outcomes are; it arises from lack of knowledge in the internal theory as to what actions are possible choices and what outcome probabilities are non-zero (i.e., what outcomes the choices can bring about); similar identifications can be made for the other five types.

Recall, now, that we previously distinguished information-seeking actions from others. These represent active efforts to elicit information from the environment and will be classified as among the choices. These particular choices have as outcomes the provision of information to the internal theory. When uncertainties are great, such a choice will in general produce a relatively large change in the internal theory. It thus provides a large amount of information and reduces uncertainty considerably, and so will have a high expected value, especially if the cost of such a choice is not great. In such cases, the decision rule will dictate an information-seeking choice. If, on the other hand, the cost of such choices is high, the decision rule may dictate a "goal-directed" choice instead. One should note, though, that it is not the mere quantity of information that provides the high expected value - it is the degree to which the additional information is of the sort that would reduce the goal-directed uncertainties in the decision-making problem at hand. It is here that the distinction between information quantity and information value (or between decision-making potential and decision-making performance) arises.

In the case where an information-seeking choice is the decision-maker's optimal one, that choice typically produces a new internal state in which the uncertainty is significantly reduced. But this choice is not what we have called a goal-directed one - it alone brings the desired goal no nearer. So the decision-maker is faced with another decision. Again, the optimal choice may be an information-seeking one. In particular, as information-seeking choices are continued, the internal theory continually grows and uncertainty is continually reduced. At some eventual point, however, further information-seeking choices elicit only messages whose information content (while still perhaps very high) no longer reduces the goal-directed uncertainty enough to justify the cost; at this point, the decision rule then dictates a goal-directed choice. This illustrates the rather paradoxical nature of information-gathering in which (1) information-gathering in near-total ignorance produces little information but is extremely valuable; (2) as more becomes known, new inputs tend to produce even more information; but (3) at some point, diminishing returns set in, so one ceases to gather information. At this point, as we have said, our information-gathering is now put to the service of the ordinary sort of goal-directed decision-making. But notice that until this point is reached, the converse is the case, and we have exactly the information-decision cycle we discussed earlier, in which the decision-making serves the information-gathering.

What we have developed here, then, is a conceptual framework for the study of information and decision, separately or together. Such a framework is a step toward the development of a full-fledged *paradigm* for a science of information. Only when we have reached this stage is our discipline a fully developed science, for it is only within the bounds of a paradigm that scientists attempt to understand and explain their subject matter - post hypotheses, develop experiments, evaluate significance, and direct new research efforts. Though this framework is only a step in that direction and does not attempt to cover the entire range of information phenomena, it does give us a primitive counterpart of a paradigm's guidance. With somewhat greater time at our disposal, we could review the various features of the framework to outline those directions in which further research is needed and to frame key questions and hypotheses to structure such research. Only if we can in such a way establish and systematically carry out an organized program of research, I believe, can our field develop the rich mixture of complexity and simplicity, of the theoretical and the experimental, which will truly justify our hopes for an information science.

References

1. Kochen, M., Adaptive mechanisms in digital "concept" processing, in *The growth of knowledge*, Kochen, M., Ed., John Wiley, New York, 1967.

2. Carnap, R. and Bar-Hillel, Y., *An outline of a theory of semantic information*. Technical Report No. 247, Research Laboratory of Electronics, Massachussetts Institute of Technology, 1952.

3. Whittemore, B. and Yovits, M., A generalized conceptual development for the analysis and flow of information. *Journal of the American Society for Information Science*, 24, 221, 1973.

4. *Ibid*, p.225.

INFORMATIONAL RELATIVITY

Satosi Watanabe

University of Hawaii, Honolulu, Hawaii, U.S.A.

INFORMATION

I usually start a course in information science with a remark: information science is not a science of information but a science of how to get rid of (useless) information. In fact, we start our day every morning by throwing away about 10 or 20 letters, keeping perhaps a short note like: movie Tuesday night, on the desk calendar. We complain about the over-abundance of information of modern life. But, since the day life was born on Earth, we have been bombarded by avalanches of physical signals. We would have been driven completely crazy if we had perceived all these signals. Nature was clever enough to shield off most of these signals allowing only a very few important ones to enter our cognizance. A good example is the mosquito-detecting signals that go from the frog's retina to its brain. The frog's eye gets rid of all the unnecessary information and perceives a very few number of particular pieces of information, including the "mosquito-ness" of the image that falls on the retina.

I do not intend to give a definition of what is information, because it would require a great deal of care in choosing proper words and proper phrase-construction. But, I will try to express my way of looking at information by giving a provisory version of a "definition".

"Information is a proposition(s) extracted deductively or inductively from raw data so as to diminish its (their) quantity and to increase its (their) usefulness for goal-seeking actions. The method of taking data and the concepts used to formulate propositions should be oriented to achieve this condition."

The reader will notice that there is a normative overtone in this "definition". Indeed, a raw datum could be correct or incorrect, whereas information could be good, bad, better or worse depending on its succinctness and its utility. The important point implied in this definition is that the methods of observation as well as the concepts used are value-dependent. This is somewhat similar to the now popular view about the theory-ladenness of observation emphasized by some philosophers of science (Feyerabeud, Kuhn). Similarity lies in what may be called relativity of information, but their contentions and ours are not the same, and as we shall see, the conclusions we draw are not the same as those of Feyerabeud, Kuhn, etc.

This definition makes clear that information is not a purely objective entity. Information is the result of an interplay between the objective and the subjective, or between the empirical and the rational. If we write a sequence: data - information - theory - world view, we can say that the participation of human intelligence becomes stronger and stronger as we progress toward the right of the sequence. But, even the left-most end of the sequence, namely the so-called raw data are not entirely independent from human choice and intention.

STRUCTURE

The history of information theory should at least be traced back to the second half of the 19th century when Maxwell, Bolzmann, (and a little later) Gibbs were trying to reproduce *Thermodynamical* Laws from atomistic laws. The properties of entropy function ($-\sum x_i \log x_i$) were well studied, and they were used to reproduce Thermodynamical entropy. This situation has not changed even in the famous papers of Szcilard who demonstrated that the increase (decrease) of thermodynamical entropy corresponded to the decrease (increase) of our information about the physical system. It was von Neumann who introduced the so-called micro-scopic entropy
$S = -\text{trace}(\sigma \log \sigma)$ (2.1) where σ is the density matrix (non-negative Hermitian matrix with diagonal sum unity). He used this quantity as a simplified version of Thermodynamical entropy. It may be said that it was my paper of 1939 in Zeitschrift für Physik which used the entropy function for the first time as a measure of ignorance (lack of information) independent from Thermodynamics.

The paper not only used this quantity as a measure of indeterminacy but also as a measure of interdependence of individuals in an assembly of individuals. The reason why this quantity can be used as a measure of interdependence can be seen by the general formula that expresses the degree of interdependence of parts in a whole:

$$D = \sum S\ (\text{parts}) - S\ (\text{all}) \qquad (2.2)$$

In the case of a quantum theoretical system S(all) becomes zero. As a result, the sum of the entropies of individuals becomes the measure of interdependence. When applied to a system consisting of similar individuals, the entropy of each individual becomes the measure of interdependence. The interdependence may be called "structure", because if every individual is independent, and has no mutual constraint, the interdependence will disappear. More than ten years later, Shannon introduced the idea of redundancy that corresponds to our interdependence.

The entropy defined in (2.1) is much more general than one used in information theory. If however all the σ's that come into our consideration commute (in the sense of matrix multiplication) then we can diagonalize all of them simultaneously. What stands in the diagonal is the usual probability. It may be useful to σ instead of probabilities in psychology. (My paper in "Information and Control" in 1971.) A σ is defined by the probabilities and the "coordinates", and the choice of coordinates add to relativity of information.

INFORMATION AND PROBABILITY

The quantity "information" can be defined only if we know what the categories are and what the probabilities assigned to them are. When we are presented with a set of raw data or the description of a set of individual cases, we usually bundle up some of them as a class or category. The probability is sometimes a simple (unconditional) probability, sometimes it is a conditional probability assigned to these categories.

The question arises whether and to what extent the information is or can be objective, in the sense of being independent of the user of the information. Relegating the problem of category to the next section, let us first take up the problem of probability. We can point out two reasons why the probabilities used in defining information cannot be "objective". First, the events that appear in different categories are not repetitive. As a result most of the time we cannot depend on the "objective" definition of probability by frequency. We have to use the subjective or personalistic definition of probability which coincides with the frequency view of probability in case the events are repetitive. This is one of the two factors that makes the probability user-dependent.

The second factor is that even when the conditional probability p(A/B) is determined by the nature of the problem or by frequency the inverse conditional probability $p(B/A) \propto P(A/B)p(B)$ depends on the "prior" probability p(B), which is often not determined objectively, i.e., independently of the information-user.

INFORMATION AND CATEGORY

When we have many items, we put similar items together in a box or category, and assign a certain probability to each category. But, is there such a thing as a class of *similar* objects? How should we define it? That was the problem to which I addressed myself.

When we say that two objects are similar, we usually mean that they share many properties, or predicates: they are both white, they both weigh between 1lb and 2lb. etc. But, I raised a doubt about this and found out a strange result which I call the Theorem of the Ugly Duckling. The Theorem says that any two objects (insofar as they are distinguishable from one another) share exactly the same number of predicates. It is therefore assumed that the total number of predicates applicable is finite. As predicates, we count all the logical functions of originally given predicates. This theorem of the Ugly Duckling seems to give a basis for a radical nominalism denying the existence of any kind of universal. But, we know that our experience can be formulated only in terms of universals or concepts. We know universals *exist*, in some sense. The reason why this is possible is that not all predicates are equally important. Two things are similar not because they share more predicates than another pair of two things but because they share more of the more important predicates than another pair. But, important for what? That depends on the purpose for which the classification is used. For a frog, a particular predicate "mosquito-ness" is an important predicate for its survival. Should we put a whale in the same box as an elephant or a tuna? The answer is that if the purpose is to make an elegant zoology it would be better to put whale and elephant together. If the purpose is to classify industries, then it might be better to put whale and tuna together.

This consideration shows clearly that the categories are value-laden things and therefore also the information, that depends critically on categories, is value-laden.

In animals, those predicates that have direct correlation with survival are more or less directly observable by sensory organs. These predicates may be compared with "innate ideas" and what I say may be interpreted in the Kantian sense. But, I am not an apriorist, and I emphasize that these "innate ideas" are created as a result of long experience through the millions of years of evolution.

In defining behavioural entropy, I mentioned (*Knowing and Guessing*) that the categories into which we classify behaviour are not determined uniquely. We said that we have to determine the categories in such a way that the inverse H-theorem holds. This is another case of relativity of categories.

EPISTEMOLOGICAL RELATIVITY

My contention with regard to the value-ladenness of concepts and the modern trends in philosophy of science about the theory-ladenness of observation (Feyerabeud, Putnam) and about the ideology-dependence (paradigm-dependence) of theories (Kuhn) have one thing in common that they may all be characterized as an aspect of epistemological relativity. But, my version of relativity has a very clear (mathematical) foundation whereas the other two versions are very vague in their argument. Moreover, my relativity does not simply remain relativity. I say that some particular concepts are chosen as required by the purpose. On the other hand, other versions tend to become anarchical and skeptical. Their conclusion sounds almost like: "everything goes". In Kuhn's theory (particularly in the first version of his book), Newton's mechanics and Einstein's mechanics are almost simple alternatives which one can choose between arbitrarily according to "ideology". But, in reality, the relationship between them is not a symmetric one. Newton's mechanics is in some sense part of Einstein's mechanics, but not vice versa. The history of theories follows a one-way evolution depending on new domains of experience opened up step by step by a new industrial development. Given a new domain of experience, scientists do not have, as philosophers imagine, many alternatives. If they find *one* way to formulate a general rule governing this new domain of experience, they should congratulate themselves. There is no room for a luxury of anarchy. Kuhnian theory does not account for this very strict "narrow gate" of evolution of scientific evolution.

INFORMATION BALANCE IN DEDUCTION

It is important to make a distinction between nomological information and contingent information. This distinction is clear in the case of obtaining *deductively* a solution of a dynamical law by an initial condition. The choice of a solution is made possible by the combination of information provided by the general rule and information provided by the initial condition. Suppose for instance we want to find a solution of the equation: $d^kf(x)/dx^k=0$ by the initial condition: $(f(o), df(o)/dx, d^2f(o)/dx^2, \ldots, d^{k-1}f(o)/dx^{k-1})$. This determines one particular curve among infinitely many possible curves. My question is, what part of information comes from the general law (differential equation) and what part of information comes from the initial condition. We cannot answer this question as it is, because it is a continuous case. But we can start with a discrete case and pass to the limit. By this calculation, one thing comes out as a conclusion that seems to be valid for many similar problems: The ratio of the information provided by the general rule to the information provided by the initial condition becomes infinite as we pass to the limit. The information provided by a law is overwhelmingly greater than the information provided by the contingent conditions.

This seems to suggest a new understanding to the genetic problem. It is true that the DNA carries a large amount of information. But, still it would not be sufficient to determine every small detail of a born offspring. Probably, the information carried by DNA can be compared to the information provided by the contingent conditions. Probably, a large part of information determining an individual, after all, derives from the physical milieu in which the fertilized egg grows.

INFORMATION BALANCE IN INDUCTION

There are two kinds of information about the general law obtained in induction. One is the information *about* the most probable hypothesis (general proposition). The other is the information that one can extract from a law. The former information obeys the rule that no information can be created by deductive inference. But, the latter is infinite, in principle, because the experimental data from which this general law was extracted are only finite, while infinitely many cases can be derived from the general law. In this sense, there is genuine creation of information. This situation will be clear, for instance, in the simplest case of induction of assigning a curve passing through given finite data points. The number of data points are finite, but the curve assigns a value of the function to each of the continuously many points of the abscissa.

Now, it is important that this curve is not determined by the data points at all. All possible curves going through the data points are permissible. This is the basic *inductive arbitrariness*. Another cause of relativity. The computer *alone* cannot overcome this relativity.

More generally and mathematically, a hypothesis H will give a probability distribution density $p(x|H)$ for individual points x. The Baysian inversion $p(H|x)$ is the inductive probability which now depends on the "a priori" probability $p(H)$ since $p(H|x) \propto p(x|H)p(H)$. The extra-evidential factors intervene in the determination of the "a priori" probability.

Pattern recognition, in this respect, can be characterized as a double induction for the following reason.

The hypothesis H and the auxilliary condition A and the experimental data D are of the following type:

H: If x belongs to class C_i, then it has probability distribution density $f_i(x)$

A: An item x belongs to class C

D: The item has position x.

The probability determined by the nature of problem is $p(x|H \cap A)=f_i(x)$.

In the training period, we evaluate the probability of hypothesis
$p(H|A\cap x) \propto p(x|H\cap A)p(H)$
In the application period, we evaluate the probability of class to which the given item belongs:
$p(A|H\cap x) \propto p(A|H\cap x)p(A)$
It is noteworthy that inductive ambiguity enters in two places.

It cannot be overemphasized that the genuine source of information is pattern recognition and induction, in which an additional factor of relativity is added through "inductive ambiguity".

PARADIGMATIC SYMBOL

There is an important question as to why pattern recognition is easy for human intelligence while it is a complicated task for a computer. To answer this question the notion of paradigmatic symbol is useful ("Paradigmatic" has nothing to do with "paradigm" in Kuhn's sense). A machine works with abstract symbols whereas the human mind works with paradigmatic symbols.

To define a concept, we can use either its intension or its extension. We can perform a classification job - pattern recognition is also a kind of classification - on a computer when the intension of the class or the extension of the class is known. However, in the usual (tutored) pattern recognition, we give paradigms (class-samples) first to a computer during the so-called training period, and then in the application period we show an object whose class-affiliation is not known. This type of paradigmatically-oriented pattern recognition is exactly what a mother would do to teach a class, say, of dogs to her child. The human child can do this kind of job very easily, but for the computer it is not an easy job. The reason is that the computer has to extract abstract features that would define the class. If the machine discovers some suitable features it can use them as the intension of the class and thereafter the job for the computer becomes easy. But, the features of the class are not uniquely determined by paradigms for the reason I mentioned already. Therefore the task becomes very complicated and remains arbitrary.

In the case of a human child, he does not abstract the features. The real dogs he was shown remain in his mind as paradigmatic symbols that is, concrete objects that stand for a class.

The paradigmatic symbol is a concrete object or a mental representation thereof that suggests a class by association. It is a special case of an elicitory symbol. For instance, a particular smell can elicit the whole memory of a hospital. When the field that is elicited is a concept, we speak of a paradigmatic symbol. A paradigmatic symbol, it is admitted, has no fixed field of elicitation. The field that is illuminated, or elicited or evoked or suggested or associated by a paradigmatic symbol depends on the circumstances. But, this flexibility is precisely what makes the paradigmatic symbol very versatile and useful. The most primitive example is the conditioned stimulus in the experiment of conditioned reflex. The highest level of paradigmatic symbols are the mental objects the scientists use in their scientific works. Such a paradigmatic symbol is the carrier of information in human information processes. It is useless to try to evaluate the numerical value of the information carried by a paradigmatic symbol.

The inductive process obviously is carried out also by the help of paradigmatic symbols in the human mind. By their capability of evoking a fringe of "influence", concrete data points as paradigmatic symbols allow us to formulate general statements about the field surrounding the data points. This, as stated, is the source of information.

I may conclude my talk by adapting the quotation that Mr.Nevitt made to express my point:

The absolute symbol defines, therefore kills
The elicitory symbol suggests, therefore creates.

NOTE

Many of the things described in this paper have their roots in Watanabe, S., *Knowing and Guessing; a quantitative study of inference and information*, John Wiley, New York, 1969. The relation between pattern recognition and induction is discussed in my article in *Methodologies of Pattern Recognition*, Watanabe, S., Ed., Academic Press, New York, 1969. An article about paradigmatic symbol is to be found in my article in the Proceedings of the International Congress on Psychology, Tokyo, 1972. A more complete article on the same subject is to be published in a forthcoming Transactions of IEEE, "Systems, Man and Cubernetics".

INFORMATION AND COMMUNICATION: A CONCEPTUAL MODEL AS FRAMEWORK FOR DEVELOPMENT OF THEORIES OF INFORMATION

Klaus W. Otten

Dayton, Ohio, U.S.A.

INTRODUCTION

It is suggested that the scientific study of information should consider the information transfer process as its basis. Building on this premise, a conceptual model is proposed as the framework for the development of theories on information. The model is general and applies to a wide variety of information systems ranging from communication systems, man-made machines, and man-machine systems to man simply interacting with his environment in the context of his cultural background and past experience. The model is modular and can be adapted and expanded to represent different information systems ranging widely in complexity.

The model is based on the functioning of the receiving part of a communication link. It assumes general validity for information systems by adding the concept of adaptive internal modification to the communication concepts of environmental relativity and systems specificity of information. Internal modifications reflect past experience, biases and task-conditioned value assignments imposed by the dynamics of the environment within which the system must function to assure survival.

BASIS FOR THE MODEL

We define a system as the recipient of "information". A system can be any information-processing structure, living or man-made: computer, living organism, man himself or an organization. "Information" may be regarded as what is needed for the control and proper functioning of a physical system in its environment that functions in the pursuit of its goals, including its survival.

Figure 1 shows a system (A) within its environment. Various forms of interactions with different components of the environment can take place simultaneously. A generally valid model has to account for all possible interactions, each within the particular context in which it takes place. Examples of interactions with the environment are: (1) reception of information from another system (e.g., from system B), (2) simultaneous exposure to various stimuli from different sources (e.g., to perceive aurally and visually what takes place around us), or (3) active interpretation of what has been sensed previously (to gain knowledge and insight), to name just a few. The major differences between the various forms of interactions are determined by the extent to which contexts play a role and by the types of context involved.

Two types of context have to be distinguished: the (historical) context of past experience; and the situational context (of the physical environment). The first reflects past interactions of a system that can adaptively learn. This context is dynamically changing and is internal to the system. For man, "past experience" includes his cultural and social background which guide his actions. Situational context is shaped by the environmental conditions in which the interactions take place. This context, which reflects the dynamic changes of the physical environment, is external to the system. The internal context of past experience and the external context of the situation form the environment in which any signal or stimulus is interpreted by the system. This interaction is governed by the overall goal-seeking control inherent in the system, with goals ranging from specific problem solving to overall systems survival.

The proposed general model is developed in two steps. First, the receiving side of the conventional model for information transfer in a communication link is expanded to include various types of context. This leads to the model for signal or stimuli, interpretation which becomes the basic building stone of the general model. In the second step, many subsystems which receive simultaneously stimuli are combined into the general model. This model is one interactive set of interpretors which can learn and which can interpret specific stimuli in the context of the historical and physical environment. This interpretation system functions according to its inherent goals aimed at survival. The general model is built on refinement, extension and multiple use of transformation functions of a signal-interpreting receiver. Signal interpretation is part of the information transfer. The understanding of the information transfer process is fundamental to the introduction of the general model and therefore must precede it.

FOUR LEVELS OF INFORMATION TRANSFER

The basis for the general model is developed by inclusion of

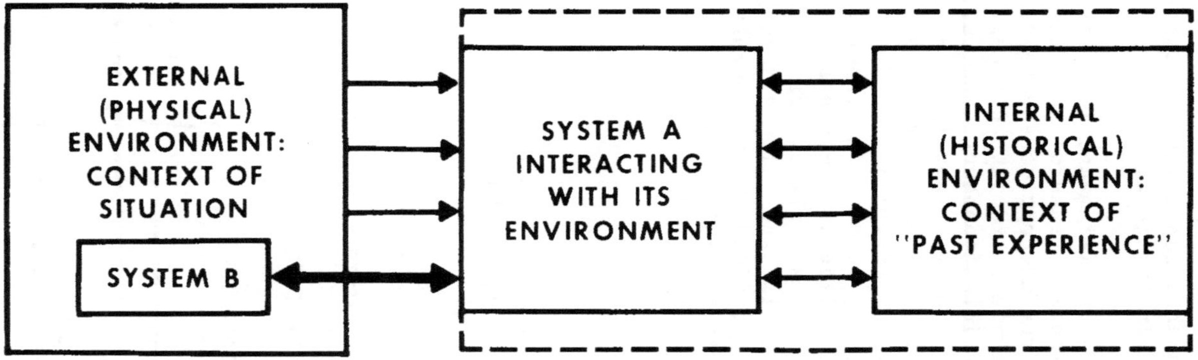

Figure 1 - System Interacting with its Environment

functional components in terms of four interrelated transfer mechanisms as illustrated in figure 2.

The bottom three transfer mechanisms are those used in modeling the one-way information transfer in a communication link. These models center around the concept of a receiver connected to a source or transmitter via a transfer channel. These three transfer mechanisms, which are interdependent, can serve to explain the transfer of meaning (in a very restricted sense) from a source B to a receiver A. Meaning in this restricted sense, disregards any context other than that inherent in the closed transfer mechanism between B and A. A fourth level of a transfer mechanism is introduced to permit the modeling of the transfer of meaning (in the general sense) in a complex context. For a human being, this context can include the social and cultural background which colors the restricted meaning defined by an otherwise closed transfer system (such as the specific meaning defined by a semantic code or a specialized language in isolation).

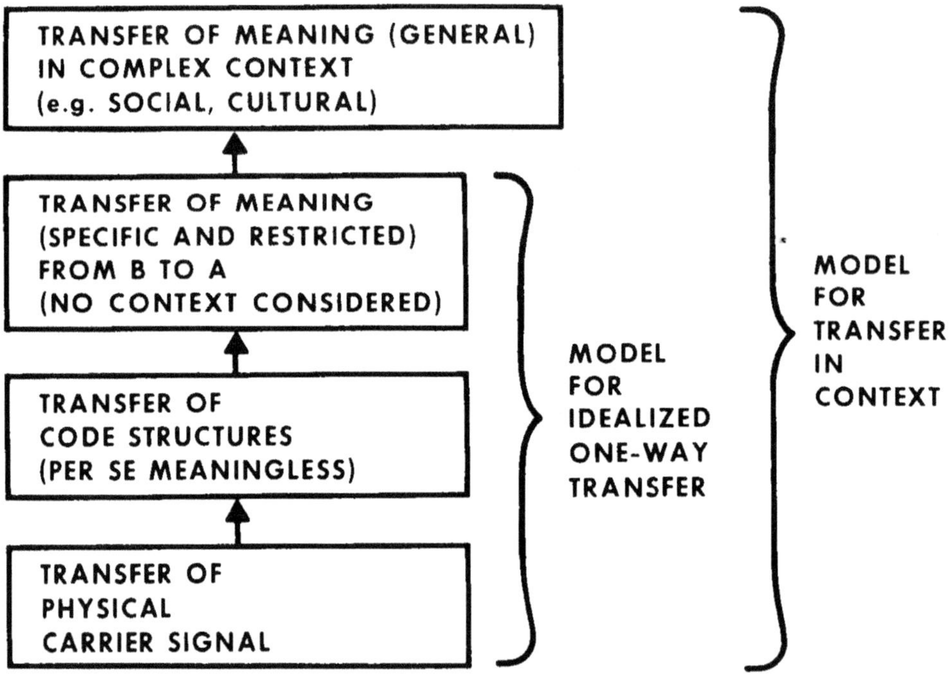

Figure 2 - Four Levels of Information Transfer

The four transfer mechanisms are interdependent: each depends on the existence of the preceding ones. Figure 3 shows the interdependencies as being hierarchical in their elementary form. The illustration indicates the increase in the scope of the system functions introduced by the addition of the fourth transfer mechanism. The three bottom transfer mechanisms (signal, code and meaning I) consider the receiving system A only in relation to the source B and only within the context of these transfer mechanisms; i.e., system B is seen through the one-way communication link as defined by the three transfer mechanisms. The fourth transfer mechanism widens the universe of the receiving system in three ways: (1) it makes it receptive to inputs from its entire environment, not only from B, (2) it interprets any input from B in the context of other signals sensed simultaneously, and (3) it interprets signals from any part of the environment in terms of "past experience". In short: System A interacts with and perceives its entire environment considering past interactions and systems goals, as limited

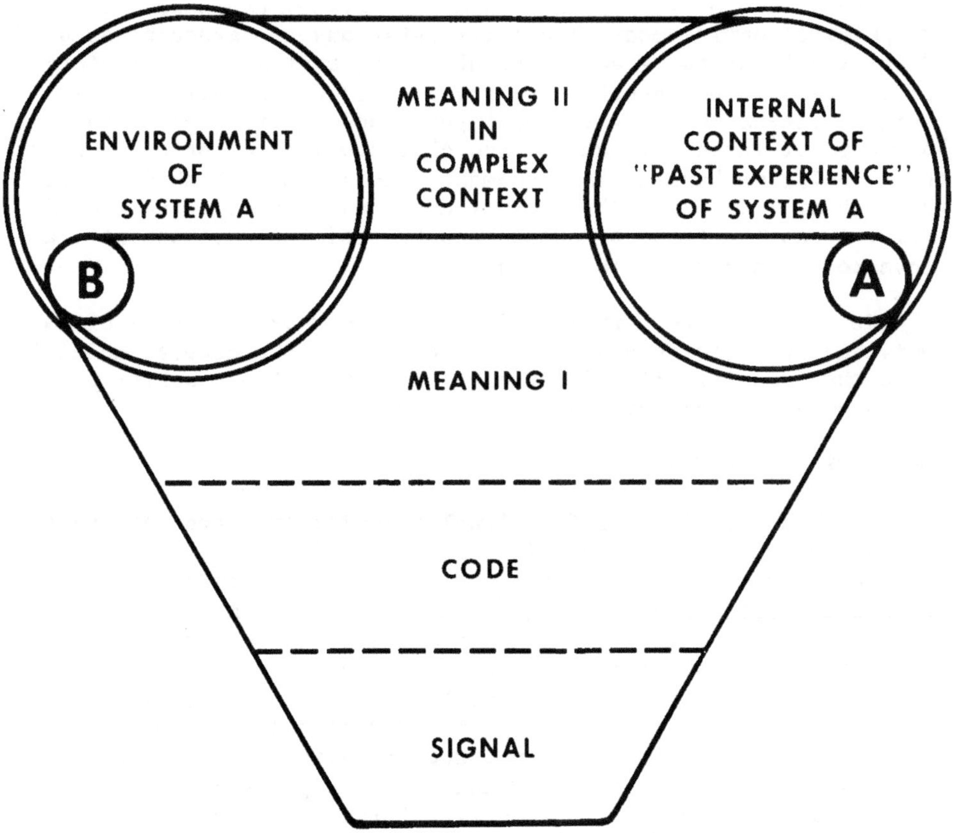

Figure 3 - Hierarchy of Interdependent Transfer Mechanisms

by the physical sensory and processing capabilities inherent in the system. It should be noted that the addition of the fourth transfer mechanism puts the entire emphasis on the signal transformation functions of the receiving system A and not on the transfer-oriented model of a communication link. Information emerges in system A as the result of its sensing and interpretation of the environment, not from the decoding of properly-encoded messages transmitted from a specific source (even though this is included in the model as a special case). Furthermore, the sensing and interpretation is conducted in the complex context of the situation and past experience and is governed by the goal-setting and survival-oriented factors inherent in the system.

The proposed model, consequently, centers around a system, which, for the purpose of explaining its operation, can be viewed as an extension of the receiving end of a communication link. The model does however not include a source and a channel and hence differs significantly from that of a communication model. The processing system of the proposed model performs transformations or interpretations of all signals which the system can sense within its physical environment. The interpretations are executed as needed for the functioning and survival of the system. The transformations are systems-specific and are limited by the performance characteristics inherent in the interdependent transfer mechanisms. Knowledge of these limitations and characteristics for each one of the transfer mechanisms is fundamental to a model which describes the mechanism by which the context-dependent, interactive and adaptive transformations of sensed inputs from the environment are converted into meaning for the system.

We proceed with a review of the models representative of each individual transfer process before describing the general model in which the transformation functions inherent in the transfer processes occur in various interactive forms.

MODEL FOR SIGNAL TRANSFER

The functional model for signal transfer is shown in figure 4. A signal source produces physical signals which are transmitted through a physical transfer medium to a signal classifier. Both the signal source and the signal classifier must know the signal specifications. To achieve signal transfer, the receiving system requires a processor (signal classifier) and a memory to store what the receiving system knows of the signal specifications. The transfer process is initiated by an external selection of the signal. With reference to the signal specifications, the appropriate signal (one of those defined by the signal alphabet) is produced by the signal source. It is then transferred through the medium. In this process it may be altered, mutilated or weakened. The signal classifier interprets each physical signal it receives based

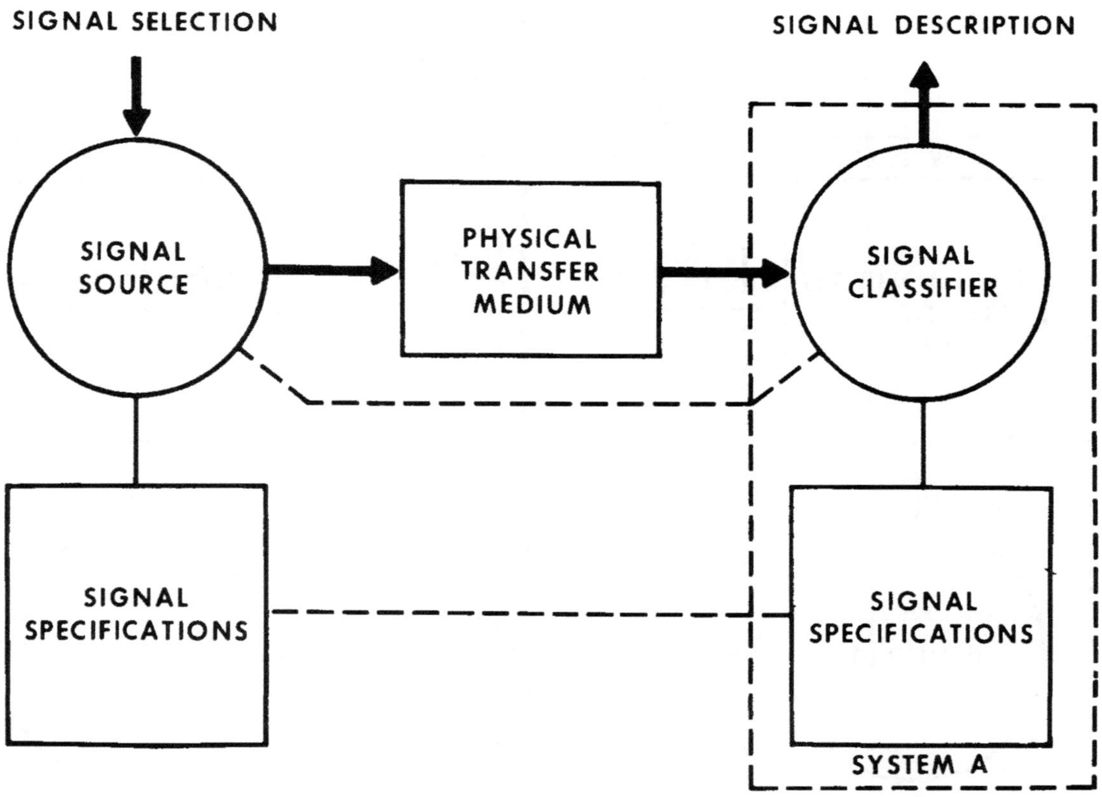

Figure 4 - Model for Signal Transfer

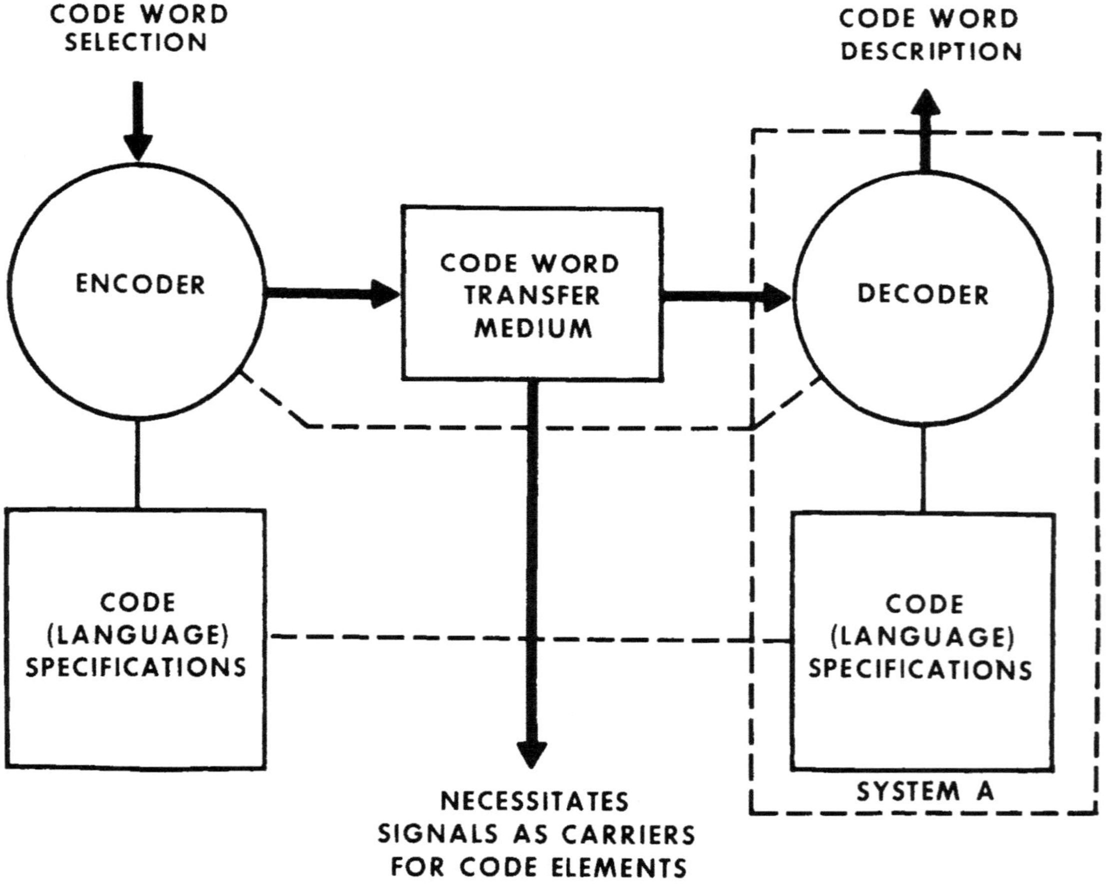

Figure 5 - Model for Code Word Transfer

on its knowledge of the specifications of all possible signals expected from such a source and then classifies it accordingly. The description of the classification represents the result of the transfer process.

MODEL FOR CODE WORD TRANSFER

The functional model for the code word transfer structurally resembles that for the signal transfer (Figure 5). Signals transferred by the signal transfer mechanism are carriers for code elements. For each code element there is a unique signal. Since the number of clearly distinguishable signals is small, communication of messages requires coding. A message can be encoded as a code word in the form of a structure, pattern or sequence of code elements. Code words are composed according to the rules of the language used or the specifications of the code.

Most languages are redundant; i.e., not all code element structures conceivable within the confines of the language are used to represent messages. Language redundancy permits, at least in principle and within limits, reliable code word transfer without reliable code element transfer. If the redundancy exceeds a certain minimum, and if the decoding receiver knows the rules of the redundant language and can apply them, the originally transmitted code words can be reconstructed from mutilated code element sequences. These mutilations occur because of imperfect signal transfer. Since all signal transfers are imperfect, the redundance of a language, the knowledge of the language by the receiver and its processing limitations play important roles in code-word transfer.

In the code-word transfer process a selected code word is encoded into a structure of code elements, the encoding being ruled by the specifications of the language. A decoder, with reference to the same language specifications, interprets the transferred code-element structures to derive the most likely correct code-word description. The transfer of the code word necessitates the use of physical signals as carriers and the existence of the signal transfer mechanism as described above. Figure 6 shows the extended model for the code-word transfer with the signal transfer model as an integral part.

MODEL FOR THE TRANSFER OF MEANING

Meaning can be associated with each code word. Code words are, per se, meaningless symbolic representations. They assume meaning within a system only if they serve as initiators or triggers of specific actions. Transfer of meaning can be viewed as the (internal or external) induction of an action. The set of correspondences between encoded messages and actions can, in its elementary form,

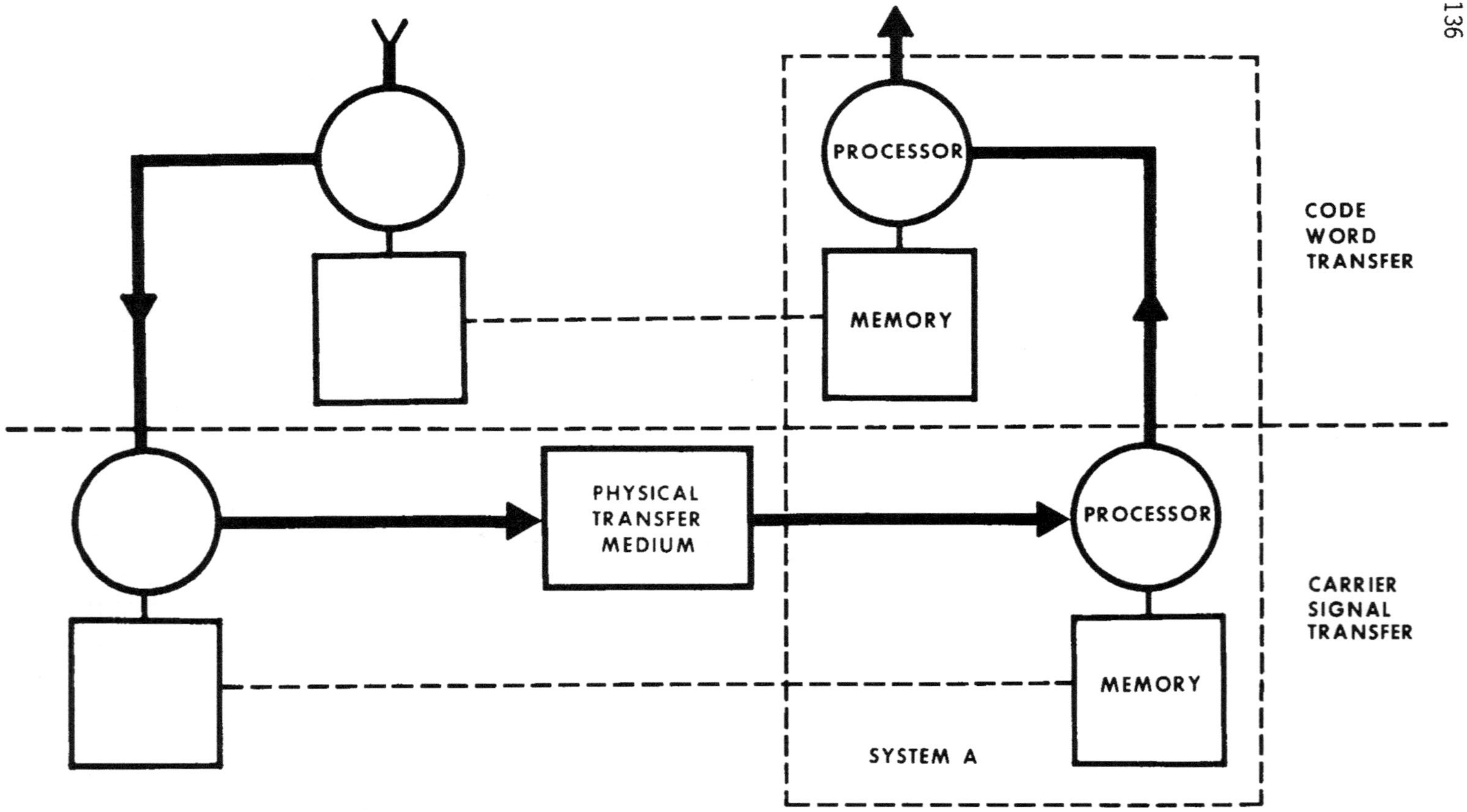

Figure 6 - Extended Model: Signal Transfer as Basis for Code Word Transfer

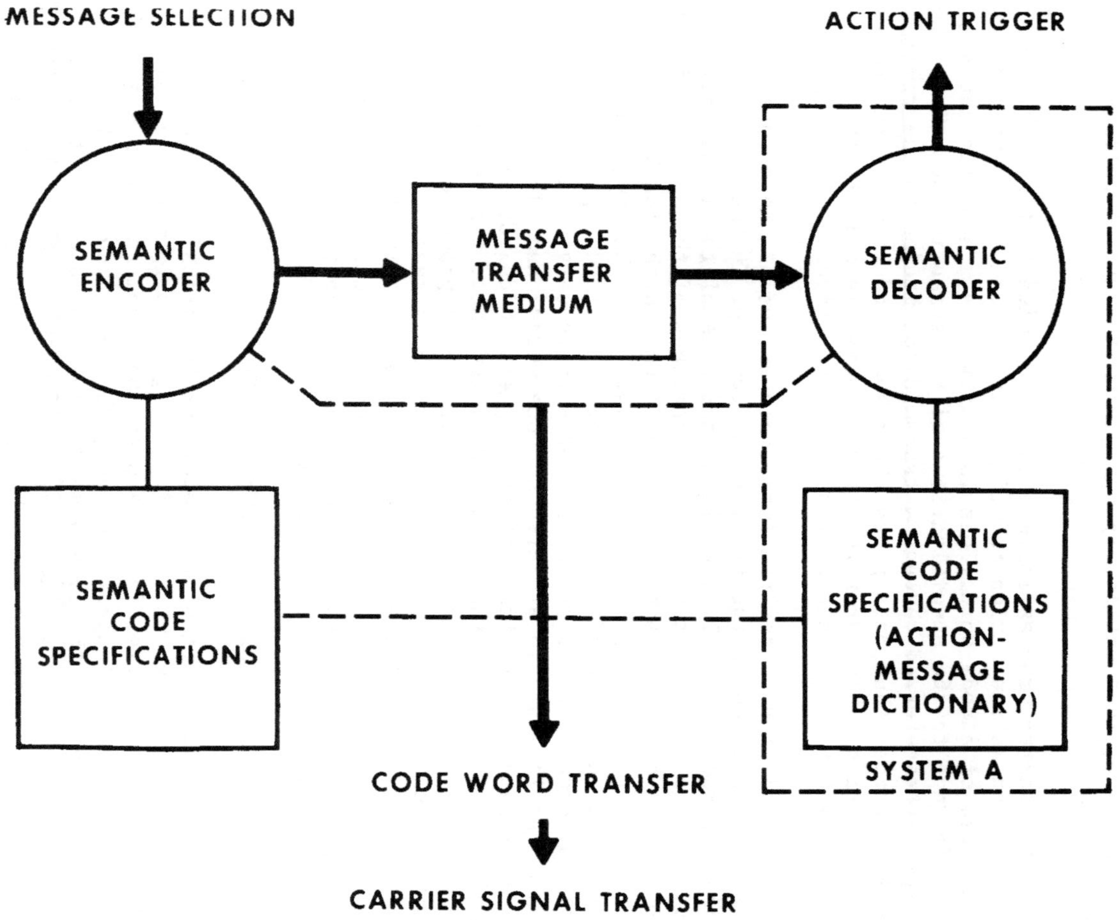

Figure 7 - Model for Transfer of Meaning

be conceived of as a dictionary containing the specifications of a semantic code.

The transfer process for meaning is modeled in Figure 7. The message, that is the action-inducing unit, is encoded symbolically in the form of a code word. This is achieved by a semantic encoder with reference to the specifications of the semantic code. The message is transferred through a transmission medium to the remote semantic decoder that interprets the message and triggers the appropriate action. This interpretation is possible only if the decoder "knows" the semantic code used in the encoding operation.

Figure 8 illustrates the interdependencies of the transfer mechanisms involved in the transfer of meaning. In the development of the general model only the right-hand side of the illustration is of interest - the interpretation functions of the system which allow system-specific meaning to be derived. In the idealized model three interpretation actions take place: (1) the physical input is interpreted as one of a small number of defined signals; (2) patterns, sequences, or structures of signals serving as code-element carriers are interpreted as code words; and (3) code words are interpreted as specific action triggers which possess meaning for the functioning of the receiving system. Note that each interpretation function requires the existence of a processor (represented by a circle) and a memory (represented by a rectangle). The memory stores the relations between all possible input structures or signals and their corresponding transformed outputs. The processor performs the necessary transformations in accordance with the relations stored in the memories.

MODEL FOR SIGNAL (OR STIMULI) INTERPRETATION

The receiving part of the model for the transfer of meaning does not have to know where its input signals originate. As long as the signal classifier interprets the physical inputs as signals which meet the signal specifications, the receiving part will try to read meaning into the sensed patterns. This interpretation of signal patterns as potential carriers of meaning takes place within the inherent limitations of the system including (1) those of the three processors and (2) the "knowledge" (of signal, symbolic code and semantic code specifications) stored in the memories which is applicable to the (possibly unknown) interpretation of this particular given source of signals (this may or may not be an information source in a communication link).

The functional model that represents the receiving part of the transfer mechanism for meaning can be viewed as a general model for signal or stimulus interpretation. Any signal is interpreted within the limitations of the processors and the reference information stored in the associated memories. This model for signal

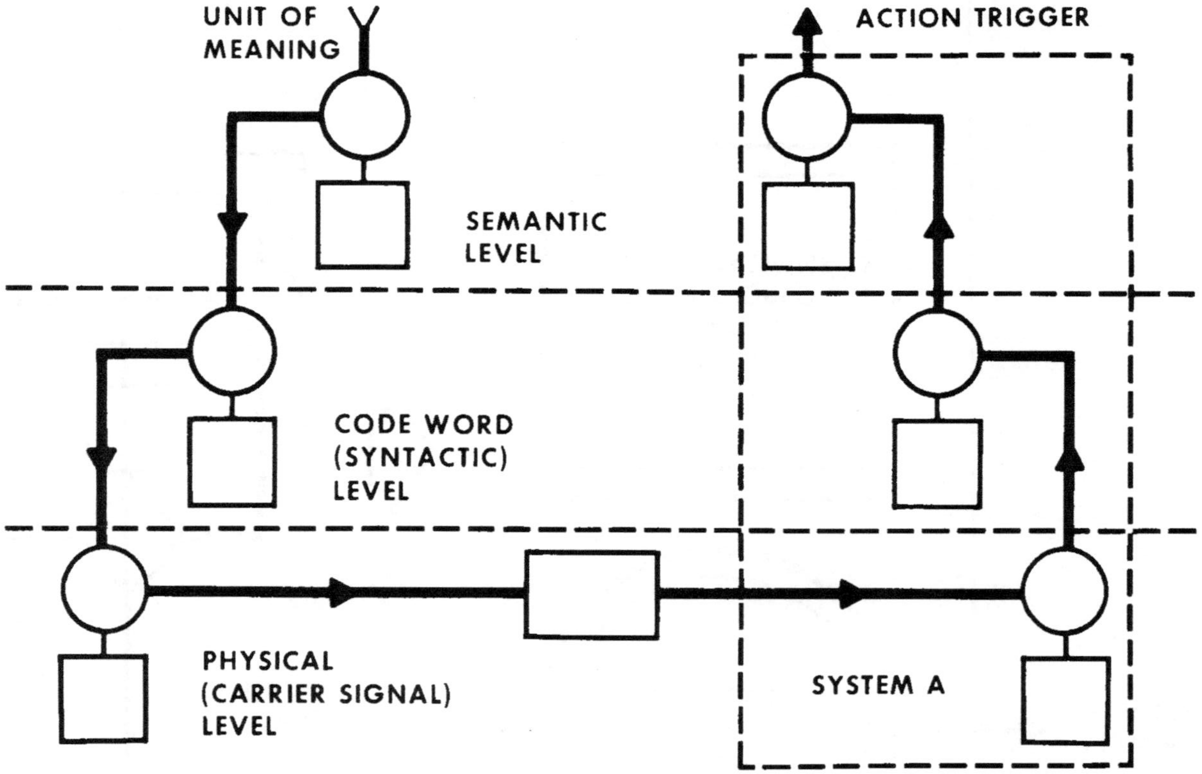

Figure 8 - Extended Model: Hierarchy of Transfer Processes for Transfer of Meaning

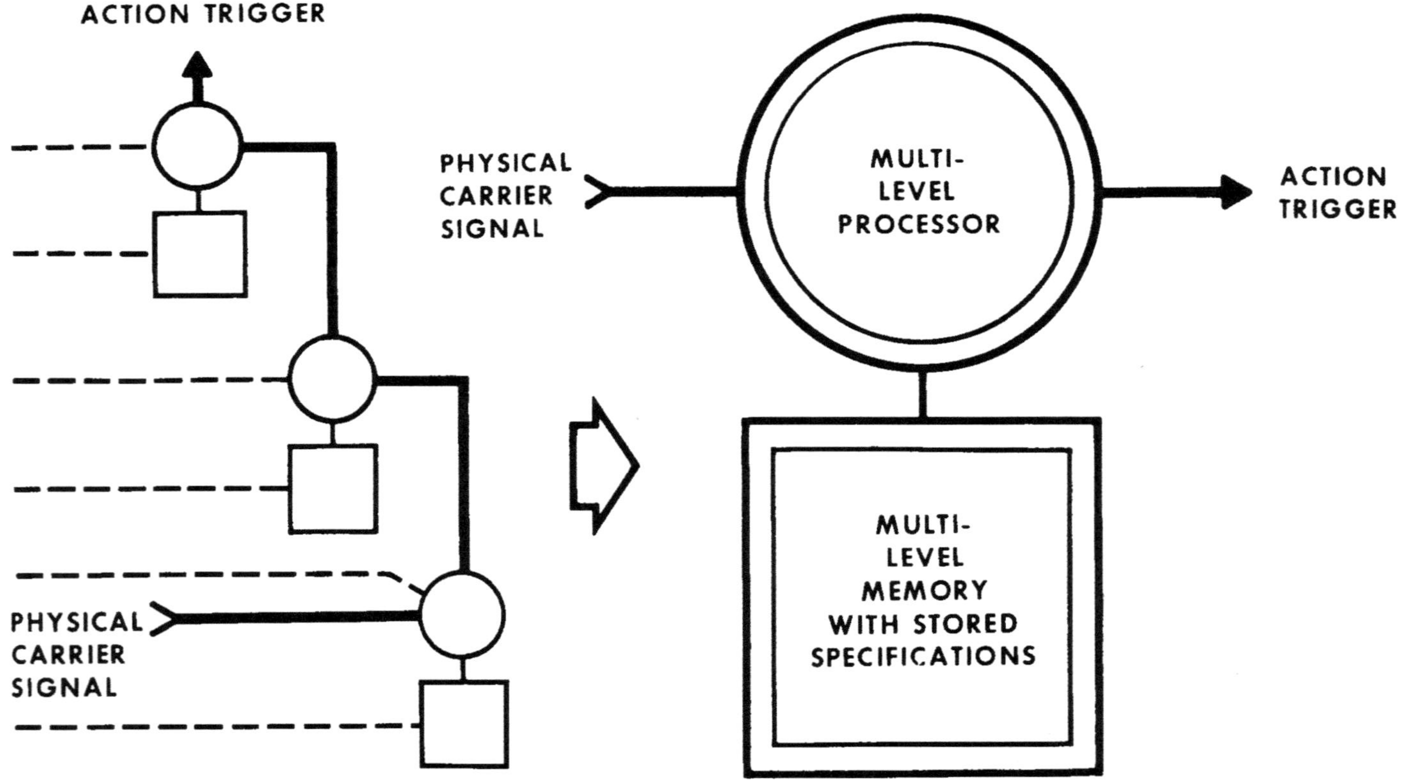

Figure 9 - Model for Signal Interpretation and its Simplification or stimulus interpretation becomes the basic building block for the proposed general model intended to serve as the framework for the development of theories on information. A simplified graphical representation of the model for signal interpretation and its more detailed equivalent are shown in Figure 9. The simplified representation will be used in all following illustrations. The dotted lines indicate that partial or complete agreement has been achieved between the stored information about the signal and the symbolic code and semantic code specifications on the one side, and actual signal, symbolic code and semantic language characteristics on the other side.

EXTENSION OF THE MODEL TO ACCOUNT FOR CONTEXT: MODEL FOR INTERPRETATION OF ENVIRONMENT

A system may have many sensors that link it to its environment by multiple means: man can see, hear, smell, taste and touch simultaneously. Input signals received from the different sensors may be interpreted differently as they may represent information in different context: acoustic signals representing speech may be interpreted with reference to the knowledge of a spoken language, similar acoustic signals originating from a bird may only be interpreted as cues for the identification of the bird or its location.

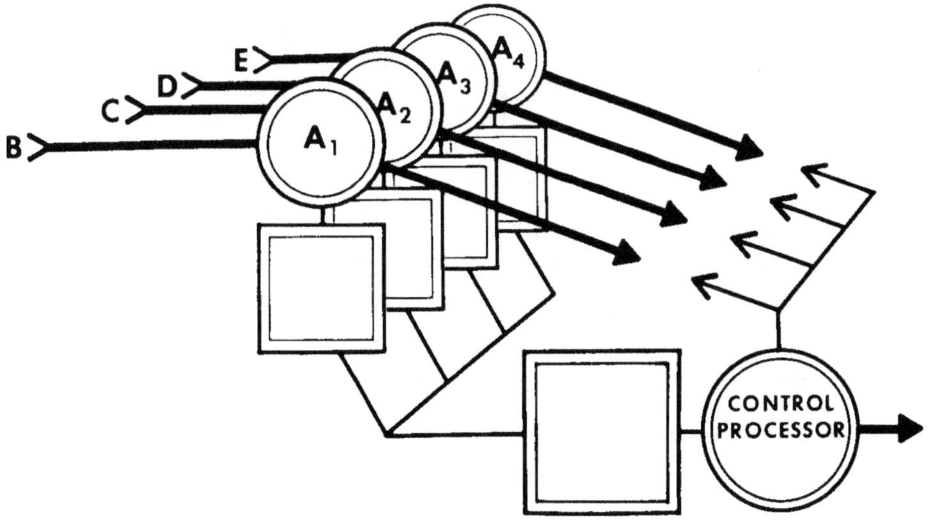

Figure 10 - Model of Multiple Signal Interpretation Mechanisms

Complex systems, such as living organisms, man himself or his organizations, are exposed to an environment which stimulates the multiple sensors almost continuously. It requires a continuous monitoring and interpretation of the signals to determine if any particular one contains "information" - something needed for the control of goal achievement or the survival of the system. Multiple signal-interpretation mechanisms, as shown in figure 10, model the "interpretation of the environment" in elementary form. Each signal-interpretation mechanism attempts to give meaning to one aspect of a sensed signal pattern which may or may not carry "information" of importance to the system. The ability to interpret is dependent on the processing capability of each processor and the knowledge of the information value of previous but related environmental stimuli for the system. The various signal-interpretation mechanisms are all potential producers of action patterns that may be competing for the control of the system. Hence, the model for the interpretation of the environment must be supplemented by a control processor which takes an overriding function in attaching weight to the interpretation functions (biasing processors and memories) according to the needs of the system.

ADAPTIVE SYSTEM FOR INTERACTION WITH THE ENVIRONMENT

The model as developed until now is a multi-input processor which transforms various input signals into action triggers that, in turn, steer the functioning of the system. Interpretation of the environment as discussed above appears to be a passive function of a system. Information systems, except some of the most elementary and many of the man-made ones, require active adaptation and learning capabilities to allow proper functioning under changing environmental conditions. Therefore, the model is extended to reflect the need of information systems to learn and adapt. This extension is known as feedback. In addition to, or separate from, its function to guide the systems responses, any action trigger can be fed back into the system and used to restructure the signal-interpretation processors and to update the information stored in the reference memories. Stimuli can be transformed into feedback action triggers that result in internal actions which influence future stimulus interpretations. These internal actions in the form of modification of processing algorithms and stored reference information, if conducted with the aim of improving and optimizing the systems functions, represent adaptation and learning. The complexity of a model representing these additional functions, as outlined, increases rapidly. Feedback and modification mechanisms can only be indicated symbolically in a graphic presentation of the model. They are shown simply as interconnections between all processors and memories involved. Interconnections are required: (1) within each single signal-interpretation model and (2) between those of all signal-interpretation models and the control processor model that establishes priorities in the integrated interpretation of the environment.

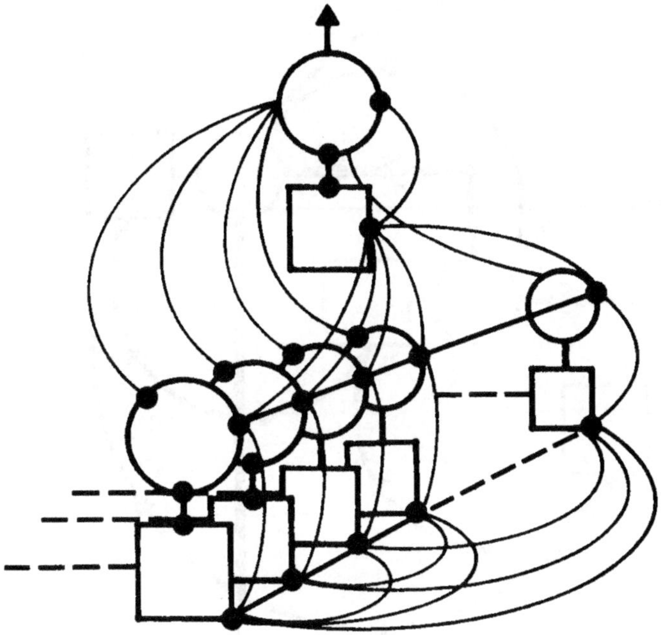

Figure 11 - Model of Adaptive System for Interaction with Environment

The general form of the model is suggested by figure 11.

REFINEMENT OF THE MODEL: DE-EMPHASIS OF HIERARCHICAL STRUCTURE

In all models of the interaction of a complex system with its environment, the elementary model for signal or stimulus interpretation finds repeated use as the basic model component. This basic model, as introduced in figure 9, utilizes a hierarchical structure at consecutive transformations: signals are interpreted as code elements, code-element structures as code words, and code words as action triggers.

This simple hierarchical structure was introduced to stress the essential need for a physical carrier signal and for symbolic coding in the transfer of meaning. The model for signal interpretation, which evolved from the receiving function in a one-way communication link, has been oversimplified and requires refinement. Interpretations of signals do not necessarily take place in the originally modeled sequence; e.g., from signal to code word to action trigger identification. More often the interpretation can be thought of as the result of an iterative process whereby, e.g., a mutilated speech signal may first be interpreted with reference

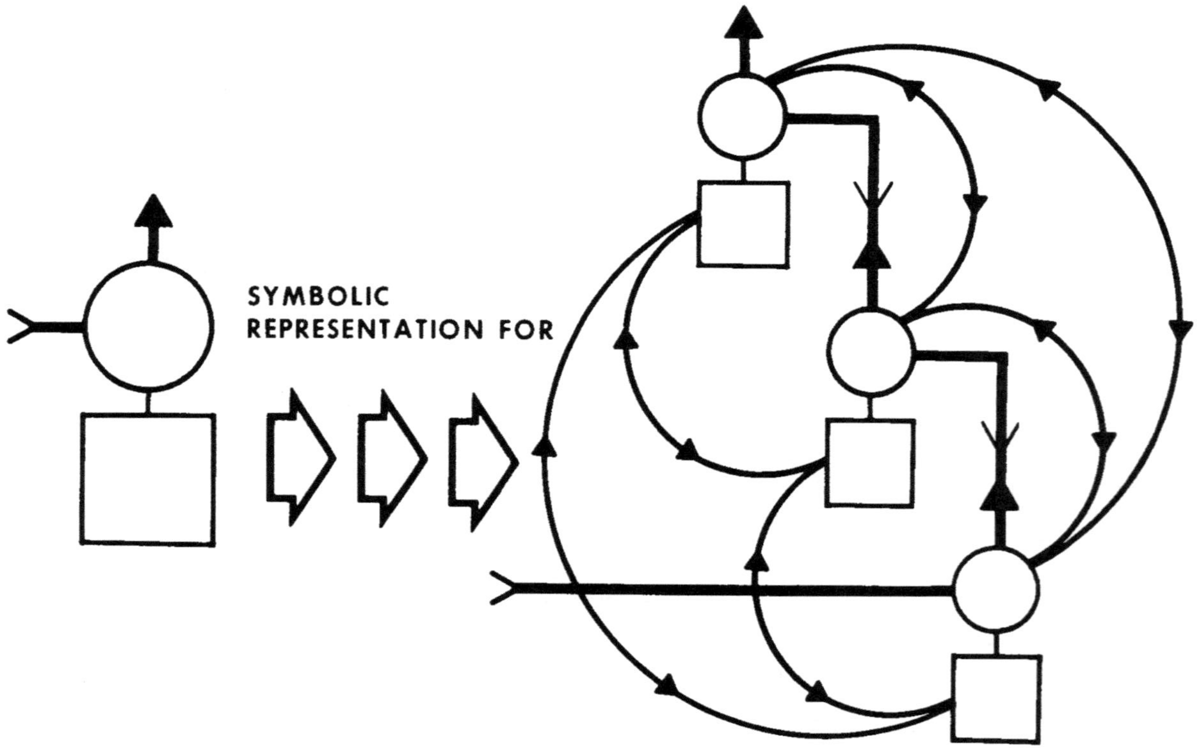

Figure 12 - Refined Model of Signal Interpretation
(Bidirectional Interpretations)

to the applicable language rules, to clarify it from a linguistic point of view, and then re-interpreted in the context of the situation to render it meaningful to the listener. If the original linguistically (formally) valid interpretation carries no operationally significant meaning for the user, the interpretation is repeatedly revised and tested until an alternate meaning is found which appears more likely in the total context of the situation. This can lead, in turn, to a downward revision of all interpretations along the hierarchical chain by the same process. In short: A more realistic model for the signal or stimulus interpretation is not hierarchical but considerably more complex. It permits bidirectional interactions between all (idealized) levels of interpretation. The refined model for signal interpretation (in isolation) is symbolically illustrated in figure 12.

GENERAL MODEL: "INFORMATION" AS RESULT OF ADAPTIVE INTERACTION WITH THE ENVIRONMENT

The general conceptual model which is proposed as the framework for the development of theories on information can now be summarized. A system interprets what data it is physically capable of sensing from its environment according to its functional needs and goals. This interpretation is limited by the physical processing capabilities (processing structure and memory capacity) and the acquired knowledge (learned or inherently provided) of the origin and meaning of environmental stimuli relative to the system itself. This interpretation transforms external stimuli into action triggers which serve either to provide the physical control and guidance of system actions or, in the form of feedback, the data necessary for the learning on the part of the interpretation system itself by modification of the processing structure and/or the reference information stored. This adaptive processing permits learning of the system. In the process of learning, new internal stimuli may be produced as a result of the adaptive restructuring without external stimuli. These internal stimuli are interpreted in the changed context of the "updated" interpretation capability and may result in new action triggers, both for external system actions and for further internal updates or restructuring.

The proposed conceptual model pictures "information" as system-specific interpretations (transformations) of stimuli. "Information", according to this concept, is manifested by an internal change in the state of the system which can (but does not have to) be reflected by external changes or actions of the system. The interpretations are system-specific in a multiple sense: (1) they serve the functioning of the system; (2) they take place in the context of the system's perception of its environment as previously interpreted and presently sensed, and (3) they are limited by the specific processing, memorization and association capabilities of the system. The model is shown abstractly in figure 13. The core

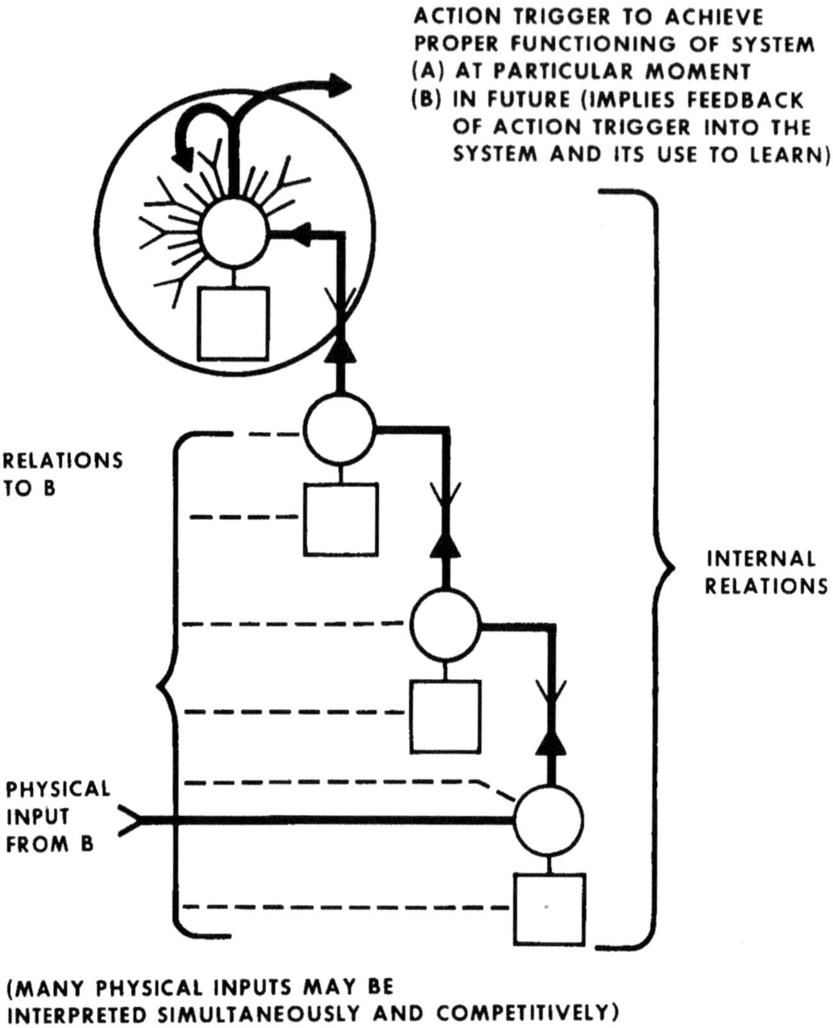

Figure 13 - General Model of Adaptive Systems Interaction
With Environment

of the system (top) interprets (transforms) stimuli from in part simultaneously stimulated environmental sensors (including receptors in communication links) and internal feedback loops (action triggers). It transforms its inputs into the action triggers needed to achieve the proper functioning of the system at the particular moment (that is, considering the integrated effect of the situational context) as well as in the future (by use of action triggers as feedback for the adaptive learning of the system). Each of the interpreted stimuli is the result of a stimulus interpretation involving interactions between signals, code words and semantic interpretations possibly at multiple levels. All interpretations require physical processing and memorization capabilities that may or may not be adaptive. Any sensed physical signal, as the potential carrier of information, provides the raw material for the formation of "information", which, according to the proposed model, is explainable only by the understanding of the system's relations to its environment. These are reflected in the states of the processors and memories, in the internal relations among these processors and memories, and by the physical ability of the system to sense the environment.

"Information" conceived as state changes within an information system, is manifested at various levels within the system's interpretation processing hierarchy and serves various specific functions. According to the level of concern one can therefore distinguish between physical (carrier signal level), syntactic (code word level), semantic (isolated action trigger level) and pragmatic (systems behavioral level) information. Accordingly, any reference to "information" should be specific and relative to the level at which the state changes serve a systems function. This qualifying statement appears necessary in addition to that which identifies the system (because information is inherently system-specific) and that which identifies the environmental relationship (the environment as perceivable and actually perceived by the system).

SUMMARY

It is suggested that the study of "information" in the widest sense of the word can be approached gradually in a scientific way, if one views information as a result of the interaction of a system with its environment. A general model is developed which is suggested as the framework for specific studies of information phenomena and processes. For the most elementary case, the interaction of a system with a single information source, the interaction can be considered as the result of a one-way communication process. This elementary case becomes the starting point for the development of the proposed model. The final and general model evolves from the functional description of the receiving end of a one-way communication link (modeling the elementary one-way interaction, referred to as signal interpretation, but disregarding context) into a complex, multi-layered adaptive processing structure. This structure transforms multiple simultaneous inputs (the stimuli to which the system is exposed and can sense) into action patterns. These transformations take place in the context of the situational environment and that of "past experience". The resulting action patterns include internal systems modifications (adaptations) that are required to learn and to maintain the functioning system in a continuously changing environment. This final and general model differs from the elementary (signal interpretation) model by the multiple use of the modified elementary signal interpretation model and the addition of feedback and adaptive capabilities. The model provides the ability to represent functionally the interpretation of any stimulus from the environment in the context of the dynamically-changing environment, of all "past experience" of the system, and of system goals that govern the relative value given to the stimulus interpretations.

REFERENCES

The paper reflects, in the terminology of the model, the "context" of my "past experience". It incorporates, in abstract and integrated form, many published or verbally communicated findings and ideas of scientists, researchers, engineers and students too numerous to mention. Any attempt to select the "most important" ones as references would produce a listing of only those sources that at the time of the writing appear most significant. This "bias" by the "situational context" would do injustice to others who contributed substantially and perhaps only indirectly during the formative phases of the model development.

INFORMATION REPRESENTATION AND DECISION MAKING

Hugh T. Smith

Sheffield University, Sheffield, England.

INTRODUCTION

It is hard to escape the impression that Information Science neglects the *detailed* study of the individual and his representation of information in the environment. Most researchers indeed appear to be content to work with perceptions of human behavior that have been derived from "field glass" rather than microscopic studies. Yet surely it is pertinent to ask where else may we learn about the substance of this stuff, this "information". My own contention is that at the present stage of the science far more attention should be directed towards descriptive analysis of human behaviour as a means of establishing "concrete" building blocks rather than towards the formulation of global theories.

My lack of faith in the normative theories of Bob Barnes[1] and Marshal Yovits[2] stems not so much from any incompleteness of the theory but from the lack of suitable directions and measuring instruments to put it to the test and thereby establish its usefulness. For example, who, in all but the most simple closed systems, has not foundered upon the twin rocks of expected probabilities and subjective value judgements? My principal reason for advancing the study of the behaviour of the decision-taker or problem-solver is in fact to be able to determine the selection of that subset of relevant theory out of the total body available.

To illustrate this point take the example of the development of ball catching skill provided by Papert[3] and shown in Figure 1. What methods does the human processing system adopt to enable the estimation of the descent path of the ball? Out of several possible hypotheses let us consider just three. Firstly the ability to ex-

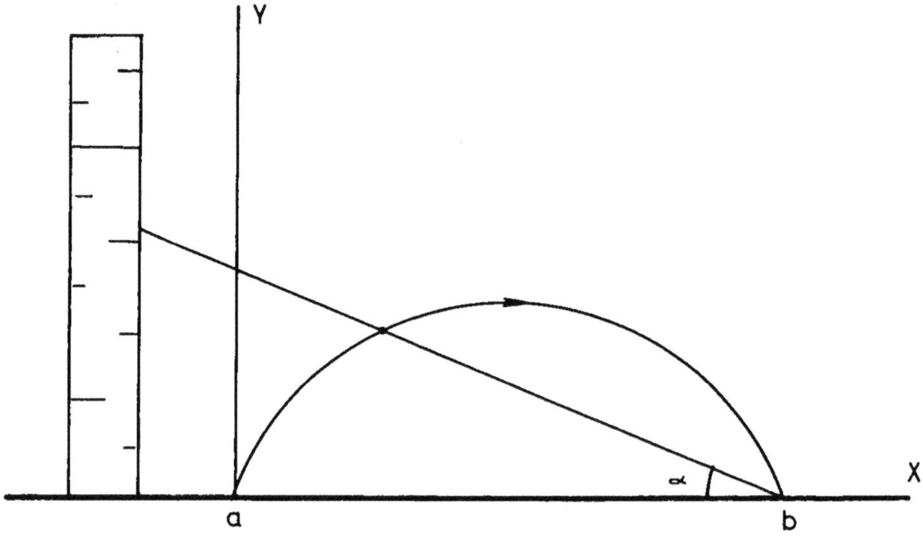

Figure 1 - Ball Catching

plicitly formulate and solve projectile equations like

$$V \text{ apparent} = b \frac{d}{dt} \frac{-ax(x-b)}{(b-x)}$$

- not really very likely. Another strategy would be to use the change in apparent size of the ball during its flight. In fact from what we know of human response characteristics the sampling time available would be too small. A third suggestion seems to fit more reasonably with the available facts - since the rate of change of vertical velocity of the ball in flight is not constant a good strategy would consist of moving along a course so that it appeared to maintain as constant a rate as possible. We might find confirmatory evidence for this hypothesis in the (empirically-established) highly-developed sensitivity of people to slight rates of change variations.

The point is not that this particular hypothesis is necessarily a correct description of the relevant thought process but that it is a good place to start in elaborating or investigating ball-catching skill because it fits some of the available data. In most problems there are many possible algorithms that might be put up to explain behaviour - it is most unlikely, however, that some would not be discarded by such references.

Another example of choosing the correct representation nearer to our technological society is the solving of real-world sequential decision problems, be it file retrieval, management forecasting or transport control. In these problems we usually have a well-specified objective, but not too much information as to how well or badly it is possible to satisfy it. (See Figure 2.)

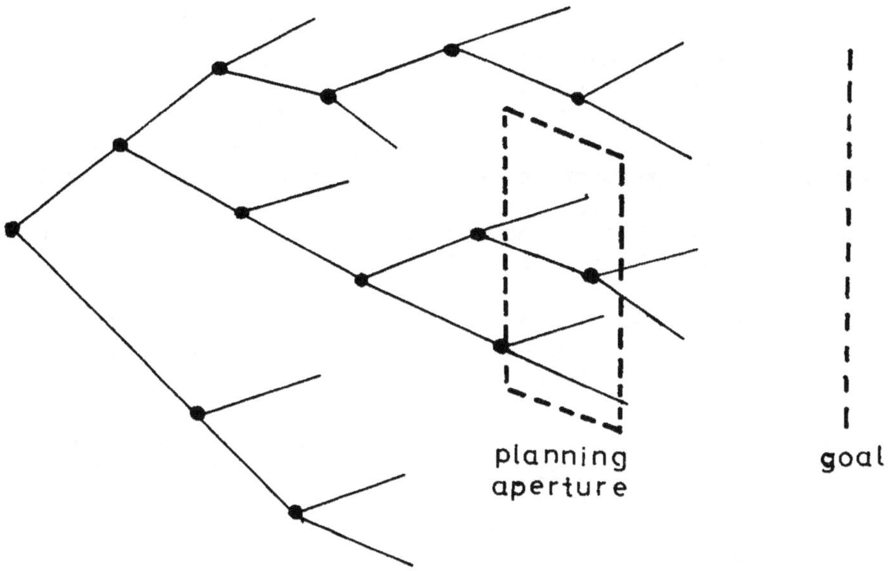

Figure 2 - Search Tree

The set of possible states that can be visited in the search for an optimal solution may be considered infinite for most practical purposes. We could not employ a machine (time and cost wise) nor expect a human to solve these problems by means of exhaustive search techniques. Pragmatically then in decision situations we can choose either rules that the machine finds easy or those human heuristics that certainly might not be so easy to find or program but often do have the distinction of working. It is this choice of solution *method* alone that determines the relevance of our model of the environment and *not* how much sophistication is built into it.

The consequences of the temptation to the system specialist to produce very detailed models of the environment in his search for validity is inevitably a host of "special rules" to accommodate specific instances. Carried too far it may suffer from the tottering instability of Ptolemeic astronomy.

PROCESS VERSUS STATE

I believe Information Science is at the same crossroads as the discipline of psychology which is struggling to sort out what are useful and what are non-useful concepts. Nowhere is this more apparent than in the developing interest in "process" models of behaviour as opposed to static or state models. The inadequacies of the state profile approach can be shown by reference to two examples:

(1) If we compare the performance of a chess master or novice in terms of such concepts as general intelligence or short-term memory capacities we will often be lucky to find other than artefactual differences, or at least differences of a much smaller magnitude than their respective skill in playing the game. Clearly the something that separates them is task specific and, if we observe a game, not immediately obvious in terms other than that the master player always chooses the "right" move.

(2) The second example concerns an apparently very useful way of looking at choice performance in decision situations that was originally put forward by Brunswick[4] - the lens model (Figure 3).

In this model the experimenter assumes a simple relationship between the experimental environment and the subject's responses. The attribute weights, criterion and judgement values can be predicted from linear combinations of the cues by simple regression equations. In fact there are several methodological reasons why it is difficult to relate human decision behaviour to the environment through this model. One of the more obvious difficulties in sorting out whether the apparent differences in attention paid to one set of attributes at the expense of others is conscious or unconscious. Often this phenomenon appears to reflect an inability of the human to simultaneously "notice" and "compare" increasing numbers of cues. Bruner *et al* and Shepard[5] have pointed out that this unconscious selection process may be performed to reduce the cognitive "strain" of complex or indeterminate situations. The relatively recent use of multidimensional scaling reflects an attempt to at least get at the subject's as opposed to the experimenter's representation of his problem. In addition no reference is made in the model to the way subjects serially select objects for inclusion in their judgement process. It is therefore difficult to remove order effects from such studies.

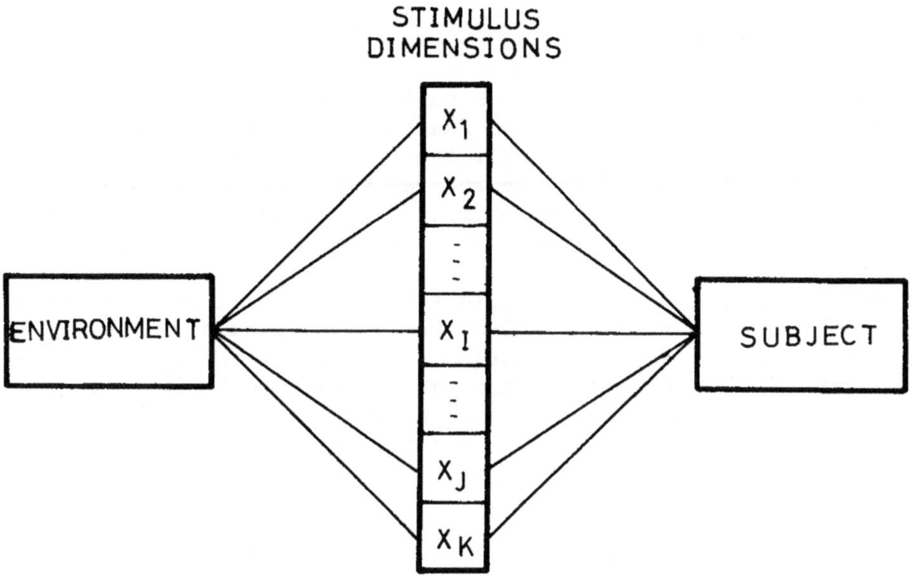

Figure 3 - The Brunswick Lens Model

A completely different approach to the study of human decision-makers is adopted by the "information processors" who stress both the importance of the task specification and the serial nature of behaviour. Much of the early research in this area was performed by Simon and Newell and reviewed in their rather discursive book *Human Problem Solving*.[6] Their general theory involves one proposition - that only a few gross characteristics of the human information-processing system are invariant over both task and problem solver - and two basic notions, the problem space and the production system. Essentially the problem space is considered to be that representation of the task inside the problem solver in which all problem-solving activity occurs. Problem-solving behaviour is considered to be the search or transformation of this space by suitable operators - selected by a strategy/method program (the production system). As the production system can only react to and lead to elements which are in the problem space its activities are a function of that space.

The crucial part of their theory depends however on the relationship between the task and the problem space. A possible criticism of Newell and Simon lies in their occasionally stated claim of a simple one-to-one mapping between the two. As was mentioned earlier however humans often appear to work with transformations or representations of minimal dimensions of the environment which do not necessarily come from a simple mapping. Much of their work on well-defined problems like chess or cryptoarithmetic is devoted to describing how the form of active search behaviour of the subjects mirrors the apparent processing limitations and advantages of the human cognitive apparatus. Examples are the influence of short term memory capacity on search patterns (like De Groot's "progressive deepening")[7] and search-reducing methods such as means/end analysis. What tends to be difficult to explain however is information storage and the relationship between the "noticing" and planning abilities of skilled compared to unskilled subjects. Experienced chess players for example appear to have integrated into their problem space patterns which they have previously observed and which therefore do *not* appear in the protocol analysis as active search. This information "chunking" appears to be intimately tied to perceptual pattern recognition. Thus we may get the substance of the observation - chess masters only "actively" look one move ahead but it is always the right one.

A recent paper by Simon and Chase[8] appears to suggest a methodology of getting at the form of these chunks which will be useful in other behavioural situations. They performed two tasks, one of which was a replication of an experiment by De Groot[7] which confirmed findings that chess masters were superior to other players in recalling the layout of a chess board only when the pieces were arranged in likely game state positions. The other, perhaps more important experiment, consisted of making subjects reproduce a board state after a brief exposure. If more pieces remained to be placed than they could recall from one such exposure they were allowed further glimpses until the configuration was reproduced fully. By timing the rate at which sub-patterns of pieces were laid out between glances they were able to speculate on the nature of the information chunks. Perhaps importantly they assert that the "number" of chunks (chunk addresses?) held in short term memory was similar for *any* level of player but the players differed in relation to the richness of these chunks.

REDUCING ERRORS IN SEARCH BEHAVIOUR

A scheme for assessing human behaviour in sequential decision tasks in terms of an environmental model, action set and evaluative function was recently proposed by Smith.[9] If we try to apportion non-optimal performance to this scheme certain interpretational difficulties immediately present themselves. For example, the existence of an explicit iterative search procedure was suggested in which the value of an action was obtained by assessing its predicted effect (measured by the evaluative function) on the system up to some predefined "horizon". If we ignore gross errors of the type caused by drifting attention, does the choice of a non-optimal action indicate the subject has been skimpy in his search procedure, used a faulty evaluative function or both? Worse still, if we allow that the model is dynamic in form (i.e., adaptive) an error in the formulation of the evaluative function might cause an error in the model.

Clearly if we are interested in improving human performance then we must try to speed up the acquisition and development of the correct cognitive "model" by distinguishing these errors. The difficulties inherent in the practical realisation of these sorts of simple feedback notions however have limited progress in the field of man-machine systems.

In recent years there have been several attempts to improve upon algorithmic solutions to large scale search problems by organising data so that a human can "see" the most promising set of possible solutions. Perhaps the most notable progress in this area has been made in the class of routing or scheduling problems (for example see Krolak and Felts,[10] Krolak and Nelson,[11] and Michie, Fleming and Oldfield.[12] Generally speaking there are two main difficulties to detailed search for humans: (1) keeping track of the nodes visited (i.e., choices made) as the search progresses; (2) estimating the present distance from the goal or subgoal in order to determine whether to extend or terminate the search branch.

One of the principle focuses of man/machine research has been in fact to overcome these difficulties by choosing a new symbolic representation of the problem. I am just beginning such an experiment which attempts to improve judgement by extending the decision horizon. A. Battersby[13] suggested that the decision maker would be greatly helped if his problem could be transformed into a conceptually easier domain allowing some use of these innate perceptual skills. He experimented with a pattern matching task in the context of a PERT problem. In our experiment the basic situation has been extended and tree search and retrieval routines added. Figures 4 and 5 show the same problem in its two fundamentally different representations. In the one case subjects try to minimize the *total* perimeter of a set of related squares by changing the shape of any square (such that the area is kept constant) and optionally moving its position. The second representation is in the more traditional form of a PERT problem where the emphasis is on the choice of the dispatch sequence for a set of interrelated activities.

What are the differences in the information content of these two environments? Although they constitute the same problem one immediately appears much easier to conceptualise than the other. Does this affect the transformations possible in the problem space? Can previously-learned associations or concepts really be transferred to other tasks? Hopefully the answer to these and other questions will be provided by empirical research.

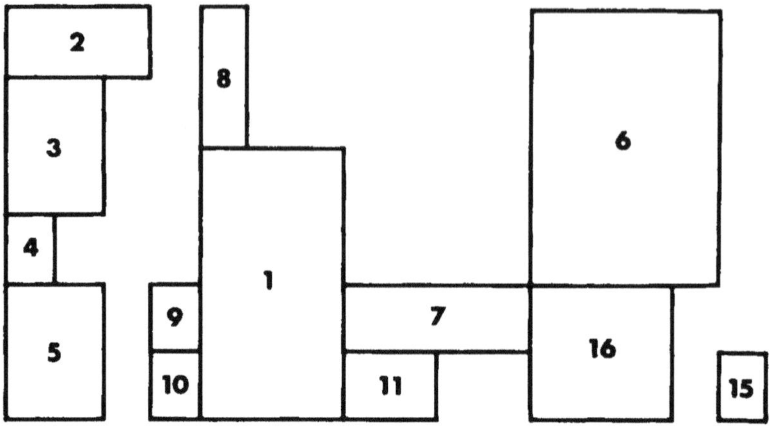

RUN 1

Figure 4 - Pattern Matching Representation

(2,4,5,) (9,10) (3) (8,1) (16) (11) (6)
1 (1,3) (1,3) (11,1) (11) (11) (6) ( )
0 1 2 4 5 7 8

RUN 1

RESOURCES 1 11 4

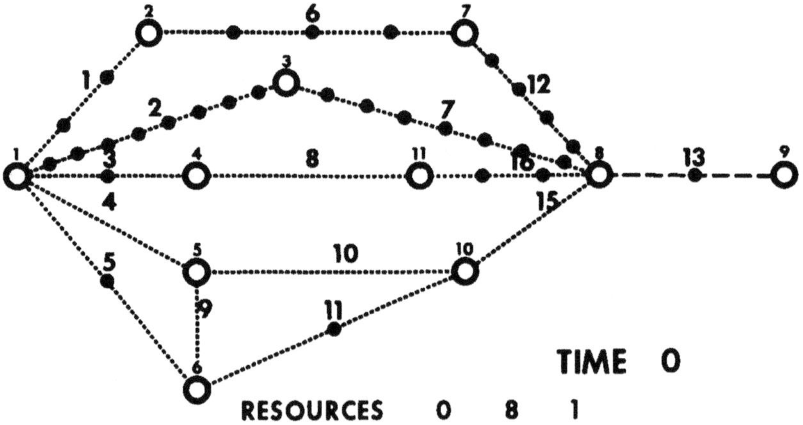

TIME 0

RESOURCES 0 8 1

DISPATCH 5 4 2

DELAY 3 1

ACTIVITY TIME RESOURCES START FINISH DELAY

Figure 5 - PERT Representation

CONCLUSIONS

Although at the beginning of this paper I criticised the theoretic "top-down" approach of information science researchers the dividing line between us is small. We all seek those tools and directions which will allow the evaluation and prescription of the role of information in our society. Perhaps the main unitary effort in this respect will come from the development of such techniques as "Fuzzy Set Theory" which attempt to formalise a more human approach to probability in judgement. Certainly a good prognosis for the future of Information Science is heavily dependent upon the development of more integrated theories.

References

1. See the paper by Barnes in this publication.

2. See Yovits, M.C., Paper to appear in *Information Science: Search for Identity*, Marcel Dekker, New York, 1974.

3. Papert, S., Theory of knowledge and complexity, in *Process Models for Psychology. Lecture notes of the NUFFIC International Summer Course, 1972*, Netherland Universities Foundation for International Co-operation, Dalenoort, G.J., ed., Rotterdam University Press, 1973.

4. Brunswick, E., *The Conceptual Framework of Psychology*, University of Chicago Press, Chicago, 1952.

5. Shepard, R.N., On subjectively optimum selection among multiattribute alternatives, in *Human Judgments and Optimality*, Shelly, M. and Bryan, G., Eds., John Wiley, New York, 1964.

6. Newell, A. and Simon, H.A., *Human Problem Solving*, Prentice-Hall, Englewood Cliffs, N.J., 1972.

7. De Groot, A., *Thought and Choice in Chess*, Mouton, The Hague, 1965.

8. Chase, W.G. and Simon, H.A., Perception in chess, *Cognitive Psychology*, 4, 55, 1973.

9. Smith, H.T., Human decision making, to appear in *Information Science: Search for Identity*, Marcel Dekker, New York, 1974.

10. Krolak, P., Felts, W. and Harble, G., A man-machine approach toward solving the traveling salesman problem, *Communications of the ACM*, 14, 327, 1971.

11. Krolak, P. and Nelson, J., A man-machine approach for creative solutions to urban problems, in *Machine Intelligence*, 7, Melzer, B. and Michie, D., Eds., Edinburgh University Press, 1972.

12. Michie, D., Fleming, J.G. and Oldfield, J.V., A comparison of heuristic interactive and unaided methods of solving a shortest route problem, in *Machine Intelligence*, 3, Michie, D., Ed., Edinburgh University Press, 1968.

13. Battersby, A. and Berners-Lee, M.L., Communication through interactive diagrams, in *Computer Graphics in Management: case studies of industrial applications*, Green, R.E. and Parslow, R.D., Eds., Gower Press, London, 1970.

Chapter 2

NATURE OF A SCIENCE OF INFORMATION

Nature of a Science of Information

Editorial Summary

Quite early in the game, the development of information systems - systems which provide man with the technical tools for decision-making - formed the basis for conceptualizing a science of information. The burgeoning of knowledge which resulted from World War II Research and Development, aided by an increased use of that knowledge in all areas of science and technology, also led to an interest in a science of information.

Recently, the interest in a science of information has increased, particularly among technologists and librarians. The technologists, primarily computer scientists, have conceived of information science as an off-shoot of the development of computers. Some librarians have perceived information science as computer-automation oriented, while others in the library camp have found it difficult to recognize a science with which they were historically aligned, but which was apparently conceptualized as a new discipline.

No statement concerning the basis for a science of information would be complete without the inclusion of philosophy and psychology - two disciplines that have traditionally been concerned with knowledge and information. These disciplines provide the formal underpinnings for a science of information, but are not concerned with the practical matters of acquiring, disseminating, and storing published knowledge, or with the media which contains such published knowledge.

In brief, the recognition of a need for a science of information emerges from three fundamentally different sources: those concerned with the development of information systems (these are primarily engineering-oriented), those concerned with the logistics of information distribution, such as librarians, and those concerned with the social and human aspects of information processes. These three elements, all reflected in the Institute, represent the belief systems which underscore the need for a science of information.

Licklider represents the technological viewpoint. He maintains that a need for such a science has become apparent because of the interdependence between computer science and communication science and their respective technologies. This interdependence is manifest in a number of pragmatic problems dealing with the acquisition, storage, and retrieval of the vast amounts of knowledge that have been recorded. Licklider does not propose to describe what such a science would look like. However, very possibly he, representing the engineer's point of view, would propose that operational analysis, simulation, and modeling be central to the interests of such a science. In this view, information science is a theoretical tool serving the information technologist.

Kesting is a librarian who perceives information science as part of an expanded concept of library science. Information science's major concern is library automation and information dissemination. Consequently, a science of information is needed to improve information acquisition, storage, and retrieval. But it would be an injustice to Kesting's position to limit his case solely to this. Kesting recognizes the importance of the work that the librarian undertakes with the user. He sees the need of a science of information that ties the humanistic aspect of library work with the technical. Better understanding of the user and his needs, and of the way technology can be applied to the needs of the user would then constitute the corpus of a science of information. Kesting expresses this view by projecting a future library as an environment dedicated not solely to meeting man's need for information, but also to the expression of the humanistic aspects of science.

The philosophical framework for a science of information is represented by Monasterio, whose paper is included in the previous chapter. To Monasterio, the basis for understanding information lies in our understanding of consciousness. Consequently, the basis for the science is the need to link experience with information and the capacity that technology provides for influencing that experience. It is not a rejection of science, but rather an incorporation of science into the philosophical construct that underlies our present concept of knowledge.

In contrast, Nevitt and Rosenberg join Tom Martin, who, in the 1972 Institute held at Seven Springs, expressed despair with the scientific method and its ability to provide any understanding of information. To Nevitt and Rosenberg, the basis for a science of information is phenomenology; it is the experience itself that counts. The method of science is counter-productive because of its inability to deal with qualitative or subjective experience. There is a need for an appreciation of the experience of information, and this appreciation is achievable through an understanding of human values and their expression.

Nevitt, a colleague of McLuhan, expresses the view eloquently. What a person perceives is what constitutes information. The perception and the information are intertwined. Fundamental to this view is transfer, more commonly referred to as communication. It starts off with an inter-relationship in which the observer interprets what he observes. From this interaction, concepts emerge. These concepts constitute information. The mechanism available to the observer to deal with the observed is language. Language is limited, Nevitt claims, because it cannot represent the real world adequately. Most observation is not encoded informal language and most communication is achieved with tools other than formal language. Nevitt emphasizes that the meaning of a word is not what it says, but rather what the word does in influencing the observer-observed interaction: "meaning is not the mechanical transfer". Meaning, therefore, is the resulting action. To Nevitt, the basis for understanding information lies in the nature of this resulting action. Nevitt's concept of information expressed metaphorically corresponds closely to Otten's abstract and functional model of information.

Rosenberg represents a view which is gaining attention because it suggests that traditional scientific methodology is insufficient to answer questions about man's state and destiny. This position questions the traditional view of the scientist, who is primarily concerned with physical laws as opposed to "universal laws" and with objectivity. The Rosenberg position questions the role of objectivity and stresses human value systems and the need to understand them as the basis for an intellectual grasp of information.

Because of the visible impact of this position on the entire effort of the Institute, it should be explored further. First, the position questions the value of conventional "hard" sciences in dealing with complex human problems. Because information is regarded as a human problem, the position seriously questions the ability of traditional science to deal with the problems of information, and claims that little is gained by looking to hard science for an understanding of information. Without developing a true critique of the position, the following counter-points might be advanced.

First, our increase in knowledge and advances in the state of the art in technology have been the result of orderly investigations generated by a discipline - normally a discipline relying on the scientific method. Although advances in man's knowledge have also been generated by imaginative insight, and through serendipity, serendipitous discovery has not been the norm in science.

Second, the assumption of randomness as the fundamental character of phenomena (as contrasted to order) will significantly influence the manner in which inquiry is pursued and the way information is viewed. The movement to randomness and subjectivity is a reaction to the complex nature of the problems with which man is presently confronted. The escape from the rigor of formal science, with its assumption of the orderliness of the universe results in an orientation to the complexity of problems which in itself may be counter-productive. By assuming such a position, we may be reducing the magnitude of the problem without accounting for the problem itself.

Despite the fact that both Nevitt and Rosenberg would like to move outside the formal realm of science to approach information, their position is positive in suggesting a paradigm for the study of information. It would appear that their formula is to derive a personalistic statement about the experience of information and then to derive the means for capturing these experiences vividly in human terms. Moving in this fashion, we can improve our understanding of communication - the process by which man gives "information" to others. An understanding of how man generates information can also be attained. With this accomplished, it would seem that the first step of formal science will have been achieved, namely, the observation of phenomena or events. A question remains as to where to go from there. On this point, Nevitt and Rosenberg are not explicit. However, their critical approach toward an understanding of information science alerts us to the dangers of failing to pursue new avenues in the search for science. This is a valid contribution. But it would be dangerous not to continue the use of proven scientific methodologies in the search for knowledge wherever it applies, as long as these open-minded parallel searches constructively suggest how we can gain knowledge by other means.

PSYCHOLOGICAL AND TECHNOLOGICAL DYNAMICS OF INFORMATION SCIENCE

J.C.R. Licklider

Massachussetts Institute of Technology, Cambridge, Massachussetts, U.S.A.

INTRODUCTION

Let me propose, at the outset, that for our purposes we reject the notion, though it may ring with truism and resound with importance, that information science is the science of information. The domain of information is too extensive for this small group to master or to govern. Why should we stake a claim, at this late date in the history of science, to territory occupied so long by logicians, mathematicians, psychologists, and artists, not to mention propagandists, spies, and many others whose subject is information.

Turning toward the other end of the continuum of scope, let me propose that we reject, also, the notion that information science is the science of the storage and retrieval of documents or data. That area is typified for too many people by narrowly envisioned retrieval and dissemination systems, computerized citation indexes, or conventional libraries, and it carries too strongly the connotation of storage warehouse or depot - too much emphasis on putting things away and then getting them out again and not enough on the dynamic aspects of organization and understanding.

The foregoing remarks pertain, of course, to the delimitation and thus to the definition, of the field of information science, the field upon which we are focusing or preparing to focus our efforts. I think that definition is an important consideration for this group and that "focus" is a key word for it. At the precursor meeting in Pennsylvania a year ago, it was evident that the ubiquity of information and the diversity among the participants' conceptions of the field were making it difficult for the group to focus on substantial yet malleable issues and significant yet solvable problems of common interest. The main aim of this paper is to contribute to such a focusing, to present some ideas that I think bear upon the selection of problems, methods, and criteria by this group - which is a group dedicated to the advancement of understanding and productivity in an area of science that is in the process of taking shape, of becoming structured, of adopting its goals and methods and criteria.

For information science, the once-witty circular approach to definition is particularly inappropriate. Psychology may well be what psychologists do more than non-psychologists, but information scientists should not be happy with such a definition of information science. The reason is fairly simple and, I believe, fundamental to our effort to develop a focus within which our work will be productive and significant. It is that, unlike other sciences, information must arise and develop and live within a technology, and that the technology, heretofore not capable of supporting the science in the way required for truly significant progress, is now rapidly achieving that capability.

When I speak of the technology as supporting the science, I do not refer only to the fact that information scientists use computers in their work. I mean, in addition, that the central problems for information scientists to attack are problems in the organization and understanding of information and that the essential methods for attacking those problems involve the "objectification", in the form of computer processes, of organization and understanding. Let me be clear that I am not suggesting that information science must wait for the day when computers are intelligent enough to create a science man cannot create. That is not the idea I am presenting. I am saying that a set of problems is beginning to form as efforts are made to apply computers in the intelligent organization and understanding of information, that the problems will become clearer as the efforts progress, and that the problems constitute the subject-matter core of what information science will or should become.

Nor do I mean to say that computer technology constitutes the only technological base for information science. Communication technology also contributes to the base.

ADVANCES IN INFORMATION TECHNOLOGY

As recently as twenty or thirty years ago, the only systems capable of processing information in ways that could give rise to knowledge or understanding were brains. They were supported, of course, by spinal cords, peripheral nervous systems, sensors, effectors, and other living organs and by such tools and systems of tools as pencils, pens, ink, paper, books, desk calculators, scale models, libraries, and telephone, radar, and fire-control systems. Nevertheless, almost all the information processing that could be called "cognitive" - the processing that organized data into knowledge and gave rise to understanding - took place in brains. Despite many years of study, no one had a sure understanding, or even a generally satisfactory theory, of cognitive information processing. Indeed, the subject was very difficult to study productively for many reasons, one of the strongest of which was that the brains in which the processing took place were protected rather well by boney skulls against all but subjective observation.

The two or three decades that constitute the operational history of digital computing have not greatly changed the state of our knowledge of cognition, but they have greatly changed the prospect of our understanding it. The opening of the horizon has taken place, actually, mainly in the last decade, and much more in the later than in the earlier half of it. The critical factors were that computers of at least marginally sufficient capability became available for use in research and that the costs of computing (per unit of processing and per unit of storage) declined to levels affordable by at least some researchers.

To appreciate what has happened and is happening, one must examine the long-term trends of performance capability and cost in the area of computer hardware. A simple "law" summarizes some of the most important facts: The size of the largest processible memory, the speed of the fastest store, and the amount of processing that can be done for a hardware dollar double every two years. There are fluctuations, of course, and some classes of equipment (e.g., consoles, small secondary stores) progress less rapidly than the doubling rule asserts, but other classes (e.g., very large stores) progress more rapidly. The important thing to realize, however, is what doubling-every-two-years means. The consequences are easier to calculate than to appreciate: A factor of 32 in 10 years; more than 1000 in 20. If it takes 8 years to move a system from research to operation, the researchers can visualize the operational system as having 16 times as much "power" as a static technology would give it.

There is, of course, no assurance that the doubling law will continue to hold far into the future. For some manufacturers, the manufacturing cost *per se* is already but a small fraction of the sales price of processors and memories. On the other hand, the computer industry is competitive, new technical bases (e.g., large-scale integration) continue to take over when older bases cease to support the advance, and at present the derivative is about as high as ever. My feeling is that the trend will continue almost in full force for at least another decade if the problem introduced in the next paragraph is solved.

There is a fact about the software world that, in the eyes of many computer users, deprives the hardware doubling law of much of its effect. The fact is that software remains time-consuming and expensive to develop. The reason is not that important software problems are not solved. The first FORTRAN compiler took 300 man-years to develop; a recent one, less than 3. The reason for the recalcitrance of software is that its domain is not only broad and deep and complex but also deceptive. Whereas in fact it requires a fundamental scientific approach combining both analytic and synthetic skills, it has deceived many administrators and managers into attacking it with masses of infantry. The present unsatisfactory state of software is therefore in fact an aspect of the dynamics fostering the rise of an information science. The hardware situation is supportive. There is a great need to master the software situation. Indeed, the need is being recognized in key areas of administration and funding.

A bit earlier, I mentioned that the cost-to-manufacture is for some manufacturers and some products only a small fraction of the sales price. That is the case, dramatically, for processors and memories. In a competitive field of business, it is of course a very dangerous situation. It creates great pressure on the manufacturer to sell more processors and more memory despite the fact that each processor and each megabit of memory do more work than their precursors. The fact is, however, that there are not enough routine informational chores needing to be done in the world to occupy all the processors and memories that industry "needs" to make. (Note the quotes, though I shall not follow the line of discussion they obviously suggest.) The pressure to sell is therefore translated into a pressure to develop new application areas. But the potential application areas require new software. Many of them require software that embodies knowledge. And that fact contributes to the need for an information of the kind that is the subject of this paper.

Another part of the technological basis for information science must be mentioned at this point. It is the technology of transmission. If storage and processing belong to the computer field, transmission belongs to the communication field, and the two fields have been interacting as though they were superpower antagonists in a cold war. In psychology and style the two fields contrast sharply: competition vs. monopoly and regulation; a new product announcement every Tuesday vs. 10 years of research, 10 years of development, and 30 years of operation. Nevertheless, during the last few years, there has been progress in the joining together of computer systems and communication circuits to form computer-communication networks.

In my assessment, computer-communication networks are vitally important, not because of what they are doing, but because of what they can do. They can provide an almost perfect matrix for the development of an important and, indeed, essential information science. They can make it possible for people and programs to interact with one another and with organized stores of information - rapidly, economically, and almost without regard to geographical distance, which is today, despite the automobile and the telephone, such a very strong inhibitory factor in so many areas of human endeavor. Computer-communication networks can deliver mail as it is typed, retrieve information as it is needed, yield answers instead of data, conduct "meetings" in which ideas interact with one another while people interact with ideas instead of one another. In short, computer-communication networks can greatly improve the intellectual-informational world in which people live and work, but, of course, computer-communication networks cannot do any of the things I have suggested without human effort, and the necessary, essentially human effort is part and parcel of the development of a technology-based information science.

One more ingredient is needed to complete the schema of this long introduction. What the situation requires is not only a dynamic information science concerned with the organization and understanding of information and a technology-based information science dealing with potentials and problems that are emerging from the revolutionary advances of information technology; what the situation requires is also a use- and user-oriented information science. That requirement stems from the fact that so much of the forward thrust of computers and networks has come from inside, that so many great potentials have been built up *by* technologists *for* users instead of by users or by technologists and users working together. As a result, the potentials are appreciated mainly inside the field, inside the industry - by the creators, by the technologists.

I am convinced that an information technologist and a potential user - a programmer and a chemist, to choose more nearly specific instances - perceive almost totally different things when they sit down at a console and log into an interactive computer system or computer-communication network. The programmer feels that, privileged with a briefcase full of keys and passports, he is entering a world of almost limitless informational resources. Unless he has had experience with the particular system, on the other hand, the chemist - or economist or administrator or businessman: there is no special reason for selecting a chemist for this comparison - probably has difficulty even in logging in. At best, he knows how to use two or three already prepared and packaged services, and the thought of exploring the system or network, of browsing around to see what is new and interesting, is kept from entering his mind by an aura of inhibition conditioned in him by past painful experiences. He plays the game strictly according to the few recipes he knows - if he plays it at all.

The use of interactive computer systems and networks is growing, but at present there are far more potential users than actual users, and great advances in human-factors aspects of information-system design will have to be made before the potential users become actual users. That thought leads me into the first of two problem areas that I want to advocate as foci for information science.

ORGANIZATION OF INFORMATIONAL RESOURCES FOR USE

The image that is foremost in my mind as I start to describe this problem (how to organize informational resources for use) is my image of the ARPA Network. The ARPANET is, as most of you know, a computer-communication network that links (and functionally includes) more than 50 operational computers at about 35 locations across the United States and (incipiently) beyond. Each operational ("host") computer is connected to a nearby communications computer ("interface message processor" or "IMP"), and the IMPs at the various locations are interconnected by transmission channels, most of which are 50,000-bit-per-second lines leased from common carriers. Interconnected with the IMPs are several TIPs ("terminal interface processors"), each of which serves as a portal to the network for several or many individual terminals. Most of the hosts are interactive time-sharing systems, and each terminal associated with a host - as each terminal associated with a TIP - has open to it, insofar as administrative and financial arrangements have been made, a wealth of informational resources that, even at this early date in the history of networks, boggles the mind.

As I suggested in the introductory remarks, the nature of the boggling depends upon the nature of the mind.

The computer buff inventories the resources of the network in terms of LISP interpreters, PL-1 compilers, text editors, assembly-language debuggers, subroutine libraries, and so on. He may be attracted, for example, by the hitherto undreamed of opportunity of comparing, in rapid succession, several BASICs and several APLs before selecting one of them for use by his introductory programming class.

The behavioral scientist with data to analyze may think of the computer buff's inventory as just a lot of paraphernalia for coders and focus his own interest on search for and exploration of available versions of TSP, TROLL, ADMINS, JANUS, and/or AQD, which are program packages that provide data-management and data-analysis services.

The meteorologist also delimits his field, insofar as procedures are concerned, to a few well used program packages, but a giant data base looms large in his specialist's view, the U.S. National Weather Data Base, about 10^{11} bits in size, available at present in full only on tape but in (increasing) part on-line.

The applied mathematician may or may not be a programmer. If he is, the ARPANET is his apple. If he is not, he should be interested in MACSYMA, which can be of very significant help to him because it can solve all kinds of indefinite integrals, simplify all sorts of rational polynomials, and so on through quite a repertoire.

But pity the poor generalist who stumbles into the ARPANET cold. He may wish merely to look up the definition of "giga-" in a dictionary, and if he is future-minded he may realize that one ought to be able to do that by typing something like "dictionary definition of giga-" on a keyboard, but in the present network world he is asking for a different kind of mind-boggling experience. There is at least one dictionary available through the ARPANET, but all our generalist's attempt will get him is a "?" or "log in, please" or a "TOO-MANY-ARGUMENTS-SUPPLIED" or a "FNF" (which the computer buff, but not the generalist, would interpret instantly as "FILE NOT FOUND"). He can cut and try at his console all afternoon, but he won't get a civil answer from the network unless he gets help from a person who has the know-how.

There is as yet no standard protocol for finding out how to locate a desired resource in the network or, if you have found it, how to use it. The programmers who created the resources and the system software of the network scarcely notice the lack. I have watched some of my M.I.T. programmer colleagues browse into, and in some instances through, interactive systems at Harvard, BBN, CMU, Stanford, SRI, UCSB, USC, and UCSD as well as other systems than their own at M.I.T. Diverse and unintuitive procedures of use affect them more as challenges to their ingenuity than as dampers upon effective discovery and use of remote resources. I can understand that psychology, and sometimes I participate in it. For example, I was pleased a few weeks ago when I managed to log into a PDP-10 at CMU without having a password. More often, however, I am pained and frustrated by the simple but myriad difficulties that arise making arrangements, finding manuals, interpreting explanations, and being confounded by programs that have been updated while their documentation has not.

The preceding paragraph leapt into the difficulties of system and network coherence - or, more accurately, of incoherence - as though it were the only problem in the sharing of resources. It is of course not the only problem, but for a very large fraction of potential users it is one of the two main remaining problems. The other one, of course, is the creation of the many additional resources that potential users will require. The creation of software resources has barely begun, I think, and it seems fairly obvious that standard guidelines or protocols should be defined as soon as possible because it is very difficult to revise a great diversity of software into a condition of coherence after the software has been brought into operation.

The word "remaining" in "two remaining problems" was intended to suggest that some problems in the sharing of resources have in fact been solved. The ARPANET is now about 5 years old. The first problem was to get diverse and separated computers into elementary signalling interaction with one another through a transmission scheme with potential for growth and economic performance. That problem was solved by the design and development of the IMP network and the guidelines (HOST-IMP Protocol) for connection of hosts to IMPs. The second problem was to move from elementary signalling interaction to program-to-program communication. It was solved by the development of HOST-HOST Protocol and network control programs for the various host computers. The third problem was to give users access to remote resources as if the resources resided in the local host. That is an open-ended problem, but important parts of it were solved by protocols called "TELNETS" for the user's computer and the serving computer, and now it is almost as easy and in some cases almost as satisfactory to use a resource 3000 miles away as to use an equivalent resource at home. (It is helpful in evaluation to have some redundant resources.) The fourth problem, and the last I shall mention in this series, was to extend the network access to users without local host computers. That problem was solved by the TIPs, which are relatively inexpensive and therefore can be located wherever there is a small community of users. Now all one needs to get onto the network is a console with a "modem" and a telephone - plus, of course, administrative and financial arrangements, introductory documents, and determination.

The present capability of the ARPANET is very great, and I believe that history will judge it to be as significant as the first automobile or the first airplane. However, "all that's past is prologue" to the two main remaining problems I have mentioned. Both problems pose challenges for the kind of information science I am advocating. Since I am focusing here upon organization of informational sources for use, I shall limit this list of challenges to that problem:

(i) To bring the resources of a computer-communication network under "bibliographic control". An important aspect of this challenge is interactive, on-line documentation.

(ii) To extend the scope of information retrieval to include the retrieval of procedures (programs, subroutines, functions, macros, etc.). This challenge includes devising ways for the program library or librarian to specify what each prepared procedure will do when called and ways for the user to specify, via some kind of retrieval prescription, what he wants to have done by the procedure he is asking the library to lend him.

(iii) To define an "interaction language" or "interaction protocol", implementable in almost any computer system, that will permit users to log in and out, prescribe access permissions, edit files, find existing procedures and data, execute commands, define complex commands (macrocommands), send "mail" messages, engage in on-line communication with other users and in "teleconferences", etc. The best interaction language from a potential user's point of view, may be based upon an interactive programming language (e.g., LISP, MUDDLE), but it will require a two-dimensional (graphical) representation as well as a linear (alphanumeric) one, and of course the computer-to-user flow will be much more important than it is in a programming language.

(iv) To develop an effective way of determining "user needs" and relating them to technical and economic feasibility. This challenge is difficult: much of what has been said and written about user needs in related fields is pap; when asked what he needed most, the old farmer replied, "A stronger mule".

(v) To develop effective ways of educating users in artful exploitation of informational resources. I think that the ultimate answer to this challenge is to meet the other challenges. A good interactive system - computer or network - should be its own best teacher.

The foregoing constitute my first suggestion of a productive focus for information science. I believe that computer-communication networks will provide challenges, tools, criteria, and reinforcement sufficient to foster a golden age if information scientists provide enough intelligence and imagination.

ORGANIZATION OF KNOWLEDGE

The second focus that I want to advocate for information science is a problem area of which parts are being attacked, with marked success, in research in artificial intelligence. I do not urge that information science try to appropriate what artificial intelligence is handling so well. Rather, it is a large problem area, and some of it seems right for information science.

It used to be presumptious to speak of knowledge outside the mind or brain. Research on relational and semantic nets and on procedural embedding of knowledge has changed the situation, however. Now one can say that organization of knowledge in computer systems is a central problem, or even *the* central problem, of information science and technology.

At present, the study of the organization of knowledge is very young, but it is alive and growing. It presents an opportunity for information science, I believe, because many of its problems present themselves within the context of computer systems of sizes and complexities that are just now becoming affordable and manageable.

An important approach to the problem, I believe, is opened by distinguishing between what I shall call "cognitively structured knowledge" and "functionally structured knowledge". The former is what scholars have in abundance; the latter, doers. A scholar knows his field, its methods, and its history, understands the relations among its concepts, can explain its phenomena, and can prove that he is knowledgeable by talking or writing. A doer knows his field, too, but in a very different way: he may be indifferent to its history; he may not understand the relation among its concepts to the satisfaction of the scholar; he may not be able to explain its phenomena but he can demonstrate some of them; and he can prove that he has knowledge by doing things that could not be done without knowledge. I shall not belabor the distinction, since it is well enough established to be the point of such familiar quotations as "Those who can do, do, and those who can't do, teach". Let me mention at the outset, however, a very insightful discussion by Sussman, contrasting "modular knowledge" and "skill", that stimulated my interest in this problem.

The most fundamental consideration concerning the distinction between cognitive structuring and functional structuring in the organization of knowledge is that the world can present a seemingly infinite variety of particular tasks - of calls for functional application of knowledge - and that it does in fact present very, very many particular tasks. No matter whether we think in terms of the potential diversity or the actually encountered diversity, we see that there are far more different calls for application than there are modules of cognitively structured knowledge to be applied. It seems reasonable as a first principle, therefore, to organize knowledge cognitively for storage until a particular task is presented, and then, when the task is at hand, to retrieve the parts of the cognitive base that are pertinent to the task and organize them into an appropriate function and apply it.

However, a second consideration presents itself immediately: Certain tasks occur repeatedly in precisely the same form, and certain other tasks appear to be variants of one another and thus to require application of the same basic function with *ad hoc* parameter values. This consideration suggests that the knowledge required to respond to tasks of these kinds be organized functionally and stored in functionally organized form so that it will be ready for use as soon as the tasks present themselves. Thus we find ourselves contemplating a somewhat redundant store of knowledge - part of it structured cognitively and part of it structured functionally. Of course there must be an arrangement for converting from cognitive to functional organization. It is difficult not to complicate matters immediately by observing that surely the basic core of this arrangement must be organized functionally - though some of it may be stored in cognitive form and converted when it is needed to convert something else.

The cognitive-functional problem arises from many different sources in the design of systems.

One of the most important sources is the desire to achieve both comprehensibility and efficiency in the same system design. A system must be comprehensible if it is to be constructed, maintained, and used. It must be efficient if its use is to be economic and satisfying. But there is, or surely seems to be, an inherent incompatibility between comprehensibility and efficiency.

To make a system comprehensible, designers tend to make it modular. For complex systems, such as programs, libraries, businesses, and armies, the favorite principle of organization is modular hierarchy. The overall system is divided into subsystems; each subsystem is divided into subsystems; ...; and at each level each sub...system is packaged to facilitate indentification, description, access, and isolation from its neighbors.

To make a system efficient, it is essential to introduce synergy, to ensure that the parts that must function together to produce each desired effect do so with good coordination and without waste of time or energy or material. To operate efficiently in a very complex task environment, a system requires many and diverse synergies, many different intimate coordinations of parts. Synergy requires and is the result of functional, not cognitive, organization.

The essential difficulty, of course, is that, no matter what basis the designer chooses for the hierarchical, modular organization of his system, it is bound to be wrong for the achievement of many of the required synergies. This difficulty manifests itself even in the layout of a supermarket, with its spatially separated departments for meats, produce, canned goods, and so on. Few shoppers buy all meat, all soap, or all paper goods and nothing else - patterns of purchase that could be achieved with fairly high efficiency within the standard supermarket paradigm. Most shoppers get a few items from each of several departments at the cost of pushing a cart all around the store. An efficient supermarket - a synergic one, to use the term I have been employing - would package each individual shopper's order and have it ready for him at the door. But going back to comprehensibility, think of the problems of taking inventory in such a supermarket:

The modularity-synergy problem is ubiquitous. The librarian deals with it in part by putting certain books "on reserve" in a shelving pattern intended to match an expected pattern of use. The business manager and the army commander deal with it in part by forming *ad hoc* task forces in which talents are grouped, not by discipline, but to meet the requirements posed by a particular challenge. The computer programmer, however, is the person who is truly confronted by the problem: He faces very many of its aspects. They present themselves on a fast time scale. Many strategies and factors are available to him as he attempts to solve the problem.

If the programmer attacks the problem of modularity vs. synergy in order to find a good design for his program, he may find the good design, but he will not gain a significant understanding of the problem. To gain a significant understanding requires a broader attack in which the problem is identified as a target for scientific study and is investigated in several or many different system environments. I think it is a good focus for an organized program of research.

In order to illustrate a few aspects of knowledge organization and especially of the problem of cognitive vs. functional organization, I shall describe some of my own encounters. For the past year, I have been using a language called "MUDDLE". It is similar in many ways to the language LISP but provides a more readable syntax, a greater variety of data types, and better graphics and network primitives and is more readily extended by users. MUDDLE is a function-application language. It has about 150 built-in functions. Users find it easy to create new functions of their own by combining built-in functions and user-prepared functions in novel combinations - almost easy enough to justify writing an *ad hoc* function to handle each new task, but not quite. It is worthwhile to prepare some functions for general applicability as well as for use in the project at hand and to save them for future reference. In my year with MUDDLE, I have amassed a collection of over 1000 MUDDLE functions. They reside in a library that is presided over by a dynamic loader which, whenever a function is applied that has not yet been loaded into main memory, notices the deficit, loads the function, and then lets the computation proceed. A MUDDLE program is just a MUDDLE function - typically one that calls subordinate functions that call subordinate functions... - applied to arguments that are either data or other functions.

One might say that, if there is any knowledge at all represented in my library of MUDDLE functions, it is organized more nearly in functional than in cognitive form. There are many degrees, however, on the scale from "extremely cognitive" to "extremely functional", and my MUDDLE functions do not occupy the extremely functional end of the scale. For one thing, I wrote them to be generally useful, and most of them could be made more efficient for a particular application if they were specialized for that application. For another, they are prepared in a form that requires interpretation by the MUDDLE interpreter, and in that form they are easy to debug, update, and modify for related but different purposes, whereas if, instead, they were compiled for direct execution, they would run much more efficiently. Finally, their layout in the memory of the computer is almost entirely haphazard, each function occupying a place in the memory map that the space allocator chanced to have ready when the dynamic loader started to load the function, and haphazard layout is not conducive to efficient performance in a paged, swapping computer system, which happens to be the computer environment in which I work.

In order to improve the efficiency of some of my MUDDLE functions, I could use MUDDLE's declaration functions. If I declare to MUDDLE, for example, that inside function ALPHA the value of variable BETA would always be an integer, the compiler could use that fact as a basis for eliminating some type checking, and the compiled ALPHA would run more efficiently. I do not want to do that, however, because I think it is more important to me to keep ALPHA more generally applicable, to preserve its ability to operate on integers or real numbers, whichever present themselves. What I really need is a utility-based monitoring and decision-making program that amasses statistics of use, knows (or estimates) the trade-off parameters, and declares and compiles when and if declaring and compiling increase the expected utility.

Another way (possible) to improve temporal efficiency would be, in functions that call functions, to replace the function calls with the bodies of the called functions. For many reasons, "melding" functions in that way can be a very complicated undertaking, but in situations of limited complexity melding is done routinely by certain compilers, and I have melded a few MUDDLE functions successfully. However, melding may cost globally more than it saves locally. If functions A and B both call C, D, and E, and if A, B, C, D, and E are all about the same size, and if total space requirement for A and B and their callees, unmelded, is 5 units, then the total space requirement for A and B with C, D, and E melded into each is just a little less than 8 units. In a complex of functions, the commonality may be very high and the space-cost of melding therefore very great. Moreover, in a heavily used swapping system, space-cost translates itself into time-cost with a vengeance, and time spent in swapping (in order to get the expanded functions into processible memory) may more than counterbalance the time saved through elimination of function calls.

Still another approach would be to mastermind the layout of functions and data in the memory map in such a way that functions that call other functions occupy the same pages. (Note that melding achieved essentially that result by eliminating subfunctions.) The trouble is that processing time and storage space are required to determine which functions call (or probably will call) which other functions, resources have to be expended to find a good layout, and the best layout is likely to require redundant representation of functions (which MUDDLE, as presently implemented, is not able to handle).

I think I have described enough of the problem as it presents itself in this relatively simple context to make it evident that, to determine even a roughly optimal position on the cognitive-functional scale, one must develop a model that involves all the important trade-offs and estimate the parameters of the model, many of which will be dependent upon the environment of the system under study. The model can be no better, of course, than the knowledge of knowledge-organization techniques embedded in it, but that field has some content now and is growing, perhaps even burgeoning.

The discussion in terms of a library of MUDDLE functions has presented a few aspects of knowledge organization but has not dealt with anything like the full scope of the problem. Indeed, the full scope of the problem is far beyond my ken and capability. But let me proceed just two more steps and then cite references that go well beyond my stopping point.

One should not have to have, already prepared and waiting in his program library, precisely the program or function required to handle a particular application. It should suffice to have a set of primitive functions (or function generators) and an "automatic programmer" capable of putting the primitives together in a (or the) required pattern. Cordell Green (personal communication) and I have independently been exploring an approach to automatic programming in which the application problem to be solved by the automatically programmed program is to transform a member of a class of simple input patterns into the corresponding member of a class of simple output patterns. The required transformation is specified by giving the automatic programmer a sample input-output pair. For example, if the programming program is given the pair

IN ⟶ (A B C D E F G H I J)
OUT ⟶ (J I H G F E D C B A)

it is being given the task of finding or composing a function that will reverse the order of the elements of (A B C D E F G H I J) or (A B C D E F) or (1 2 3 4 5) or (ALPHA BETA GAMMA) or (JOE BILL HARRY JACK) or any other list of (atomic or numeric) elements. Inasmuch as there is in the library a function, REVERSE, that will do just that, the task of the program in this case is merely to find REVERSE and return it. There is also a function in the library, ROTATE.LEFT, that will turn (A B C D E F G H I J) into (B C D E F G H I J A) and (A B C) into (B C A). When the sample pair

IN ⟶ (A B C D E F G H I J)
OUT ⟶ (I H G F E D C B A J)

is presented, what is desired is the concatenation of ROTATE.LEFT and REVERSE. My automatic programmer limits its search to component functions that, on the basis of a few descriptors, seem appropriate as candidates, and it looks for a combination that, in terms of a limited model of the function-application process, does the right thing. In this case it synthesizes and returns the definition of a function, which - unless the user supplies a name - it calls

"AP.RESULT". In MUDDLE notation, the definition is:

```
<DEFINE AP.RESULT (S) <ROTATE.LEFT <REVERSE .S>>>
```

The "S" is a dummy name for the argument to which the function is to be applied, and the "." in the ".S" indicates that the local value of S is to be supplied. The angle brackets are parentheses, each pair of corresponding brackets enclosing the name of a function and the specification of its argument(s), if any.

Actually, there is also ROTATE.RIGHT in the library, and a trivial change in the ordering would cause the program to yield:

```
<DEFINE AP.RESULT (S) <REVERSE <ROTATE.RIGHT .S>>>
```

There is a bit more to the automatic programmer, but the foregoing account is enough to illustrate the idea of automating some of the process of synthesizing functions. (Functions usually embody functional knowledge, knowledge organized for application to a task.) The final step, which I can illustrate as yet only weakly with the automatic programmer, shows that it is convenient for storage, maintenance, and retrieval purposes to organize the component modules on a cognitive rather than on a task-oriented basis. ROTATE.LEFT, ROTATE.RIGHT, and the more general function, ROTATE, are members of one subset of the library; SUBSTRUCTURE.MODULO.2, SUBSTRUCTURE.MODULO.3, SUBSTRUCTURE.MODULO.4, and the more general function, SUBSTRUCTURE.MODULO.N, are members of another subset; and TRANSPOSE.M.2, TRANSPOSE.M.3, TRANSPOSE.2.N, TRANSPOSE.3.N, and the more general function, TRANSPOSE.M.N, are members of a third. Those groupings make sense in the context of preparing, descriptorizing, updating, and retrieving the modules, but the members of a given subset tend *not* to be used together (more than they tend to be used together) in synthesizing a function to perform a given task.

At present, the automatic programmer itself is mainly just a moderately complex MUDDLE function, but, in future versions, increasing amounts of knowledge about programs and programming will be represented in a cognitively structured information base. (I avoid the term "data base" here because the distinction between cognitively structured and functionally structured is not entirely the same as the distinction between data and procedure.) Several models are available to guide the organization, and there is much interest in exploring new ones. Among the existing models are the document-descriptor model familiar from conventional information-retrieval systems, the predicate-calculus model used in several computer-based question-answering systems, and the models of PLANNER, CONNIVER, and QAS-4, which represent significant recent advances in both the tool and the concept areas of artificial intelligence.

Among the best examples I know of the organization of knowledge on a significant scale within computer systems are MACSYMA, a convincingly capable mathematical assistant and, indeed, in a few areas a mathematician in its own right; DENDRAL, an expert at interpreting mass-spectrometric and related data; MOONROCKS, a program that retrieves, from a data base, in response to queries presented in natural English, facts about rocks brought back from the moon in the APOLLO project; and, of course, PROGRAMMAR, Winograd's well known program that, in response to natural-language commands and queries, carries out fairly complex operations in a small world of toy blocks and explains why it performed them the way it did. To these I would add two quite remarkable new programs, an automatic programmer developed by Martin and several of his colleagues and an automatic debugger developed by Sussman. Martin's program prepares report-generation and related programs for a simulated chain of supermarkets, composing executable and even rather well optimized PL-1 programs replete with job-control cards. Sussman's program debugs programs that operate in Winograd's "Blocks World" and revises the code to correct errors that it observes in its own performance. Study of those programs and other more or less like them, reported mainly in the *Journal of Artificial Intelligence*, will develop better and more fully the concept of knowledge organization that I have attempted to introduce here from just one point of view.

TOWARD A HOLISTIC CONCEPT OF KNOWLEDGE

Jean Gideon Kesting

University of South Africa, Pretoria, South Africa.

In a recently published book, *Surviving the future*,[1] a dialogue is recorded between a Japanese academician, Professor Wakaizumi of the Kyoto Sangyo University, and the renowned British historian, Arnold Toynbee. The dialogue comprises concisely formulated, pregnant questions by Professor Wakaizumi concerning profound issues in contemporary man's existence on this planet, and wise commentary by Dr. Toynbee, articulated in a free-flowing, conversational style. The first of the seven questions posed, touches upon the most searching philosophical inquiry for all humans: What is the purpose of life? Wakaizumi poses this question in a specific contextual framework:

> Science applied to technology has been producing changes in our way of life that are diverse, complex, and revolutionary. Our difficulty in dealing with these changes is being aggravated by their rapidity. A society that has been agricultural, pastoral, and rural for thousand of years is being transformed into one that is industrial and urban. The consequent confusion, pressure, and strain from which we are suffering today are moving us to reconsider the fundamental question of the meaning and aim of life. What should man live for?[2]

Toynbee's able response to the challenge is to present his article of faith in life. 'I would say', he replies, 'that man should live for loving, for understanding, and for creating. I think man should spend all his ability and all his strength on pursuing these three aims, and he should sacrifice himself, if necessary, for the sake of achieving them.'[3] Wakaizumi's profound question has elicited a profound reply. Both question and reply reflect wisdom.

The other questions raised relate to man's growing anguish about the ultimate destiny of the manifestations of modern evils, such as the consequences of an over-populated, underfed earth, of man's spoliation of his environment, of mankind's terrifying tyrannies. What is the place of religion in the scheme of things? What are the goals embracing the education of man? As has been hinted, the essential quality of the answers and their response is wisdom. A general definition of this quality, formulated in the *Shorter Oxford dictionary*,[4] states that it is "the capacity of judging rightly in matters relating to life and conduct; soundness of judgement in the choice of means and ends; sometimes, less strictly, sound sense, especially in practical affairs". If this definition can be accepted as a concise explication of the essence of wisdom in its least doctrinaire sense, then it is clear that *sound judgement* and *common sense* imply the ability to discriminate qualitatively among different manifestations of thought and among varieties of objects relevant to particular applications. The content of thought and the substance of an object can both be related to the phenomenon of information (the latter term again used in its most widely-accepted sense). If wisdom presupposes selection, decision and synthesis, it stands to reason that the sage is aware of isolated and synthesised elements with which he has to reckon. In other words, he has to possess information of some kind concerning such elements. He may not be a learned man, he may not have the innate ability, the inclination or the education to know in any *exact* terms what measurements are involved in his selection or decision-making. Like many truly wise men, he may be educationally unsophisticated. Yet, by means of a mental and psychic process largely unknown to the conscious investigator, he is guided by a sense as true as instinct.

Such a man was Solomon, or Blake or Walt Whitman. A student of Whitman will recognise the following statement as an utterance of truth and not of mere unbridled speculation on the part of the poet: "I feel and know that death is not the ending, as we thought, but rather the real beginning - and that nothing ever is or can be lost ..."[5]

At this point the traditional natural scientist will raise his hands in horror. He has been trained to evaluate every theoretical statement in terms of a precise definition acceptable to the consensus of his contemporary scientific community, ordaining that the statement relates in precise manner to a science, i.e. a body of knowledge. This knowledge, he will insist, must be verified through controlled experimentation and should be based on laws or paradigms pertaining to its peculiar nature. These paradigms, which unify the phenomena of which the science in question is a subject, constitute the framework of that science, such as, for example, entropy as the second law of thermodynamics. The scientist will claim, furthermore, that effective research in any field is dependent on factors such as a firm knowledge of the fundamental entities of which the universe is composed; how these entities interact with each other and with the senses; and what questions may legitimately be asked about such entities and what techniques employed in seeking solutions.[6] More rigid terms for qualification under the notion referred to by Kuhn as 'normal science' abound, casting the logical fabric of Whitman's wisdom aside on the assumption that the embodiments of wisdom are suspect because the source itself is unverifiable.

In so far as this type of verdict represents the current scientific consensus, we are faced with a cultural dilemma. If scientific mandates - and here the concept is used in its Germanic connotation, i.e., referring to a systematic approach to all phenomena, whether of nature itself or emanating as articulated thought from the depths of the human mind or psyche - rule out the reflective and aesthetic in scholarship as a valid field for disciplined academic investigation, the scientist is in fact suggesting that a large body of human knowledge, experience and intuitive manifestations is unfit for inclusion in the proper domain of sophisticated mental activity. A logical conclusion is that the inspirational in the creative fine arts and literature, in philosophy as intuitive speculation, and all that is "soft" in the humanities and social sciences are considered by scientists to be somewhat beyond the pale. What this means is that C.P. Snow's initial conception of the origin and widening of a schism between the "two cultures",[7] or the melancholy reflection by Michael Yudkin[8] that the only way out of the modern predicament is the tacit acceptance of the final conquest of the humanities by the "hard" sciences is to be taken at face value as a sound probabalistic theory. This extreme view, in my opinion, is fallacious and unnecessarily pessimistic. First, I wish to postulate a modification of Snow's revised,[9] more tempered diagnosis of a recognisable social disorder, viz. that a cogent case may be made for the conception of the sciences and the humanities, not as "two cultures" subsisting in barren isolation from each other, but as poles which derive their vital solidarity from the continuum of our total human culture. Second, I have argued that wisdom implies a systematic substructure of a certain type of knowledge, which I shall define more closely in the course of this essay. Third, the assumption that a systematic order of any kind (i.e., viewing science as a study of order in the universe) provides a framework for evaluating the "soft" fields of knowledge within the broader context of scientific investigation. Finally, it must be pointed out that the experiences of man as a total being encompass not only physical and sensory stimuli, but thought and psychic processes as well.

For the full man, all experience is real, and therefore worthy of evaluation and of scientific investigation. There is no absolute criterion prescribing that his thoughts and conclusions be recorded in symbolic language, which is understood by others operating in the same cultural milieu to have a commonly understood meaning, although truly effective communication would naturally make it desirable. Therefore, the exclusion of recorded or unrecorded information from certain fields of knowledge, amounts to an exclusion of reality itself, since reality in its ultimate essence is synonymous with the idea of wholeness.

An extension of the view that the two cultures represent opposite poles in the concept of a unified culture, is the observation that the so-called "hard" sciences (Kuhn's[10] notion of "normal science") are characterised by precision and exactness as prerequisites for arriving at conclusions based on logically directed empirical research, whereas the "soft" sciences, or "soft knowledge", if you wish, is fundamentally reflective by nature and is characterised by the prevalence of elusive variables. It has been suggested that the enthusiasm elicited among some physical scientists, in particular, to dedicate themselves with such incredible devotion to "miniscule" excursions into the undiscovered microscopic world of their chosen speciality may be attributable to the high probability of success in problem-solving situations.[11] The "scientific revolutions" which force them to reject old paradigms for new ones in accordance with their code of loyalty to the fixed principles of laws newly uncovered, are obviously of a much more minute order than the comparable premises of philosophy which are seldom *completely* superseded. It is justifiable to add that the methods of scientific investigation into the physical sciences in accordance with the dictates of evolving paradigms naturally presuppose that objectivity and scientific consensus are infinitely more easily attainable than in the case of the "speculative" sciences.

It seems likely that the disproportionate incidence of subjectivity as a characteristic of analysis and synthesis at the reflective end of the spectrum of knowledge and the corresponding limitations of indisputable success in problem-solving, has been a major factor which has rendered humanists suspect as worthy colleagues of natural scientists.

We should now pause for a while to reflect on the typical characteristic which accompany scientific investigation as far as the "soft" or predominantly speculative sciences are concerned. These characteristics - by implication scientific flaws - have been identified as a tendency toward subjectively conceived conclusions, lack of closely-knit sets of rules, absence of a significant degree of anticipation of ensured success in problem-solving, the prevalence of seemingly unmanageable variables and the synchronous existence of diverse and chronologically unsuperseded paradigms available optionally to the "soft" explorer of the universe. All the alleged flaws, I suggest, emanate from one single, but immensely potent factor: the enigmatic nature of the human mind and psyche. The human mind in the exploration of the physical sciences is present only in the posture of critical observer and experimenter. In the physical sciences there are few factors which appear to be capable of interference, in the form of crudely developed or sophisticated intelligences. Rapoport,[12] expressing scepticism with regard to the claims of mathematicians that they had discovered an analogy in the principles governing information and the formula for entropy in statistical mechanics, finally rejects these assertions on the grounds that a fundamental principle distinguishes a mathematical analogy and an ordinary "metaphorical" one, and elaborates his point in further complex argumentation. The main point he wishes to make is that "arguments based on ordinary analogies are seldom conclusive".[13] The most significant conclusion arrived at by Rapoport[14] centres on the speculation whether a useful analogy can be drawn between the concept of entropy as originally introduced in the context of physical thermodynamics and entropy as a paradigm in chemistry and biochemistry. However, in experiments conducted to test the reactions accompanying the synthesis of sugar from water and carbon dioxide by green plants, it was observed that the ordinary processes of the entropic principle were reversed. The logical conclusion was that "interference" had taken place and that this interference had something to do with an assumed built-in force of self-preserving equilibrium present in any living organism, which, in this case, was identified as one of a low order of assumed intelligence.

Assuming that plant life is capable of exercising a suggestion of some undiscovered manifestation of an "intelligence" principle, there is little difficulty in imagining the potential of intelligently-directed actions among higher forms of living. Ethology, the study of behavior of animals in their natural environment, provides much information of this phenomenon. From a generalist's point of view, observations made by the highly gifted South African poet, nature-lover and natural scientist, Eugène Marais, make stimulating reading. After the end of the second Anglo-Boer War, Marais and a companion sharing his friendship and his interest in animal life spend three years (1902 to 1904) of lonely existence on a deserted farm in the Waterberg district of the Northern Transvaal, studying the habits of a troop of some 300 chacma baboons inhabiting a large area in the folds of the semi-dry hills. In a foreword to Marais' *The soul of the ape*,[15] Robert Ardrey, the well-known American ethologist, asserts that if Konrad Lorenz is to be accorded the status of father of ethology (a science established some fifty years ago) "then Eugène Marais was its prophet".[16] Ardrey further claims that Marais was the first man in the history of science to conduct a prolonged study of one of man's primate relatives in a state of nature, but qualifies his praise with the remark that Marais' status as a scientist at that stage should not be overemphasised.[17] To demonstrate the relevance of *The soul of the ape* in the argument, attention must be drawn to the central hypothesis of this book, namely the evolutionary origins of the unconscious in man. This was a deduction drawn from Marais' intimate study of the large troop of baboons in the Waterberg. But the author deserves special credit for his invention of two simple phrases, *phyletic memory* and *causal memory*. (The word *phyletic* is a derivative of the biological term *phylogeny*, which means the history of a species and its antecedent species.) *Phyletic memory* refers to the notion of instinctive knowledge, the origin of which remains unknown, since the animal's memory itself has been embedded in his genetic make-up and has indelibly recorded evolutionary crises in the far distant past of his ancestry. Marais could identify the instinctive behaviour of the chacma baboon and distinguish this with great clarity from its conscious actions. In the baboon the division between the instinctive and the conscious appeared to be more clearly demarcated than in man. But Marais, a largely self-trained behavioural scientist, who had studied and contemplated on Darwin's notions of evolution, discovered natural analogies between the levels of consciousness in the baboon and those of selective man. Eugène Marais' study (the manuscript, revised and retyped several times during his lifetime, had been unaccountably mislaid until it was discovered in 1968, 32 years after the author's death) is of extreme importance in so far as his isolated ingeniously developed methods of research in a behavioural science yet to be born, resulted in findings which anticipated the great hypotheses of Jung, especially in regard to the obvious parallels between *phyletic* consciousness among the baboons and Jung's idea of the collective unconscious, and archetypal realities hidden in the depths of man's genetic structure.

It may appear that I have strayed from the main argument, and it is perhaps necessary to recapitulate briefly. The narration surrounding a remarkable South African who is slowly beginning to emerge from undeserved obscurity as an ethologist (and not a poet, as he is thought of primarily in the country of his birth, since he wrote in the relatively isolated language of Afrikaans) was introduced to elaborate on Rapoport's assumption of the presence of an implied intelligence in living organisms, by way of illustrating that higher forms of living, known to possess infinitely superior degrees of intelligence, behave in a manner which refuses to be shackled by such a beautifully conceived symmetric formula as entropy. Nevertheless, the intricate and complex behaviour of the baboon as a single organism as well as a member of a troop, revealed a *system* containing elements of regularity, consistency and predictability. These "paradigms" which Marais had discovered among animals living in unison as a troop in undisturbed natural conditions are taken to be of a higher order than those conceived of by the early physicists. The study of the behaviour of living organisms, culminating in the phenomenon of man, implies the structuring of the requisite sciences in increasing complexity, and the concomitant realisation that the rewards of problem-solving are ever so much smaller than those promised by the study of the "hard" sciences.

At this point, it might be helpful to summarise the foregoing by reducing my view of the structure of knowledge and the place of the "hard sciences" in the scheme of things, by suggesting the mould of a receptacle as a metaphor representing the container of all notions of knowledge, information, and discrete facts. The mould is that of an inverted cone, with its broad, open end reaching upwards into infinity. The increasingly sharpening lower end of the cone is designated as the base, or proper domain, of the "exact" end of the knowledge spectrum. Therefore, lines drawn vertically from the sharp tip of the cone at its lower extremity to its ever-widening open end would span the total routes of interaction between the two opposite poles. The very fact that the lower end of the cone constitutes a relatively confined and rigorously demarcated space of potential *Lebensraum*, earmarks it an eminently suitable area of operation for the characteristically "hard" sciences. "Normal science" wants to know from what fixed premises it should proceed; it wants to make quite sure exactly what its "spatial" limitations are at any given time in the evolution of its fundamental sets of rules; it experiences comfort, rather than a sense of claustrophobia, in its spatially compressed state; there is probably nothing it abhors so much as the seemingly chaotic conditions prevailing in a void of immense space abounding in fluxes of imponderables. This lower end of the cone will comprise an orderly arrangement of the atoms and molecules of "exact" particles of knowledge, structured in accordance with compatible mathematical precepts.

This end of the cone would provide the kind of base that a physical science such as chemistry ideally requires. Chemical technology will be superimposed on the basic branches of classical chemistry, take root there and then move upward along the route that ultimately leads to the upper lip of the cone from which height it extends into the immeasurable space of infinity. This single example dramatizes the origins of our dilemma: we know today that the uncontrolled advancement of chemical technology is the prime cause of at least some major aspects of environmental pollution and of an upheaval in the equilibrium of insect and animal life. So we have turned full cycle and find ourselves participating in Toynbee-Wakaizumi speculations of attempted clairvoyance and wisdom, not merely on the configurations of remedial technology, but rather on philosophical reassessments of the ultimate place and purpose of man as a biological being.

Apart from the interactions taking place along the vertical routes linking a world of factual knowledge, which has a common code of symbolic language to a world of reflective dynamism, there are also horizontal interactions which have lately begun to culminate in the emergence of such interdisciplinary natural sciences as biochemistry and physical chemistry. This line of reasoning will no doubt be appreciated. It can only lead to further elaborations on the same theme, which at this point, will neither provide more conclusive proof of the validity of the premise of the indivisibility of knowledge, nor present more perceptive insight into the phenomenon with which we are dealing.

However, it seems useful to cite two examples relating to such fields as management information systems and applied economics, if for no other more meaningful reason than to shed light on the relevance of the conception of interdisciplinary relationships beyond either the natural sciences or the humanities, viz. those affecting branches of the social sciences. In a paper delivered by Professor J.L. Livingstone to the Northern Transvaal Branch of the South African Society of Accountants in June this year,[18] it is apparent that the principle of intrinsic differentiation between the "exact" and "speculative" poles of knowledge as elements in a dynamic interplay of facts and forces is analogous to the general postulate expounded above. First of all, Livingstone illustrates his point by drawing three concentric circles, representing, respectively, the role of three dimensions influencing organisational levels within management information systems. The inner circle represents technical information, which is roughly of a precise and reasonably predictable kind. The two outer circles, referred to respectively as managerial and institutional components of information, are subject to the penetration of imponderable configurations of influences emanating from the unpredictability of precisely-conceived problems and influences. Finding that business management labours under the burden of being largely uncertain of the reality of the effect of reports (which, in his analysis, can range from those based on accurate information to those obscured by error and conflict, error and irrelevance, to supremely inaccurate information) he concludes that it is better to be roughly right than to be exactly wrong, and from there onwards to proceed with the arduous task of identifying the problem and to devote all efforts toward its (still uncertain) solution, i.e., judged from static and dynamic viewpoints. The most significant characteristic of this type of approach to management information systems, however, is the speaker's insistence on the relevance of non-profitmaking issues, such as a due consideration of environmental influences and effects, and of laws of administrative complexity, of complexity as a contributory factor to staff passivity and of increasing pressure on the information system in proportion to the rate of dynamic changes in business and commerce.

A matter of deeper principle is the recent observation that inflation need not necessarily be accompanied by a stagnant economy, a fact generally accepted until today. De Villiers[19] claims, for example, that "an important implication of the cost-push explanation of inflation is that prices can rise in conditions of declining demand and increasing unemployment, giving rise to so-called *stagflation*". That a phenomenon as odd as *stagflation* could manifest itself as a new feature of economics 37 years after the publication of the incisive and provocative writing in 1936 by Lord Keynes, *The general theory of employment, interest and money*[20] seems astounding. Keynes' hypotheses must naturally have benefited from knowledge gained against the background of the fluctuations during and after a major world war and globally the most crippling depression civilisation has ever known. This can only serve to underscore the essential principle of the need to revise old paradigms and to replace them with revolutionary new laws. These laws apply with increasing intensity to those fields in which the factual is interwoven with an intricate fabric of imponderables in knowledge, and even more so in demarcated disciplines in the pure humanities, characterised by subjective and qualitative issues which preclude conclusive results at any time, and yet are the very stuff of wisdom. If we have difficulty in controlling the elements and the unity of knowledge to control the ills of the material world, are we not at a greater loss when we deal with the demands and promises of equilibrium in a meta-earthly dimension of consciousness?

In *Man in search of himself*,[21] Jean Charon, the French natural scientist, postulates a general theory of man as a fundamentally three-dimensional being. Like his evolutionary predecessor, the highest of the primates, his nature is claimed to be divided by a threshold between his conscious and his unconscious self. There is, of course, nothing new in this idea. However, Charon introduces the novel premise that man's faculty for science or rationalism operates purely in the domain of the conscious, well above the neutral equator of his total being. His artistic sense, on the other hand, dips below the equator in order to derive inspiration from the springs of the unconscious, but returns to the realm of the conscious, to give expression to man's aesthetic revelations in an idiom or symbolic form generally understood to have the same meaning among his peers. Then a third dimension is seen to complete man's three levels of knowledge, namely his religious awareness. This awareness exists in a realm almost totally submerged in his unconscious: a minute apex protrudes beyond the equator which comprises his total ability of translating his religious experiences into a kind of interpersonal, symbolic language,[22] so as to be shared among the esoteric and the exoteric levels of human consciousness in desperate need of wisdom drawn from the deepest wells of the spirit.

It does not matter whether or not a person declares himself a supporter of Charon's principle. In as much as Rosenberg[23] implicitly expresses reservations concerning Bronowski's "main contribution" postulating a broad division of the history of science (see *The common sense of science*)[24] into "the idea of order" (cf. Linnaeus, taxonomy in botany and zoology), "the idea of cause" (e.g., Newton's theory of gravitation) and "the idea of chance" (the assumption that the law of probability shares honours with the notion of causal principles),[25] it is reasonable to expect that such a broad classification may be considered to be too simplistic for the mind initiated in the mental and working climate of the natural sciences.

Be it as it may, there is something significant in these bold attempts at reducing complex sets of circumstances, events, objects and natural or human manifestations of any kind into their most essential qualities. Even if these attempts result in erroneous conclusions, they often pave the way for greater clarity in man's natural desire to see the wood for the trees. The type of mind that has this inclination, reveals the urge to synthesis. Alexis Carrel, in his *Man, the unknown*, uses this theme as a central key to a deeper understanding of man and nature and claims.

> ... that there is just as much need for minds that can synthesise as for minds that can analyse. If only people would recognise the superiority of this bent of mind and take steps to encourage its development, specialisation would cease to be so dangerous.[26]

Carrel, of course, had been conducting a private crusade against the trend of specialisation in medicine, where specialists tend to treat the parts of the human body as fragments operating almost in isolation. He may have overstated his case, and it may be claimed that he put too low a premium on the value of the dissective and the analytical as indispensable processes in the scientific and systematic advancement of a world bedevilled with ignorance and crude superstition. The most meaningful omission, in my opinion, is the neglect on his part to emphasise the complementary relationships, the fundamental interdependence, of his explorations, the methodologies and illuminations derived from the two poles of knowledge. Carrel's main point, however, is one that can no longer be dismissed without further investigation and discussion.

Let us take a look at a familiar case in point. As every student of British history knows, the Wars of the Roses represent a series of events (of observed, interpreted and recorded "facts"), so intricate that the casual reader is likely to end up more confused than before he had read the first broad outline of this vital phase in the evolution of British history. It is well-known that Sir Winston Churchill's account of these events in the first of his four-volumed *A history of the English-speaking peoples*[27] was hailed by historians as the clearest account of the convolutions in the struggle for supremacy between the houses of Lancaster and York yet written at the time of its publication, Churchill certainly had the sweep of a mind that was able to discern patterns larger than the traces of structure that appeared to emerge under microscopic observation of isolated facts, events and objects. He appeared to be a generalist, as the achievements of his many-sided career would suggest. Nevertheless, there were times, especially when he was at the height of his political power, when it was demonstrated that, in order to be accorded the status of a synthesist, he had succeeded in gaining intimate knowledge of the relevant details and had paid due attention to the meaningful foregoing analysis of a situation.

Such a man also was Jan Christiaan Smuts, the South African soldier, statesman, natural scientist and philosopher. While studying law at Cambridge in the years immediately prior to the outbreak of the Anglo-Boer War at the end of the nineteenth century, he began to develop a concept of the nature of an evolving universe which remained an abiding interest throughout his life. He derived the word "holism" from the Greek *holos* and in 1926 published *Holism and evolution*,[28] having revised an original set of manuscripts at a time when he was temporarily relieved of his political responsibilities. A practical and rational mind, he saw holism as an attempt at synthesis, not a system of philosophy. However, it is difficult to draw the line between the concept as a philosophy - a layman's philosophy - and as a pragmatic blueprint for his interpretation of the major phenomena of the world and the events in his personal and public life. It seems more acceptable to view the concept as a guiding theory and an article of faith bolstered by the cumulation of wisdom in the course of his rich and diverse existence. The idea of holism was linked with a dynamically modified version of Darwinian evolution. Darwin's supposition of the origin of species by natural selection implied an acceptance that mutations had occurred in an orderly way, but guided principally by instinct and animistic affinity. Smuts, however, not only thought, along with a number of geneticists, that such a supposition was in conflict with general scientific laws of probability, but became convinced of the principle of a divine ordering of the universe and saw teleology as an evolutionary factor. Smuts' holism as a principle of life was expressed in such terms as these:

> 'There is a process with a persistent trend (in nature) which cannot possibly be the result of mere accident. In fact, the idea of Chance arises from a too limited and abstract view of the facts'[29]

Ancient philosophers and modern scientists alike either inspired Smuts to arrive at his remarkable conclusions or supported him directly and independently. Plato developed his notion of *Kosmos* (order out of chaos) and concluded that, where the universe provided evidence of creative evolution, it was divinely guided. His famous remark, "God geometrises", must be seen as a blend of scientific observation and philosophic speculation in a bygone age, when the study of philosophy was inseparable from that of science. Heracleitus (c. 540 - 475 B.C.) discerned a general law operating in the universe which he called its *Logos*. (It is interesting to note that the phrase "In the beginning was the *Word*", appearing in the fourth Gospel, is considered by Greek scholars to be an unsatisfactory translation for *Logos*, a concept which implies that there is a divine plan for the universe.) Support for belief in a divinely ordered universe came from Einstein himself, in words astonishingly explicit in their import:

> Quantum mechanics (as a theory) is very impressive. The theory yields much, but it brings us no nearer to the secret of the Creator. I am convinced that He does not run the world by chance or by throw of dice.[30]

Similar views are expressed by such notable scientists as Sir Basil Schonland in *The atomists*,[31] by Jeans, Robert Broome and the famous German mathematical physicist, Max Planck, who made a plea for the inclusion of wholeness into the field of physics, as well as into the field of biology, in order to enable man to understand and formulate the laws of nature.[32]

It is not possible to conduct a full pro and con debate on the subject of the validity of the concept of wholeness as one of the most fundamental properties of the physical universe, its living organisms, and the mind of man. It has been chosen as a point of departure in this essay, and I am committed to pursue this theme and to show its relevance to the problems of information theory in general. A final, concise, definition of holism seems necessary before the next step can be taken: the *SOD* defines it simply as "The tendency in nature to produce wholes from the ordered grouping of units."[33]

As may be expected, the absence of an attempt to define the concepts of importance to a discussion of the nature, problems and potential of information science thus far, will undoubtedly have raised questions in the mind of the critical listener. The concepts referred to are "information", "knowledge", "science", "communication", "learning", "wisdom", "facts", "data" and the like. I do not propose to deal with the problem of definition and differentiation beyond a reasonable level of clarity. Briefly, the central concept of information I have taken to be related to each of the terms referred to. The heart of the matter, then, is that the preceding argumentation is designed to show that a discipline of "information science" is committed to organise itself in such a way that it can deal effectively with the storage, retrieval, and dissemination of knowledge in all its facets: first, by its very explicit designation and, second, by the profound reality of the *existence* and the *need for the dissemination* of information in the complete spectrum of knowledge, i.e., a corpus extending from the most discrete facts abounding at its exact extreme to its fine reflective thrust into infinity.

This, however, does not go far enough to emphasise my concern with current trends. One centrally important implicit accusation is that for historical, strategic and other pragmatic considerations, information science, under its many assumed names in many different countries and in all its recognisable archetypal guises, has been claimed to be the exclusive province of the exact base of our metaphoric cone. This, I suggest accounts (at least partially) for the preoccupation among information scientists in hitherto - perhaps unconsciously - attempting to impose the criterion of compatibility with mathematical modelling upon all serious efforts at analysing and synthesising problems and experimentation relating to the sharing of knowledge as enlightenment and as a gift of grace, as if this were to be the established First Law of Information Science, and the exclusive domain of the sciences that lend themselves *par excellence* to preciseness in research.

This narrow view is totally incompatible with reality. In reality, information is generated from all fields of knowledge. Those fields cannot be hermetically sealed in packages - for the express purpose of devoting our attention methodologically to selected clusters, and ignoring the remainder of the world of reality as if it did not exist. Knowledge, as we know, is essentially indivisible. Its nature is organic; it vibrates with mental, emotional, aesthetic and religious energy. Knowledge issues from an eternally evolving human intellect and psyche. Mathematic modelling and excessive concern with formalistic logical regimentation - devoid of the interpenetration of a logic inherent in a dynamic, orderly universe - have driven us to a sterile plateau, where we can move about but not elevate ourselves to significantly higher octaves. On this plateau we imprison ourselves in Icelandic closets, indulging in psychological-warfare chess-playing games, blind to the bounty and fullness of organic life outside which is a property of the awareness of life as a whole. Information does not exist in a vacuum. Generated by humans, it is designed to fulfil the needs of full humanity, individually and collectively.

It matters little if this conclusion confronts us with the greatest predicament imaginable to information science, as long as the premise is prescribed by the dictates of reality. It is quite conceivable that the main factor which has thus far inhibited the systematic moulding of a worthy discipline out of the phenomena of information as a science is attributable to the staggeringly meagre foundations we have prepared with such sincere intentions and devotion in the last two decades. It matters little if the phenomenal widening of the scope implicit in a concept of wholeness appears to render control unmanageable. It is the force of reality - *not* that of a conscious or unconscious conspiracy of a select brotherhood of man - that will decide ultimately what nature a science in the making will assume.

Our reality has two fundamental determinants: (a) the principle of the indivisibility of knowledge; and (b) the principle of the full man. It goes without saying that we are greatly indebted to those pioneers from a multitude of disciplines (such as mathematics, physics, psychology, computer science, linguistics, logic, cybernetics, management information, etc.) for their immensely valuable scientific activity in the past 20 years and more in achieving progress in understanding some of the phenomena of information and to make it work where the need has seemed most apparent. As Louis Vagianos has stated, scientists operating at the exact extremity of the spectrum of knowledge have built their *raison d'être* "on a mighty foundation, a foundation for all ages, forged in the firmament of time, continually evolving, and invulnerable; a foundation distinguished by one common denominator: dedication to the understanding of the physical universe through systematic investigation and careful measurement, through the harnessing of curiosity, specialised training, rigid discipline, exceptional instruments, and perhaps through intuition and luck!"[34] Having acknowledged this, I wish to reiterate the need for a fusion of the two poles of knowledge to re-emphasise the need for a recognition of their innate complementality.

In conclusion, I wish to make a plea:

(1) that the reality of the indivisibility of knowledge and the dynamic continuum of our total human culture be adopted as a first principle in approaching the problems of communication and information science;

(2) that a consensus be evolved in which emphasis is placed on laws of correspondence rather than on laws of division and in terms of which the reality of the co-existence of the exact and the reflective poles of knowledge as a prerequisite for further purposeful action is acknowledged;

(3) that a code of ethics be formulated among information scientists as professional people, to serve as an essential condition for assuming their responsibility toward mankind, based upon the principle that the quality of life is the most important goal to which man can aspire.

References

1. Toynbee, A., *Surviving the Future*, Oxford University Press, London, 1971.

2. *Ibid.*, 1.

3. *Ibid.*

4. *The Shorter Oxford English Dictionary on Historical Principles*, 3rd ed., Onions, C.T., Ed., Clarendon Press, Oxford, 1968, 2436.

5. Quoted by Henry Miller in his *The Books In My Life*, Peter Owen, London, 1961, 231.

6. Kuhn, T.S., *The Structure of Scientific Revolutions*, 2nd ed., University of Chicago Press, Chicago, 1970.

7. Snow, C.P., *The Two Cultures and the Scientific Revolution*, Cambridge University Press, London, 1959.

8. Yudkin, M., Sir Charles Snow's Rede Lecture, in *Two Cultures? The Significance of C.P. Snow*, Leavis, F.R., Windus, London, 1962, 44.

9. Snow, C.P., *The Two Cultures; and a Second Look*, Cambridge University Press, Cambridge, 1965.

10. Kuhn, T.S., *op. cit.*, 35.

11. *Ibid.*

12. Rapoport, A., What is information? Reprinted in: *Introduction to information science*, Saracevic, T., Ed., R.R. Bowker, New York, 1970, 11. (Appeared originally in *ETC: a review of general semantics* 10, 1953.)

13. *Ibid.*

14. Marais, E., *The Soul of the Ape*, Anthony Blond, London, 1969.

15. *Ibid.*, 38.

16. *Ibid.*, 1.

17. Ryke, P.A.J., Eugène Marais, natuurkenner en -navorser, *Tydskrif vir geesteswetenskappe*, 12, 1972, 219.

18. Management information systems; a paper delivered in Pretoria on June 23, 1972.

19. De Villiers, H., Investment counselling in an inflationary climate, in Natal Society of Chartered Accounts, 4th Autumn School, Hilton College, Natal, April 7-10, 1973.

20. Keynes, J.M., *The General Theory of Employment, Interest and Money*, Macmillan, London, 1936.

21. Charon, J.E., Anderson, J.E., trans., *Man in Search of Himself*, Allen & Unwin, London, 1967, 131.

22. *Ibid.*

23. Rosenberg, V., The teaching of information science to non-scientists, *Journal of Education for Librarianship*, 13, 97, 1972.

24. Bronowski, J., *The Common Sense of Science*, Vintage Press, New York, n.d.

25. Rosenberg, V., *op. cit.*

26. Carrel, A., *Man, the Unknown*, Penguin Books, West Drayton, 1948, 54.

27. Churchill, W.S., *A History of the English-speaking Peoples*, I, Cassell, London, 1956, 334.

28. Smuts, J.C., *Holism and Evolution*, Macmillan, London, 1926.

29. *Ibid.*., 182.

30. Haarhoff, T.J., *Smuts the Humanist; a Personal Reminiscence*, Basil Blackwell, Oxford, 1970, 82.

31. Schonland, B., *The Atomists*, Oxford University Press, Oxford, 1969.

32. Planck, M.K.E.L., Prof. Planck and the principle of causality in physics, A. Ferguson, *Nature*, 130, 45, 1932.

33. *The Shorter Oxford Dictionary ... op. cit.*, 911.

34. Vagianos, L., Information Science: a house built on sand, *Library Journal*, 97, 153, 1972.

A NOTE ON PARADIGMS IN INFORMATION SCIENCE

Xavier O. Monasterio

University of Dayton, Ohio, U.S.A.

It is my understanding that one of the basic purposes of this Institute, though certainly not the only one, is to discuss the foundations, the prospective shape, and the methods of a science of information capable of providing a unifying theoretical framework to the increasing number of disciplines in which information appears to play a central role. This paper is conceived as an attempt to contribute to that specific goal.

I will endeavor to show that there are two fundamentally different views on information. One of them, whose origin is to be found in our everyday experience, sees information in terms of meaningful reference and purposive action. The other, which has emerged more recently within the context of strictly technical research, sees information in physical terms as a mathematically formalizable function. Obviously, the shape that the science of information should take depends on the extent to which informational phenomena can be subsumed under the latter view, that is, on the extent to which the science of information can dispense with the common-sensical categories of meaning and purpose. The position I will take is that the science of information cannot dispense completely with those categories without losing its own identity as a discipline and without disregarding severely the complexity of its own object of study. As a consequence, I will advocate for a science of information that, instead of attempting to reduce the semantical and the teleological character of some information systems to the merely physical order, it should explicitly recognize their at least provisory non-reducibility and try to cope with it accordingly.

I am not unaware either of the highly controversial character of my position or of the extreme complexity of the issues at stake. Unavoidably, I have had to over-simplify them in order to make some sense within a short period of time, which makes my views even more controversial. Perhaps this is not a defect, after all. Our last year's NATO Institute showed how far we are from having reached a clear idea of what information is, and a sufficiently broad basis of agreement as to the shape that the science of information should take. A controversial position has at least the advantage of being a position from which a better position can be reached through discussion. Even if this paper failed in all other regards, I am confident that it will contribute to clarifying some of the reasons for our basic disagreements.

THE NON-TECHNICAL VIEW OF INFORMATION

For reasons whose importance will appear later in this paper I will start with a very summary analysis of the fundamental characteristics of the view of information that we work with in our everyday life.

"Information" is a word that has been in Western languages for a long time. Long before becoming a technical word - that is, a word used with a particular meaning in the context of an explanatory theory in certain disciplines - it was and still is, a word of our every-day language. Now, there is no doubt that we use that word in a rather loose way, since we speak sometimes as if information were an object, sometimes as if it were an event, sometimes as if it were an operation, sometimes as if it were a process, and so on. Hoeever, one characteristic of the common usage of the word "information" in every-day language is that it is employed primarily, not to say exclusively, in reference to people. For instance, we treat information as a thing in the sentence "Here you have this piece of information", accompanied by the gesture of presenting a newspaper or a document, but we would never dream of offering a piece of information to a door or to a typewriter. Speaking of information as an event, someone would say "I informed my house-keeper that I had not found my mail". The informed one may be my house-keeper or the President of France, but is a human. Speaking of information as an operation, one would say "I obtained such information reading this book", but no one would expect the book itself to have obtained any information by being printed. Speaking of information as a process, one would say "on this point I am better informed now than last year", but there is always an I or a he or a you or a she who has become better informed - never an it.

Why this constancy? Unless I am very badly mistaken, the reason is that information, as understood in every-day life, is a concept closely tied to the experience of being or becoming aware of something, that is, the experience of being conscious. Indeed, in the framework of every-day language, only those beings supposed to possess the quality of being conscious can be informed. Humans are the primary examples of such beings, and thus the word "information" is primarily used in reference to them. Sometimes, though not often, we speak of information in reference to animals, for animals are conceived to be gifted with awareness, but we simply do not speak of things as subjects of information except in obviously metaphorical senses.

Another trait manifest in ordinary usage of the vocabulary of information is that information is conceived of as a semantical reality, that is, a reality of the order of signs, not of the order of mere objects. This is not to say that information has nothing to do with objects, for it certainly does, but that it has to do with objects inasmuch as they function as sign or are signified. The semantical order is precisely constituted by objects that function in that particular capacity. Another way of saying this is that the semantical order is the order of reference constituted by that which refers, that to which there is reference, and the subject who is referred from one to the other. In order not to make a well-known tale too long, I shall content myself with only a few superficial remarks about a subject on which much has been written.

Imagine a person who is carrying a message but who does not understand the message he is carrying. Would we speak of information in this case, in every-day language? Yes and no. Yes, because the message means, that is, it refers to something besides itself; if properly decoded, it will yield information about that to which it refers. However, there is no information in the sense that the person who carries the message is in a similar position to the book which contains information, since the person is not informed of anything by carrying the message. In other words, there is no information because there is as yet no awareness of the meaning of the message, but there is information because the message refers, though it is not actually functioning in that capacity for the person involved. We call it a message because we know that it can function as a referring reality. Should we discover that it refers to nothing, we would not call it a message anymore, for it would have lost all its semantical capacity. In short, the way in which we talk about information shows that we conceive of it as a reality which is neither independent of nor unrelated to physical objects but which is not fully reducible to them.

There is a third and last characteristic of the non-technical view of information which I want to underline: its connection with action and, therefore, its practical usefulness. Though information is often spoken about divorced from action as if its function were exhausted by dispelling the ignorance or the merely theoretical uncertainty of the subject who becomes informed, it is really appreciated mostly because of its practical usefulness. Or, what amounts to the same, information is not conceived merely as something to be enjoyed and possessed but rather as something to be used. For the fact is that information results in adapted action. When we are informed about something we are able to act in an adapted way, that is, we are able to move in a certain field, to act upon certain objects, to react to a certain environment in a way that is conducive to obtaining the effects we desire. On the other hand, when we lack the required information we hesitate to act, we act in a disfunctional or non-adapted way and we obtain the desired effects only by chance, if at all. The connection between information and adapted action is conceived to be so close that we call the information that results in non-adapted action misinformation. And, in the same vein, we detect the information of people in a certain field by observing them in action. The method appears to work so well that it can even be refined into a sort of scientific tool. For instance, that is what Piaget did in his study of the development of intelligence in infants.[1] Since infants lead, by definition, a pre-verbal existence, asking them questions was unthinkable. But inferring the quality and degree of their information about their environment from the increasingly adaptive way in which they acted upon it and reacted to it proved extremely valuable.

Without entering into the intricacies of purposive behavior, it seems rather obvious that the every-day or non-technical notion of information as being useful for action involves the conviction that behavior based on information is purposive, that is, behavior organized in view of a certain effect to be obtained in preference to others. Indeed, the usefulness of information vanishes as soon as all the effects produced by acting are considered as indifferent. Only in a context in which one particular effect is intended among various possible effects, and therefore a particular course of action is selected among various possible courses of action, can information be viewed as useful.

To recapitulate, then, the three basic traits of the non-technical view of information which have been considered are: (a) that information belongs in the semantical order, that is, in the order of meaning, which is not identical with the physical order; (b) that information refers to consciousness, in the sense that only beings gifted with consciousness can be informed subjects; finally (c) that information is useful in that it makes possible the success of purposive action.

THE TECHNICAL CONCEPT OF INFORMATION

In striking contrast with the preceding, a technical view of information has progressively taken shape in the last forty years within the context of scientific research. A very quick review of the break-throughs which proved to be influential on the technical view of information will be useful.

To start with communication technology, the first name, chronologically, is Hartley's. In 1928 he introduced into the field his famous formula for measuring information.[2] That was the initial core around which progressively emerged a whole theoretical set which has been named sometimes information theory and sometimes communication theory. Of that theoretical set the most influential formulation seems to have been the one published by Shannon and Weaver in 1949.[3] Writing in a descriptive but certainly not uninformed vein, Michael J. Apter made the following characterization of information or communication theory:

> Information theory is a branch of probability theory. It can be used in situations involving sequences of events in which a probability value can be attached consistently to the occurrence of each event. When this can be done, the amount of information in a given event at a given moment can be estimated, as can the combined amount of information in a series of events. Thus if we take it that the probability of each face of a die is one-sixth on each throw, we can measure the amount of information contained in a series of throws. The more random, and therefore unpredictable, a sequence of events or symbols constituting a message is, the more information in the message. So information has an inverse relation to probability or expectancy on the part of the receiver of the message: the more probable the message, the less information.
>
> Just like in control-theory, information theory can be used to refer to both living and non-living systems: the set of probabilistic events could be the letters of the English alphabet, the frequency of English words when used in a given context, the alternative responses available to a subject in a psychology experiment, the 'coding' molecules of DNA, the different morse code signals, the notes of the piano, a set of numbers to be operated on by a computer, etc., etc. It should be made clear, however, that information theory says nothing about the meaning of a message. So in information theory terms, where for example the probabilistic events are taken to be the letters of the English alphabet, the phrase 'Visitors are kindly requested not to feed animals' would contain exactly the same amount of information as 'kindly animals are requested not to feed the visitors'.[4]

To turn our eyes now to the field of self-controlled machines, the significant break-through seems to have come about during World War Two. Working in the area of self-directed weaponry, Norbert Wiener realized the central importance of the concept of feedback and of feedback mechanisms. Moving along that line, he reached the conclusion - expressed in a paper published by Wiener, Rosenblueth and Bigelow in 1943 - that the purposeful behavior of humans could be replicated by computing devices using feedback mechanisms and, therefore, that the idea of a goal-seeking consciousness could be dispensed with completely.[5] The road from there to a unified theory of self-steering systems was paved. Indeed, in 1948 Wiener himself published a book characteristically entitled *Cybernetics: or Control and Communication in the Animal and the Machine* in which he developed the theory of self-controlling systems.[6] Needless to say, as suggested by the very title of the book, the old distinctions between conscious and non-conscious, meaningful and un-meaningful, intentional and non-intentional were viewed as irrelevant within this context of cybernetics.

As is well known, computer technology has profited immensely from the afore-mentioned break-throughs represented by an information measurement, a probabilistic approach to information, a generalization of the feedback concept, and a general theory of self-steering systems. Moreover, by successfully implementing the theories, computer technology has undoubtedly contributed to consecrating them. Perhaps the most striking illustration of this point is to be found in the growing, and nowadays fairly common, cultural conviction that humans are nothing but highly sophisticated computers, a conviction apparently well-supported by the actual and projected achievements of computers in the replication of the human as information-processor and adaptive agent. In short, whether or not the systematical disregard of meaning and consciousness of meaning in communication theory, and of purpose and consciousness of purpose in self-controlling systems theory were meant originally as purely methodological, that is, as merely convenient approaches to problems related to information, the fact is that they have become a view of reality - a view of reality considered to be <u>the</u> scientific one. It is difficult indeed to imagine that Turing's famous proposal for an empirical test of the fundamental identity between humans and computers could have been conceived in a different climate of scientific theory and technological achievements.[7]

A FUNDAMENTAL ALTERNATIVE

If we now ask whether the foundational concepts for the science of information should be derived exclusively from what I have called the technical view of information, or whether the science of information should also include among its foundational concepts the commonsensical categories of consciousness of meaning and of purpose, the answer is, at first sight, obvious. The impressive theoretical and technological progress realized thanks to the reduction of information to a physical function specifiable in purely mathematical terms speak for themselves. On the other hand, such hazy categories as consciousness, meaning, and purpose are, as such, scientifically intractable. As for myself, I will argue - however startling or naive my point may sound - that there is no hope for a coherent science of information if we decide to treat information as a merely physical function. Indeed a science of information of this type would literally find itself without any object of study, the reason being that the only means we have to determine whether a given system belongs or not in the field of study of the science of information is by reference to our everyday experiences of meaning and of purpose.

What, for instance, makes the electrical current of a telephone line an object of study belonging in the field of information, while so many other electrical currents do not? The answer, I think, is that that particular electrical current conveys information, while many other currents do not. Were it not for that property, the information scientist would not concern himself with the electrical current of the telephone line any more than he concerns himself with the current of many other circuits. Now, how is it determined that the telephone line in question conveys information? Because the final output of the line can be decoded in terms of meaning, that is, because the final output of the system can be converted into awareness of a certain meaning. To make the same point in another way, imagine, let us say, a computer whose final output would be a string of symbols that could in no way be decoded, that is, that could in no way be converted into awareness of meaning. The result would be that the symbols would not be symbols, and that the system, as long as it is not meant to produce meaningful symbols, does not fall within a field of study of information science any more than a dynamo or a car.

Of course, there is the other aspect of information which we discovered in the non-technical view of information: that it results in adapted action. Now, if we conceive of information from this point of view, we seem to be able to dispense completely with the idea of meaning and consciousness of meaning, for information becomes simply the independent variable upon which adapted behavior depends. We can, therefore, identify the object of study of information science without any reference to meaning and consciousness of meaning, thereby cleansing from it all anthropomorphical adherences.

One problem with this approach is that it would eliminate from the field of study of information all those systems which convey, process, or store information without themselves applying it to action, for, as we just saw, those systems cannot be identified as belonging in the field of information without some reference to consciousness of meaning. On the other hand, I personally do not see that this approach would be any less anthropomorphical. For how are we going to determine whether or not a system belongs in the field of information? Because it is self-governing? Well, I think that the planetary system is, in a sense, a self-governing system, but no one, as far as I know, has so far suggested that it belongs in the field of study of information science. When we speak of self-governing systems as being part of the field of information science, we mean, I think, that their behavior can be read as goal-directed because, among alternative courses of action, they pick with significant consistency the ones that will best serve a determined purpose. Now, the similarity between this description and the experience we have when, according to our everyday experience, we act on the basis of information, is too close and striking not to suggest that we determine whether or not a system belongs in the field of study of information by implicit reference to the way in which we act in view of a purpose. In other words, those systems are considered to be information systems which, like informed humans, adapt their actions to a goal. This, of course, does not mean that all information systems work on the basis of goal consciousness. It simply means that the field of the science of information cannot be delimited, and perhaps not even conceived, without reference to the central elements of the non-technical view of information. Thus, it seems to me, a science of information attempting to work exclusively with a physical view of information would find itself, if it were consistent, with no object of study.

This may be taken as a proposal that the science of information move backwards to metaphysics. Why should we encumber the theoretical set of the science of information with such notions as consciousness of meaning and of purpose when we can dispense with them so successfully? After all, the fact is that it makes no practical difference whether or not a system of information is conscious, whether or not a message is meaningful, or whether or not behavior is purposive. Such being the case, information can and must be treated not as a semantical or teleological function but as a physical function.

The idea is tempting, and its advantages are many. Among them are its elegance, and the fact that it introduces perfect unity and exactness into the field. However, it postulates the radical homogeneity of all information systems, for only if they are radically homogeneous can they be subsumed under one and the same set of explanatory concepts. What is more, the postulate is beyond experimental proof. Indeed, as soon as one assumes that all information systems are homogeneous, one is bound to treat all differences in performance between systems of information as non-significant, that is, as mere differences in degree of sophistication whose explanation does not demand any fundamental changes in the theory. This amounts to saying that no experimental evidence can be taken as a proof that the postulate is false.

To state the point more concretely, suppose that I am as convinced that the performance of human information systems is in certain cases to be explained in terms of consciousness of meaning and of purpose as you are convinced that it is to be explained in purely physical terms. If, trying to prove our respective points, we compare the performance of a human to that of a computer in a particular field, it is clear that I will read the differences in performance between the human and the computer as a proof that they work under different principles and that their performances must be explained differently, while you will interpret the differences as a matter of degree of sophistication of two homogeneous systems. Our discussion could go on fruitlessly forever, for we are interpreting the same data according to different fundamental views.

Those of you who are familiar with Kuhn's idea of paradigms in science may already have guessed that what is at stake here is a matter of paradigms. If I cannot prove my point to you and you cannot prove your point to me, it is because we are working with different paradigms. For, as Kuhn pointed out, paradigms cannot be proven false by any experiment, for they determine implicitly what the experiments can prove or not prove.[8]

At this point of the game and whatever the future may bring, it is generally admitted that there are important differences in performance between humans and computers. Whether we consider the case at the level of mathematical computation, of chess playing, of pattern-recognition, of language-translation, of information retrieval, and so on, the performances of the human and of the computer are not the same by any standards. For some tasks the computer performs more efficiently than the human, while for other tasks the reverse is the case. Now, if the differences are recognized by all but explained differently according to the demands of different paradigms, this means that what makes a significant difference is the kind of paradigm with which scientists work. What is at stake, then, is the kind of paradigm that should be adopted for the science of information. This seems to be a matter of choice. But such a decision cannot be purely arbitrary, for decisions have consequences. My point is that the practical consequences of adopting as a paradigm the merely physical view of information are such that they should constrain us to adopt a different paradigm. A couple of instances of those practical consequences will illustrate my point.

If we take the view that there is no significant difference between human and non-human systems of information, a view which follows from the idea that all systems of information are to be explained in the same way, then we should conclude that human and non-human systems of information should be treated in the same way. This, however, makes perfectly unintelligible the issue of the uses of non-human information systems to meet the needs of the human - an issue which, by the way, we are supposed to consider in this Institute. Why not the other way around: how can the human systems of information meet the needs of the non-human systems? Fortunately for us, no scientist concerned with information and information systems is ready to go to that extreme. I suppose that the reason is that it makes a great practical difference, after all, whether a system of information is human or not. But, then, why go on clinging to a paradigm that constrains us to consider as non-significant the differences between human and non-human systems of information?

Perhaps the following illustration of my point will strike home more directly, for it sounds more plausible. In a paper presented in our last year's NATO Institute, Hugh Smith advocated a trend of practical research that would start capitalizing on the differences of performance between humans and computers.[9] According to him, the optimistic hopes of the late 50's and early 60's for a growing co-operation or 'symbiosis' between man and machine have not been fulfilled. Instead of that, we have an increasing substitution of the machine for the human. Is this failure a mere accident? My contention is that it is not. It is, rather, the consequence of a certain view of information and information systems which tends to consistently dismiss as non-significant the differences between different systems of information. Indeed, given the fact that machines happen to be more enduring, predictable and tractable than humans, and given the principle that the machine can perform as well as the human because there is no significant difference between them as systems of information, the normal attempt has to be to substitute the machine in place of the human. Is this to be pragmatic? Not at all, if we listen to Smith, who finds that machines are far more efficient than humans for certain tasks while humans are far more efficient than machines for other tasks, for the simple reason that they operate on different principles, among which consciousness of meaning and of purpose need to be counted. Smith argues for a more pragmatic approach that capitalizes on the differences between different information systems instead of ignoring them. The approach is pragmatic because it leads to better practical results. In the name of the same pragmatic interest I have been arguing for a science of information that does not preclude systematically such results by adopting a theoretical framework that leaves no place for the recognition of significant differences between different information systems.

Is there an alternative paradigm for the science of information? It seems to me that the alternative to a paradigm that systematically reduces the semantical and the teleological to the physical is a non-reductive paradigm. In other words, the science of information, in my opinion, should work on the explicit assumption that information is not an homogeneous but an heterogeneous phenomenon, that the large variety of systems of information cannot be reduced to a single and homogeneous set of explanatory concepts, and that quantification and mathematical formalization are neither the only nor necessarily the best adapted tools for the study of information at every level of the phenomenon.

I am the first to acknowledge that the proposed change of paradigms does not simplify the issues. But the issues are not simple anyway. That is why the science of information has had, and is still having, such a difficult time in finding its own identity. It is my hope that the views expressed in this paper may have contributed to exposing some of the fundamental reasons for our lack of agreement and, perhaps, to suggesting a direction whose possibilities remain to be explored.

References

1. Piaget, J., *The Origins of Intelligence in Children*, Cook, M., Trans., W.W. Norton & Company, New York, 1963.

2. Hartley, R.V.L., Transmission of Information, *Bell System Technical Journal*, 7, 535, 1928.

3. Shannon, C.E. and Weaver, W., *The Mathematical Theory of Communication*, University of Illinois Press, Urbana, 1949.

4. Apter, M.J., *The Computer Simulation of Behavior*, Harper & Row, New York, 1970, 36.

5. Rosenblueth, A., Wiener, N. and Bigelow, J., Behavior, Purpose and Teleology, *Philosophy of Science*, 10, 18, 1943.

6. Wiener, N., *Cybernetics; or, Control and Communication in the Animal and the Machine*, The Technology Press and John Wiley & Sons, New York, 1948.

7. See Turing, A.M., Computer Machinery and Intelligence, *Mind*, 59, 433, 1950.

8. Kuhn, T.S., *The Structure of Scientific Revolutions*, The University of Chicago Press, Chicago, 1970. See particularly 131-135, and ch. XII.

9. Smith, H.T., *Human Decision Making and the Computer* (Medical Research Council, Social and Applied Psychology Unit, Department of Psychology, The University, Sheffield. Prepublication Version. Without date).

INFORMATION THEORY AND HUMAN EXPERIENCE

Barrington Nevitt

"For your information, let me ask you a question"
(Sam Goldwyn, *Metro-Goldwyn-Mayer*)

Aphorisms, representing a knowledge broken, do invite men to inquire farther; whereas *Methods*, carrying the show of a total, do secure men, as if they were at farthest.
(Francis Bacon, *The Advancement of Learning*)

We had the experience but missed the meaning,
And approach to the meaning restores the experience
In a different form, ...
That the past experience revived in the meaning
Is not the experience of one life only
But of many generations -
(T.S. Eliot, *Four Quartets* - "The Dry Salvages".)

MEANING IS DETERMINED THROUGH INCLUSIVE RELATIONSHIP, NOT EXCLUSIVE DEFINITION

Nothing has any meaning alone. A single note does not make music nor does an isolated word constitute a language. The meaning of anything springs from its dynamic relationship as a *figure* to its context or *ground*. The meaning of a word is not what it *says*, not its definition which is *exclusive* (kept within bounds), but what it *does*, its resonance which is *inclusive* (beyond bounds).

When we fail to perceive the active *ground*, which normally remains hidden like water to a fish, we miss the meaning. The experience of anything is through interplay which may be unconscious, but the meaning is conscious reply and recognition. Pre-literate children and tribal adults alike learn the meaning of words by their *effects* in their social context. While children play with words like toys and learn the powers of language through probing their audience, poets use their knowledge of language powers deliberately to "put on" their audience.

As humorist Mark Twain remarked, *Le Mot Juste* is "the difference between lightening and the lightening bug". And poet Richard Aldington explains that:

> The exact word does not mean the word which exactly describes the object itself. It means the exact word which brings the *effects* of that object before the reader as it presented itself to the poet's mind at the time of writing the poem.[1]

To achieve his desired *effect* and communicate his experience, the poet shares his poem with an audience who participate in *making sense* of his poetic *figures* in the *ground* of their common culture. Communication is change that transforms its participants.

Every word originated as a breakthrough in human perception - an invention that created a perceptual bridge to new patterns of thought which reshaped its users. For language is the hidden *ground* that mediates all human thinking - it figures! Through constant use words become clichés or "software" tools, which accumulate untold human perceptions beyond any possible definition. These perceptions are unleashed by the resonance of poetry and pun. Symbolist poet Stéphane Mallarmé perceived *that* when he said: "To define is to kill, to suggest is to create".

In contrast, scientific usage aiming at "clarity" by defining "one word with one meaning" methodically strips natural language of its cultural *investment*. Pushed to extreme specialist language and concepts alike are reduced to binary bits by computer programs. And Robert Theobald points out that:

> Computer systems, not men, will first realize humanity's old dream of universal language, and the subtleties and nuances of human thought will risk being mediated through the restricted and standardized symbols of computer communication.[2]

The boundless dialectic of existence itself is transformed into the Yes-OR-No logic of computer wit. This wit without humor shuffles concepts minus percepts, abstract *figures* divorced from their actual *grounds*.

CONCEPTS ARE PRECEDED BY PERCEPTS AND BYPASSED BY WITS

Percepts are the direct sensory experience of our encounter with existence, whereas concepts are derived from repeated percepts, fossilized as it were, and packaged for convenient use. Perception as such is a proportion among proportions apprehended in our sensory lives. It is an analogical proportionality or a *figure/ground* relationship. For no sense can operate in isolation, but only in relation to other senses which are constantly transforming each other in the process of *making sense* through *synesthesia*. E.H. Gombrich notes that:

> What is called "synesthesia", the splashing over of impressions from one sense modality to another, is a fact to which all languages testify. They work both ways - from sight to sound and from sound to sight. We speak of loud colors or of bright sounds, and everyone knows what we mean. ... There is touch in such terms as "velvety" voice and "cold light", taste with "sweet harmonies" of colors and sounds, and so on through countless permutations. ... Representation is never a replica.[3]

Jokes are percepts that bypass concepts. Perception inspires levity rather than gravity. Wits choose their jokes and puns deliberately to break down conceptual barriers and break through to fresh insights. Newton did not discover "gravity" but "levity", for everybody knew that apples fell downward to the earth, but only Newton perceived that apples also tend to fall upward to the stars. His contemporaries who could not conceive action without contact called him a "Gothic Ape" for implying "action-at-a-distance" as he perceived the gravitational "resonant bond". This is still the hangup of those who can conceive only *visual* connections. If they cannot see what is resonating they cannot believe it. Today, "hardware" products are being constantly converted into "software" information or design to do even more with ever less in what Arnold Toynbee calls the process of "etherealization".

> There are, in fact, no connections in the material universe. Einstein, Heisenberg, and Linus Pauling have baffled the old mechanical and visual culture of the nineteenth century by reminding scientists in general that the only physical bond in Nature is the resonating interval or "interface". Our language, as much as our mental set forbids us to regard the world in this way. It is hard for the conventional and uncritical mind to grasp the fact that *"the meaning of meaning" is a relationship: a figure-ground process of perpetual change*. The input of data must enter a *ground* or field or surround of relations that are transformed by the intruder, even as the input is also transformed. Knowledge, old or new, is always a *figure* that is undergoing perpetual change by "interface" with new environments. Thus it is never easy to divorce knowledge and experience. In the same way that knowledge and experience are continuously modifying each other, the relation between "hardware" and "software" is not fixed but is in a perpetual state of metamorphosis.[4]

"Software" is the knowledge which shapes and transforms its "hardware" embodiment. Both "hardware" and "software" are human artefacts which create environments of services and disservices which transform their users.

SCIENCE ORGANIZES KNOWLEDGE THAT DISPLACES PERCEPTS BY CONCEPTS TO MATCH THE OLD

The use of concepts to program rather than to probe "where IT's at" can lead only to "hindsight" projects. For "IT" is normally the hidden *ground* that only direct perception can reveal. Concepts block perception, while classification as such rejects the breakthrough of process pattern recognition.

Thomas S. Kuhn has already exposed the hidden "paradigm trap" of *normal* science:

> One of the reasons why *normal* science seems to progress so rapidly is that its practitioners concentrate on problems that only their own lack of ingenuity should keep them from solving ... Discovery comes with the awareness of anomaly, i.e., with the recognition that nature has somehow violated the paradigm induced expectations that govern *normal science*.[5]

Science organizes conceptual knowledge, for quick retrieval and repeated use, whereas art uses concepts as probes to organize ignorance for invention and discovery.

Every culture finds its bearings in terms of some previous culture or Golden Age. Renaissance Europe harked back to the Roman Empire and to ancient Greece. Today, we still think of nineteenth-century mechanical "progress" with its ultimate "goals" and "missing links". Such metaphors or figures of speech bias the current modes of thought by creating its cognitive climate, for every metaphor (Greek *metapherein*: carry across or transfer) translates one form of being into another.

Today's dominant metaphors are of electric circuitry. The Shannon-Weaver information theory is based on a "telegraph model". This information theory deals with the practical problem of designing telecommunication circuits that can *match* their outputs to defined inputs with predictable error at minimum cost in bandwidth and transmission power. The theory is not concerned with the semantic *meaning* either of its "signals" or "noise", and it ignores all psychic and social effects - its *message*. It treats information or "software" like a "hardware" product transported through an information "pipeline". Norbert Wiener extended this theory to cybernetic systems, but biologists aware of its limitations are now turning to the DNA molecule as a new metaphor for explaining the genetic structure and behaviour of stable living cells - a more sophisticated form of *matching*. General systems scientists are following a broader course but still via concepts stripped of human percepts - abstract *figures* minus their existential *grounds*, stranded on the "hard facts" of nineteenth-century assumptions.

There is no information in a telephone directory - only data. And *relevant* data is not merely what is *given* but what is *taken* to create information for use. Language itself is a metaphor, a universal booby-trap, for it is the hidden *ground* that shapes perception which determines choice. Radioastronomy was discovered in the "noise" of radio circuits. And radio "signals" like all electromagnetic and gravitational phenomena are propagated through "empty space" where the action is, not through "void" where nothing can happen by definition. For "action" is not through connection but through the interaction of interpenetrating processes at their interfaces.

Unlike machines, people always choose the events to *make* the news. The historian first selects the events to report - *figures* abstracted from the *ground* of existence itself; he next transforms the chosen events via media; then he interprets the result. *History* is a human artefect - a triple fake. *Truth* is not something to *match*; it is not a label. *Truth* is something people *make* in a process of remaking the world as the world remakes them and their "human nature". Is there anything really old under the sun?

THE CONFLICTING GROUNDS OF WESTERN *EYE* AND EASTERN *EAR*

Some 2500 years ago the Greeks invented *Nature* and created new *visual* "order" out of the old *acoustic* "babble" of total existence. In his *Preface to Plato*,[6] Eric Havelock describes this process which Harold Innis had previously noted in his *Empire and Communication*.[7] Pre-literate tribal Greeks, who identified their technologies with gods as services, put the disservices mythically into Pandora's Box. The literate culture of the Western world, pushed to civilized extremes by Guttenberg, scrapped mythic involvement and created Western philosophy, science, and industry alike. As René Descartes perceived: "*L'occidental est condamné a la sequentialité*" (Western man is doomed to sequentiality).

Western civilization substitutes the EYE for the EAR and reduces all order to visual order. *Visual space* assumptions are the hidden hangup of Western man, who *imposes* a visual structure of separate centers with fixed boundaries, clear perspectives, and private points of view upon all existence. *Visual space* either contains or excludes definable components or specified properties. It demands YES-OR-NO logic for categorical *matching* that reduces diversity and uniqueness to the uniformity of common denominators. In visual space everything is logically connected and sequentially ordered in its proper place and time. It is the space of Euclid, Descartes, and Newton - the mind's EYE.

Adam Smith and the classical economists equated the laws of the market with the laws of nature itself. Whereas the road to hell had been paved with *good* intentions, henceforth the "hidden hand" of the market would ensure that the road to heaven could be paved with *bad* intentions. Today, cost-benefit economists kill what we love by striving to "quantify the quality of life".

Ever since the Golden Age of Athens we have continued to study the "causal connections" of our own technological extensions as abstract *figures* while ignoring their "resonant interplay" - the new hidden *grounds* of services and disservices created by these technologies which transform everybody ECO-logically. By failing to consider and to anticipate their psychic and social *effects* in remaking "human nature", we have abdicated the power of human choice to the "technological imperative" that leads to *Ultima Laputa*.

In his "Essay on the Chinese written Character", Ernest Fenollosa contrasts the psychic bias of Western visual culture and Eastern oral culture created by their communication media and exemplified by their differing modes of "explanation":

> In Europe, if you ask a man to define anything, his definition always moves away from the simple things that he knows perfectly well to abstractions. Thus, if you ask him what red is, he says a colour. If you ask him what a colour is, he tells you it is a vibration or refraction of light, or a division of the spectrum. And if you ask him what vibration is, he tells you it is a mode of energy, or something of that sort, until you arrive at a modality of being or non-being ... beyond your depth and his depth. But when the Chinaman wanted ... to define red ... he put together the abbreviated pictures of ROSE, CHERRY, IRON RUST, FLAMINGO ... The Chinese ideogram for red is based on something everyone knows.[8]

Since all Chinese words have many meanings which depend only upon their tone and context, every Chinese statement is loaded with puns and poesy. Fenollosa's essay recalls Francis Bacon's understanding of both "Aphorisms" and "Methods" during the intensified shift from *audile-tactile* to *visual* sensory bias in the age of Guttenberg. Today, Mao's aphorisms *replayed* in the Chinese ideographic culture never lose their home spun quality, while any aphorism *repeated* in the print-oriented Western world soon becomes threadbare. But aphorisms like aristocrats never stoop to exposition. Aphorisms and myths encapsulate and re-present ancient and timeless experience of tribal village life in *acoustic space*. Totally involving like a musical surround, *acoustic space* has no perspectives, no goals and no points of view whatever. It is inclusive and wholly related, never connected. In *acoustic space* there is no framework but only resonant interfaces between interpenetrating processes having neither inside nor outside. The structural pattern of *acoustic space* is that of centers everywhere with boundaries nowhere. It is the mind's EAR.

COMMUNICATION WITH PEOPLE IS BY MAKING NOT MATCHING

"What we cannot speak about we must confine to silence." (L. Wittgenstein)

Metaphors like all communication media are languages that foster either *visual* or *audile-tactile* patterns of psychic bias. Every human artefact as a technological extension of man creates an environment of services and disservices as *ground* for human consciousness as *figure*. Thus every innovation, whether of "hardware" or "software", becomes a medium of communication.

In communicating with people, the users are always the *content* of the *medium* which shapes or massages and transforms them unconsciously. The consciously contrived *program* is a hybrid of media which may have effects quite opposite to those intended. The *message* is always the totality of effects, psychic, and social, both of "hardware" and "software" whether intended or not. Each individual involved in the communications process *makes sense* or *meaning* out of the total process by using all his wits and senses. In *The Silent Language*, Edward T. Hall[9] exhibits the contrasting *effects* of differing uses of "time" and "space" in tribal and civilized cultures, and how ignorance of such *figure/ground* relations can reverse the intent of any communication. Charles F. Berlitz goes on to explain "How to insult everyone regardless of race, color, creed, or national origin" by wordless gestures.[10]

The problem of communication is to translate the intentionality of the participants into a *message* of desired effects. Whereas the scientist *conceives* communication in terms of matching desired effects to program content, the artist *perceives* communication as sharing experience that transforms all who participate in the process. Information is determined by communicating with people in the process of making the new rather than matching the old. The Oxford English Dictionary illustrates the parallel process of enrichment and impoverishment that occurs in defining the word "information" itself as a *figure*, while the *ground* of European culture changes from oral to visual stress. The question is being asked, whether there is a "Law of Conservation of Information" like other conservation "laws". Such a question reveals its visual bias of concepts for matching abstract Nature, out of *this* world. What theories of communication are implicit in the "software" artefacts of cultures past and present?

ART ORGANIZES IGNORANCE AND DISPLACES PERCEPTS TO MAKE THE NEW

"I talk in general terms because the particular has no language" (T.S. Eliot, "The Family Reunion")

In the Western world, artists like William Shakespeare, James Joyce, and T.S. Eliot learned to use language conceptually for exposition and perceptually for exploration to dis-cover hidden grounds that shape visible figures. Symbolist poet Edgar Allan Poe's story of *The Descent into the Maelström*[11] is a classic example of using the problem as *ground* to reveal its own solution as *figure*. Poe's hero, swept to "sure death" in *that* giant whirlpool, saved himself by ignoring his pre-conceptions and studying the process patterns of the Maelström itself. *This* approach bypasses concepts and models alike to deal with existence in its own terms. Understanding is neither a point of view nor a value judgement, but the *simultaneous* apprehension of the dynamic *figure/ground* relationships in any situation whatever - the recognition of its process patterns. By instant replay, we can have the meaning but miss the experience.

In symbolist statement, the normal *ground* is suppressed in order to highlight the *figure*. For example, a chair on a highway is a symbol whereas a car is not. The medium or *ground* can reverse the roles of writer and reader, or of producer and consumer, from merely passive to active co-makers of the experience. The "medium is the message" is a symbolic statement, because it suppresses the fact that the user or audience is both the *content* and the maker of the experience, in order to highlight the effects or *message* of the medium as the hidden *ground* of experience. In symbolist art, logical connections are deliberately removed in order to involve the public in a creative role.

Depth awareness is created by parallel suggestion not by connected statement. In his essay on the "Emotion of Multitude", W.B. Yeats explains the process of creating depth awareness by parallel suggestion:

> I have been wondering why I dislike the clear and logical construction which seems necessary if one is to succeed on the modern stage.... The Greek drama has got the emotion of multitude from its chorus ... to witness as it were, some well ordered fable.... The Shakespearean drama gets the emotion of multitude out of the sub-plot which copies the main plot, much as a shadow upon the wall copies one's body in the firelight. We think of King Lear less as the history of one man and his sorrows than as the history of a whole evil time.[12]

The role of the artist in any field is not merely to react but to *anticipate* the effects both of new knowledge and of his own actions. He does so by displacing percepts to sharpen awareness of human experience in perceiving the hidden environments of his own times. He learns to remake the old eternally new by juxtaposing ordinary things to dislocate human perception for attaining extraordinary vision. His art transforms old trash into new treasure. Similarly, the scientist of comprehensive awareness may use old models, old metaphors, and old concepts consciously as *probes*, rather than unconsciously as *programmes*. Whereas the old science is of concepts with visual cultural bias, the new *Scienza Nuova*[13] is of percepts and comprehensive awareness. The comprehensivist can thus learn to "predict" what has already happened by being the first to perceive IT. For IT is the hidden *ground* which shapes the visible *figures*. "Future Shock" is failure to perceive where IT's at - that the future of the future is the present.

INFORMATION TRAVELLING AT ELECTRIC SPEED IS THE HIDDEN GROUND OF THE TWENTIETH CENTURY

> *World War 1* - A railway war of centralization and encirclement
> *World War 2* - A radio war of decentralization concluded by "the bomb"
> *World War 3* - A TV guerilla war with no division between civil and military fronts.
> (Marshall McLuhan and Barrington Nevitt)

Hot war overkill has been cooled by electric media. Hardware "fighting" pushed to extreme has retrieved the ancient software of "flyting" - the names that really hurt are the latest reality. Today, only small wars are fought with hardware. But obsessed with old *visual figures* the specialist, whether in "warfare" or "welfare", is unable to perceive the new *acoustic ground*. For the specialist, unaware of the hidden visual bias of his own culture, demands to *see* the "vibes" before believing them. Meanwhile all rigid bounds of matter, time, and space have vanished in the boundless resonances of electric media - WE are *there* and THEY are *here* instantly, as discarnate minds "etherealized" by telecommunication. At electric speeds, all times merge in NOWNESS. Change itself has become the only constant. We can no longer afford to "wait and see" what effects our actions will have. Today, we must "know in advance" what the effects will be *before* taking any major action. In the contrast and comparison of developing situations, we can now learn to perceive the relevant information in process patterns that lie inside our problems themselves. We can thus *anticipate* "inevitable fate" by *understanding media*.

As hardware products are "etherealized" into software information at the speed of light, the old ground rules collapse or reverse their effects, like the old *Spitfire* as it went through the "sound barrier". Today's fact and fiction and fake unite, for the process of *making the news* now involves more people than the events reported. Information overload as breakdown leads to process pattern recognition[13] as breakthrough. We return to human scale, and to *making sense* with everybody's wits and all our senses. Today's REALITY is *making the new*.

The conflict of EYE and EAR becomes complementarity - the comprehensive resolution of polarity in paradox. For paradox is the technique of seizing the conflicting aspects of any complex process in a single instant. And Rosalie L. Colie points out that:

> Paradox exists to reject such divisions as those between "thought" and "language", between "thought" and "feeling", between "logic" and "rhetoric", between "logic" "rhetoric" and "poetics", and between all these and "experience". In paradox, form and content, subject and object are collapsed into one, in an ultimate insistence upon the unity of being.[14]

Niels Bohr's classic example of complementarity in physics is discussed by Thorlief Boman, who also observes that, "sense-physiologists do not seem to be aware of the essential difference and the incommensurability of optical and acoustical perceptions".[15] More recently Gordon G. Globus has explored the same theme for resolving the "mind-matter problem".[16]

IN TUNE WITH THE FINITE

"The Keys to Given" (James Joyce)

Through comprehensive awareness we can now make "bridges" of reason, rhyme, and rhythm between the worlds of EYE and EAR. Questions not answers are our main resource, when we TAKE TODAY in its own terms:

> The familiar idea of "making the news" now yields to making the world itself. For the best part of a century, we have been programming human consciousness with retrievals and replays of the tribal unconscious. The complementary of this process would seem to be the "natural" program for the period ahead: *programming the unconscious with the recently achieved forms of consciousness*. This procedure would evoke a new form of consciousness radically different from former consciousness. Everybody becomes a voluntary participant in creating diversity without loss of identify. Man is the content of the environment he creates, whether of "hardware" or "software", whether of consciousness or unconsciousness. There is therefore no technical alternative to "humanism", even though for many this would include the divine grace of the superhuman. INNOVATION IS OBSOLETE. SO IS OBSOLESCENCE, AS

INFORMATION SPEED-UP TRANSFORMS MAN AND HIS WORLD INTO ART FORM.[17]

Understanding bypasses all media and mediators alike. There is nowhere for elites to hide in the global electric theatre, where audience and actors merge. Now everybody is on stage. As perceptual knowledge replaces conceptual experience Everyman becomes the measure of all things for *Anticipatory Democracy.* In the *Age of Understanding Media* we shall choose effects to *precede* causes, for thought travels much farther and faster than light.

References

1. Richard Aldington cited by: Isaacs, J., *The Background of Modern Poetry*, Dutton paperback, New York, 1952, 45.

2. Theobald, R., Social Impact of Cybernetics, in *Master Minds: Portraits of Contemporary American Artists and Intellectuals*, Richard Kostellanetz, ed., Macmillan, New York, 1969.

3. Gombrich, E.H., *Art and Illusion; a study in the psychology of pictorial representation* [2nd ed., rev.], Pantheon Book, New York, 1961, 366 and 370.

4. McLuhan, M., and Nevitt, B., *TAKE TODAY; The Executive as Dropout*, Harcourt, Brace, Jovanovich, New York, 1972, 86.

5. Kuhn, T.S., *The Structure of Scientific Revolutions*, University of Chicago Press, Chicago, 1962, 37 and 52.

6. Havelock, E.A., *Preface to Plato*, Harvard University Press, Cambridge, 1963.

7. Innis, H., *Empire and Communications*, University of Toronto Press, rev. by M.Q. Innis, 1972.

8. Ernest Fenollosa cited by: Pound, E., *ABC of Reading*, No. 89, published in New Directions, Norfolk, Conn., 1951?

9. Hall, E.T., *The Silent Language*, Doubleday, New York, 1959.

10. Berlitz, C.F., How to insult everyone regardless of race, color, creed or national origin, *Horizon*, American Heritage Publishing, New York, Winter 1969, 120.

11. Poe, E.A., *Great Short Works*, Thompson, C.R., ed., Harper & Row, New York, 1970, 313.

12. Yeats, W.B., *Essays and Introductions*, Macmillan, London, 1961, 215.

13. Vico, G.B., *Principii d'una Scienza Nuova*, Italy, 1725.

14. Colie, R.L., *Paradoxia Epidemica: The Renaissance Tradition of Paradox*, Princeton University Press, Princeton, 1966, 517.

15. Boman, T. *Hebrew Thought Compared with Greek*, Westminster Press, Philadelphia, 1960, 207.

16. Globus, G.G., Unexpected Symmetries in the 'World Knot', *Science*, 1973, 1129.

17. McLuhan, M. and Nevitt, B., *op. cit.*, 297.

18. Nevitt, B., Computer at wits end leads to process pattern recognition, *The First Computer Communication Conference, October 24-26, Washington, D.C.*, IEEE publication.

THE SCIENTIFIC STUDY OF INFORMATION - ITS NATURE AND IMPACT

Victor Rosenberg

School of Librarianship, University of California,
Berkeley, California, U.S.A.

Throughout my years of work as an information scientist I have been plagued by a personal and professional sense of doubt with respect to the field. A central theme of every conference that I have attended related to "what is information science" or "is information sçience a science". In this paper I hope to take these questions head on. I must begin by saying that I *do* take information science seriously as a science. I see it as the quest for understanding of the nature of information and man's interaction with it. That we lack so much in this quest for understanding is the greatest challenge of the science. I intend to explore the tradition or world view bearing on the scientific study of information. The currently prevalent world view is the scientific tradition which extends from the Enlightenment to the present. I also intend to look critically at what I perceive to be the premises underlying most of our current efforts to understand the phenomenon of information. The criticism will by necessity be speculative. I intend to stick my neck out, not because I can prove my assertions, but because I believe these ideas must be discussed.

One of the most important essays on the nature of science is Thomas Kuhn's *The Structure of Scientific Revolutions.*[1] In this essay Kuhn develops the notion of the scientific paradigm and the idea of the scientific revolution. Kuhn's notion is that a paradigm in science is the world view within which scientists see their environment. The notion is imbued with a fundamental relativism. It suggests that what one sees is due in large measure to what one is conditioned to see. The paradigm serves as a structure for organzing knowledge, and organizing perceptions of the world. This notion is, of course, not new, but is elegantly developed by Kuhn as he looks at the Western scientific tradition. The idea of a paradigm is, broadly, that of the *weltanshauung*, the way in which one sees the world. Kuhn's idea is radical in that he goes so far as to say that different scientists with different paradigms see different worlds. I believe this is true.

The closest thing to a definition of a paradigm is found in Dudley Shapere's review of Kuhn's essay. "Most fundamentally, ... , Kuhn considers [paradigms] as not being rules, theories, or the like, or a mere sum thereof, but something more 'global', from which rules, theories, and so forth are abstracted, but to which no mere statement of rules or theories or the like can do justice. The term 'paradigm' thus covers a range of factors in scientific development including or somehow involving laws and theories, models, standards, and methods (both theoretical and instrumental), vague intuitions, explicit or implicit metaphysical beliefs (or prejudices). In short anything that allows science to accomplish anything can be part of (or somehow involved in) a paradigm."[2] Shapere goes on to criticize Kuhn for being so vague about the exact definition of a paradigm. I hope to turn the vagueness to advantage in using the notion of paradigms to look at information science.

Kuhn's ideas regarding scientific revolutions are equally provocative. He suggests that what one age regards as scientific truth is, in fact, truth only within the world view prevailing at the time. He challenges the notion of scientific change as a linear process of ever-increasing knowledge and the search after ultimate truth. The scientific truth of a given age is that which conforms to the prevailing paradigm. A scientific revolution occurs when the paradigm begins to break down under constant testing and observation. Kuhn supports his view with examples from the history of chemistry, physics, and astronomy. He shows how Ptolomaic astronomy was considered truth within its time, and how Newtonian mechanics was considered truth within its time. In each case as new observations were made, anomalies developed. These anomalies were observations which did not conform to the theoretical structure which was understood at that time. As these anomalies continued to pressure the prevailing paradigm, forces began to develop which suggested the need for a new paradigm. As the pressure increased the old paradigm ultimately gave way to a new paradigm. Kuhn's ideas suggest that the world seen by Einstein was a different world than that seen by Newton. Scientific truth was different with each world view. Kuhn argues that science does not proceed toward a unified truth, but rather that it is a process of explicating nature within the structure of the adherents to a given paradigm.

I believe it is worthwhile to explore the basic premises of what we today call information science in the context of a paradigm. Whether or not Kuhn's notions of scientific paradigm and scientific revolution are true, they do serve as a useful metaphor for discussion of information science. What then are some of the characteristics of the scientific paradigm of information science?

Most of the research done to date in information science has been done in what we can broadly call the tradition of Newtonian mechanics. In this tradition, the world and man are perceived to be essentially mechanistic. All phenomena are understood in the context of logical and ultimately mechanical processes. Central to this world view is the notion of causality. Although some phenomena must be viewed as probabilistic, there is an underlying fundamental logic. Both causal and probabilistic laws, or explanations, are derivable from observation. The most important element of this scientific tradition is the belief system based on the notion that behind every phenomenon there is an ultimately-discoverable cause or explanation.

The method for discovering the mechanism or cause is to reduce all phenomena to their basic component parts, to dissect, to simplify. Also essential to the methodology is the notion of objectivity - the removal of the observer from that which is observed. This scientific tradition and its methods have demonstrated at least qualified success in most scientific disciplines.

The modern digital computer represents a high point in the development of deterministic electro-mechanical devices. In many ways it also represents the high point, at least to date, of the scientific tradition just described. In information science, as well as other disciplines, the computer has become more than a tool or machine, it is a way of looking at the world. The computer has emerged as a cultural phenomenon. I call this the "gestalt of the computer". Although the world view derived from the computer is important in broad cultural terms, it is particularly important in the field of information science. Because information science has been so closely linked to the computer, the device has thoroughly colored our view of what information is and how people use it. Broadly speaking, the computer has caused us to view human information processing as analagous to machine processing. The success of this approach is similar to the success that Kuhn describes with regard to obsolete paradigms (such as Newtonian mechanics). Historically scientific paradigms have been successful and "true" for the period during which they prevailed. In a limited way the "obsolete" paradigms remain useful in describing certain phenomena. A paradigm becomes obsolete, however, when the number of questions that it cannot answer increases and, additionally, when phenomena are observed which the paradigm cannot explain.

There is no doubt that a large portion of human behavior associated with information is deterministic, even computer-like. Further, there is little doubt that human communication is in fact achieved by symbols which can be incorporated into a computer. For example, it is quite clear that there is a direct relationship between an index term and the document it indexes, and in linguistics there is a direct relationship between the symbols in the written sentence and the meaning of that sentence. There is no doubt in my mind, nor should there be doubt in anyone's mind who has worked with computers, that computers can perform tasks which very closely resemble human behavior. Much of the fundamental work in information science has been done using the tools of logic and mathematics. Most retrieval strategies currently use at least some form of Boolean logic or Boolean operators to compose queries for these information systems. Many efforts have been made, and still are being made, to explore the fundamental logical nature of the information retrieval process.

Beyond the logical components of the information retrieval process or other processes involving man and information, there is the basically human element. The human and social side of information science has been explored using the techniques developed over the past century for the study of sociology and psychology. Although many scientists are beginning to challenge the premises of the behavioral sciences, much of the scientific work in information science is based on principles largely in the tradition of behaviorist psychology. It is not my intention to explore the nature of behaviorist psychology or the theories challenging it, but rather to link it with the notion of Newtonian mechanics and the "gestalt of the computer".

The Newtonian or deterministic scientific tradition reaches full flower in the study of artificial intelligence. In many ways the field of artificial intelligence epitomizes that loosely connected set of scientific ideas which I call the paradigm for information science. Whether wittingly or unwittingly scientists studying artificial intelligence have developed a vocabulary, and have given rise to popular ideas, that suggest a basically anthropomorphic nature of the computer. In the June 28, 1973 issue of the *Wall Street Journal*,[3] Herbert Simon is quoted as saying, "I think you can say that the computer is now showing intuition and the ability to think for itself". In the same article is the statement "in giving birth to intelligent machines man is designing them in his own image. Some will have a voice like humans and the ability to understand speech". Our own language is full of statements like "computers can retrieve information", computers can "read", or can "write". Perhaps the most far-reaching statement here attributed to Herbert Simon is "some of us don't see any principle or reason that would prevent machines from becoming more intelligent than man."

Although the paradigm I have described has many ramifications, I believe the basic question is: what is the relationship between man and his computer? There are two ideas current.

One suggests, either directly or indirectly, that the nature of man's mental processes, social and personal interactions, and physical behavior is machine-like. By machine-like, I mean that there is a fundamental order beneath the observable phenomena and that this order is ultimately discoverable by scientific study. This notion further suggests that man's brain is essentially a mechanical device, and that intellect and thought are ultimately mechanical processes. From this stems the belief that ultimately man will be able to create a computer as intelligent or more intelligent than himself. Most scientists are humble enough to admit that the current state-of-the-art of computer technology will not permit the replication of advanced human processes. But the belief persists that eventually a device can be built to replicate the human mind. This notion of man as an ultimately mechanical and understandable device has broad acceptance in our culture.

It is important at this point to reveal the crux of my argument. First, although I believe that much of man's intellectual and physical behavior is indeed mechanical, I do not believe it is *essentially* mechanical. I believe that the notion of man as a mechanical device and the notion that human behavior can be perfectly replicated by computers are fundamentally wrong. Note that I said I *believe* it is fundamentally wrong. There is no adequate way of *proving* the assertion at this time. What is important here is not whether the idea is right or wrong, but what research and what social consequences flow from this idea and what alternatives might be possible. I believe that the essentially reductionist view of man which emerges from the "gestalt of the computer", is ultimately demeaning to man, is scientifically counter-productive, and it is arrogant. Nevertheless, I am not suggesting that all the work that has been done in replicating human intellectual behavior using computers is of no practical value. Remember, at the outset I said that I am taking information science seriously as a science. In the same way that Newtonian mechanics, although proven false as a scientific theory, still has enormous usefulness in every-day life, similarly the use of computers to process information has and will continue to have enormous value. However, as a basic principle for understanding, scientifically understanding, the nature of information and its use, the paradigm is of extremely limited value.

Since I have just stated, with an overweening arrogance of my own, that the fundamental premises on which information science is currently based are all wrong, I must support this conclusion. I will support this conclusion in essentially three ways. First, I will attempt to evaluate critically the nature of scientific endeavor to understand information and its relationship to human beings. Second, I will attempt to explore pathogenic social and political consequences of the mechanistic view. And third, I will attempt to sketch the beginnings of an alternative and contrast the two types of research that might develop from each paradigm.

Characterizing what is now almost a generation of research in information science since World War II requires the broadest of generalizations. As with all such generalization, there will be many exceptions and qualifications necessary. Nevertheless, we can trace much, if not most, of our recent scientific heritage to the scientific developments which came to flower during World War II. Vannevar Bush's article "As We May Think" contains many of the seeds of what followed in information science research.[4] As Dr. Bush said "the instruments are at hand which, if properly developed, will give man access to, and command over, the inherited knowledge of the ages". Dr. Bush and others who followed him, hoped to reduce thought, at least thought associated with information retrieval, to logic; and further to reduce logic to machine processes. There was, and perhaps there still is, a shared belief that eventually the essential processes of information retrieval could be reduced to machine instruction and processes. In the process of developing mechanical analogs of human behavior, the true nature of the processes could be discovered.

Since generalization is so difficult to deal with, let me use a specific example with which I am somewhat familiar. Many researchers have attempted to develop an understanding of the semantic relationship between concepts, or between index terms, or between documents. The basic notion of cross referencing in library systems, and all the more advanced notions of relatedness used in retrieval systems are based on the development of a definition and ultimately a quantification of relatedness. Yet, in my own work I attempted to develop measures of relatedness between index terms and to evaluate the measures' effectiveness for indexing and retrieval. The basic premise is that there exists a fundamental logic, and a fundamental mathematically-derivable relationship between terms. Bar-Hillel comments on this search for measures of relatedness by saying,

> "The failures of these attempts, I would say, are not due to the insufficient stature of those who made them, but rather to the fact that they attempted the (almost) impossible. I do not hesitate at all to state that... the establishment of a concept of 'relatedness or connectivity of documents in a collection along with a suitable metric to define the degree of the relatedness', is either unattainable at all or at least a devilishly difficult task, no approach being in view that has any chance of leading to a function that will give us a useful measure of distance between topics, and still less between documents."[5]

Bar-Hillel's criticisms, although written more than ten years ago, are as true today as they were then.

Bar-Hillel, in his essays, provides us with similar devastating criticisms of almost every aspect of the use of mathematics or logic in the area of information retrieval. I can find no instance in which the intervening years have proved him wrong. Here I must again reiterate my distinction between a fundamental understanding of information science and the development of useful practical tools for handling information mechanically. There is no doubt that large computer systems can effectively handle information as data. But the development of even the most sophisticated information retrieval systems has not enhanced our fundamental understanding of the nature of information. In fact, I would argue that the development of automated systems has inhibited this fundamental understanding. The demands of the computer, and the computer industry, have reduced much of our fundamental scientific effort to very sophisticated paper-pushing. If, as I believe, the nature of human information processing is fundamentally different from machine information processing, then the development of digital computer systems becomes an obstacle to the understanding of information and its use.

What I am attempting to articulate, is that the digital computer represents more than just a machine but rather part of a world view or paradigm. The computer carries with it a set of values, scientific values. These values are basically deterministic, reductionist and mechanical. The paradigm specifically inhibits serious consideration of concepts that are social, cultural, or spiritual. In this argument I am asserting that my notion of the existing basic paradigm for information science should be rejected and an alternative sought because it is fundamentally counterproductive.

Let me turn now to the second of my arguments for rejecting the deterministic paradigm. It is my firmly held belief that many of the premises which make up the currently held paradigm of information science are socially and politically pathogenic. These premises are shared with most other scientific and technological disciplines. My notion of pathogenic premises comes from a paper by Willis Harmon, *Alternative Futures and Educational Policy.*[6] Harmon's definition of a pathogenic premise is one which is problem-generating even though in other ways it may produce useful consequences. He lists the following:

(1) The "technological imperative" that any technology that can be developed, and any knowledge that can be applied, should be.

(2) The premise that the summed knowledge of experts constitutes wisdom.

(3) The reductionist view of man, a premise associated with the development of contemporary science, that lends sanction to

dehumanizing ways of thinking about and treating man.

(4) The premise that men are essentially separate, so that little interest and responsibility is felt for the effect of present actions on the remote individuals or future generations.

(5) The premise that man is separate from nature, and hence that nature is to be exploited and controlled rather than collaborated with.

(6) The disbelief that "what ought to be" is a meaningful concept and is achievable.

To these I would add my own which deal more directly with information science.

(1) The premise that useful information is that which can be handled as data.

(2) The premise that human information processing is ultimately mechanically reproducible and understandable.

(3) The premise that there exists a direct relationship between human thought and the symbols of communication.

Let me attempt to illustrate what I mean by the results of pathogenic premises within the context of information science. Among the many different information retrieval systems are those which are designed to store and manipulate police-record information. The information involves the arrest and conviction records for individuals. In my work with police-record systems, I have discovered that the systems tend to homogenize criminals. That is, they treat alike all criminals arrested for a given specific offense. Criminal records systems do not provide extensive detail on the nature of the crime, and in many cases not even the disposition of the arrest. If an individual is arrested for a specific crime and is found guilty of that crime, he is often treated just like the individual who is arrested for the same offense in a situation which was later determined to be a case of mistaken identity. I only wish to stress the important point that these systems reduce individuals to items of data. I can generalize and say that any information system reduces an individual or an individual's writing, to relatively homogenous data. The computer simply cannot deal with the infinite variety of human experience.

Most information scientists will go along with me when I assert that human values, and human dignity, are often trampled upon by information systems. They will argue that great care must be taken to avoid this possibility when designing an information system. But, where most will get off, is taking the next logical step.

And that is to indict the logic and substance of the scientific endeavor which led to the development of such a system. I take the further step which suggests that the pathology and the unfortunate consequences of information systems are inherent in the philosophy underlining their design. When we design an information system we have a set of premises in mind, whether implicitly or explicitly. When these premises, which I have previously articulated, collide with pressing social problems, such as crime, the result is most often not a solution of the problem but a profileration of problems. When we apply the knowledge derived from the scientific study of information we find a situation that does not fit solutions to problems so much as it fits problems to solutions. Although most consider the development of an information system as a partial solution to the problem of crime, the reality of the situation more often is that the problem is redefined such that an information system can be useful. The reason for this is simple and obvious. The problem of crime is intractable. The solution of providing a data processing system is not. The problem here is not the direct tangible harm that the information system does to a specific individual. Rather it is the image of man inherent in it. The inclusion of any individual in a data system is inconsistent with the inherent dignity of every individual. The data system reduces the individual to something he is not. This ends up as counterproduction scientifically and also counterproductive socially.

I do not wish to belabor the example too long. It is the subject for another paper at another time. What I do wish to stress is that the social and political consequences of the thinking derived from the premises of science, and more particularly of information and computer science, may well be disastrous. Again I must stress that the picture is not simply black and white. The "gestalt of the computer" and the notion of a mechanistic universe do have their place. But in the search for scientific principles, we should not be misled by our short-term successes nor can we afford to ignore our obvious failures.

One could, of course, argue that social and political consequences are irrelevant to the development of a scientific theory. Is it not objective reality that determines the validity or falseness of scientific theory? Here I must return to Kuhn's notion of paradigms. If one believes, as I do, that objective reality is largely colored by the prevailing paradigm, the cultural values that every scientific community shares, then the social and political questions become very real and in fact determine the nature of our reality and our scientific theory. The essence of Kuhn's argument is the demonstration that, historically, scientific theories and cultural values have always been inextricably related, and that, as cultural values are superseded by newer ones, so are scientific theories. Why, then, should our age be different than earlier epochs?

Theodore Roszak in his book *Where the Wasteland Ends*, states:

> "we lose sight of the arrogance that lies hidden in the... 'humble practical men' of science, who simply happen to know the difference between what is reliably so and what isn't and who possess the only means of telling the difference,... because their claim makes such perfect sense to us. All the metaphysical and psychological premises of that claim have become the subliminal boundaries of the contemporary mind-scape; we absorb them as if by osmosis from the artificial environment that envelopes us and which has become the *only* environment we know. The scientists simply work within this universally endorsed reality. They search for what that reality defines as knowable and - with strict professional integrity - feel obliged to entertain no alternative realities. Scientific knowing becomes,..., the orthodox mode of knowing; all else defers to it. Soon enough the style of mind that began with the natural scientists is taken up by imitators throughout the culture... we find ourselves surrounded by every manner of scientific-technical expert, all of them purporting to know as the scientist knows: dispassionately, articulately, on the basis of empirical evidence or experiment, without idiosyncratic distortion, and if possible by the intervention of mathematics, statistics, or a suitably esoteric methodology."

Now I turn to the third point in my argument, the alternative. To criticize is not enough; one must attempt to develop an alternative. Kuhn states, "To reject one paradigm without simultaneously substituting another is to reject science itself. That act reflects not on the paradigm but on the man. Inevitably he will be seen by his colleagues as 'the carpenter who blames his tools'".[7] The development of an alternative scientific paradigm is always an enormously difficult task. But then that's the stuff that science is made of.

A paradigm is never developed by one man, but rather by a community of scientists. All the critics of science seem to have one common failure, they cannot provide an alternative. I can only suggest a few avenues of research which might be productive.

First, information science must abandon its deterministic approach and must recognize the computer as perhaps an historical accident rather than a scientific organizing principle. We must get out from behind the computer. A more holistic approach is needed. In this the field is perhaps most closely related to a field such as ecology. We must begin to look at the interrelationships between various parts of the information environment.

We must begin to pay more attention to the social, cultural, and spiritual aspects of human communication. We must recognize that what a man says or writes is not simply the additive sum of the phonemes or the morphemes, the words or the sentences he utters. To deal effectively with the transcendent qualities of human communication we must admit as scientific evidence the intuitive, the subjective, and the experiential.

The paradigm crisis that I have presented is by no means unique to information science. Similar problems appear in all the social sciences, and to a lesser extent, perhaps, in the physical sciences. In his essay "Aspects of Science, Rediscovering History"[8] Stephen Toulmin outlines what may well prove to be the new direction of science, and by extrapolation the new direction of information science.

Toulmin sees the last fifty years as dominated by logical models and analyses which he calls "synchronic" issues. By synchronic he means issues which are seen out of an historical context in one slice of time. According to Toulmin, the focus on synchronic issues "delayed the moment at which [scientists] were prepared to tackle once again the increasingly urgent (but diachronic) problems of social and cultural change".

The shift from synchronic to diachronic concerns is seen as a return to a more consistent trend in the history of science. The synchronic approach tends to be historical and concerned with logical structure. The new concentration will be on conceptual change. As Toulmin says:

> "The problem of conceptual change became inescapable, the moment philosophers recognized that the *rationality* of any natural science involves something more than the *logical coherence* of the propositions forming its current intellectual cross-section. For - as a few of us had been insisting all along - scientists display their rationality as much through the manner in which, and the circumstances in which, they are prepared to give up one set of concepts and hypotheses in favour of another, as they do by their ability to judge the formal consistency of their current beliefs. This being the case, we should be paying just as much attention to the intellectual procedures by which - in historical succession - different families of scientific concepts displace one another in the course of time, as our immediate predecessors paid to the logical relationship within particular intellectual cross-sections of science. So the philosophy of science must come down from its formal quasi-mathematical perch and recognize that the *historical* development of the natural sciences has a genuine relevance to the *philosophical* understanding of scientific procedures."

If my reading of Toulmin is correct, the most productive avenue of research in information science will be the examination of conceptual change in the field, i.e., it will look at the concepts forming the basis for research. Once new concepts are developed new understandings of phenomena will follow.

References

1. Kuhn, T., *The Structure of Scientific Revolutions*, The University of Chicago Press, Chicago, Illinois, 1962 (second edition 1970).

2. Shapere, D., (Review of) *Structure of Scientific Revolutions*, in *The Philosophical Review*, 63, 383, 1964.

3. Brand, D., Latest machines see, hear, speak and sing - and may outthink man, *Wall Street Journal*, 88, No. 25, 1, 1973.

4. Bush, V., As we may think, *Atlantic Monthly*, 176, 101, 1945.

5. Bar-Hillel, Y., *Language and Information: selected essays on their theory and application*, Addison-Wesley, Reading, Massachusetts, 1964, 351.

6. Harmon, W.W., *Alternative Futures and Educational Policy* prepared for Bureau of Research, U.S. Office of Education by Stanford Research Institute, ERIC No. 6747-6, 9, 1970.

7. Kuhn, *op. cit.*, 79.

8. Toulmin, S., Aspects of Science, Rediscovering History.

Chapter 3

INFORMATION SYSTEMS: NATURE, DESIGN, AND USE

Information Systems: Nature, Design, and Use

Editorial Summary

The components of an information system may be understood through the presentations of Barnes and Otten discussed earlier in these proceedings. But Mitroff attempts to make explicit the basic nature of such systems - and particularly their purpose. Mitroff's view of information systems is similar to Churchman's view, which defines information systems essentially as inquiry systems. To Mitroff and Churchman, information or inquiry systems assume the nature of knowledge systems represented in philosophical discourses throughout the centuries. In this sense, these systems represent specific modes of dealing with reality, or the event world. Mitroff specifically alludes to the Leibnizian system as a formal deduction system, the Lockean as an experiential induction, the Kantian which gives equal emphasis to deduction and induction, the Hegelian as a dialectic system, and finally the Singerian-Churchmanian as the integrative-interdisciplinary system. Mitroff's major point is that each of the users of an information system brings his own view or "revolutionary shock" to bear on the data. Consequently, the design of an information system should provide enough flexibility to accommodate the idiosyncratic characteristics of the user.

The nature of information systems is such that they reflect the manner in which the data can be assembled to support or provide evidence for perception, and, ultimately, action. Mitroff's position assumes that the manner in which scientists defend their theories can be directly related to a man's mode of viewing the world. Royce adds support by postulating personality dimensions to epistemic styles. To Royce there are numerous psychological processes which tend to influence the world view. These psychological processes can be identified through factor analysis. What does all this mean for understanding the nature of information systems and their design? What Mitroff and Royce attempt to make clear is that there are idiosyncratic styles for executing the transformation of data to information. The styles are well organized and in many ways inter-related. This position gives considerable support to the emphasis on the user in the total system design process.

Bar-Hillel assumes a different perspective as to the nature of information systems, discussing them in terms of communication. He is particularly concerned with language structure and the function of semantics. Bar-Hillel believes that an information processing system requires a natural language and that the prospects of developing such a language are not good. Bar-Hillel proposes an experimental program on interactive languages, while assuming a modest position in claiming the existence of realistic information processing systems.

Given the nature of information systems, one can postulate models for their design. An attempt is made by Maher and Kabrisky, utilizing a visual process model to simulate pattern recognition. Pattern recognition has received considerable attention by psychologists, computer scientists, electrical engineers, and others. The thesis advanced in bionics underlies this position: an understanding of the brain mechanism involved in pattern recognition could lead to a better understanding of human information processing and subsequently to principles for the design of information systems. The findings of the Maher-Kabrisky experiments provide evidence that recognition of form by the human can be mathematically described. The significance of Maher's effort for information science remains to be acknowledged or confirmed through further experimentation, but it clearly represents an attempt to provide some understanding as to the underlying mechanisms responsible for the transformation of data to information. Conway, however, reminds us of the need for modesty in our dependency on modelling the underlying processes. Conway discusses the numerous aberrations to which the visual system is susceptible. Conway uses the distortion of visual space, particularly some of the "impossible figures" presented by M.C. Escher to alert us to the complexities of perceptual phenomena. In particular, he addresses the problems encountered by the information seeker in the acquisition of sensory data.

Returning to design considerations, the two basic configurations discussed were the establishment of *utilities* and the creation of *networks*. Saracevic has given considerable thought to the question of information systems as public utilities which could be studied for developing university libraries, etc. He states the properties and conditions usually associated with utilities and analyzes his environment considering these. He concludes that much more understanding is needed to establish criteria by which to operate an information system as a utility.

Another design concept is networks - the combining of one or several information and/or computer systems in an integrated communication organization. Samuelson claims that information networks provide the opportunity for enhancing the free flow of information, as opposed to the passage of static, serial data through computer networks. Samuelson suggests that it would appear best to design the global network from the top down rather than from the bottom up. Communications and service stations should be provided at an earlier stage than data retrieval or computer processing. Thus, priority would be given to two types of transfer, multipoint telesatellites for person-to-person communication, and multi-media displays.

A critique of a design concept is presented by Mullin, who maintains that the one critical deficiency in design is the lack of ability of a system to adapt to changing conditions. Mullin says that most retrieval systems tend to be too large. He cites several common design errors, namely, the predetermination of record structure and record size, the predetermined generalized descriptions of records, and the tendency of the designers to tell the users how to use the system. Design decisions for implementation are based on expected usage of the system, but the usage is estimated with insufficient data.

The acceptance of the system by the user is fundamental, as suggested by Mullin. However, Friedman analyzes the problems associated with assessing the success of storage and retrieval systems in general. His proposed approach to systems development would improve user satisfaction. Instead of merely specifying a system in terms of performance parameters, designing it, and later testing it for agreement with the system specification, a development is suggested in which the user of the system contributes to the conceptual design. He proposes a methodology for testing the functional value of complex storage and retrieval systems which uses a management information system in a form permitting feedback to improve the use of the system.

Orin Taulbee explains some of the problems facing the operators of information systems in making the service available to the public at a fair price. He points out that one should distinguish between information access systems, and information support systems (information handling and processing services). With increasing technological sophistication, information systems are enjoying wider use and the financial base is changing. While systems initially were operated on a budget provided from outside to serve those in need of the service, the trend is towards self supporting systems. This raises the questions of the value of information to the user and the public and of establishing a price which is considered fair by the user and will provide enough revenue to the operator. The problem of economics is further complicated by the nearly monopolistic environment in which most systems operate because of the uniqueness of the data base or the service. Economic criteria may change the technology applied in systems; more economical ways of duplicating large data bases are required. Taulbee outlines the problems associated with resource sharing and gives us food for thought on the question of considering information systems as public utilities.

FROM APOLLO TO INFORMATION SCIENCE: INFORMATION AS ORGANIZATION THEORY, ORGANIZATION THEORY AS INFORMATION - A WHOLE SYSTEMS CRITIQUE

Ian I. Mitroff

University of Pittsburgh, Pittsburgh, Pennsylvania, U.S.A.

INTRODUCTION

We live in a society where the production and handling of information has become increasingly more important. For this reason alone it behooves us to ask what the relation if any there is between the structure of an organization and the kinds of information activities it typically is able to support.[1] If we go a bit further and conceive of information in its broadest and most generic sense as the product of some basic process of inquiry,[2] then the most general question we can ask is what the philosophy of science can contribute to the study of organizations, i.e., to the development of organization theory and its relationship to information science broadly conceived. The purpose of this paper is to raise precisely this question.

More specifically the author feels that the current state of organization theory is in great need of some fundamentally new kinds of questions and challenges. The same applies to information science. Put slightly differently, the author believes that the future course of organization theory is in need of some radical new directions. Almost by definition, these "new" directions can only be gotten by stepping outside of the usual modes of thinking by going to fields of inquiry, like the philosophy and social psychology of science, which traditionally have been conceived to be far removed from the concerns of organization theory. If the success of any field is ultimately tied to the radical ideas and paradigms that it is willing to entertain, then the concern of this paper is with the most speculative and wide-ranging hypotheses that can be formulated by looking at organization theory from some radically different perspectives.

THE PHILOSOPHY OF SCIENCE AS THE MANAGEMENT OF INQUIRY[3]

In a number of previous papers,[4] the senior author has described a number of fundamentally differing inquiry systems. These systems basically derive from the history of Western epistemology. In particular they derive from the recent efforts of C. West Churchman[5] to cast some of the traditional ideas of Western epistemology in modern systems terms so that epistemology might thereby become more directly relevant to the design needs of the practicing systems analyst. Since most of these systems have already been extensively described elsewhere, we shall, for the most part, confine ourselves here to an extremely brief description of them. One aspect in particular of one of these systems has not been dealt with previously, or at least not in the context of organizational studies, and it is thus one of the main purposes of this paper to focus on this aspect here.

Before describing these systems, it is important to appreciate that none of them is "true" in the literal sense. Each system represents a characteristic way of "representing" or viewing reality rather than reality itself.[6] That is, each of these systems does not describe reality *per se*. Rather, each of them describes how those who are committed to each system view reality, and hence, the various frameworks into which the proponents of these systems attempt to squeeze our knowledge of reality. In a word, each system expresses the value commitments of its proponents. And in fact it can be shown[7] that *some* set of normative or value judgments *always* precedes the collection of *any* set of scientific data or the construction of *any* scientific model. Scientific description is not value-free because one always has had to presuppose some normative model of the world in the first place in order that one can collect observations in the second place. The very act of scientific observation expresses a normative commitment, i.e., the metaphysical belief that the world is basically orderly enough so that observation of the world is possible.

Leibnizian Inquiring Systems (IS) (as they are termed by Churchman[8]) are the epitome of formal, deductive reasoning systems. No matter what the problem, Leibnizian IS always strive to reduce it to a completely pure mathematical, logical, or formal representation of some sort. Unless one can do this, those who advocate Leibnizian inquiry do not feel we have adequately understood or captured the "true essence" of the problem situation.

It is of course a trivial matter to note (at least for those who are not Leibnizian in their basic outlook) the implicit premises on which this view of the world rests. One premise certainly is the belief that the structure of the world is inherently logical or mathematical so that it can be explicated in logical or mathematical terms. Another premise is that all problems are best studied through the medium of formal mathematical systems. Still another premise is that mathematical inquiry is superior to other forms of inquiry. To be sure this hardly even begins to touch on the many properties associated with Leibnizian IS, let alone the many attitudes of their proponents. There is however one additional attitude or property of Leibnizian IS that this paper is particularly critical of and hence that we wish to challenge more directly later. This is the notion that Leibnizian IS not only regard the activity of formal theory building *as superior to* the activity of data collection, but that they regard theory building as an activity which can proceed *independently of* the activity of data collection. Given the inevitable errors to which human data processing (perception) is liable, Leibnizian IS are extremely distrustful of any claims to knowledge which rest on empirical considerations of any sort. They prefer instead conclusions which are based on the impersonal dictates of formal reasoning.

If Leibnizian inquiry represents a pure, abstract theory-before-data view of the world, then Lockean inquiry represents the exact reverse of things. *Lockean* IS are the epitome of inductive, experiential systems. No matter what the problem, Lockean IS always strive to break it down into its "basic experiential components" on which one can make some "direct observations". Unless one can gather some data that directly pertains to the problem, those who advocate Lockean inquiry do not feel we have adequately understood or captured the "true essence" of the problem situation. In short, if Leibnizian inquiry says that we haven't understood something until we have a good formal theory of it, Lockean inquiry says we haven't understood something until we have collected some good data on it. Although this position as well gets immensely complex and rests on many sophisticated premises, there is one assumption in particular that we especially wish to challenge. This is the notion that Lockean IS not only regard the activity of data collection *as superior to* the activity of theory building but that they also regard data collection as an activity which can proceed *independently of* theory building, i.e., that one need not presuppose *any* prior theory in order to engage in the activity of scientific observation. Finally, it should be noted that *Lockean IS are consensually-based IS*. That is, Lockean IS do not regard a datum, a conclusion, or a proposition as objective unless two or more experts can be found who agree strongly with respect to their observations, conclusions, etc.

If Leibnizian inquiry represents a view of the world that emphasizes the primacy of theory to the detriment of data, then Lockean inquiry represents a view that emphasizes the primacy of data to the detriment of theory. By contract, *Kantian* IS emphasize a view of the world which gives equal emphasis to both. Kantian IS emphasize that it is impossible to collect data on a phenomenon without having to presuppose some theory, no matter how implicit that theory may be, with respect to the phenomenon being investigated. Put somewhat differently, Kantian IS stress that every Lockean IS has unwittingly presupposed some Leibnizian IS in its operation. In short, scientific observations are not theory-free.[9] Different scientific theories can not only lead us to the collection of different sets of data but even when they direct us to collect the "same" set of data, more often than not they give it different interpretation.

In effect, a Kantian IS is an information system which presents multiple Leibnizian views of any problem so that the decision-maker can witness explicitly how the "nature" of his problem changes as the underlying Leibnizian or theoretical representations of it are varied. This is in sharp contrast with Leibnizian and Lockean IS which typically tend to present only one view of a problem situation for a decision-maker's elucidation, acceptance, or rejection. Most important social problems like abortion, pollution, the busing of children to achieve racial quotas, etc. call for Kantian inquiry, at a very minimum. The nature of important social problems is such that there is no single "well agreed upon" view of the problem so that Leibnizian or Lockean inquiry is sufficient to treat them, i.e., to "represent" them. The very thing we want to witness with respect to important social problems is precisely how they change as we view them from different perspectives. Kantian IS are designed to accomplish exactly this.

If Liebnizian and Lockean IS represent the antithesis of one another with respect to issue of what comes first, data collection or theory building, then Lockean IS and *Hegelian* (or Dialectical) IS are the antithesis of one another with respect to the place of agreement in the conduct of inquiry. If Lockean IS are founded on the principle of total agreement, then Hegelian IS are founded on the principle of total conflict. Further, if Kantian IS are designed to present multiple images of a problem, not all of which may necessarily be in conflict or disagreement with one another - that is, the multiple images of a Kantian IS may actually overlap and compliment one another - then Hegelian IS are designed to actively seek the strongest possible conflicting views of a problem situation.

No matter what the problem, Hegelian IS are designed to present the strongest possible debate on the nature of the problem between two or more sharply contesting views of the problem. Contrary to Lockean IS, Hegelian IS emphasize that by themselves data are not information. Information is that which results from the interpretation of data. Data without interpretation are meaningless in themselves.[10] Hegelian IS attempt to illustrate the crucial nature of diverse views of the world by applying two or more opposing views *to the same data set*. It is hoped that as a direct result of witnessing the conflict and the discrepancy between opposing interpretations of the *same* data set that the decision-maker will be in a better position to observe for himself the crucial role that assumptions play in the analysis of problems. Hegelian or Dialectical IS are founded on the presumption that conflict is a better principle than agreement in exposing the hidden assumptions on which every representation of the world depends. If agreement is a strength and benefit, then it can also be a weakness and a liability. It can obscure precisely those aspects of a problem that should be most severely scrutinized but are hidden precisely because of the agreement between experts. On the other hand, conflict is no virtue in itself. If Lockean IS are designed to construct a view of the world that is founded on agreement even when there may be little agreement "naturally" occurring in the problem situation, then Hegelian IS are designed to construct a disagreement view of the world even when there is little disagreement "naturally" present in the situation. Hegelian inquiry would seem most appropriate - indeed an absolute necessity - for those social problems which naturally present the most intense division regarding the "true nature of the problems", e.g., abortion, pollution, the busing of children to achieve racial quotas, etc. Hegelian IS are deserving of increasingly more serious attention in a culture that is giving increasingly more serious attention to the idea of placing computerized information systems in the homes of its citizenry. It is disturbing, to say the least, to think of the consequences of such systems if they are not *explicitly* designed to present a variety of opposing views on important issues.[11] The dangers are only too well illustrated by recent political events. One can only speculate as to whether President Kennedy's decision with respect to the Bay of Pigs invasion might have been drastically different if he had been presented with a strong counter-position which challenged every one of the assumptions on which the position he ultimately acted on was based.[12]

The last IS that will be dealt with is the most complicated to treat in brief terms. The reason is that it is a complex function of all the previous IS. For this and other reasons too extensive to go into here, we must confine our treatment to a single aspect of that is most pertinent for the present discussion. It should also be borne in mind that there are many more IS than the few we have touched on in this paper. Further, there are an infinite number of ways of combining these IS. Leibnizian, Lockean, Kantian, and Hegelian IS represent pure or extreme types.

Singerian-Churchmanian IS are the epitome of integrative interdisciplinary systems. In a word, Singerian-Churchmanian (SC) IS are anti-reductionistic.[13] Contrary to each of the previous IS, SC-IS do not believe that the process of inquiry can be reduced to a single set of fundamental entities on which knowledge can be shown to depend. Thus, Leibnizian IS believe that inquiry can be reduced to formal, mathematical entities; Lockean IS, to experiential elements combined with consensus; Kantian IS, to possibilities; Hegelian, to conflict. SC-IS believe that each mode of inquiry can be shown to depend on each of the others in subtle and not so subtle ways. Thus, for instance, we have already pointed out that all Lockean IS have unconsciously presupposed the existence of some Leibnizian IS in their design. Hegelian IS only work because they presuppose some minimal level of agreement. Likewise, Lockean IS presuppose the fundamental existence of some conflict, e.g., that between theory and data. In a similar fashion SC-IS are devoted to showing how each of these IS constantly interact, presuppose, and strengthen one another. From the point of SC-IS, each of the previous IS are nowhere near as independent or as pure as their proponents would have us believe.

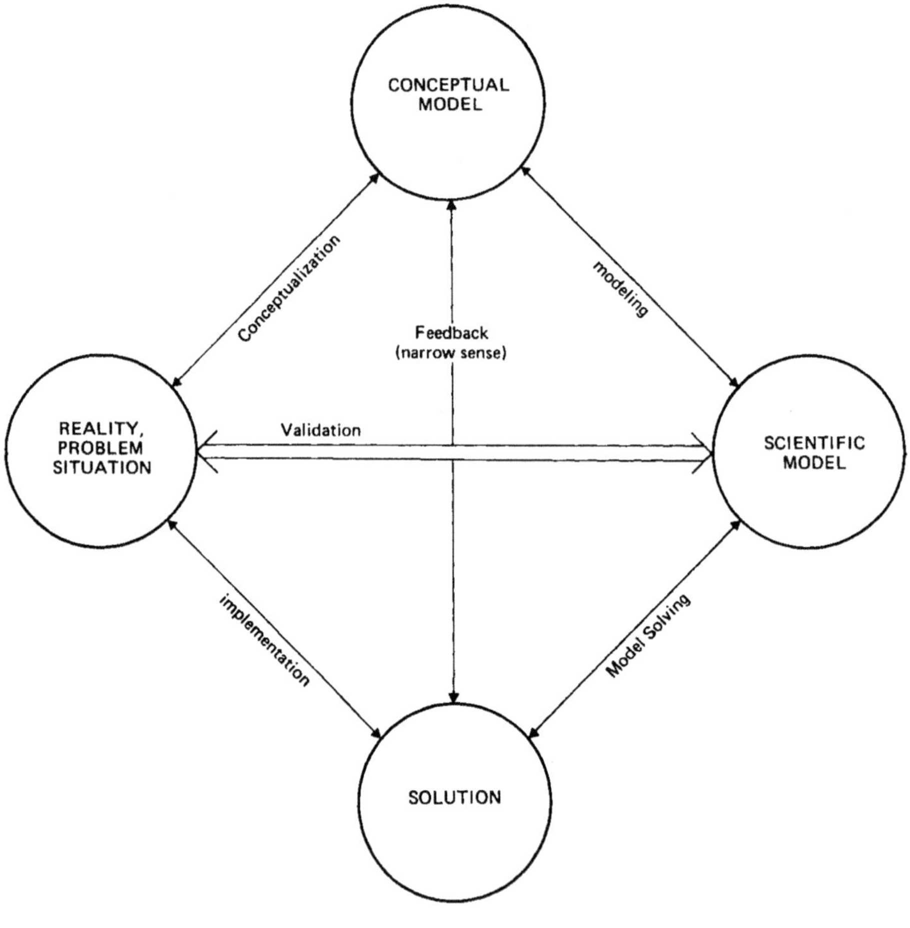

Figure 1 – A Systems View of Problem-Solving

A diagram may help to bring out some of the more distinctive features of SC inquiry (Figure 1). Figure 1 represents a simple "whole systems" view of the activity of problem solving.[14] For the moment let us suppose that every inquiry starts with the extreme left-hand circle, the existence of a problem-situation. (Actually, if one takes a systems view of things, there are no simple "starting" or "ending" points.[15] One can begin the "process" at any point in the diagram.) The arrow or path from the circle labelled "Conceptual Model" is meant to indicate that the "first phase" of problem solving consists in formulating a conceptual model of the situation. The conceptual model sets out in broad terms the definition of the particular problem that will be solved; it specifies the field variables that will be used to define the nature of the problem and the level to which the variables will be treated, for example, whether from a micro or macro point of view. Once a conceptual model of the problem has been formed, a scientific or formal model can be formed, and from this a solution, if one is possible, can be derived. If the solution is then fed back to the problem for the purpose of taking action on it, we have the case of implementation. In this way, any one of a number of iterations can be performed on the problem. To complete the diagram, the arrow or path from "Reality" to the "Scientific Model" corresponds to the "the degree of correspondence" between "reality" and the model in the traditional meaning of the notion of correspondence. The vertical path between the "Conceptual Model" and the "Solution", we have labelled "feedback in the narrow sense". The reason for this will be made clear shortly.

One purpose, among many, of the diagram is to draw out further distinctions between the IS. For instance although a proponent for each of the IS might contend that they were interested in the whole system, i.e., all the phases of problem-solving, from the perspective of the SC-IS, each of the IS have typically restricted themselves to studying one or more phases of the diagram. Thus, Leibnizian and Lockean IS have shown an extreme preoccupation with the right half of the diagram to the almost total neglect of the left half. Leibnizian and Lockean IS typically act as if the cycle from "Conceptual Model" to "Scientific Model" to "Solution" back to "Conceptual Model" was all there was to problem solving - that the goal of problem solving is to iterate on deriving better scientific solutions to better scientific models. This is what we mean by "feedback in the narrow sense". This form of activity never gets back to "reality" to question its initial assumptions; i.e., it takes its conceptualization of the problem for granted. Kantian and Hegelian IS, on the other hand, are extremely good at questioning the initial assumptions, first principles, or conceptualization on which every scientific model depends. Unfortunately, they often never move beyond this stage, and for this reason, remain locked in the path between "reality" and the "Conceptual Model" exploring conceptualization after conceptualization, or the act of conceptualization for its own sake.

SC-IS represent a very different attitude toward problem-solving. Under a Singerian-Churchmanian view of the world, no single aspect (component, stage, path) of the problem-solving process is more important than any other. Indeed, under a conception of the world which treats reality as a system of highly interconnected entities, the components of such a system can only be separated in one's imagination, not necessarily in reality.[15] A whole systems view of the world implies that in the strict sense there are no such things as "components" which have an existence independent of other components and of the whole system itself.[16] Needless to say, this view leads one to be immensely critical of the ways that we have typically studied the process of problem-solving in the past. Under this philosophy, it is wrong to perceive any single component as more critical than any other, and hence, it is wrong to allocate all of one's inquiry efforts to studying one phase of the system over any other. It is fair to say that we live in a culture that has given its major scientific allegiance to Leibnizian and Lockean modes of inquiry. As a result, we have a preponderance of research dealing with scientific modeling and model solving but comparatively little research dealing with conceptualization and implementation. We have sophisticated models and treatises on modeling and model solving, but comparatively speaking, only the most rudimentary treatises on conceptualization and implementation.

In order to correct for the overemphasis which has been given to formal modeling and model solving, SC-IS stress the importance of studying the implementation phase of problem-solving.[17] In part this is due to the fact that Singerian-Churchmanian inquiry arises directly out of the tradition of American pragmatism.[18] For the pragmatist the "truth" of a model is not to be identified solely with its formal structure or with formal truth tests of the model but rather with the ability of the model to effect significant social action, i.e., implementation. For this reason, SC-IS are often called "implementing IS" although it should be strictly understood that SC-IS do not say that implementation is thereby the most important aspect of inquiry. Were they to contend this then they would be reductionistic in their own way. They would commit the very kind of error they are so critical of the other IS for committing. They would be saying that all inquiry reduces to a single aspect, in this case, implementation.

The whole notion of IS raises some fundamental questions and challenges for organization theory. For one, if we admit that each IS is actually most appropriate for a certain class of problems, and since all inquiry takes place within a certain social context that is necessary to support that inquiry, then are different kinds of organizations (structures) needed to support different kinds of IS? If so, what are they? Further, are they different from traditional organizational structures?

These kinds of questions become particularly critical for the Hegelian and the SC-IS. First, with respect to the Hegelian IS, dialectical thinking is so foreign to our culture that experiments[19] reveal that it cannot be supported and sustained without extensive training and continual reinforcement. Another way to put this is to say that one cannot bring in naive untrained subjects off the street, subject them to the arguments of strongly opposing experts, and then expect them to form policies which are richer than those of the original opposing experts. However, with training and reinforcement, laboratory experiments and field studies[19] give tentative support to the hypothesis that decision-makers who are able to think in a dialectical mode or who are able to make use of Hegelian inquiry systems are able to form long-range strategic policies and plans which are superior to those formed in accordance with other modes of inquiry.

The most critical challenge may be that which is raised by the SC-IS. It is a safe generalization to say that past research in organization theory has been largely compartmentalized; i.e., the vast majority of research efforts have proceeded on a largely piece-meal and fragmentary basis. There are few studies in organization theory or few theories of organization which sweep around the whole of Figure 1. As a general rule, we find that either by training or by preference most researchers have confined themselves to working on one or two branches of the diagram at most and at separate points in their careers. There are extremely few researchers who have simultaneously worked on more than two aspects of the diagram: Virtually non-existent are those whose interests and efforts have spanned the entire diagram.

Notable exceptions are Russell L. Ackoff, C.W. Churchman, and Herbert A. Simon.[20] Each of these outstanding researchers has done significant work on nearly every phase of the diagram at some point in their careers. The aim of their efforts is twofold: (1) the development of programs of research on complex systems and organizations devoted to working on *all* the phases of Figure 1 *simultaneously*, and (2) the development of comprehensive theories of organization able to account for every phase of Figure 1 considered as activities that are performed within the context of some organization. Notice that this involves two distinct but nonetheless related uses of Figure 1. In the first sense Figure 1 points to the phases that any comprehensive organization theory will have to go through in order to be developed. In the second sense, Figure 1 points to the very phases considered as organizational activities that any comprehensive theory of organizations will have to explain. In this sense, the figure is doubly prescriptive, or evaluative, in that it can be used as a standard against which to measure current efforts.

THE SOCIAL PSYCHOLOGY OF SCIENCE - A BRIEF REPORT OF A STUDY OF THE APOLLO MOON SCIENTISTS

> ... I think many scientists are genuinely unaware of the extent, or even of the fact, of this personal involvement, and themselves accept the myth of impersonal objectivity. This is really very unfortunate. It is true that only a man who is passionately involved in his work is likely to make important contributions, but the committed man who knows he is committed and can come to terms with this fact has a good chance of getting beyond his commitment and of learning how to disassociate himself from his idea when this is necessary. There is little in the traditional education of scientists to prepare them for this necessity, and there are many who are still unaware of it.
>
> Ann Roe, "The Psychology of the Scientist"[21]

Thus far we have outlined a program of research for organization theory that has drawn its main inspiration and justification from considerations derived from the philosophy of science. We want now to turn to another series of considerations that lend additional justification to the program that has been outlined. These considerations derive from a recent study performed by the senior author into the psychology of the Apollo moon scientists.[22] The main object of the study was to identify, if possible, the differences in psychology, if any, between (1) those scientists who exhibited strong prior commitments to their pet theories regarding the nature of the moon and as a result showed an extreme reluctance to part with their pet theories in the face of the data returned from the moon, and (2) those scientists who were perceived by their peers as more willing and able to part with their pet notions. The differences in psychology between these two "types" of scientists are not only interesting and striking in themselves but they are even more interesting with regard to what they imply about the nature of science. The results suggest that we have been extremely narrow and limited with respect to much of our thinking regarding science. As it has been throughout this paper, our argument is that a different conception of science allows us to visualize and open up new kinds of research. The biggest limiting factor in the kinds of research we have undertaken in the past may have been the very images we have had of science itself.

Almost three months to the day of Apollo 11's landing, a series of extensive interviews were begun with over 40 of the most eminent scientists who studied the moon rocks. Each scientist was interviewed four times over a span of three years. The interviews were generally conducted in the period between the completion of one Apollo mission and the start of another. Thus the scientists were interviewed between Apollo 11 and 12, 12 and 14, 14 and 15, and for the last time, between Apollo 15 and 16. The style of the interviews ranged from open-ended discussions on the opening round to written questionnaires on subsequent rounds. The open-ended discussions were designed primarily to explore the range of issues connected with the lunar missions. The written questionnaires, which were given in person for the express purpose of encouraging the scientists to talk about and even criticize the items on the questionnaires and not merely fill them out, were designed to focus in on specific attitudes towards some of the issues that had been identified from the opening discussions.

The first interviews struck a dominant theme that was to be repeated time and again. No matter what the topic that was being discussed - for example, the status of some highly technical physical theory - the scientists steadfastly moved the discussion toward highly personal matters. It seemed fundamentally impossible for them to discuss the status of a physical theory and the evidence pro and con with respect to that theory in *purely* impersonal or so-called "objective" terms. For every theory, it seemed that there was some scientist or group of scientists who were clearly associated with that theory and who served as its personal advocates and defenders. As a result, it was impossible to react to a theory without reacting simultaneously to the proponents of that theory. Little wonder then why the image of science that one gets from the interviews is so greatly different from the image one gets from other sources. Scientists may be the epitome of cold, calculating, impersonal, rational creatures as portrayed in "the standard accounts of science", but they are not like that in doing science, in talking about it, and especially, in talking about their colleagues.

One of the most surprising outcomes of the whole study was the reactions of the scientists to the abstract concept of the "disinterested observer". No single notion seems more basic to the idea of science. At the very least the notion of the disinterested observer implies that scientists should adopt a neutral or uncommitted attitude towards their hypotheses. If not, they may so bias their observations and experiments that they may end up unwittingly confirming what they should have rejected as false. For this reason, if scientists should develop any kind of an extreme attitude, it should be one of extreme skepticism. Indeed, there are whole schools of contemporary philosophy of science which are founded on the idea that scientists should do everything in their power to *disprove* their pet hypotheses, not affirm them.[23] Only after this has been done repeatedly and only if the evidence keeps mounting in favor of the proposed hypothesis are we then justified, if even then, in accepting the hypothesis as a "provisional truth of science" that some future inquiry may well rescind.

It was thus somewhat surprising when *out of the 42 scientists interviewed on the first round there was not one scientist who in one way or another did not indicate that he thought of the notion of the purely objective, disinterested science naive.* The vocal and facial expressions that accompanied the verbal responses were the most revealing of all. They ranged all the way from mild humor and guffaws to extreme annoyance and clear expressions of anger. It was generally felt that the only people who took the idea of the purely objective, disinterested scientist literally and seriously were the general public or beginning science students. Certainly no working scientist, to quote from one of the respondents, "believed in that simple-minded nonsense". Because they actually did science and because they had to live with the day-to-day behavior of some of their more extreme colleagues, they knew better.

What was so surprising about the reactions of the scientists was not that they rejected the notion of the "disinterested scientist" as an accurate *descriptive* account of the workings of science but that they even rejected it on the stronger grounds as a desired *prescriptive* ideal or normative standard. Not only was it recognized that in point of actual conduct that the good scientist was more often than not highly committed to a point of view - at the very least to his pet theories and hypotheses - but even more interesting and important, strong reasons were evinced why this *ought* to be the case, that *ideally* scientists *ought not* to be without strong prior commitments. Even though the general behavior and personality characteristics of some of their more extremely committed colleagues absolutely served to infuriate them, as a rule the scientists still came out in favor of scientists having strong commitments. The following are typical quotes from the respondents:

> Scientist A - Bias has a role to play in science and it serves it well. Part of the business [of science] is to sift the evidence and to come to the right conclusion, and to do this you must have people who argue for both sides of the evidence. This is the only way in which we can straighten the situation out. I wouldn't like scientists to be without bias since a lot of the sides of the argument would never be presented. We must be emotionally committed to the things we do energetically. No one is able to do anything with liberal energy if there is no emotion connected with it.
>
> Scientist B - The uninvolved, unemotional scientist is just as much a fiction as the mad scientist who will destroy the world for knowledge. Most of the scientists I know have theories and are looking for data to support them; they're not sorting impersonally through the data looking for a theory to fit the data.
>
> You've got to make a clear distinction between not being objective and cheating. A good scientist will not be above changing his theory if he gets a pre-

> ponderance of evidence that doesn't support it, but basically he's looking to defend it.
>
> Without commitment one wouldn't have the energy, the drive to press forward sometimes against extremely difficult odds. Trying to collect data on the moon is not the easiest thing in the world [this may be the biggest understatement of the entire study]. There are not only physical problems but there are bureaucratic problems as well to fight.
>
> You don't consciously falsify evidence in science but you put less priority on a piece of data that goes against you. No reputable scientist does this consciously but you do it subconsciously.

It should be clearly understood that it is not being said that the moon scientists did not critically test their ideas or those of their colleagues. Likewise, it is not the case that they have abandoned all of the critical and usual methods of science. There was ample testimony to the pursuit of normal scientific activities. Further, there is no reason to believe that these scientists are any more or less critical, objective or subjective, than their colleagues in other fields of science. If anything this group may have merely been more candid in describing some of their underlying motives. Thus, it is not that these scientists do not test their ideas but that the real process of doing science is much more complicated and filled with subjective and even irrational elements than we have acknowledged and have dared to study in detail. To remove commitment and even bias from scientific inquiry may be to remove one of the strongest sustaining forces for both the discovery of scientific ideas and for their subsequent testing. And in fact, we would even argue that it should not be our goal to eliminate bias and commitment from science altogether but to understand them better so that we could account for their influence.[24] Unfortunately, those views of science which only see commitment and bias in a negative sense are the least helpful in this regard. They hinder our recognizing the influence of bias and commitment in the first place and they keep us from studying them in the second place.

It is unfortunately beyond the scope of this paper to discuss in any great detail the model of science that the results of the study of the moon scientists suggests. It is likewise beyond the scope of this paper to discuss in detail the differences in psychology between some fundamentally differing kinds of scientists. Of necessity we must confine ourselves to some general remarks.

The model that emerges is one that suggests that science is governed by deep internal conflicts. It would appear that not only has science learned how to live with these conflicts - if only to tolerate them - but that in an even more fundamental sense it has learned - albeit unconsciously - how to profit from them. It would appear, in other words, that these conflicts are vitally necessary for the rise, growth, and sustenance of science.[25] Conflict is not tangential to science; it lies at its very core.

While the conflicts that operate in science are many and varied, they manifest themselves on at least two levels: (1) the individual, and (2) the social or institutional. On the individual level, conflicts arise because of the sharp personality and stylistic differences between scientists. In this regard, the IS we described earlier do not merely describe and represent abstract conceptual systems. They also describe and represent some very concrete and real personality types. They certainly represent some very different attitudes toward the nature and practice of science. In the study of the moon scientists, significant examples of specific scientists were identified for each of these styles of inquiry. More specifically, the single quality or dimension which best served to differentiate between the scientists was "willingness to speculate beyond the data available at hand".

In any social grouping, there are always those who prefer to stay close to "the facts" and those who prefer to venture out beyond them even at the risk of ignoring them. Type I scientists, as they were referred to, were distinguished by the essential defining quality that they excelled at extrapolating from data. Although they were often fine, detailed experimenters themselves and at times enjoyed experimental results or numbers for their own sake, theorizing was obviously their most pleasurable and exalted task. Type I scientists were willing and even relished the bold intuitive and theoretical leaps always required in making inferences from incomplete data to a comprehensive and encompassing theory. Type III scientists, as they were referred to, represented the other end of the spectrum. Here, numbers are often relished primarily for their own sake. There is a preoccupation, even an obsession, with data gathering. There is often an extreme disdain of theorists who deal with highly inferential and abstract concepts. Speculation or extrapolation from data is valued little and only engaged in when the data clearly warrants such extrapolation and then only with extreme cautiousness. Where type I's more readily tended to see the positive advantages of speculation in science and to speak glowingly of it, type III's tended to disparage it. They tended to equate speculation with wild theorizing and refer to it as "finger-painting in the sky".

The significance of these differences is that if the type I scientist most nearly corresponds to the Great or extraordinary scientist,[26] then in the judgments of the moon scientists themselves, the type III scientist most nearly corresponds to the Average or typical scientist. Furthermore, if the results of the study are representative of scientists everywhere, then they imply that, much more than his type III counterparts, the type I or Great scientist is more likely to be creative in the production of bold and speculative ideas. They are also more likely to be interested in the "big", the unsettled and ill-defined questions and problems. However they are also the kinds of the scientists who are most likely to become most rigidly committed to their ideas once they have produced them. Contrary to popular misconception, it is the "lesser" not the "greater" scientist who is more likely to have an "open mind". The greater the scientist the more likely he is to develop a line and to push it for all it is worth; also the more likely it is that he will be able to defend this line although this does not mean that he is more likely to be ultimately proven "correct" in the long run. In a word, the greater the scientist, the more he is likely to belie the myth of the "disinterested, uncommitted" scientist.

In a similar fashion it can be shown that these same differences manifest themselves on the social or institutional level of science.[27] For example, it can be shown that if "disinterestedness" is a legitimate social norm of science (which it *is*), then there is an opposing norm on the other side - call it "interestedness" - which can also be construed as a legitimate norm of science. In fact, it can be argued[28] that there is an opposing counter-norm for every one of the norms of science that have been proposed by the sociologists of science[29] and that it is the dialectic or Hegelian tug of war between the norms and the counter-norms which runs the institution of science.

CONCLUDING REMARKS

We have argued throughout this paper for a different model of science. We have argued for a conception of science that takes as its starting point several facts or premises: (1) the fact that scientific observers are not neutral, untainted observers but have strong prior commitments; (2) the fact that scientists differ with respect to one another in the degree and kind of their prior commitments; i.e., that there exist different kinds of scientists; (3) as a result of these differences, conflict arises between scientists; i.e., there is a differential toleration and appreciation of the necessity for these differences between scientists; and (4) that the differences between scientists and the conflict engendered as a result of them are *necessary* (but not sufficient) conditions for the rise and the continuance of science. If the results and conclusions of this paper are generally valid, then the Hegelian IS may be even more important than we have hypothesized. Not only is conflict an important principle in the development and explanation of organization theory, but it is an important principle in the development and explanation of science, the very thing we use to explain other things.

In a culture that relies more and more on science for the solution to all its ills, it is more and more important to examine the nature of the beast that pretends to effect a cure. The question therefore is how we can better study science by means of science itself. The question for organization theory is how it can aid in this effort. The question is whether it can choose not to and still remain viable.

References

1. Mason, R.O. and Mitroff, I.I., A program for research on management information systems, *Management Science,* 19, 475, 1973.

2. Churchman, C.W., *The Design of Inquiring Systems: Basic Concepts of Systems and Organization*, Basic Books, New York, 1971.

3. Betz, F., On the management of inquiry, *Management Science*, 18, B-117, 1971.

4. Mitroff, I.I., A communication model of dialectical inquiring systems - A Strategy for strategic planning, *Management Science*, 17, B-634, 1971.

   Mitroff, I.I. and Betz, F., Dialectical decision theory: A meta-theory of decision-making, *Management Science*, 19, 11, 1972.

   Mitroff, I.I., Williams, J. and Rathswohl, E., Dialectical inquiring systems: A new methodology for information science, *Journal of American Society for Information Science*, 23, 365, 1972.

   The following articles are in the press:

   Mitroff, I.I., A Brunswik lens model of dialectical information systems, *Theory and Decision*, to appear.

   Mitroff, I.I., Epistemology as general systems theory: An approach to the conceptualization of complex decision-making experiments, *Philosophy of the Social Sciences*, to appear.

   Mitroff, I.I., Systems, inquiry, and the meanings of falsification, *Philosophy of Science*, to appear.

5. See reference 2.

6. See reference 2 and the last two items in reference 4.

7. Churchman, C.W., *Challenge to Reason*, McGraw-Hill, New York, 1968

Churchman, C.W., *The Design of Inquiring Systems*, Basic Books, New York, 1971.

Churchman, C.W. and Ackoff, R.L., *Methods of Inquiry; an Introduction to Philosophy and Scientific Method*, Educational Publishers, Saint Louis, 1950.

Churchman, C.W., *Prediction and Optimal Decision, Philosophical Issues of a Science of Values*, Prentice-Hall, Englewood Cliffs, New Jersey, 1961.

8. See reference 2.

9. See last three items in reference 7.

10. Churchman, *op. cit.*, and Hanson, N.R., Observation and interpretation, in *Philosophy of Science Today*, Morgenbesser, S., Ed., Basic Books, New York, 1967, 68.

11. Churchman, C.W., *Challenge to Reason*, McGraw-Hill, New York, 1968.

12. Whyte, W.F., Organization for the 1980's, paper delivered at Georgia State University on October 17, 1972.

13. Churchman, C.W., Concepts without primitives, *Philosophy of Science*, 20, 257, 1953.

Churchman, C.W., *The Design of Inquiring Systems*, Basic Books, New York, 1971.

14. Feyerabend, P., Problems of empiricism, in *Beyond the Edge of Certainty; Essays in Contemporary Science and Philosophy*, Colodny, R.G., Ed., Prentice-Hall, New Jersey, 1965, 145.

15. Ackoff, R.L. and Emery F., *On Purposeful Systems*, Aldine, Chicago, 1972.

16. Ackoff and Emery, *op. cit.*, and Churchman, C.W., *The Design of Inquiring Systems*, Basic Books, New York, 1971.

17. Churchman, C.W. and Schainblatt, A.L., The researcher and the manager: a dialectic of implementation, *Management Science*, 11, B-69, 1965.

18. Singer, E.A., *Experience and Reflection*, Churchman, C.W., Ed., University of Pennsylvania Press, Philadelphia, 1959.

19. Mitroff, I.I., Williams, J., and Rathswohl, E., Dialectical inquiring systems: A new methodology for information science,

Journal of American Society for Information Science, 23, 365, 1972.

20. Simon, H.A., *Models of Man: Social and Rational; Mathematical Essays on Rational Human Behaviour in a Social Setting*, Wiley, New York, 1957.

21. Roe, A., The Psychology of the scientist, *Science*, 134, 456, 1961, 457.

22. Mitroff, I.I., *The Subjective Side of Science: A Philosophical Enquiry into the Psychology of the Apollo Moon Scientists*, Elsevier, forthcoming.

23. Popper, K.R., *The Logic of Scientific Discovery*, Harper, New York, 1965.

    Popper, K.R., Normal Science and its dangers, in *Criticism and the Growth of Knowledge*, Lakatos, I. and Musgrave, A., Eds., Cambridge, 51, 1970.

24. Churchman, C.W. and Ackoff, R.L., *Methods of Inquiry*, Educational Publishers, Saint Louis, 1950.

    Mitroff, I.I., *The Subjective Side of Science: A Philosophical Enquiry into the Psychology of the Apollo Moon Scientists*, Elsevier, forthcoming.

25. Kuhn, T.S., The essential tension: tradition and innovation in scientific research, in *Scientific Creativity, Its Recognition and Development*. Selected papers from the proceedings of the first, second and third University of Utah Conferences: "The Identification of Creative Scientific Talent", Taylor, C.W. and Barron, F., Eds., Wiley, 1963, 341.

    Mitroff, I.I., *The Subjective Side of Science: A Philosophical Enquiry into the Psychology of the Apollo Moon Scientists*, Elsevier, forthcoming.

26. Kuhn, T.S., Logic of discovery or psychology of research? in *Criticism and the Growth of Knowledge*, Lakatos, I. and Musgrave, A., Eds., Cambridge University Press, 1970, 1.

    Kuhn, T.S., *The Structure of Scientific Revolutions*, Second Edition, University of Chicago, 1970.

27. Merton, R., *Social Theory and Social Structure*, Enlarged edition, The Free Press, New York, 1968.

28. Mitroff, I.I., On the norms of science - A report of philoso-

phical case study into the psychology of the Apollo moon scientists, paper submitted to *Science*.

Mitroff, I.I., *The Subjective Side of Science: A Philosophical Enquiry into the Psychology of the Apollo Moon Scientists*, Elsevier, forthcoming.

29. Merton, R., The ambivalence of scientists, *Bulletin of the Johns Hopkins Hospital*, 112, 77, 1963.

Merton, R., Behavior patterns of scientists, *American Scientist*, 57, 1, 1969.

Merton, R., *Social Theory and Social Structure*, Enlarged edition, The Free Press, New York, 1968.

Storer, N.W., *The Social System of Science*, Holt, Rinehart, and Winston, New York, 1966.

EPISTEMIC STYLES, INDIVIDUALITY, AND WORLD-VIEW

Joseph R. Royce

The University of Alberta, Edmonton, Alberta, Canada.

INTRODUCTION

As a result of certain life-long interdisciplinary interests and involvements I have been struck by how difficult it is for specialists to interact and communicate across disciplines. While there is nothing new in this observation, I found, when I pressed the inquiry beyond the obvious answers of technical jargon, different interests, different content, and different methods, that I was unable to come up with satisfying alternative hypotheses. Some years and one book later[1] I came up with the idea that attempts at communication between specialists fail because of insufficient overlap of reality perspectives, and further, that such non-overlapping is due to differences in either implicit or explicit epistemological commitments. Stated another way, specialists don't communicate well with each other because they have different notions of what's "really" real. For example, the empirically-oriented laboratory scientist places his faith in facts, the mathematician bets on logical consistency, and the poet sees meaning in metaphors. I further concluded that knowledge specialists have very limited images of reality, and that even the most broad and learned men have a limited perspective of "the way things are". After still further analysis and study, it became increasingly clear that such limitations in looking at the world were not restricted to academicians; rather, the limited vision of the specialist was merely more extreme, more exaggerated, and more obvious. In short, my analysis suggests that all men are "encapsulated"; and that our models of the totality of life are dependent upon idiosyncratic psycho-biological histories. During the ensuing years I slowly came to the realization that I was being engulfed by what we might refer to as epistemological psychology,[2] and that the problems of special interest to me might be called *the psychology and philosophy of the Weltanschauung.*[3] This is the problem to which I will address myself in this paper.

More specifically, I will deal with the psychological basis for differences in world-view. The answer I come up with involves a synthesis of two heretofore independent projects - one involving the development of a general theory of individual differences,[4] and the other involving the development of a psychological theory of knowledge.[5] This synthesis is accomplished by:

(1) conceptualizing the previously identified three ways of knowing as epistemic styles,

(2) including style constructs as an integral part of the multi-dimensional, multi-system, structure of individuality,

(3) regarding epistemic styles as higher order personality integrators,[4]

(4) identifying the cognitive processes and epistemologies which correspond to a particular world-view.

We begin with a brief summary of the relevant portions of the theory of individuality.

THE STRUCTURE OF INDIVIDUALITY

The problem at hand is to come up with a general theory of individual differences. Such a theory must accommodate *all* psychological differences, ranging from molecular differences such as reaction time and conditionability to molar differences such as values and world-views. Furthermore, it must include all organisms from amoeba to man, and it must accommodate the total life span, contributions of hereditary and environmental sources of variation, and biological (i.e., neural, biochemical, hormonal, etc.) correlates. It goes without saying that this is an enormous task, and that a full exposition of such a theory would require considerable space. Thus, while a presentation of the complete theory is neither feasible nor desirable in this context, it is possible to present a condensation of the most relevant portions of it. I will, therefore, summarize the structural-dimensional aspects of the theory. The reader who also wants the dynamic-functional aspects is referred to a recently completed monograph-length paper and a forthcoming book.[4]

The essence of the proposed theory is that one's psychological makeup is an organized, multi-dimensional system. Highly oversimplified, the key idea is that each of us can be characterized by a profile on a large but finite set of traits or dimensions.

These dimensions are organized in some unknown way, and, of course, the way they are organized is important in determining the behavioral outputs of the system.

Thus, the theory postulates an overall system under the rubric of *psychological structure*, which is defined as *a multi-dimensional, organized system of processes (subsumes mental structure) by means of which an organism manifests behavior and mental phenomena.* Further, the theory defines *mental structure as a multi-dimensional, organized sub-system of processes (subsumes cognitive and affective structure) by means of which an organism manifests mental phenomena*, where mental phenomena refer to internal (i.e., neurologically coded) representations. Parallel to the previous two definitions, I will define *cognitive structure as a multi-dimensional, organized sub-system of processes (subsumes perceiving, thinking, symbolizing) by means of which an organism produces cognitions*, where cognitions refer to those mental phenomena which are products of cognitive processes (i.e., perceiving, thinking, symbolizing). It should be apparent from these definitions that we are describing a general system and its sub-systems, and that a given structure is nested or embedded within a more inclusive structure (see Fig. 1A). This means, for example, that cognitions (e.g., percepts, ideas, insights; see Fig. 1B) evolve primarily from cognitive structure, but that such cognitions are also influenced by the underlying mental structure, which, in turn, is influenced by the organism's total psychological structure.

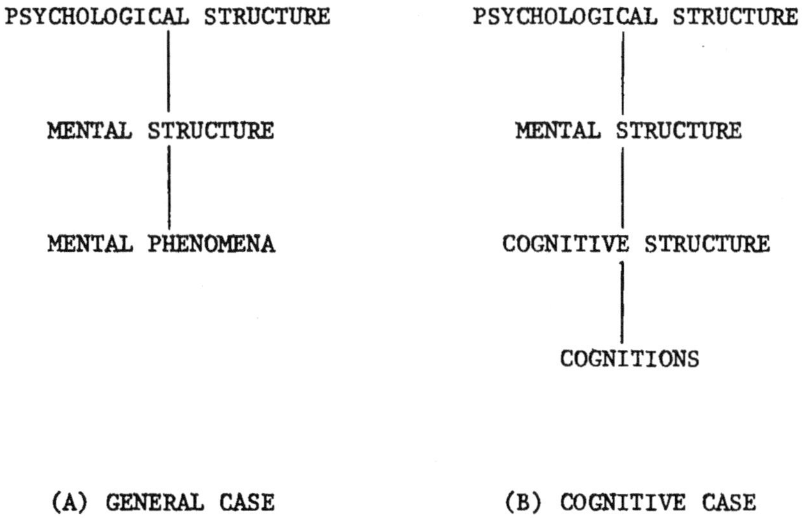

Fig. 1. The hierarchical embeddedness of cognitive structure (Royce, 1973)

Let us now define *affective structure as a multi-dimensional, organized sub-system of processes (subsumes temperament and emotionality) by means of which an organism manifests affective phenomena.* Thus, the implication is that affect emerges out of affective structure, which is directly influenced by mental structure, which, in turn, is influenced by the organism's total psychological structure. Note that affective structure is parallel to cognitive structure - that is, affective structure is *not* subsumed by cognitive structure. A more complete statement of relationships between cognitive and affective structure will be summarized later when we bring in the linkage taxonomies of style structure and evaluative structure.

What are these cognitive and affective dimensions, how might they be organized, and where did they come from? So far the theory includes some 50 cognitive dimensions, plus around 25 affective dimensions which appear to be hierarchically arranged. And they have emerged inductively from the thousands of empirical reports in the factor analytic literature. Thus, we have relied upon the factor analytic model to provide us with the basic taxonomy of dimensions.

Factor analysis is, of course, uniquely suited for such a task - the task of identifying dimensions, and the task of generating a plausible taxonomy of dimensions. It begins with a large number of measurements, n, and a correlation matrix, R, from which one derives a factor matrix, F, such that the number of factors (m) in F is less than the original n measurements. This first step, called factoring, can be summarized in matrix algebra terms as

$$R = FF' \qquad (1)$$

F' is the transpose of the factor matrix, F, such that when it is pre-multiplied by F, it reproduces the original correlation matrix, R. The second step, transformation, involves a rotation of the orthogonal reference vectors in F to a non-arbitrary location. This step, which is usually taken in accordance with the criteria of simple structure, gives us a matrix V, which is used for "interpretation" purposes. Thus, in matrix terms, we have

$$F\Lambda = V \qquad (2)$$

where F is the orthogonal factor matrix, Λ is the transformation matrix, and V is the rotated (usually oblique, but, in special cases it can be orthogonal), to-be-interpreted matrix.

The main point thus far is that, following the principle of parsimony, we have moved from a large set of measurements, n, to a smaller subset of underlying theoretical constructs[6] or factors, m. Furthermore, if the factors are correlated (i.e., an oblique V matrix), which is usually the case, one can then factor these (lower order) factors in search of second order and still higher order factors. Thus, in the summary figures below, such a hierarchical arrangement of factors emerges in both the cognitive and affective domains. The reader who wants a more complete exposition of this particular model is referred to basic works on factor methodology, such as Harman[7] and Thurstone,[8] the thousands of empirical reports, and my more elaborate statements.[4]

Because of the complexity of these hierarchical arrangements, let us begin by focusing our attention on the first and second order factors of Fig. 2. Here we see 24 first order factors which are grouped under six second order factors. There are, for example, five first order reasoning factors, four first order fluency factors, and two memory factors. If we now broaden our scope to include the more peripheral information processing of sensory inputs and motor outputs, we would add additional factors, such as the first and second order psychomotor factors summarized in Fig. 3. And if we now elaborate vertically as well as horizontally, we get the hierarchical cognitive system summarized in Fig. 4. A similar analysis leads to the hierarchy depicted in Fig. 5 for the affective domain.

Having presented both cognitive and affective structures, it now becomes possible to explore the possibility of cognitive-affective linkages which provide for integrating certain aspects of these two broad domains. I will make use of two classes of linkage constructs, i.e., style factors and evaluative factor.

Style Structure

A *style* is a characteristic mode or way of manifesting cognitive and/or affective phenomena. I will define *style structure as a multi-dimensional, organized sub-system of processes (subsumes cognitive, affective, cognitive-affective, and epistemic styles) by means of which an organism manifests cognitive and/or affective phenomena.* Thus, style structure is a sub-system of psychological and mental structure which serves a synthesizing role either *within* a sub-domain, such as cognition (e.g., cognitive style) or affect, or *between* cognitive and affective structures (see upper half of slashed line sections of Fig. 6).

When the style construct is linked primarily to cognitive phenomena we will refer to it as *cognitive style*. When it is linked primarily to affective phenomena we will refer to it as *affective style*. When it cuts across both domains we will refer to it as *cognitive-affective style*. And when the style construct simultaneously invokes a valid truth criterion (i.e., leads to a justifiable knowledge claim in addition to being a characteristic mode or way of interacting with the environment) we will refer to it as *epistemic style*.

While it is epistemic style that is of particular importance

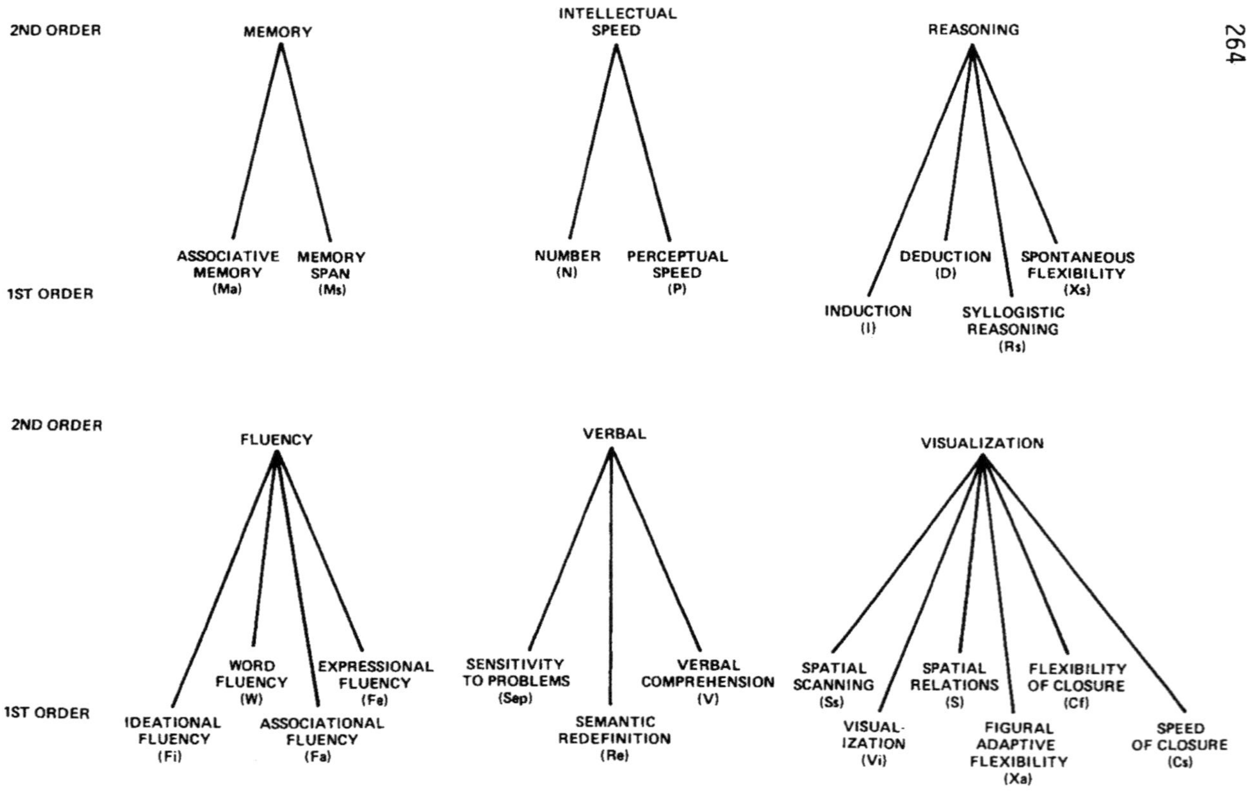

Fig. 2. Hierarchical structure of intelligence factors (Royce, 1973).

(N.B. Due to space limitations the two level hierarchy has been split and displayed in two rows rather than one)

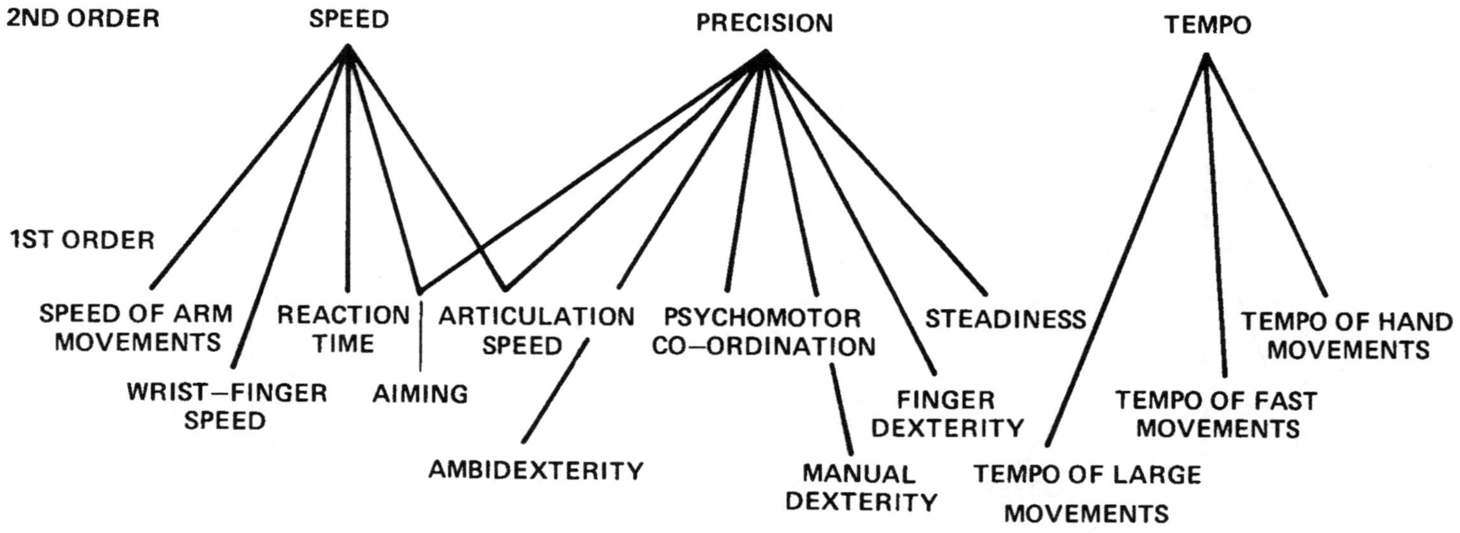

Fig. 3. Hierarchical organization of the psychomotor factors (Royce, 1973).

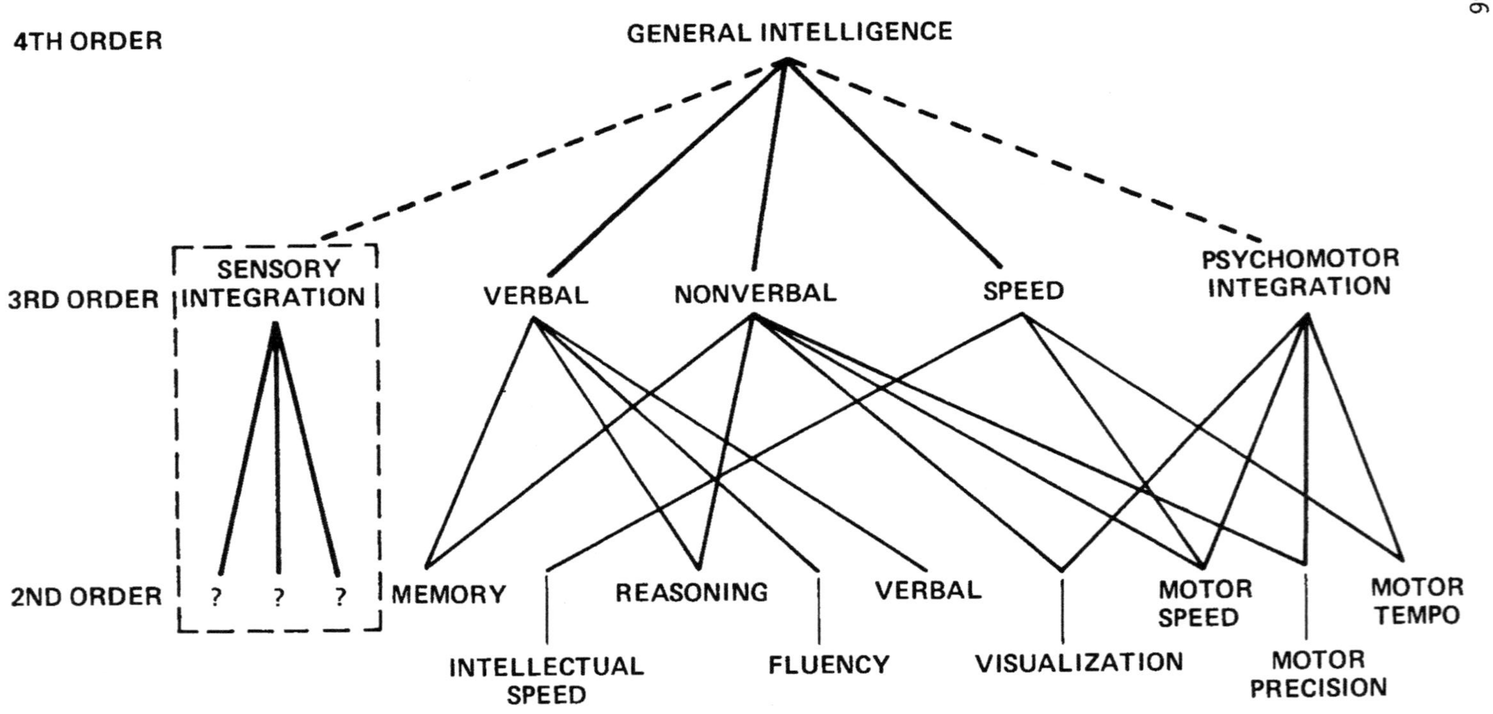

Fig. 4. Tentative synthesis of cognition (Royce, 1973).

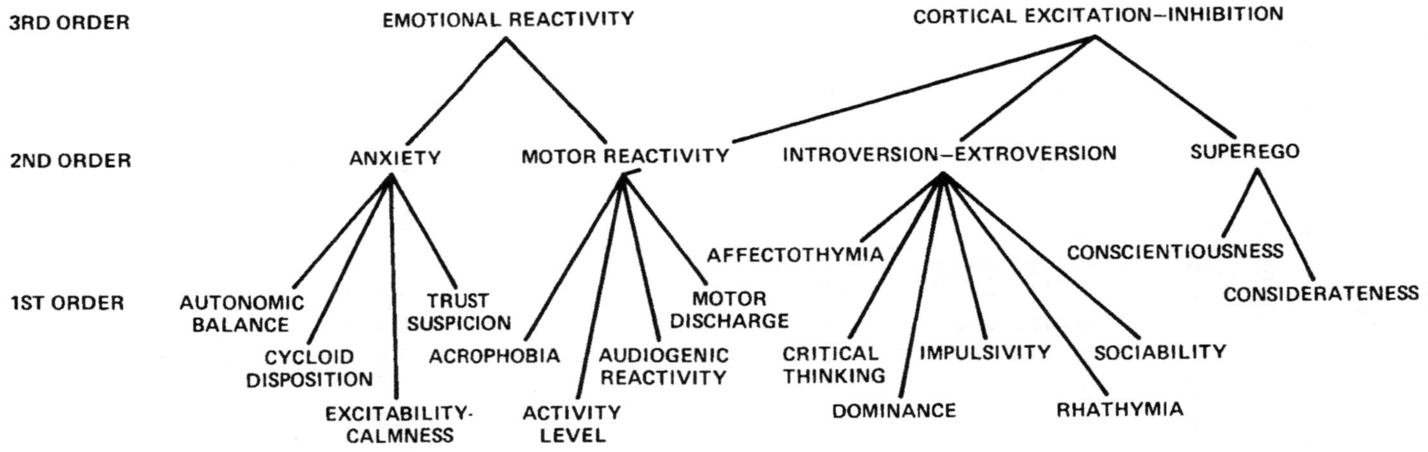

Fig. 5. A conceptual synthesis of temperament and emotionality structure (Royce, 1973).

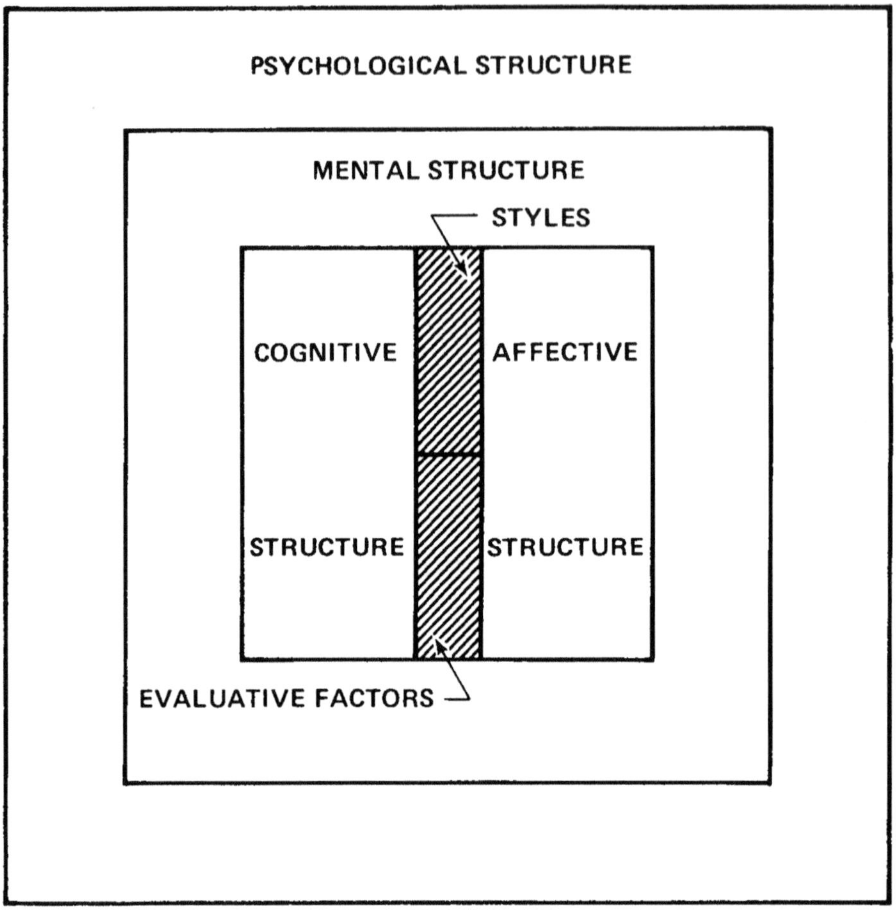

Fig. 6. Relationships between style and evaluative factors (shaded areas) and various structures (Royce, 1973).

in this context, I will postpone further elaboration of it until later. The immediate task is to present a summary of style structure. Unfortunately this cannot be done adequately for the simple reason that the empirical evidence on the factorial definition of these constructs is so limited that it is premature to spell out a comprehensive structure at this time. However, we can present the details for selected styles as examples of the empirical-conceptual linkages we envisage in this domain. These are indicated in Fig. 7.

Extensiveness of scanning, reflection versus impulsivity, and field articulation are examples of cognitive style, affective style,

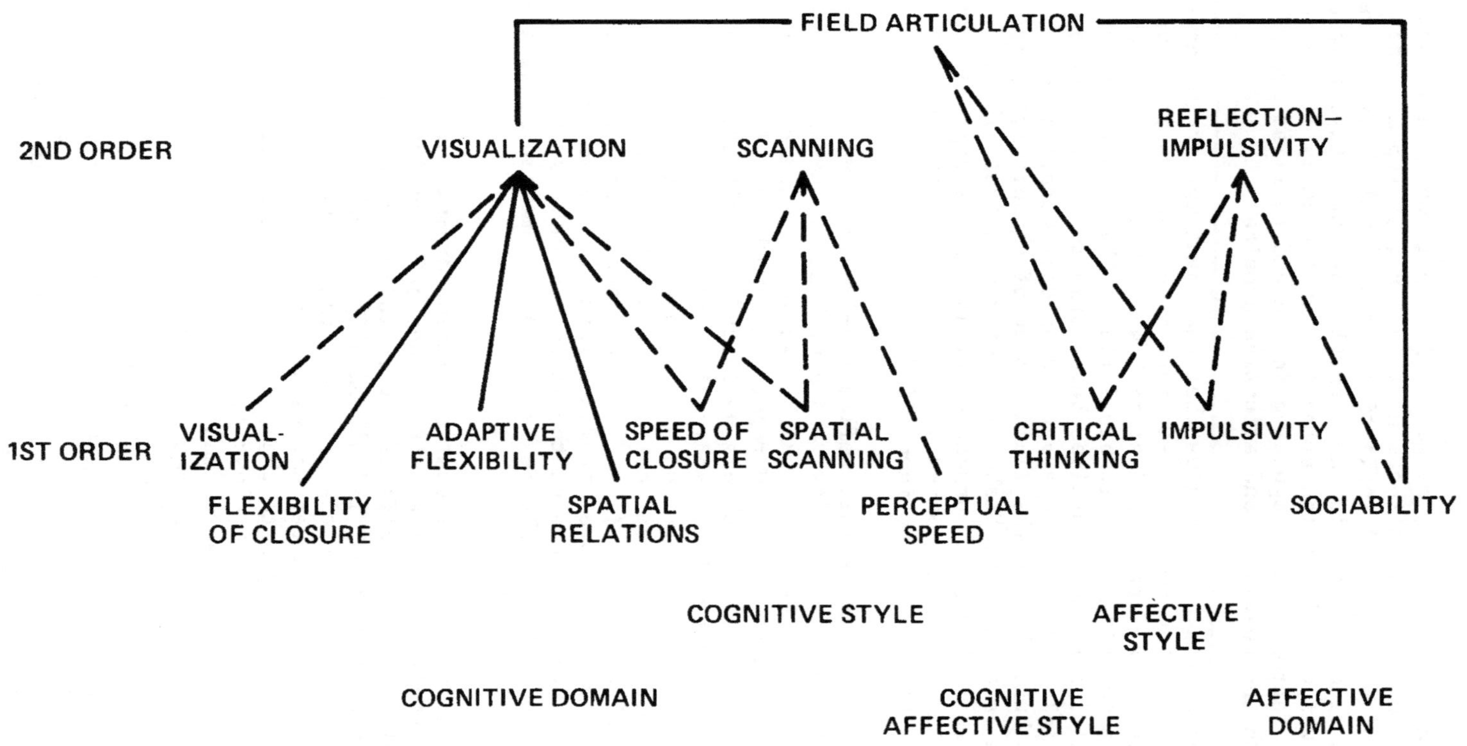

Fig. 7. Postulated linkages of style factors to cognitive and affective structure (solid lines indicate empirically based linkages) (Royce, 1973).

and cognitive-affective style, respectively. While we cannot specify order level of style constructs at this time, it appears that extensiveness of scanning and reflection versus impulsivity are related to first order factors in the cognitive and affective domains, respectively, while field articulation is simultaneously linked to a higher order factor in the cognitive domain and to first order factors in both the cognitive and affective domains.

Inspection of Fig. 7 indicates that the cognitive style factor, scanning, is linked to the three first order factors of speed of closure, perceptual speed, and spatial scanning. All three abilities involve scanning the perceptual field. The most important of these, spatial scanning, involves being able to quickly survey a complex spatial field to find a particular configuration representing a pathway through the field. Speed of closure requires that the field as a whole be "taken in" quickly so that likely material can be "filled into" unseen portions. And perceptual speed requires that the field be scanned quickly so that specified small elements of the pattern can be identified.

The affective style factor, reflection-impulsivity, is linked to critical thinking, inpulsivity, and sociability. The essence of this construct is "the degree to which a subject considers alternative hypotheses in constrast to reporting hypotheses with minimal consideration of their probable validity". The affective factors of critical thinking (i.e., meditative, reflective) and impulsivity (i.e., minimal consideration of alternatives) are obvious temperament counterparts of reflection-impulsivity.[9] Although the relevance of the sociability (i.e., meeting people, active social life) factor is not as obvious, it makes intuitive sense to suggest that it is the impulsivity end of the style continuum which correlates with high sociability. There is, in fact, some empirical evidence that this is the case.[10]

Field articulation is the most convincing example of a cognitive-affective style. Cognitively it focuses on an analytic versus a global way of perceiving; affectively it is concerned with the extent to which an individual is autonomously guided and the degree to which he is affected by social and emotional stimuli. Field articulation is primarily a perceptual rather than a conceptual style, as evidenced by its linkages to the six first order factors of visualization, flexibility of closure, adaptive flexibility, spatial relations, speed of closure, and spatial scanning.[11] These linkages occur via the second order factor of visualization, a general ability concerning the structuring and restructuring of spatial elements. The primary perceptual manifestations of the field articulation style occur via flexibility of closure, adaptive flexibility, and spatial relations (solid line connections in Fig. 7). All three of these draw heavily on abilities required to imagine changes in spatial arrangements and to select out or identify parts embedded in "distracting" stimulus configurations. The affective manifestations occur via the temperamental factors of critical thinking, impulsivity, and especially sociability. The implication is that the field articulate person is autonomous, socially uninvolved, and relatively deliberate.[12]

Many other linkages between styles and affective or cognitive constructs have been found (in varying degrees of adequacy), but much more research is necessary before a definitive style structure becomes apparent. However, it appears reasonable to say that styles are embedded throughout affective and cognitive structures, and probably provide higher order linkages between them.

In addition, we can anticipate a hierarchy of style factors. This might occur in two ways: (1) when style constructs link higher order cognitive and/or affective factors, and (2) when style constructs link other styles. Thus, while we do not know the order level of the three style constructs summarized in Fig. 7, it is probable that scanning and reflection-impulsivity are second order constructs, and that field articulation is a third or fourth order construct.

A similar analysis showing the relevance of the value system is also possible, but, since it is somewhat distant from our primary focus, I will merely allude to this area. Thus, we can define *evaluative structure as a multi-dimensional, organized sub-system of processes (subsumes interests and values) by means of which an organism manifests evaluative phenomena.*

THREE WAYS OF KNOWING

In several recent publications[13] I have outlined a psycho-philosophical scheme for ways of knowing. In particular, I posit three ways of knowing: empiricism, rationalism, and metaphorism. These three "isms" are regarded as basic because of their dependence upon various cognitive processes on the one hand, and their epistemological justifiability on the other hand. This view can be briefly summarized by reference to Fig. 8. The implication here is that each of these "isms" represents a legitimate approach to reality, but that different criteria for knowing are involved. Rationalism, for example, is primarily dependent upon logical consistency. That is, this approach says we will accept something as true if it is logically consistent, and we will reject something as false if it is illogical. Empiricism says we know to the extent we perceive correctly, and metaphorism says that knowledge is dependent upon the degree to which symbolizing cognitions lead to universal rather than idiosyncratic awarenesses. While each of these cognitive processes may lead to error, the implication is that each is also capable of leading to truth. The possibilities of perceptual error, for example, are readily apparent. The errors of the thinking pro-

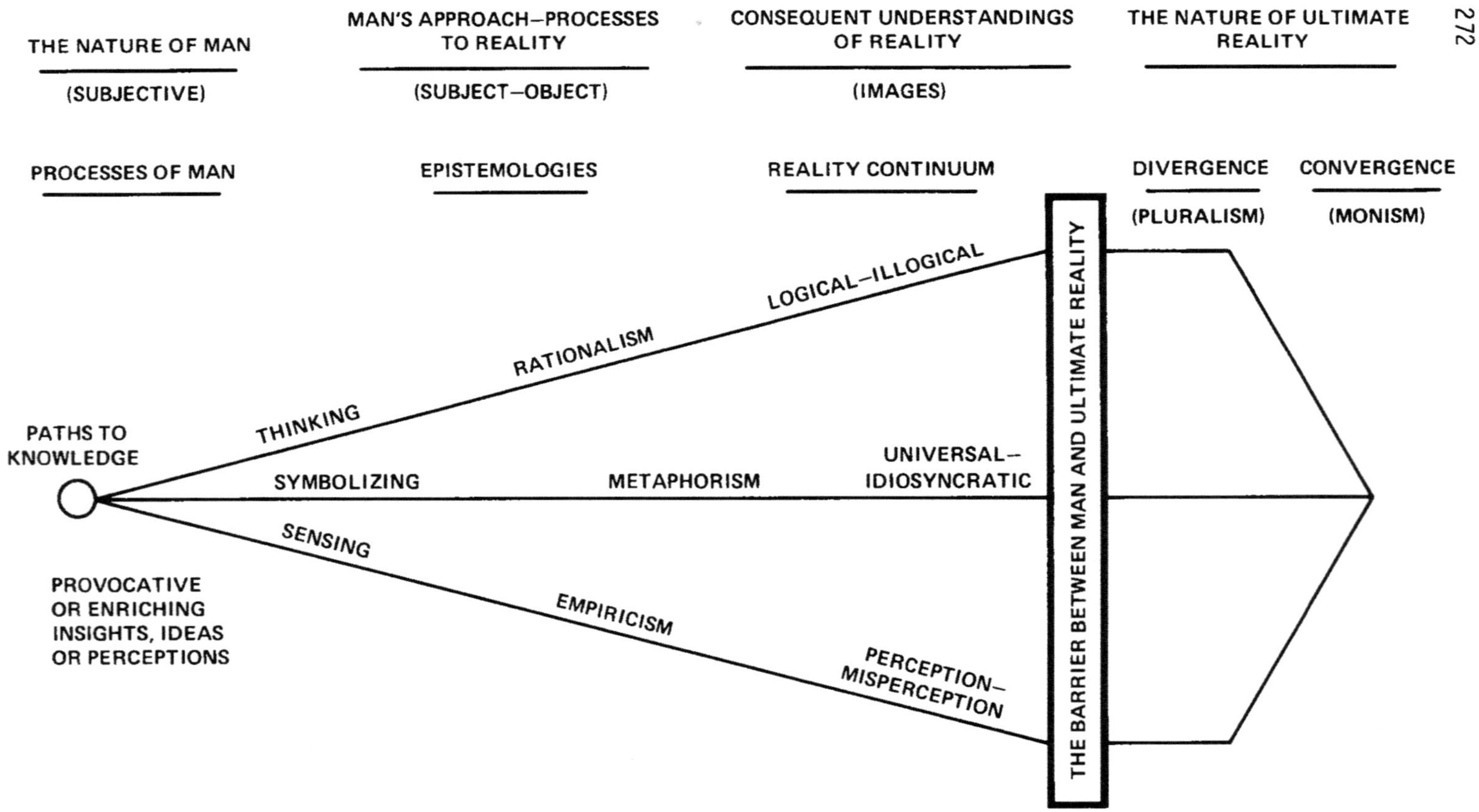

Fig. 8. The basic paths to knowledge (Royce, 1974; modified version of Royce, 1964).

cess are probably more subtle, but I have been led to believe that they have plagued the efforts of the logicians and mathematicians. And the errors of symbolizing are even more elusive, primarily because of the sheer difficulty of providing an adequate articulation of metaphoric knowing (e.g., the problem of what symbols "mean"; what qualifies as "universal"?). Furthermore, we recognize that none of these psychological processes operates independently of the others. That is, one does not think independently of sensory inputs and the process of symbol formation; nor do we perceive independently of thinking. In short, although the correspondences indicated in Fig. 8 are oversimplified for purposes of analysis and exposition, they represent the best fit between a given cognitive class and its parallel epistemological criterion. Although psychologists have just begun the task of evolving an adequate cognitive psychology, it is incumbent upon me to at least indicate what I am alluding to when I employ such terms as thinking, perceiving, and symbolizing. Working definitions of these terms follow.

Thinking: Cognitive processes which focus on concepts - their formation, elaboration, and functional significance to the organism; more deductive than inductive; focus is on the logical consequences of information currently available to the organism.

Perceiving: Cognitive processes which focus on observables - sensory inputs and their "meaning"; more inductive than deductive; focus is on the processing of sensory information.

Symbolizing: Cognitive processes which focus on the formation of symbols - "constructed productions" offered as representations of reality; analogical rather than deductive or inductive; focus is on the processing of "new-formation" (i.e., internally generated forms) rather than "information".

In Fig. 8 I have attempted to show the relationships between man, the knower, and the nature of reality via three ways of knowing. The two columns to the extreme right are separated from the other three columns by a barrier between man and ultimate reality. That which is epistemologically untestable lies to the right of this barrier and constitutes unknowable ultimate reality. That which is testable by some criterion for knowing lies to the left of the epistemological barrier and leads to "reality images" which are "true" or "real". Despite the efforts of great thinkers to somehow circumvent the epistemological limits involved, the only valid assessments open to finite man necessarily lie to the left of the barrier.

EPISTEMIC STYLES AS HIGHER ORDER PERSONALITY INTEGRATORS

It will be recalled that we defined *epistemic styles as those style constructs which simultaneously invoke a valid truth criter-*

ion (i.e., lead to justifiable knowledge claims in addition to being characteristic ways of interacting with the environment). At this juncture it should be apparent that ways of knowing take on psychological-theoretic significance when seen as styles. Although I had used the terminology "ways of knowing" from the outset (around 1955), it was not until recently that I saw the conceptual connections to styles. The long delay in coming to this insight was due primarily to the fact that I saw my early interest in epistemology as a philosophic side trip which had little or nothing to do with psychology. When it slowly dawned on me that specifiable cognitive processes correlate differentially with different epistemologies we had overcome the first big conceptual hurdle. However, the big jump in conceptual awareness came about in the context of another project, the development of a general theory of individual differences. More specifically, I came upon the conception of epistemic styles as part of our elaboration of how styles provide higher order linkages between cognitive and affective dimensions. Thus, ways of knowing were then conceived as those cognitive styles (i.e., epistemic styles) which also lead to valid truth claims. The rational mode or style therefore typically involves thinking processes, whereas the empirical mode or style typically involves perceptual processes, and there may be typical affective process correlates as well.

Given that ways of knowing should now be conceptualized as epistemic styles, the most obvious question to be asked is "How many epistemic styles are there?" I have argued for three,[14] but, of course, this is a contentious issue. There are proponents who argue for less than three - for example, the super-rationalists or the super-empiricists who see only one reality, that of logical consistency or veridical perception respectively. Or, take the typical 20th century victim of "scientism" who will not grant validity to metaphorism. I have dealt with these instances of excluding valid epistemologies as manifestations of psycho-epistemological "encapsulation".[15]

Then there are those who argue for additional ways of knowing - intuitionism, authoritarianism, and mysticism for example. It has been my experience that, whereas the erroneous claims of the "something less" variety (i.e., denying a place for an epistemically valid claim) are usually due to excluding an "alien" cognitive process, the erroneous claims of the "something more" variety are usually due to including an invalid epistemology. Let me briefly elaborate on the examples of intuitionism, authoritarianism, and mysticism by way of making myself clearer. I have rejected intuitionism as an epistemic style because it does not have its own valid epistemology. I do, however, see it as a valid, even though not well understood, mode of cognizing.[16] The equating of cognizing with knowing has been dubbed "psychologism" by the epistemologists. This is a common error of the psychologist. Thus, in my judgment, what happens when an intuition turns out to be correct is that it receives confirmation from some other source - e.g., from the factual source of empiricism, or the logical source of rationalism.[17]

Authoritarianism as a separate and distinct epistemic style fares less well, for it neither has its own cognitive process nor its own epistemic criterion. The authoritarian "epistemology" at its best simply rests its case on some finding which has already been validated (e.g., the facts in a scientific handbook), but it doesn't have a corresponding cognitive process (i.e., "believing" per se is not a cognitive process separate and distinct from thinking, perceiving, and symbolizing, etc.).

The problem with mysticism is that it simply hasn't put forward a viable epistemology. It is typical for the mystic to say that such and such an experience constitutes the epistemic criterion. While the antagonist need not question the validity of experience, it certainly seems reasonable to question the claim that experience is synonymous with knowledge. Experience, after all, is involved in all forms of knowledge. In short, experience is a necessary, but not a sufficient, condition for knowledge.

The question of epistemic style structure now arises. Since it is premature to put forth a style structure at this time, it follows that the situation is even more tenuous in the case of epistemic styles. In fact, all I can do at this juncture is speculate on the possible relationships between epistemic styles and other style constructs. I have done this in Fig. 9.

Let us now shift our attention to higher order factors as personality integrators, with particular attention to the role of epistemic styles. The notion that higher order factors might play an integrative role seems to be a reasonable one, although there is little empirical evidence available to back it up. Let me, however, briefly summarize the hypothesized role of introversion-extroversion as the best established case in point.[18] First of all, it is influenced by a third order factor of cortical excitation-inhibition whose biological underpinnings are thought to be the reticular-activating system. It is hypothesized that introverts have an excess of cortical excitation, whereas extroverts are dominated by cortical inhibition. The theory is that under identical stimulating conditions introverts would be operating in a relatively higher state of cortical arousal than extroverts. Thus, introverts, in general, will not seek out additional external stimulation. Thus, in terms of first order factors (refer to Fig. 5), we would expect introverts (high on cortical excitation) to have low scores on affectothymia, rhathymia, impulsivity, and sociability, and extroverts to show the opposite pattern.[19] Similar pervasive effects could be elaborated in the cognitive domain - in the case of the

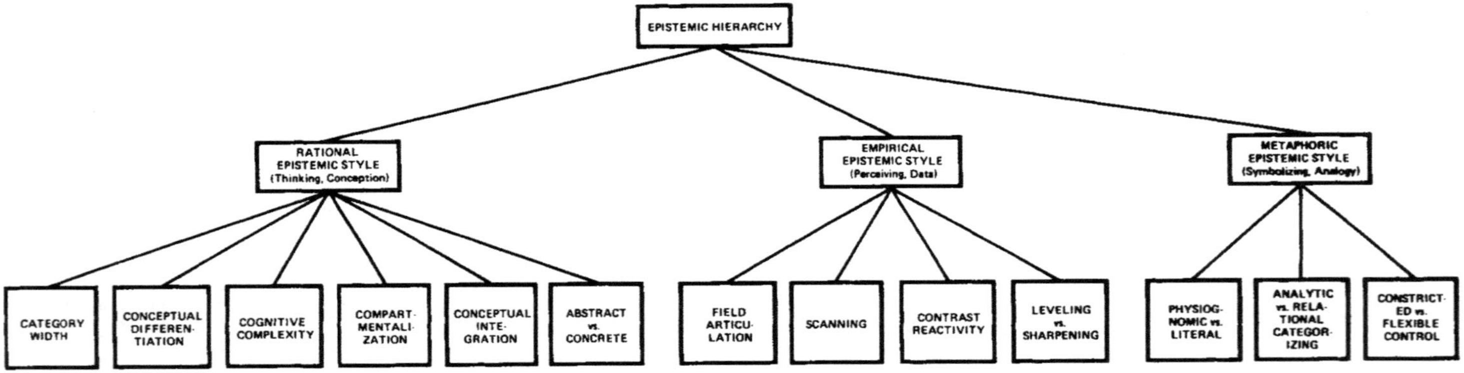

N.B. THE SPECIFIC LINKAGES OF EACH EPISTEMIC STYLE TO COGNITIVE ABILITIES AND AFFECTIVE TRAITS REMAINS TO BE SPELLED OUT (see Figure 7 for examples). HOWEVER, IT IS ANTICIPATED THAT EPISTEMIC STYLES WILL BE PRIMARILY COGNITIVE IN NATURE (i.e., with few affective linkages).

Fig. 9. Hypothesized epistemic style structure.

third order verbal or speed factors, for example, as well as the obvious role of psychomotor integration in coordinating the lower order psychomotor factors.

Even more provocative, however, is the synthesizing potential of the style factors, especially those which carry the potential of linking the abilities and affective traits. Witkin's field articulation construct carries this kind of potential with it. The point is that this style manifests itself both cognitively and affectively, thereby suggesting that such a style also exerts a pervasive influence on behavior.

Thus, I am suggesting that the closer a higher order construct is to the apex of a system hierarchy the greater its potential influence on that class of behavior on the one hand, and the greater its integrative role on the other hand. Furthermore, the deeper a theoretical construct lies in the nomological network,[20] as in the case of style constructs, the greater its synthesizing potential. Thus, when we consider epistemic styles we are looking at theoretical constructs which can be characterized as both higher order and cognitive-affective linkage constructs. And, if we also remind ourselves of the fact that epistemic styles carry the potential of leading to knowledge, we can immediately see its preeminent relevance to an account of world-view.

We can best see these and related relationships by reference to Fig. 10. This figure provides an overview of psychological structure, with the two lower strata of the sub-structures of cognition, styles, values, and affect not shown but implied at the bottom, higher order personality integrators indicated at level III, system typologies given at level IV, the system integrations of world-view and ideologies indicated at level V, and integrated personality indicated at the apex. If we now proceed from the top down, the implication is that one's self concept or overall cognitive-affective balance has *the* most pervasive effect on behavior since all six levels of the dimensional hierarchy are affected by it. At the next level we have the essence of one's cognitive integration indicated in terms of world-view, and the integration of one's affective life linked to the concept of ideology. Thus, ideology will affect one's entire value system as well as affective structure. Similarly, world-view both affects and is affected by cognitive structure (via general intelligence) and style structure (via epistemic hierarchy).

The next level, level IV, constitutes the apex of each of the four sub-structures, and is thereby a very crucial level of psychological integration. For, what is integrated here is nothing less than an entire sub-structure. Let us take the sub-structure of values as an example. Value analysis at level IV refers to a subject's profile of the entire value domain. Since such a dimensional

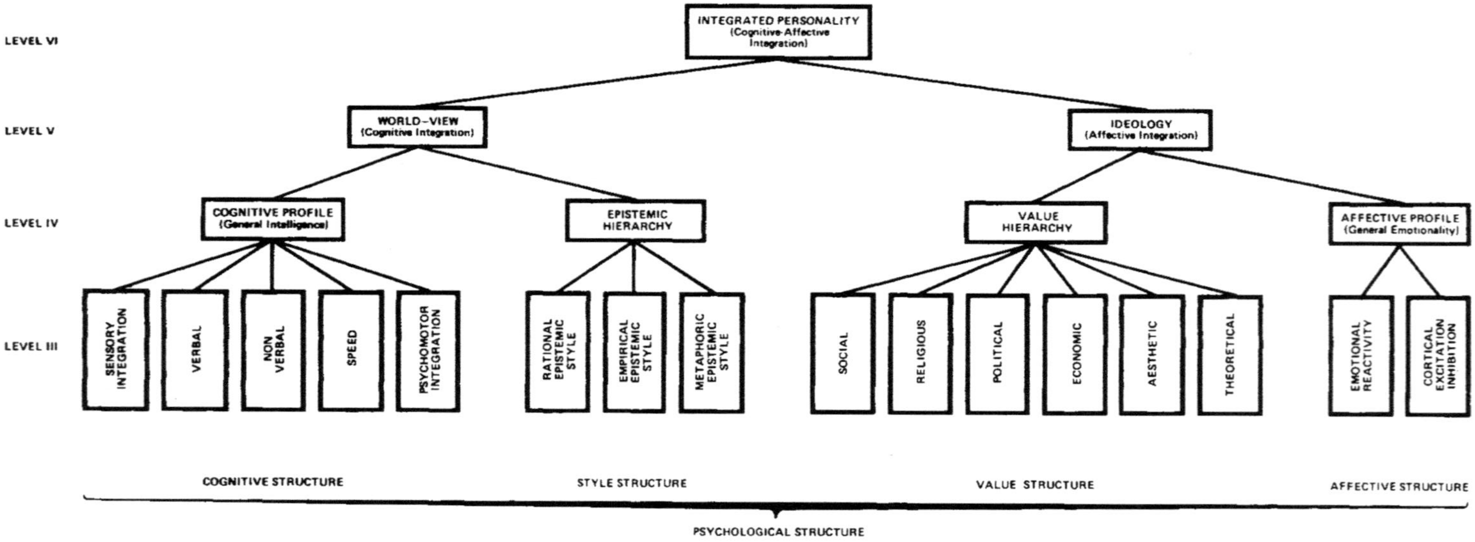

Fig. 10. The hierarchical structure of personality.

profile covers the full spectrum of scales, it also constitutes an ordered series, that is, a hierarchy of values. Further, those subjects who share the same value profile constitute a common class or set, which is one definition of a type. Hence, my designation of this level as the level of *system types*.

In the case of cognition the profile would cover the higher order factors at levels II and III and the primaries at level I, and we would have, in addition, a "general intelligence" score in the form of an I.Q. or its equivalent. How shall we look at the concept of "general intelligence"? For example, is it a pervasive integrator like the factors at levels II and III? Yes and no. In my opinion, the integration that is crucial here takes the form of a hierarchically ordered cognitive profile which constitutes a cognitive type. Furthermore, "general intelligence" turns out to be merely an average of all the components. I now refer the reader to Fig. 11 in order to bring out these points by way of a concrete example. In this figure we show two simplified (i.e., only seven primary factors) cognitive profiles, A (solid line) and B (dotted

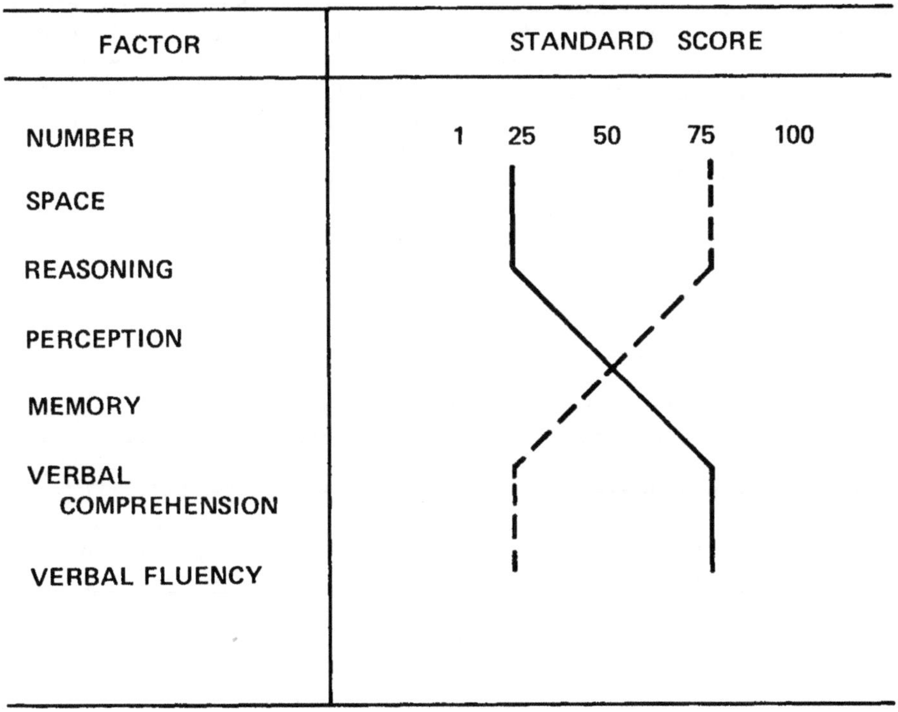

Fig. 11. Showing two persons, A (solid line) and B (dotted line) with the same I.Q., but with opposite mental ability profiles (Royce, 1957).

line), with the same I.Q. of 100 (i.e., standard score of 50), but with opposite mental ability profiles. Thus, from this profile we would conclude that person A is a verbal type and that person B is a quantitative type, and that both of them show the same overall intelligence average. While it is true that an intelligence average is of some value as a reflection of the general intellectuality of a person, it clearly is *not* a unitary theoretical construct or process which has a pervading causal effect on other cognitive constructs. Rather, it is merely an average of one's cognitive profile or type. The *hierarchy of traits* in such a profile, however, is of considerable integrative significance, as it reflects cognitive strengths and weaknesses. Such strengths and weaknesses of cognitive type indicate which cognitive dimensions are optimally available to the subject in his interactions with the environment. Thus, the verbal type will do better with "verbal" tasks and the "quantitative" type will do better with "quantitative" tasks.

However, in this context we are more concerned with style structure and epistemic hierarchy. By epistemic hierarchy we mean a profile of scores on all style constructs, but especially the three epistemic styles indicated at level III. The implication is that there will be a rational epistemic type, an empirical epistemic type and a metaphoric epistemic type. That is, the "rationalist" will manifest a typical style profile, as will the "empiricist" and the "metaphoric". And, because there are also (unknown) relationships between styles and cognitive dimensions, it is to be expected that different sub-sets of these dimensions will also be part of the total psychological profile of the "metaphoric", the "rationalist", and the "empiricist". Thus, the main point is that it is this epistemic hierarchy, in coordination with cognitive profile, which determines world-view.

EPISTEMOLOGY, COGNITION, AND WORLD-VIEW

Further exposition will be facilitated by reference to Fig. 12. The key problems are indicated down the center of the figure, with psychology and its approach indicated on the left, and philosophy and its approach indicated on the right. Knowing is indicated as the first problem. The psychologist's rubric for this class of problem is cognition, and his mode of attack is to provide an empirical description of knowing processes. The philosopher, on the other hand, regards this as epistemology or theory of knowledge, and his mode of attack is to elaborate on truth criteria. Although the approach I take puts greater weight on the cognitive processes of psychology, it is consistent with the philosophic approach in requiring epistemological criteria. Thus, my definition of *knowledge is that it refers to those cognitions of an organism's psychological structure (especially the cognitive and style sub-structures) which are epistemologically justifiable.* If we move to the next level of Fig. 12 we see that several ways or modes of knowing

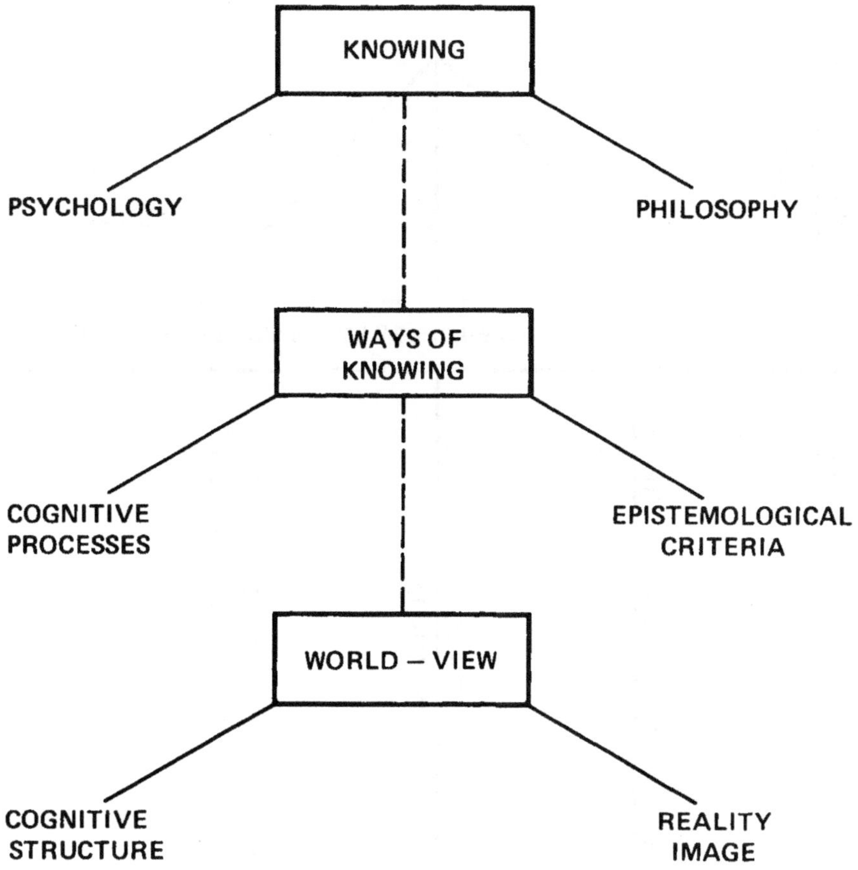

Fig. 12. **Psychological and philosophical aspects of the problem of knowing (Royce, 1974).**

are involved - this implies that knowing in the arts, for example, is not the same as knowing in the sciences. How might they differ? On the philosophic side there may be different truth criteria. Does it make sense, for example, to demand empirical repeatability in evaluating truth claims in literature? And psychologically one might expect differential involvement of cognitive styles and abilities. The psycho-epistemological basis for this state of affairs is depicted in Fig. 13. It is understood that all three epistemic styles or epistemologies are involved in each of the three representative disciplines of knowledge, but it is also clear that each

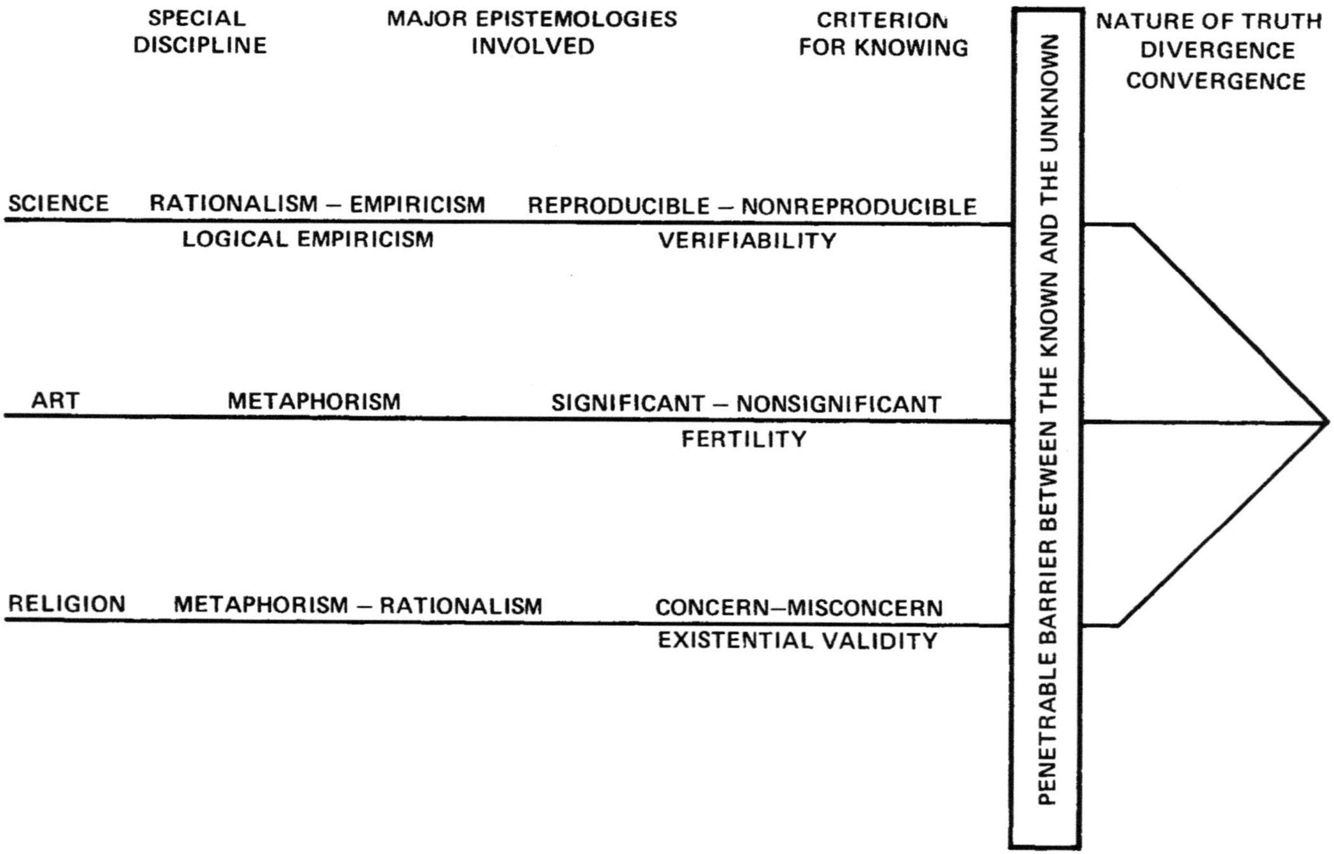

Fig. 13. Representative special disciplines of knowledge (modified version of Royce, 1964).

discipline gives greater credence to one or more of them. In doing science, for example, the scientist "thinks", "symbolizes", and "perceives", but he maximizes the rational and empirical ways of knowing and minimizes metaphoric symbolizing as final judge. Conversely, the artist, who also invokes his entire cognitive repertoire, maximizes the symbolizing process at the expense of the thinking and perceptual processes. There are, of course, wide variations in the possible combinations of epistemological profiles; this brief exposition should be taken as relative and typical rather than as absolute and general.

Although it is minimal, there is empirical evidence to show that people do, in fact, combine these three epistemic styles in different preference orders.[21] For example, the data summarized in Table 1, based on administration of an inventory known as the Psycho-Epistemological Profile,[22] show that empiricism is the dominant characteristic of the Chem-Biology group, that metaphorism is highest for the Music-Drama sample, and that rationalism is highest for the Math-Theoretical Physics group. Furthermore, according to Figs. 8 and 10, differences in cognitive ability profiles also contribute to different world-views, where world-view is defined as "an organism's organized set of personal cognitions which constitutes a model or image of reality (i.e., the way things are)".[23] More explicitly, in Fig. 2, first order factors subsumed under the second order factor of reasoning are likely to be most relevant to the epistemic style of rationalism, first order factors subsumed under the second order factor of visualization are most likely to relate to the epistemic style of empiricism, and the four first order fluency factors are most likely to relate to metaphorism. We are presently conducting several factor analytic studies on epistemic styles which should throw light on the relationships between the three ways of knowing and styles, abilities, and affective traits. In addition, the long awaited 367 variable factor study by Messick and Dermen[24] should greatly enhance our general understanding of style structure.

Thus, the position herein explicated is that personal cognitions (which refer to that subset of cognitions or concepts which tend to verify or systematize reality for a given person) are directly based on cognitive ability and style structures, and indirectly based on affective and value structures. It is further implied that world-views tend to be "encapsulated", i.e., less than the totality of reality. The precise form that encapsulation takes depends primarily upon psycho-epistemological profile; that is, whether rationalism, empiricism, or metaphorism is at the apex of a person's epistemological hierarchy. If rationalism, for example, is dominant this means the resulting world-view is based on an essentially thinking cognitive structure (see Fig. 14A). And if we are considering a case of ultra-empiricism, then we have an example of perceptual dominance (see Fig. 14B). This means the result-

GROUPS	EMPIRICAL SCALE	METAPHORIC SCALE	RATIONAL SCALE
Chem—Bio N = 48	1.223	1.141	1.337
Mu—Drama N = 50	1.102	1.489	1.203
Math—Phy N = 44	1.162	1.142	1.452

Table 1. Group Means on Empirical, Metaphoric, and Rational Scales of the Psycho-Epistemological Profile (Revised Experimental Form II) (Royce, 1974).

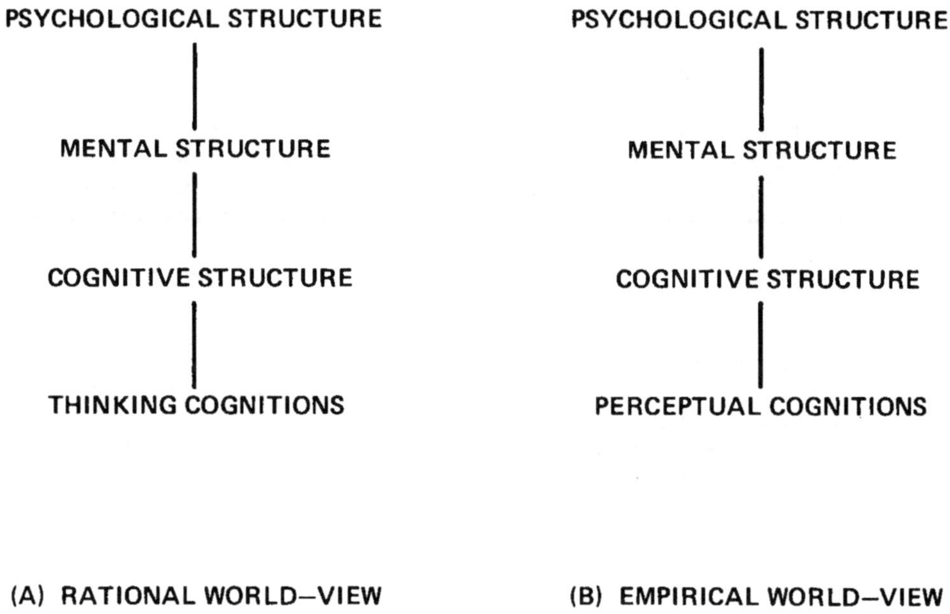

Fig. 14. Cognitive structure and epistemological hierarchies (Royce, 1974).

ing world-view is based primarily on the processes of perceiving, even though some cognitions of the total set emerge from the processes of thinking and symbolizing as well. And we epistemically validate world-views in terms of the appropriate epistemic rule. Thus, in the hypothetical case of "pure", empirical knowing, perceptual cognitions of a world-view are checked against the epistemic rules for accurate perception (e.g., yes, that is a chair). "Pure" rational knowing involves checking thinking cognitions of a world-view against the epistemic rules for logical thinking (e.g., yes, that deduction follows from this particular logical system). And metaphoric knowing involves checking symbolizing cognitions of a world-view against the epistemological requirements of universality (e.g., yes, Mr. X's play, written 500 years ago, and still produced in various parts of the world, transcends time and space to a relatively high degree).

In conclusion, it seems reasonable to say that we get variations in world-view because of variations in ways of knowing (epistemic styles), and variations in the trait profiles of knowers (individuality). Could this variety of world-views also be due to variations in what is to be known (i.e., a pluralistic ontology)?

Appendix A. Brief Definitions of Cognitive Factors

First Order Intelligence Factors

S: Spatial Relations (CFS). Ability to put together by visual imagination parts that are out of a customary place in a visual pattern and to identify such "out of place" percepts.
P: Perceptual Speed (EFU). Speed in identifying specified, small elements of a visual pattern.
I: Induction (CSS & CSC). Ability in forming and testing hypotheses directed at finding a principle of relationship among elements and applying the principle to identify an element fitting the relationship.
N: Number (MSI & NSI). Speed and accuracy in doing the basic operations of arithmetic.
V: Verbal Comprehension (CMU). Facility in understanding of English words, sentences, and paragraphs.
Fw: Word Fluency (DSU). Facility in producing words in accordance with structural restrictions but without regard to meaning.
Ma: Associative Memory (MSI). Upon presentation of one part of previously associated but otherwise unrelated material, ability to recall another part.
Cs: Speed of Closure (CFU). Ability to "take in" a perceptual field as a whole, to "fill in" unseen portions with likely material and thus to coalesce somewhat disparate parts of a visual percept.
Cf: Flexibility of Closure (NFT). Ability to "hold in mind" a particular visual percept (configuration) and find it embedded in distracting material.
Vi: Visualization (CFT). Ability to manipulate visual percepts (to imagine change in forms) and thus to "see" how things would look under altered conditions.
Fi: Ideational Fluency (DMU). Ability to quickly produce ideas and exemplars of an idea about a stated condition or object.
Fe: Expressional Fluency (DMS). Facility in finding an appropriate word or set of words to make a proper English expression.
Fa: Associational Fluency (DMR). Facility in producing (English) words having somewhat similar meanings.
R: General Reasoning (CMS). Ability in organizing the relevant aspects of problems having an algebraic quality and reasoning through to find solutions for the problems.
Xa: Figural Adaptive Flexibility (DFT). Ability to try out in imagination various possible arrangements of the elements of a visual pattern and thus to converge on one arrangement which satisfies several stated criteria.
D: Deduction (EMR & EMI). Reasoning from the general to the specific; the ability to test the correctness of a meaningful conclusion by applying general principles to the individual case.
Ss: Spatial Scanning (EFI). Ability to quickly survey a complex

field to find a particular configuration representing a pathway through the field.

Re: Semantic Redefinition (NMT). Ability to imagine different functions for objects or parts of objects and thus to use them in novel ways to accomplish stated purposes.

Rs: Syllogistic Reasoning (NMI). Ability in formal reasoning from stated premises to rule out nonpermissible combinations and thus to arrive at necessary conclusions.

Sep: Sensitivity to Problems (CMI). Ability to imagine problems associated with function or change of function of objects and to suggest ways to deal with these problems.

Xs: Spontaneous Flexibility (DMC). Facility in imagining diverse functions and classifications for objects.

O: Originality (DMT). Facility in conceptualizing phenomena in ways that in our culture are judged to be unusual and clever.

Ms: Memory Span (MSS). Capacity in number of distinct elements that can be maintained within the span of immediate awareness.

First Order Psychomotor Factors

Speed of Arm Movements - Rate of arm movement after it has been initiated.

Wrist-Finger Speed - Speed of simple movements of the wrist and/or fingers particularly repetitive movements which involve little, if any, eye-hand coordination.

Reaction Time - Speed of simple reaction to auditory or visual stimuli.

Aiming - Ability to execute quick and precise hand-finger movements with emphasis on good eye-hand coordination.

Articulation Speed - Rate of reading and speed of repeatedly articulating the same consonant.

Ambidexterity - Ability to use the nonpreferred hand.

Psychomotor Coordination - Ability to achieve degree of adjustment of limb (finger, hand or foot) motions in response to visual stimuli.

Manual Dexterity - Speed and coordination of arm-hand movements in skillful manipulation of objects.

Finger Dexterity - Speed and accuracy in manipulating small objects with the fingers.

Steadiness - Ability to perform very precise arm-hand movements with little to no emphasis on speed.

Tempo of Large Movements - Natural speed with which motor activities of limbs and trunk are carried out.

Tempo of Fast Movements - Natural speed with which quick motor activities are carried out.

Tempo of Hand Movements - Natural speed with which hand movements are carried out.

Second Order Intelligence Factors

Memory - Ability to recall previously associated related or un-

related material that is meaningful or nonmeaningful, verbal or nonverbal.
Intellectual Speed - The speed with which each correct item is solved, length of time devoted to each abandoned item (persistence or continuance) and the number of wrong items and time spent on each.
Reasoning - Ability to organize a divergent or convergent problem that presents itself in a symbolic or nonsymbolic form.
Fluency - Facility in retrieving language elements, i.e., words or ideas, from long-term storage to immediate awareness.
Verbal - A facileness in acquiring and utilizing meaning.
Visualization - Ability to imagine the way objects change as they move in space, maintain spatial orientation, perceive the Gestalt in a field of elements, maintain flexibility in perception concerning structuring and restructuring elements in space.

Second Order Psychomotor Factors

Motor Speed - Rate of initial bodily movement or movement after it has been initiated.
Motor Precision - Ability to hold the body in static positions or a prescribed sequence of movements as defined by a task.
Motor Tempo - The natural speed with which motor activities are carried out.

Third Order Cognitive Integrative Factors

General Verbal - Ability to acquire and utilize language elements, i.e., words, symbols, ideas, meanings, etc.
General Nonverbal - Ability to interact and structure nonverbal elements of the environment within the context of a defined problem or goal.
General Speed - Ability to respond quickly and efficiently to the demands of a task.
General Psychomotor Coordination - Ability to coordinate movement of parts of the body to overt or covert visual stimuli.
General Sensory Integration - Ability to integrate impinging stimuli across sensory modalities.

Fourth Order Cognitive Integrative Factor

"g" - A broad, diffuse, integrative capacity to sense, cognize, and manipulate stimuli. Not the same as Spearman's "g" since that factor is postulated as one which is common to all measures of intelligence.

Appendix B. Brief Definitions of Affective Factors

First Order Affective Factors

Acrophobia - Emotional reactivity where the stressful stimulus involves height.
Activity Level - Locomotion in a variety of situations where the external stimulus is not particularly stressful and internal stimuli are major contributors (e.g., hormone influence).
Affectothymia - Warmhearted, outgoing, easygoing, participating.
Audiogenic Reactivity - Emotional reactivity where the stressful stimulus is of an auditory nature.
Autonomic Balance - A response pattern under stressful stimulation where a major role is attributed to gland and smooth muscle functioning. A typical manifestation in rodents is "emotional elimination".
Conscientiousness - Persistent, moralistic, responsible, dominated by a sense of duty.
Considerateness - Quiet, nonaggressive, socially retiring, considerate of the feelings of others.
Critical Thinking - Reflective, meditating.
Cycloid Disposition - Moody, moods or energy level shift suddenly and without apparent reason.
Dominance - Assertive, outspoken, socially aggressive.
Excitability - Easily given to emotionality or worry, as opposed to calm, emotionally mature.
General Arousal - Tempo with which individual engages in an activity, stamina.
Impulsivity - Tendency toward acting without stopping to think about the consequences; acting on the spur of the moment.
Masculinity-Femininity - Unsentimental, self-reliant, tough-minded, practical; as opposed to sensitive, dependent, tender-minded.
Motor Discharge - A more specialized variety of emotional reactivity, describing response patterns under stress involving primarily skeletal muscles. This pattern in rodents is typically referred to as "freezing".
Rhathymia - Carefreeness, happy-go-lucky disposition.
Sociability - Enjoyment of social contacts, a preference for much social activity.
Trust-Suspiciousness - Accepting of behavior of others and of events, unsuspecting of hostility; as opposed to jealous, suspicious of interference.

Second Order Affective Factors

Anxiety - Feelings of discomfort, emotionality, and troubledness, often accompanied by feelings of inferiority; mood swings, lack of concentration, etc.
Introversion-Extraversion - Retiring from social contacts, preference toward relative solitude; as opposed to seeking out contacts

with people, enjoying being the "life of the party", leading an active life.
Motor Reactivity - Any response or set of responses involving the skeletal muscle system; of a more voluntary nature than elicited responses involving glands and smooth muscles.
Superego - Conscientiousness as a second order factor.

Third Order Affective Factors

Emotional Reactivity - Any response or set of responses to a stressful stimulus. The response may be elicited (e.g., UCR) or it may be of a more voluntary nature (e.g., escape-avoidance).
Cortical Excitation-Inhibition - A gross, molar influence stemming from the reticular activating system which either facilitates (excitation) or suppresses (inhibition) certain cognitive and affective processes.

Appendix C. Definitions of Styles

Field Articulation - An analytical, in contrast to a global, way of perceiving; entails a tendency to experience items as discrete from their backgrounds, and reflects ability to overcome the influence of an embedding context.

Extensiveness of Scanning - Extensive scanning describes a relatively stable disposition to attend to tasks intensely and in a focused manner, yet with extensive coverage of relatively incidental aspects of the field.

Conceptual Differentiation - "In contrast to category-width tests, in which each item assesses the limits of one conceptual realm, free sorting tests require the spontaneous differentiation of heterogeneous items into a complex of more or less related groups". Thus, this style refers to differentiation between concepts, while category width (or equivalence range) refers to distinctions made within a concept's dimensionality.

Leveling versus Sharpening - A tendency to be hypersensitive to minutiae, to respond excessively to fine nuances and small differences, to exaggerate change, and to keep adjacent or successive stimuli from fusing and losing identity. Levelers are characterized by maximal assimilation effects, and by memory organizations in which the fine shades of distinctions among individual elements are lost.

Tolerance for the Unconventional - Acceptance of experiences which do not agree with what one knows to be true. Reflected on the Rorschach in reluctance to project or fantasize, in refusals to attribute to the blots qualities that were known not to be there, and in difficulty in adopting an "as if" attitude ... in the apparent movement test.

Constricted versus Flexible Control - Terms ... to describe differing reactions to stimulus fields containing contradictory or intrusive cues ... constricted-control subjects resorted to counteractive measures in their attempts to overcome the disruptive effect of intrusive cues ... Flexible-control subjects seemed relatively comfortable in situations that involved contradictory or intrusive cues.

Category Width or Equivalence Range - The degree to which subjects are impelled to act on or ignore an awareness of differences. A narrow equivalence range seemed to imply detailed categorization of certain aspects of experience. "Narrow-range" subjects had relatively exact standards for judging similarity. "Broad-range" subjects, on the other hand, were less finicky about fine stimulus differences, and grouped stimuli into broader categories.

Reflection versus Impulsivity - Refers to the degree to which a subject considers alternative hypotheses in contrast to reporting hypotheses with minimal consideration of their probable validity.

Cognitive Complexity - Persons of great "cognitive complexity" presumably make more, and more complex, distinctions between persons and events ... Conceptual differentiation in categorizing can

be achieved as of an aspect of cognitive complexity.
Analytic versus Relational Categorizing - Whether one forms "analytic-descriptive" concepts (based on similarity in objective elements within a stimulus complex, that were part of the total stimulus) or "inferential-categorical" concepts ("involve an inference about the stimuli grouped together") as opposed to "relational concepts" ("based on a functional relationship between or among the stimuli grouped together").
Compartmentalization - Measured by the number of miscellaneous objects left ungrouped in a sorting task, and identified as "a tendency to compartmentalize ideas ... in discrete categories ... a possible limitation in the production of diverse ideas".
Conceptual Integration - The relating or hooking of such parts (concepts) to each other and to previous conceptual standards.
Physiognomic versus Literal - (For the physiognomic person) percepts are often subtly suffused with emotional or expressive qualities. Inanimate objects or events seem to move, become motivated and assume expressive and "human" areas ... All these experiences involved a preference for the dynamic and emotive rather than for the static and literal.
Abstract versus Concrete - The more concrete, the more the structure is assumed to be restricted to, or dependent upon, physical attributes of the activating stimulus. The ability to form concepts.
Contrast Reactivity - The degree to which individuals appreciate and respond to the contrast between stimuli that vary greatly.

References

1. Royce, J.R., *The Encapsulated Man: An Interdisciplinary Essay on the Search for Meaning*, D. van Nostrand, Princeton, N.J., 1964.

2. Royce, J.R., Kessel, F.L., Coward, H., Egan, E. and Mos, L., *Epistemological Psychology: A Critical Review*, 1975 (in preparation).

3. Royce, J.R., Cognition and knowledge, in *Handbook of Perception*, Carterette, E.C. and Friedman, M.P., Eds., Academic Press, New York, 1974 (in press).

4. Royce, J.R., The conceptual framework for a multifactor theory of individuality, in *Multivariate Analysis and Psychological Theory*, Royce, J.R., Ed., Academic Press, London, 1973 (in press); and Royce, J.R. and Buss, A., *A Multi-Factor Theory of Individuality*, 1975 (in preparation).

5. Royce, J.R. and Razeboom, W., *The Psychology of Knowing*, Gordon and Breach, New York, 1972.

Royce, J.R., Cognition and knowledge, in *Handbook of Perception*, Carterette, E.C. and Friedman, M.P., Eds., Academic Press, New York, 1974 (in press).

Royce, J.R. and Tennessen, H., Eds., *Inquiries into a Psychological Theory of Knowledge*, 1975 (in preparation).

Royce, J.R., Kessel, F.L., Coward, H., Egan, E. and Mos, L., *Epistemological Psychology: A Critical Review*, 1975 (in preparation).

6. Royce, J.R., Factors as theoretical constructs, *American Psychologist*, 18, 522, 1963.

7. Harman, H., *Modern Factor Analysis*, 2nd ed., University of Chicago Press, Chicago, 1967.

8. Thurstone, L.L., *Multiple Factor Analysis; A Development and Expansion of the Vectors of the Mind*, University of Chicago Press, Chicago, 1947.

9. Kagan, J., Moss, H.A. and Sigel, I.E., Psychological significance of styles of conceptualizations, *Monographs of the Society for Research in Child Development*, 28, 73, 1963.

10. Kagan, J., Impulsive and reflective children: Significance of conceptual tempo, in *Learning and the Educational Process; Selected Papers from the Research Conference on Learning and the Educational Process*, held at Stanford University, 1964, Krumboltz, J., Ed., Rand McNally, Chicago, 1965.

11. Gardner, R.W., Jackson, D.N. and Messick, S.J., Personality organization in cognitive controls and intellectual abilities, *Psychological Issues*, 2, No. 4, 1960.

Witkin, H.A., Dyk, R.B., Faterson, H.F., Goodenough, D.R. and Karp, S.A., *Psychological Differentiation*, Wiley, New York, 1962.

12. Witkin, H.A., Dyk, R.B., Faterson, H.F., Goodenough, D.R. and Karp, S.A., *Psychological Differentiation*, Wiley, New York, 1962.

Elliot, R., Interrelationships among measures of field dependence, ability and personality traits, *Journal of Abnormal and Social Psychology*, 63, 27, 1961.

Evans, F.J., Field dependence and the MPI, *Perceptual and Motor Skills*, 24, 526, 1967.

13. Royce, J.R., *The Encapsulated Man: An Interdisciplinary Essay on the Search for Meaning*, D. van Nostrand, Princeton, N.J., 1964.

Royce, J.R., Ed., *Psychology and the Symbol: An Interdisciplinary Symposium*, Random House, New York, 1965.

Royce, J.R., Metaphoric knowledge and humanistic psychology, in *Challenges of Humanistic Psychology*, Bugental, J.F.T., Ed., McGraw-Hill, New York, 1967.

Royce, J.R., The present situation in theoretical psychology, in *Toward Unification in Psychology*, Royce, J.R., Ed., University of Toronto Press, 1970.

Royce, J.R., Cognition and knowledge, in *Handbook of Perception*, Carterette, E.C. and Friedman, M.P., Eds., Academic Press, New York, 1974 (in press).

14. Royce, J.R., Cognition and knowledge, in *Handbook of Perception* Carterette, E.C. and Friedman, M.P., Eds., Academic Press, New York, 1974 (in press).

15. Royce, J.R., *The Encapsulated Man: An Interdisciplinary Essay on the Search for Meaning*, D. van Nostrand, Princeton, N.J., 1964.

16. Westcott, M.R., Toward a contemporary psychology of intuition; a historical, theoretical and empirical inquiry, in *The Psychochology of Intuition*, Holt, Rinehart and Winston, New York, 1968.

17. Royce, J.R., Cognition and knowledge, in *Handbook of Perception*, Carterette, E.C. and Friedman, M.P., Eds., Academic Press, New York, 1974 (in press).

18. Eysenck, H.J., Ed., *Readings in Extraversion-Introversion*, 1-3, Wiley-Interscience, 1971.

19. Royce, J.R., The conceptual framework for a multifactor theory of individuality, in *Multivariate Analysis and Psychological Theory*, Royce, J.R., Ed., Academic Press, London, 1973 (in press).

20. Royce, J.R., Factors as theoretical constructs, *American Psychologist*, 18, 522, 1963.

21. Royce, J.R., *Test Manual for the Psycho-Epistemological Profile*, University of Alberta, 1970.

Royce, J.R. and Smith, W.A.S., A note on the development of the psycho-epistemological profile (PEP), *Psychological Reports*, 14, 297, 1964.

Smith, W.A.S., Royce, J.R., *et al.*, The development of an inventory to measure ways of knowing, *Psychological Reports*, 21, 529, 1967.

22. Royce, J.R., *Test Manual for the Psycho-Epistemological Profile*, University of Alberta, 1970.

23. Royce, J.R., Cognition and knowledge, in *Handbook of Perception*, Carterette, E.C., and Friedman, M.P., Eds., Academic Press, New York, 1974 (in press).

24. Messick, S. and Dermen, D., *Sex Differences in Cognition and Personality Structure*, (in preparation), 1974.

THE IMPACT OF THE ESSENTIALLY PRAGMATIC CHARACTER OF NATURAL LANGUAGES ON LINGUISTIC INFORMATION PROCESSING

Y. Bar-Hillel

The Hebrew University of Jerusalem, Jerusalem, Israel.

Some of what I shall be saying will remind you of what has already been said by other participants in our Institute, whether in lectures or during discussions, but I hope you may discern some difference in tenor. For instance, you have already been reminded of the three medieval *artes sermonicales*, that formed the *trivium* of classical education: Dialectics, Grammar and Rhetorics, the respective arts of sound reasoning, correct talking, and efficient convincing. In due time rhetorics deteriorated into a bag of tricks for demagogic persuasion, on the one hand, and a collection of figures of speech, on the other, and receded from its central position in the education of gentlemen. Dialectics and grammar merged in the 17th century into the dream of a universal characteristic, the ideal philosophical language, with its foolproof grammar, its self-explanatory vocabulary, and its mechanical means for testing validity of arguments - the perfect tool for replacing thinking by calculation. (Many of you must have dreamt this intoxicating dream yourselves, and some information scientists I know never awakened from it.)

I could have gone on and told you some more about the awakening of the linguists, in the beginning of the 19th century, from their sweet dogmatic slumber, about the fateful disdain in which logic (the modern term for dialectics) was held by them until the mid-twentieth century, but I have told this fascinating, sad story elsewhere,[1] and you can find it now in every good book on the history of logic or linguistics.

Almost exactly a century ago, the American philosopher Charles Sanders Peirce revived the medieval trivium, giving it a new twist, and combined it with his revival of the Stoic science of *semiotics*, the science of signs and signification. Thirty-five years ago, another American philosopher, Charles Morris, re-revived Pierce's scattered thoughts and gave them a modern systematization on behavioristic foundations, coming up with his famous tripartite division of semiotics into *pragmatics, semantics* and *syntactics*, in this order of abstraction. To repeat in slogan terms what is surely well known to everyone of you: in pragmatics, the language-users, the communicants, the speaker/hearers of Chomskyan fame, are fully and explicitly taken into consideration, in semantics they are abstracted from, still explicitly countenancing the relationships between signs and objects, while in syntactics, finally, even these relations are put aside, and only properties of, and relations between, the signs are considered.

Pragmatics is clearly a messy business, and one can easily understand why tidy thinkers should have wanted to keep away from it and perform the abstractive step into semantics at their earliest convenience. It is somewhat more difficult to understand why linguists and logicians alike, simultaneously though independently, should have been so eager to perform the second abstraction and try to move as quickly as possible into what seemed to them to be the greener pastures of syntactics, in the twenties and early thirties of our century. After all, if you disregard the relations between word and object, between language and word, don't you take the soul away from the linguist's enterprise, or the logician's, for that matter?

Well, for reasons which it would take too long to describe here in detail, but which again can be looked up in the literature, the problem of the relations between language and word was regarded in those days as metaphysics-laden, dangerous, and seductive, from which the sober, positivistic scientist should therefore keep away.

In 1935 it became known that Alfred Tarski, the Polish logician, had been able to provide a solid, sober, seemingly metaphysics-free foundation for a central part of semantics, the theory of truth (though only for *constructed formalized language systems*, not for natural languages).[2] Rudolf Carnap, among others, eagerly accepted Tarski's revelations and immediately began to develop logical semantics. Only the year before Carnap had published, in German, *The Logical Syntax of Language*,[3] in which he had tried to show in addition to many other insights, how semantics can, without loss, be replaced by (logical) syntax - again with regard to constructed language systems. In the same year, Morris visited Carnap in Prague, and, in addition to successfully persuading him to leave Nazi-threatened Czechoslovakia for the U.S.A., convinced him of the basic role of pragmatics in natural communication systems. Still, initially Carnap paid not much more than lip-service to pragmatics, and never really wanted to deal himself with natural languages as such, with their chaotic messiness;[4] but in his later life he not only welcomed my attempts[5] to convince linguists (and information scientists, for that matter) of the vital importance of logical semantics[6] (as developed by Tarski, Carnap, Reichenbach, and Quine, for instance) but even himself published two papers dealing with pragmatic matters.[7] True enough, my crusade of 1954[8] (opposed at the time by none other than Noam Chomsky, in his second publication ever[9]) remained almost unnoticed for more than a decade, when all of a sudden many linguists and logicians all over the world (though in particular in the U.S.A., U.S.S.R., and Germany) became increasingly convinced of the vital role logic is bound to play in the development of linguistic semantics. Again, I shall refrain from discussing here this interesting (and, to me, highly gratifying) development and turn to our topic proper.

In the late fifties, I became more and more convinced that substantial advances in the utilization of computers in the information processing field (as well as in the transformation of data into information - as this process has repeatedly been called in our Institute, i.e., the organization of these data in some substantial theoretical framework) can only be achieved if and when the computers are equipped to understand the content of the data they are supposed to handle, whether for indexing, abstracting, extracting, or - most difficult of all - for translating. At that time I thought that the major obstacle to obtaining this goal resided in the fact that we are unable to formulate and formalize the procedures by which humans are able to disambiguate a syntactically and/or semantically ambiguous sentence, when uttered in a given (linguistic) cotext and a given (situational) context. My 1961 "proof" of the non-feasibility of high-quality fully automatic translation[10] made a considerable impression in its time and has since been often quoted (and misquoted) as having decisively exemplified "What computers can't do".[11]

During the past decade, no serious argument has been proffered to make me change my evaluation (though I would by no means agree with all the "humanistic" conclusions that have occasionally been drawn from it). On the contrary, I think I now see much clearer the difficulties that any substantial computerized information processing and retrieval system will have to overcome. It now seems to be generally recognized that such a system will require, among others, a man-computer language that will have approximately the same expressive power as a natural language. But it is by no means generally recognized that this expressive power is based on the essentially pragmatic character of natural language communication, on the constant and boundless reliance of such communication on cotext and context. Nor is it generally recognized that this expressive power is based on the constant utilization of channels, - to use, for once, communication-theoretical terminology - additional to the auditory or visual ones, supposedly exclusively used in linguistic communication. Nor that it is based on a constant reliance on so much background knowledge in the hearer (or reader), and on a substantial reliance on his goodwill.

In order to drive this point home as effectively and quickly as possible, I know of no better way than to ask you - as I often have before on similar occasions[12] - to consider the following thought experiment. Imagine that in 1984, one of the first actions of the newly established world government will be to prohibit any use whatsoever of cotext-dependency (e.g., of such *anaphoric* words as "he", "it", "this", "the former", etc.) of context-dependency (e.g., of such *deictic* words and phrases as "I", "you", "we", "here", "then", "yesterday", "five minutes ago", "at the far end of this room", tensed verbs, etc.) and, for good measure, also of goodwill. In other words, only such languages will be allowed thereafter as will look like the computer languages developed until very recently. (The only computer language known to me that allows for substantial cotext and context-dependency - though still to an extremely restricted degree only - is the one used by Terry Winograd in his dissertation of 1970,[13] which has been mentioned by Licklider in his presentation.) Since I am afraid that it must take a few minutes, if not hours, to think through what these limitations really mean and, put positively, how much additional power is gained by being free of them in natural languages, over the customary constructed language-systems, whether invented by logicians or computer-people - artificial languages like Esperanto belonging in this respect, of course, in the same league as natural languages - let me draw the conclusions for you: *communication*, as we know it, *would come to a complete standstill; humanity would cease to exist as such.* True, textbooks could perhaps be written in this Newspeak without using deictic and anaphoric expressions - though I wouldn't dare to estimate the effort it would cost to do this - but can you really envisage textbooks, not to mention journals, newspapers, and ordinary conversation, that would not presuppose, heavily and irreplaceably, vast amounts of background knowledge and goodwill?

It was Winograd's memorable achievement that he was able to equip his computer with an eye and a hand, or rather to simulate this equipment on the computer, turning it into an intelligent proto-robot, thereby enabling the computer to understand some of the simple spatial deictic expressions; to have it memorize preceding sentences of the discourse, thereby enabling it, together with a reasonable set of appropriate rules, to handle simple anaphora, even when they exceeded sentence boundaries, and to program into it certain rules of goodwill, thereby enabling the computer to reply to certain inherently ambiguous commands with "I presume you meant me to do A" and to carry out A, rather than to react with a sullen "Sorry, your request was ambiguous; could you please specify?" Winograd's program equipped the computer not only with a certain linguistic competence - this had been done before, though he was able to improve considerably on previous programs even here - but also with certain rudiments of what we might call "communicative competence". The price he had to pay for this substantial increase in the interactive competence of the computer was, as he fully realized, enormous. The field of interaction was in infinitesimal part of what a serious interactive information system - we can no longer call it a *linguistic* information system - would be supposed to handle.

As may be expected, there will be differences in the estimation of the difficulties that will pile up when Winograd's system is extended to handle additional parts. I, for one, regard Licklider's optimistic evaluation - which you heard last week - to be totally unrealistic, even, I would not hesitate to say, naively so. Not only are we still very remote from a reasonably adequate grammar of some natural language, i.e., syntax and semantics, or even from an agreement on what should be incorporated into such a grammar, but we cannot even agree on which framework this should be done within for computer processing, whether, say, in a standard generative-transformational grammar, an extended such grammar, a generative semantics of any of the two dozen or more variants known to me, a Halliday-type grammar as done by Winograd himself, or any of the countless other grammars that are sprouting all over the world. (I shall later briefly discuss one of the more recent and more important kinds, the so-called *Montague-grammars*.) The difficulties will include the programming into the computer of the equivalent of the background knowledge we would normally assume of a human interlocutor, or, alternatively, of an ability to pick up this knowledge itself, by observation and handling of its environment. The difficulties will also include the programming into the computer of goodwill and, last but certainly not least, of the equivalent of sound judgment as to the relevance of the stored background knowledge - say, as contained in the latest edition of the *Encyclopaedia Britannica*. All this seems to me a dream relative to which Leibniz's dream was child's play.

Before I turn to consider from a practical point of view where this leaves us, let me add two more theoretical comments. First, it would certainly be inappropriate here, in Great Britain, to disregard the British version of pragmatics, the so-called *speech act theory*, originating with the later Wittgenstein,[14] further developed by J.L. Austin,[15] and now actively pursued by such philosophers as Paul Grice[16] and John Searle[17] in Berkeley, Cal., Jonathan Cohen[18] in Oxford, Jürgen Habermas[19] in Germany, and by a growing number of linguists, such as George Lakoff[20] in Berkeley, Dieter Munderlich[21] in Düsseldorf, and many, many others. Their basic insight, by no means a sensationally new one, is that there exists no one-to-one correlation between the syntactic form of sentences and the speech acts performed by their utterance, their main claim for distinction being only that they take this rather trivial insight seriously. More importantly, perhaps, they take the pragmatic notion of speech act as fundamental - as Morris and, not too seriously, we remember, Carnap had already done - and attempt to derive the semantic relations from overtly pragmatic ones by explicit procedures rather than by "abstraction". Let us notice that this conception has some far-reaching implications for information systems. Information scientists have generally totally disregarded any results of speech acts, performed by uttering or inscribing the sentences in the documents they dealt with, but have looked only at the *statements*, the truth-value bearers, and have simply identified the information contained in a document with the set of statements made therein. They have, of course, also treated *questions*, a quite different kind of speech art, but only questions *about* the documents and their content and, to my knowledge, never the questions raised *in* those documents, nor, say, *suggestions* contained in them as to further research. I don't think that there is any explanation for this fact but laziness and thoughtlessness, sometimes covered up by such catch-phrases as "hard facts", and based, of course, on the age-old overestimation of the *constative* function of linguistic communication over its myriads of other functions.[22] Needless to say, philosophers, logicians, and linguists alike were responsible for this development, and information scientists just took their cue from them without giving it much thought. Don't you think that processing a document should not only present the facts stated therein, but also the conjectures proposed, the theories presupposed, the queries raised, the suggestions made, and perhaps a few more things?

Secondly, just a few words about Montague grammars. In the early fifties, Richard Montague, one of the most gifted students of Tarski and an excellent mathematical logician and model theoretician, decided to apply his great technical abilities to the study of the semantics of natural languages. As a matter of fact, he did more than that: he came up with a formal system which was not only far more precise and powerful than anything known before - though, as may be expected, it dealt again with a very small sublanguage only - but it dealt also with some pragmatic aspects of communication via natural languages, viz., some of the context-dependencies connected with speaker, hearer, and time of utterance. What he did can be formulated in slogan terms (which, I am afraid, is not very illuminating) as a *semanticization* of these aspects of *pragmatics*. Montague himself[23] and some of his followers, such as David Lewis,[24] extended this semanticization in various directions in subsequent work. (Let us recall that what Carnap did in 1934 may be called a *syntactization* of *semantics*.) I don't think that this semanticization can be pushed much further and, in particular, do not at all think that Montague and his associates can satisfactorily handle all the issues raised by the speech act theorists.[25]

What are the practical conclusions one should draw from this state of affairs for information science systems? Since the establishment of a powerful interactive man-machine information processing system requires natural language or something very close to it as the language of communication between man and computer, and since natural languages are, if I am right in my analysis, not only pragmatic but essentially so, it seems that the chances for the establishment of such a system are rather dim for the foreseeable future. On the other hand, while the pragmatic features of natural languages are just there, for interactive languages you can experiment at your ease, introducing one feature at a time, checking how much the power of the system is thereby increased and how the various features reinforce each other or, perhaps, interfere with each other, and, as a result, perhaps decide to drop one of the features.

I have no idea how long this experimentational stage might last and when, if ever, it will result in a general interactive information processing system worth its name. For the time being, I suggest that most information scientists lower their aim or perhaps, more realistically, try to improve their available systems piecemeal rather than on the basis of some grand, but utopian model. This may sound rather disappointing but, if this is any consolation, the person most disappointed by this state of affairs is - me.

References

1. Though, unfortunately, only in Hebrew, in a lecture before the Israel Academy of Sciences and Humanities, Bar-Hillel, Y., The Betrayal by the Logicians, *Iyun*, 14, 120, 1964.

2. For the English version of this paper, see Tarski, A., The concept of truth in formalized language, in *Logic, Semantics, Metamathematics*, Clarendon Press, Oxford, 1956, chapter 8.

3. For the English version, see Carnap, R., *The Logical Syntax of Language*, Harcourt, Brace & Co., New York, 1937.

4. For the story see, for instance, Bar-Hillel, Y., Remarks on Carnap's *Logical Syntax of Language*, in *The Philosophy of Rudolf Carnap* (P.A. Schilpp, ed.), Open Court, La Salle, Illinois, 1963, and reprinted as chapter 11 of Bar-Hillel, Y., *Aspects of Language; Essays and Lectures on Philosophy of Language, Linguistic Philosophy, and Methodology of Linguistics*, North Holland Publishing Co., Amsterdam, 1970.

5. See Carnap's reply to the article mentioned in the previous note, *ibid.*, 940.

6. See Bar-Hillel, Y., Logical Syntax and Semantics, *Language*, 30, 230, 1959, reprinted as chapter 2 of Bar-Hillel, Y., *Language and Information: Selected Essays on Their Theory and Application*, Addison-Wesley Publishing Co., Reading, Mass., 1964.

7. See Carnap, R., *Meaning and Necessity; a Study in Semantics and Model Logic*, 2nd ed., University of Chicago Press, 1956, supplements D and E.

8. See reference 6.

9. Chomsky, N., Logical syntax and semantics: their linguistic relevance, *Language*, 31, 36, 1955.

10. Published originally as Appendix III to: The present status of automatic translation of languages, in *Advances in Computers*, I, (F.L. Alt., ed.), Academic Press, New York, 1960, 158, reprinted as chapter 12 of Bar-Hillel, Y., *Aspects of Language; Essays and Lectures on Philosophy of Language, Linguistic Philosophy, and Methodology of Linguistics*, North Holland Publishing Co., Amsterdam, 1970.

11. See, for instance, Dreyfus, H., *What Computers Can't Do; A Critique of Artificial Reason*, Harper and Row, New York, 1972.

12. See, for instance, Bar-Hillel, Y., Communication and argumentation in pragmatic languages, *Linguaggi Nella Societa e Nella Tecnica*, Edizione di Comita, Milano, 1970, 271.

13. Winograd, T., *Understanding Natural Language*, Edinburgh University Press, 1972.

14. Wittgenstein, L., *Philosophical Investigations*, Basil Blackwell, Oxford, 1953.

15. Austin, J.L., *How to Do Things with Words*, Oxford University Press, 1962.

16. Grice, P., *Logic and Conversation*, to be published.

17. Searle, J., *Speech Acts: An Essay in the Philosophy of Language*, Cambridge University Press, 1969.

18. Cohen, J., Speech Acts, *Current Trends in Linguistics*, 12, (Linguistics and Adjacent Arts and Sciences, Tome 12/1), 1973.

19. Habermas, J., Vorbereitende bemerkungen zu einer theorie der kompetenz, in *Theorie der Gesellschaft oder Sozialtechnologie*, Habermas, J. and Luhmann, N., eds., Suhrkamp, Frankfurt/M., 1971, 101; cf. Bar-Hillel, Y., A critique of Habermas' philosophy of language, *Synthese*, 25, 1973.

20. See, for instance, Lakoff, G., Linguistics and natural logic, *Synthese*, 22, 151, 1970; reprinted in *Semantics of Natural Languages*, Davidson, D. and Harmon, G., eds., Reidel, Dordrecht/Holland, 1972.

21. See, for instance, *Linguistische Pragmatik* (D. Wunderlich, ed.), Athenaeum, Frankfurt/M., 1972.

22. Cf. Austin, J.L., Performative-constative, *The Philosophy of Language*, Searle, J., ed., Oxford University Press, 1971, 1.

23. Let me mention only Montague, R., Pragmatics, in *Contemporary Philosophy*, Klibansky, R., ed., La Nuova Italia Editice, Firenze, 1968; Universal grammar, *Theoria*, 36, 373, 1970; Pragmatics and intensional logic, *Synthese*, 22, 68, 1970, reprinted in *Semantics of Natural Languages*

24. Lewis, D., General semantics, *Synthese*, 22, 18, 1970, reprinted in *Semantics of Natural Languages*.

25. Cf. Bar-Hillel, Y., Malino, J. and Margalit, A., Logic and theoretical linguistics, *Current Trends in Linguistics*, 12 (Linguistics and Adjacent Arts and Sciences, Tome 12/1), 1973.

A PATTERN PROCESSING MODEL OF THE HUMAN VISUAL SYSTEM

Frank A. Maher and Matthew Kabrisky

Dayton, Ohio, U.S.A.

INTRODUCTION

Basically there are two approaches to simulation. The first type involves the analysis of the "real" system, so as to identify the critical processes performed by the system. The second type is not concerned at all with the processes performed by the "real" system but is solely interested in the input and output of the system. The objective of this presentation is to introduce a model of the human visual system, based upon the first form of simulation. More specifically, the model is directed toward the pattern recognition, discrimination, and classification processes of the human visual system.

The development of the model was initiated by an electrical engineer: Matthew Kabrisky.[1] The basic assumption underlying the development of the model is that an extensive investigation of the human system would provide the knowledge necessary to build hardware which would be capable of performing pattern processing tasks with an efficiency similar to that of the human system. The model then was developed so as to be consistent with the physiology of the human system, as it is currently understood. The model was then implemented via computer simulation so that the efficiency of the output could be determined and the relationship between the model and human, in performing pattern processing tasks, finally confirmed.

THE PHYSIOLOGY

The pertinent characteristics of the human visual system considered in the model development were drawn from the study of the physiology of the system. The transmission of the pattern from the fovial area of the retina to the lateral geniculate body of the thalamus and the ultimate mapping of the pattern directly on the primary visual area of the cortex is of special significance, especially considering that the retina - thalamus - cortex transmission channel maps the retina homeomorphically on the functionally flat cortical sheet. Assuming the human system is basically an efficient system, it appears logical to suggest that the pattern data would not have been carefully retained if the processing occurred before being mapped upon the cortex. The cortex than appears to be the anatomical area of interest as far as pattern processing is concerned. The cortex is functionally structured to operate in small integral and independent units. The cortical sheets are roughly analogous to the structure of honeycomb with the separate and distinct cells' coherence forming a sheet-like structure.

The basic connectivity within the cortex was first described by Cajal in 1898, based purely on histological studies. The flow of data, as shown by the general orientation of axonal direction, is perpendicular to the plane of the sheet of cortex: input data are received through axons which are effectively perpendicular to the sheet, roughly like hair growing on the scalp, and the output corresponding to a particular input departs from the cortex at the same point at which the data went in. There is virtually no transmission of data across the sheet - at least so far as concerns the identification game. Functional verification of the fact that the cortex operates as a collection of distinct cell-like units has been communicated by Mountcastle,[2] and by Hubel and Wiesel.[3] This cylindrical array of neurons is apparently the generic and basic computing entity in the cortex. Most recently, Pollen[4] found neurons in the visual cortex within a mammalian, that respond to stimuli that suggest that they are the output neurons of two-dimensional Fourier component filters of data imposed on the animal's retina.

Since visual sensory data are mapped homeomorphically onto the primary sensory cortex, and since at this stage of their processing the localized nature of the cortical computation denies them any sort of computation extending over a significant portion of any interesting visual pattern, one seeks to know where the data are relayed following their initial reception in the cortex. It can only be after this first cortico-cortical relay that significant areas of an input pattern can be made to interact. Regrettably, little is known about systems and methods of data transmission between various portions of the cortical sheet. What little is known of the scheme suggests that data from local cortical areas are broadcast widely to succeeding cortical areas by means of multitudinous axonal pathways originating from the discrete localized computing regions. McCulloch, for instance, described this phenomenon in terms of the strychnine neuronography experiments he was using to deduce cortico-cortical and thalamo-cortical connectivity patterns, as follows: "When Dusser de Barenne and I strychninized a pinhead spot on the area striata strychninized spikes appeared at many points in the secondary visual areas as if the output from each spot in the area striata were scattered at random in the secondary visual area."[5] Such a statement seems to summarize basically what is known about the destination of data following its neat mapping on the primary sensory cortex and it is necessary to try and deduce what sort of computation - in at least the sense of pattern recognition - can be performed by a device which broadcasts data widely and, seemingly, randomly to successive cortical sheets.

Curiously enough, optical systems (which can easily be set up to perform interesting computations with two-dimensional patterns) fit McCulloch's description. For instance, the scheme of light-ray paths in a slide projector fits the diagram above; and the projector, if excited with a coherent light source, generates as an inadvertent concomitant of its projection of the slide, the two-dimensional Fourier transform of the slide image in an optical place.

It produces this transform because of the interference and reinforcement of various portions of sinusoidal electromagnetic (light) fields propagating along pathways of different lengths within the system of lenses. The difference in path lengths - and hence time of propagation in a homogeneous medium - depend upon the refractive properties of, and the mutual physical relation among, the lenses in the system. In an optical homologue using fibers, or in a digital computation simulation, it is necessary to control propagation times in the individual pathways to mimic the cancellation and reinforcement that occurs intrinsically in the spatially continuous optical system. For a neuron homologue of such a system it would probably be necessary to maintain consistent relative propagation times in axonal pathways and provide sufficient overall synchronization of activity to allow the possibility of excitory post-synaptic potential and inhibitory post-synaptic potential interaction in individual neurones to provide the equivalent of reinforcement and interference found in the optical system.

THE MODEL

The key to the model was the assumption that the process being performed in the cortex was a discrete Fourier transform of the pattern data. Extending the model development, Radoy[6] and later Tallman[7] developed a computer simulation based upon the Fourier transform but with the addition of two other processes which are essential to the model: spatially filtering and cross correlation within the transform domain.

The Radoy-Tallman work showed that a large class of pattern identification tasks previously solvable only by the human visual system could be solved by pattern recognition algorithms based upon correlating the low spatial frequency components of test data with the corresponding spatial frequency components of the stored prototypes.

The advantages of a pattern recognition system based upon correlation of low spatial frequency components of the Fourier transforms of the test data should be cited:

(1) There is a potential economy of storage; identification of font-independent English and cyrillic alphabets require storage and correlation of no more than 49 Fourier components of the individual characters.

(2) The data thus stored are distributed over multiple memory locations providing immunity to error produced by discrete errors. This is a useful property, one suspects, in biological hardware and would certainly be useful in certain man-made situations.

(3) Operating in the Fourier domain enables conceptually easy implementation of search routines to find specific patterns midst cluttered backgrounds. That is, implementation of the Wiener autocorrelation theorem is applicable once the decision is made to compute Fourier transforms of data. This technique has been applied to the problem of reading Chinese,[8] aerial reconnaissance,[9,10] and speech recognition.[11]

PSYCHOLOGICAL SUPPORT

An initial attempt at validating the model was undertaken by Maher.[12] A pattern discrimination task involving 10 animal forms was performed by the computer simulation of the model and by a group of 46 human subjects. The degree of relationship between model and human output was determined using statistical correlation methods. A. Pearson correlation coefficient of .961 was calculated, thereby supporting the model. Subsequently, the experiment was replicated with a different pattern set, resulting in similar results.

Ginsburg investigated the Gestalt principles of proximity, similarity, closure, figure and ground, as well as several optical illusions in an attempt to understand these phenomena through the structure of the model. Figure 1 illustrates the principles of the whole, proximity, similarity and proximity/similarity. The first image box of each row is that of the original image. The second and fourth columns of boxes are the filtered Fourier transforms of the image, and the remaining (columns 3 and 5) boxes depict the restructured patterns which would be analogous to the images as they

Gestalt Principles of Perceptual Organization (square 4x4): Figures provided by Arthur Ginsburg and are the product of his research.

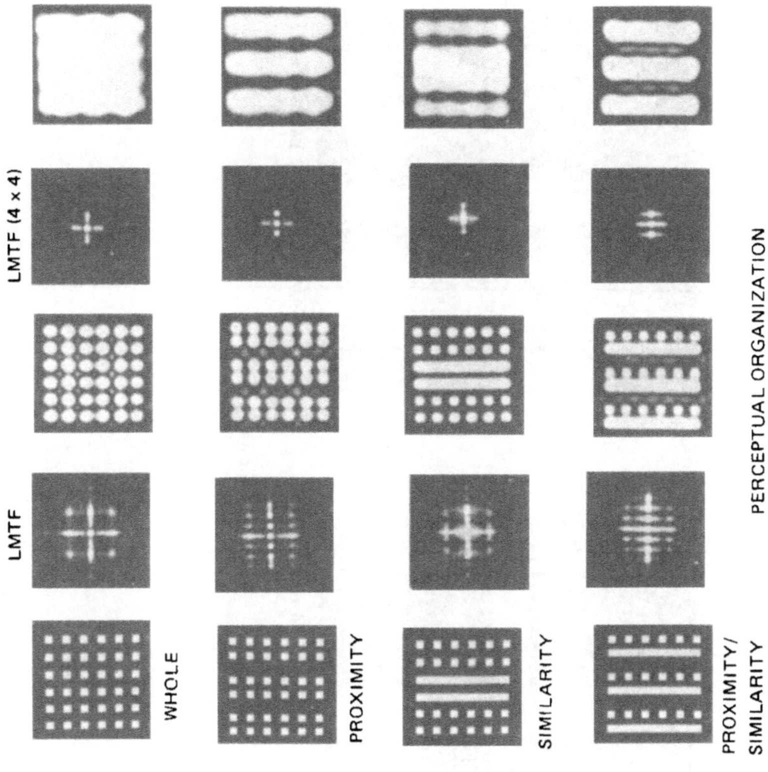

Figure 1

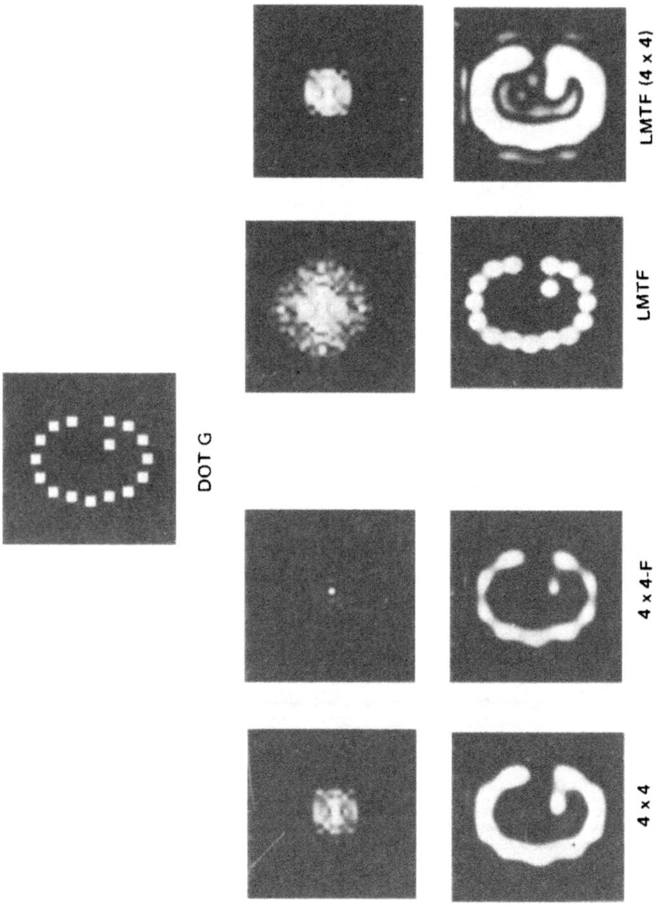

Figure 2

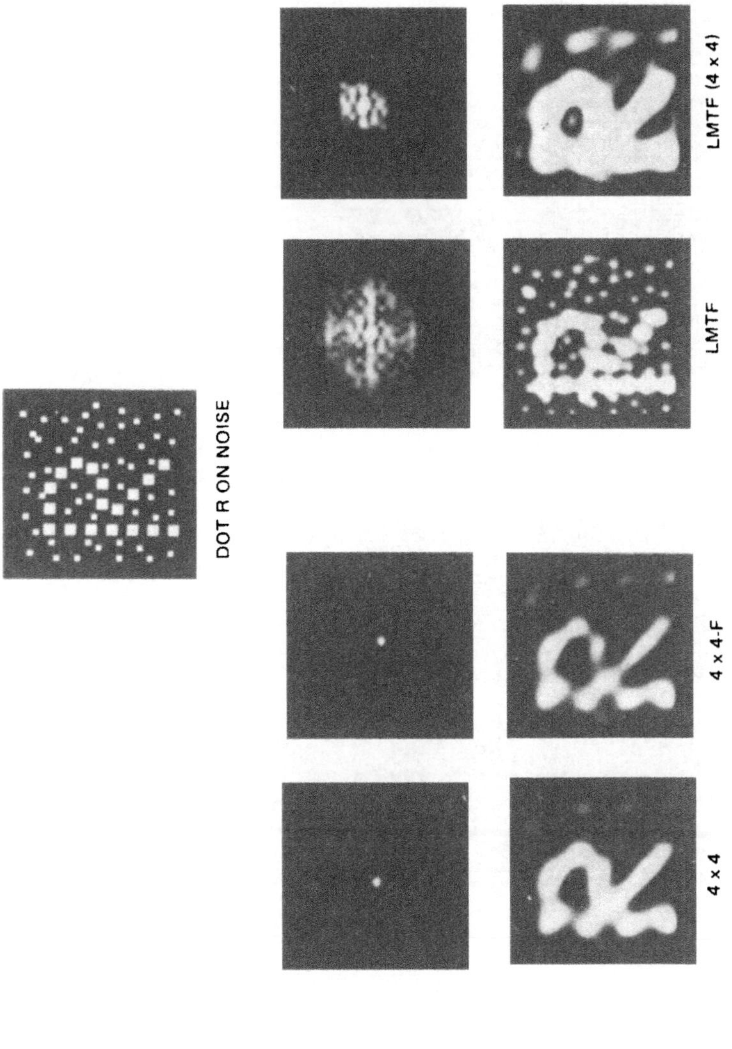

Figure 3

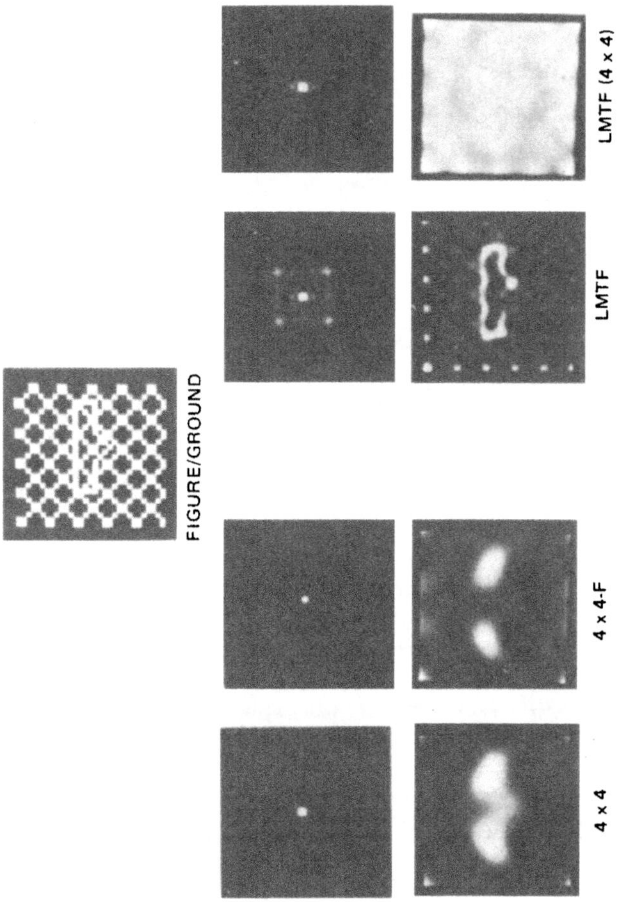

Figure 4

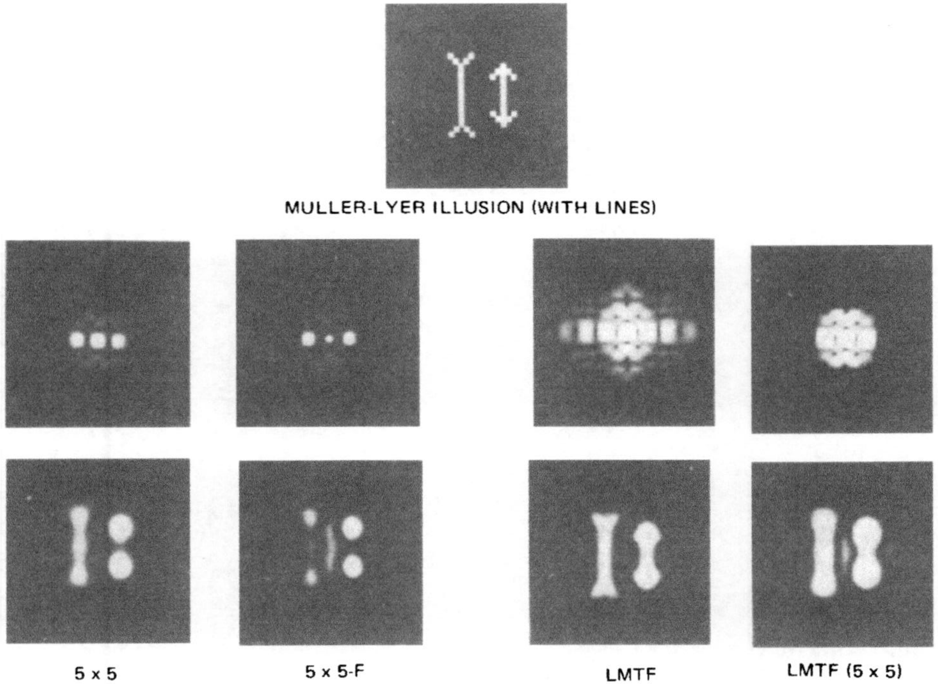

Figure 5

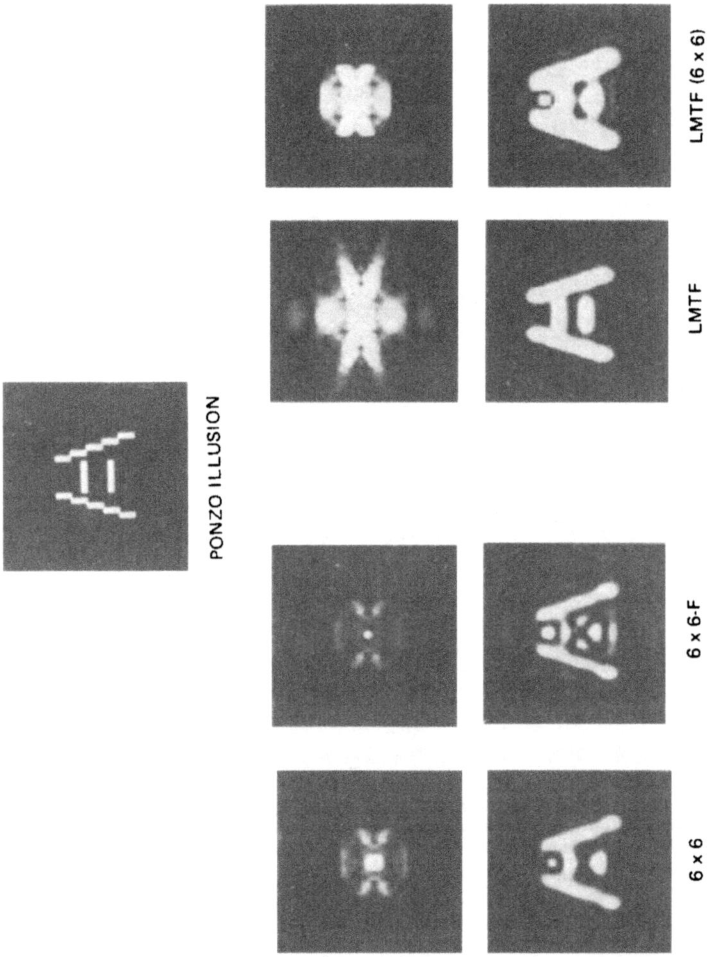

Figure 6

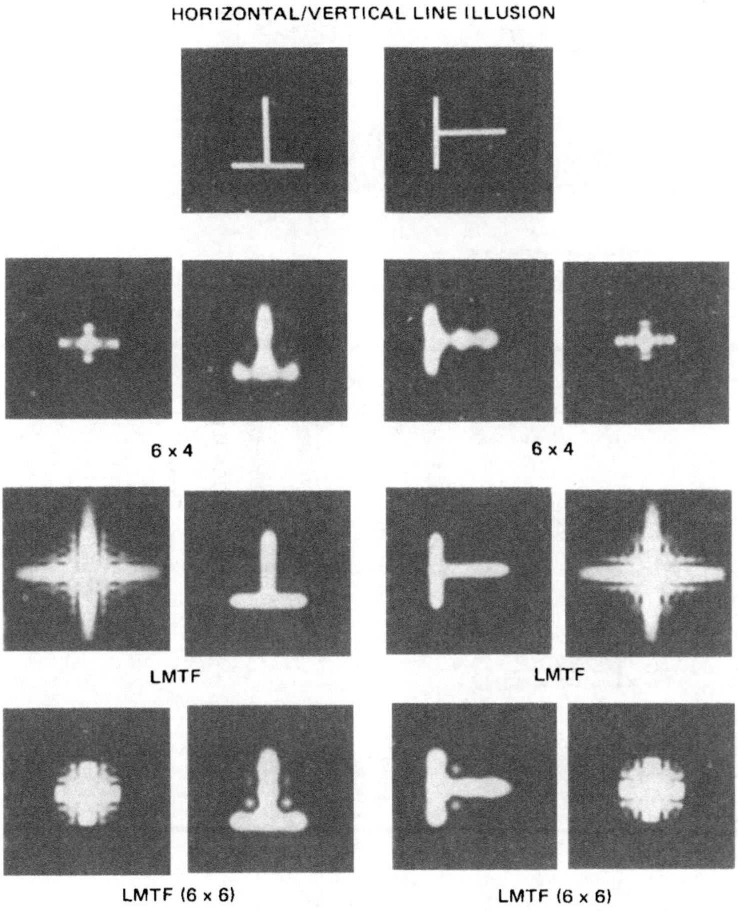
HORIZONTAL/VERTICAL LINE ILLUSION
6 x 4
6 x 4
LMTF
LMTF
LMTF (6 x 6)
LMTF (6 x 6)

Figure 7

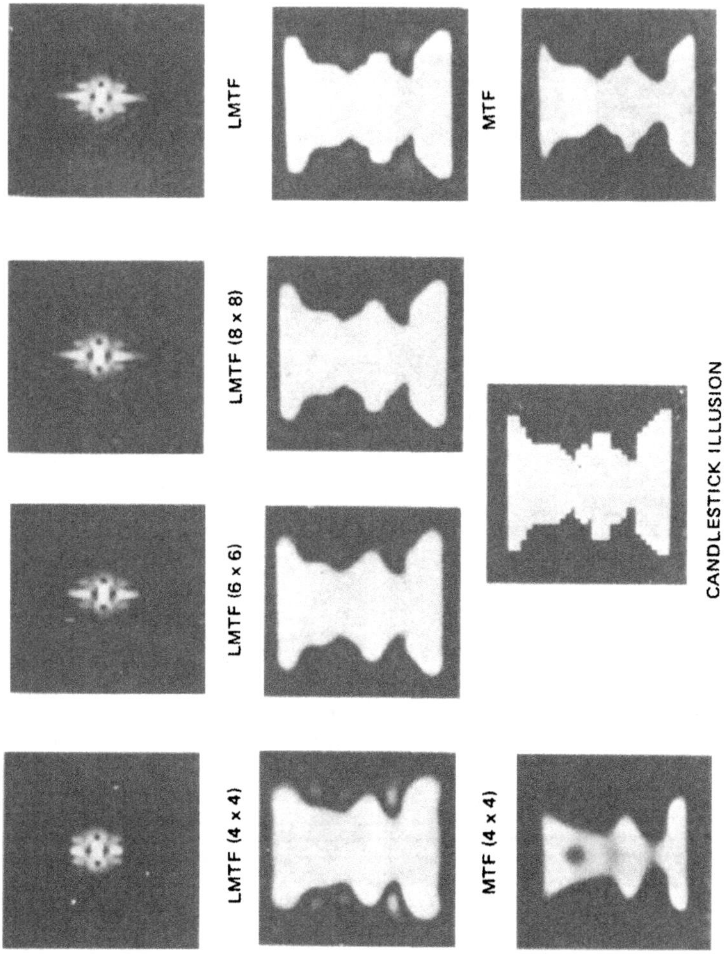

CANDLESTICK ILLUSION

Figure 8

are present in Area 19 of the cortex. In all cases the Gestalt principles are demonstrated to exist as an intrinsic function determined by the structure of the model. The other figures (2 through 7) are also formated to depict the source image and the filtered Fourier transforms in the Fourier domain and as restructured images that would be analogous to the image in Area 19 of the cortex. Figures 2 through 4 depict various perceptual capabilities of the human system that the model demonstrates to be realizable by the intrinsic characteristics of the system as it processes visual patterns. The remaining figures (5 through 7) demonstrate that optical illusions like the Muller-Lyer, Ponzo, and horizontal/vertical internally are not illusions at all, in that in area 19 of the cortex the lines are in fact unequal in length. The candlestick illusion (Figure 8) demonstrates that the perception of either the candlestick or the two profiles appears to be a function of which spatial filter is utilized by the system.

SUMMARY

A model assuming that the form of recognition function of the human brain is based upon computing the two-dimensional Fourier transform of input data has been derived from selected biological data. The model has been simulated and shown to be applicable to pattern recognition problems in a wide variety of complex data manipulating tasks. Finally, the actual behaviour of the model has been shown to be isomorphic to the behaviour of the intact visual system in:

(1) a variety of experimental psychological tests; and

(2) certain preliminary *in situ* psychological experiments.

The impact of the model upon Information Science lies in the basic approach underlying a type 1 simulation. The human system is one system which is known to deal with data and information. This is not to suggest that it is the only system that deals with information, merely that it is one such system. By an analysis of the ways in which the human system manipulates and processes data and information some basic understanding of the nature and structure of information may become available.

References

1. Kabrisky, M., *A Proposed Model for Visual Information Processing in the Human Brain*, University of Illinois Press, Urbana and London, 1966, and Kabrisky, M., Visual information processing in the brain: models for the perception of speech and visual form (edited by Wathen-Dunn) (Cambridge, Mass: Massachusetts Institute of Technology Press), 1967.

2. Mountcastle, V.B., Modality and topographic properties of single neurons of cat's sensory cortex, *Journal of Neurophysiology*, 20, 1957, 408.

3. Hubel, D.H. and Wiesel, T.N., Receptive fields, binocular interaction, and functional architecture in the cat's visual cortex, *Journal of Physiology*, 160, 1972, 106.

4. Pollen, D.A. and Taylor, J.H., The striate cortex and the spatial analysis of visual space, *The Neurosciences*, Worden, F. and Schmitt, F.O., Eds., MIT Press, Boston, 1973.

5. Dusser de Barenne, H.G., *et al*, *Isocortex of a Chimpanzee*, University of Illinois Press, Urbana, 1951.

6. Radoy, C.H., *Pattern Recognition by Fourier Series Transformation*, Air Force Institute of Technology M.S. Thesis, GE/EE/67A-11, 1967, AD 651-801.

7. Tallman, O.H., *The Classification of Visual Images by Spatial Filtering*, Air Force Institute of Technology Ph.D. Thesis, 1969, AD 858-866.

8. Ankeney, L.A., *The Classification of Chinese Characters by Spatial Filtering Techniques*, Air Force Institute of Technology M.S. Thesis, GE/EE/71-2, 1971, AD 722-852.

9. Roberson, R.S., *Application of an Image Classification Model to the Reconnaissance Problem*, Air Force Institute of Technology M.S. Thesis, GE/EE/71-23, 1971, AD 886-111.

10. Ashbaugh, J.B., *The Detection and Identification of Man-Made Objects from Aerial Reconnaissance Photographs*, Air Force Institute of Technology M.S. Thesis, GE/EE/73-1, 1973.

11. Dailey, K.G. and Sutton, F.S., *An Automatic Speech Recognition System Using a Vocoder Input*, Air Force Institute of Technology M.S. Thesis, GE/GGC/EE/72-18, AD 742-441.

12. Maher, F.A., A Correlation of human and machine pattern discrimination, (IEEE Transactions on Aerospace & Electronic Systems, 70-C-14-AES), *IEEE NAECON 70 Record*, 1970, 260.

13. Ginsburg, A.P., *Psychological Correlates of a Model of the Human Visual System*, Air Force Institute of Technology M.S. Thesis, GE/EE/71S-2, 1971, AD 731-197.

VISUAL, COGNITIVE AND BEHAVIOURAL TWISTS (SOME DYNAMIC ASPECTS OF INFORMATION

Derek Conway

JC Engineering Design Lt., Bristol, England.

To begin with I would like to allow myself the luxury of attempting a definition of Information. While participating at the Seven Springs NATO Advanced Study Institute on Information Science in 1972 I derived my own personal definition of Information which is:

DEFINITION: "Information is a perceptual-cognitive shell in which 'fundamental' ideas or notions interact"

It is tempting to try and add more, or to enlarge upon this definition, but I would like to keep it as stated above. The closeness of this definition to what we could mean by Consciousness is, I think, evident. It was not intended to represent Consciousness but was derived whilst pondering the problem of information generation.

Many of the papers presented at the Seven Springs Institute provided a framework for several theoretical ideas and experiments which I hope eventually to develop. For my work the importance of these papers, particularly those of a theoretical nature, cannot be overestimated. However, I do not necessarily agree with the point of view made by Dr. Hugh Smith in his paper "Human Decision-Making in a Sequential Task".[1] The unstated, yet implicit philosophy behind this kind of experiment which no more represents decision-making, to my mind, than describing a very ordinary chain of events, puts forward a two-value system of the experimenter always being non-involved while the subject, so to speak, is on the tread-mill. When applied to real industrial situations the implications are far too real to be taken lightly. Although the man is at work let's hope his mind still is, although work demands may not always be under his control.

This paper is an attempt to illustrate the paradoxical nature of information in relation to the information seeker[2] and to demonstrate that things are not all they appear to be. To my mind there is a relationship here between the biological/physical and psychological aspect of information-seeking as it relates to the individual and aspects of mythology, particularly Greek mythology and drama.

The term "feedback" has, of late, been used extensively and is still a very useful concept to relate the effect of "inputs" and "outputs" upon a system. It is well known that feedback can be positive or negative depending upon whether it is desired to increase or decrease error. The schema for this is well known and so I do not propose to reproduce it here.

However, it does seem to me that there are times when the feedback element neither leads to an increase or a decrease in error, but creates confusion, particularly when the system is not well structured, that is "fuzzy". In this case it could be said that "confusion feedback" exists and I'm referring here, of course, to a system in which a human element is involved. Perhaps another way of putting the case is to say that which is intended to be fed-back and what is actually fed-back are different. Different not only in degree but different in kind. (See figure 1)

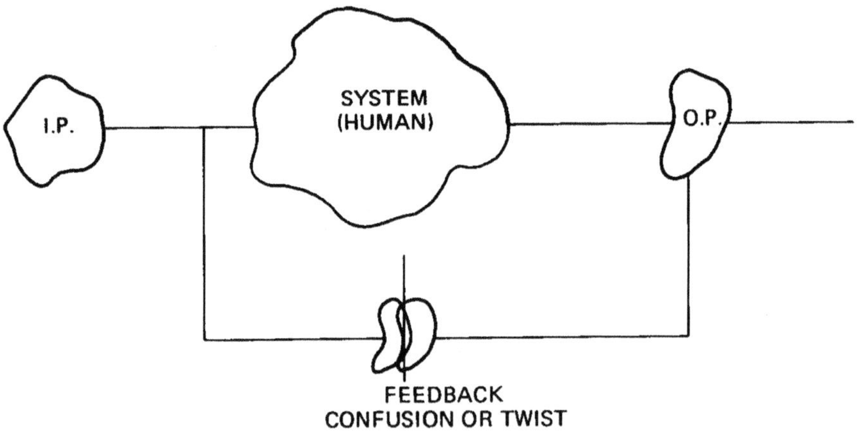

Figure 1

For the moment I would think that there are essentially two different ways of conceptualising any system, as an observer or as a participator. This could be expressed in another way, by saying, there are those systems which do not involve the individual and there are those where the individual finds himself, very much, in the feedback loop.

The two states, observer/participator are interesting in themselves and often one finds the individual switching from one state to another, or from one state to a meta-state, in quite a reasonable and controlled fashion, much I imagine like the dual function of Actor/Producer in drama.

The subject of observer/participator and hybrid or multiple states are, to my mind, extremely interesting and related to Information Science. I would like to develop this aspect another time, more fully. For now I would like to bring to attention those mythological hybrid words of Minotaur and Centaur and this raises the concepts or meaning behind the word, perhaps the "picture" the word creates. Just as Sherlock Holmes was a master of disguise so too, it seems, we may use words to disguise the meaning as we either wish or wish not to convey. This is not quite the same as saying the meaning of the words is in the gap between the words,[3] although I am in sympathy with that point of view. The point I wish to make is that the meaning of the word may lie behind the word, much like the face behind the mask. Instead of looking at the gaps, use another dimension, so to speak, to probe *behind* the words.

There is an element of the dual state in the Orpheus myth, the World - the transition - the underworld, and I think this myth is a good introduction to the point I wish to develop of visual transitions, twists or paradoxes.

It should be remembered that Orpheus in his quest for Eurydice entered the Underworld and that eventually, having found Eurydice he was allowed to lead her from the Underworld provided that he never turned to look upon her during the journey. We know that as the story goes he did turn to look upon her and that she instantly vanished for ever.

In this particular story it can be said that the feedback element of the "turning-to-look", perhaps for Orpheus to reassure himself, produced confusion to say the least. A visual pragmatic effect.

Now, I would like to move on and direct attention to some visual paradoxes as found in the Impossible Pictures by the Dutch artist, M.C. Escher.[4,5,6] There are many which indicate some kind of loop or oscillatory effect but in particular there are two which demonstrate the element of movement in some kind of "closed-loop", or feedback system.

In the one,[7] the waterfall, the painting is so put together that the water can be seen to run first downhill (the waterfall) and eventually flows back to the top of the waterfall, apparently flowing uphill.

In the second painting, The Impossible House, the element of measurement is not quite so marked but the figures climb the ladder to the top of the house, a ladder which is both inside and outside the house at the same time.

Both of these paintings bring out for me, the transition from one state to another, although where precisely this transition takes place is not a particularly well-defined area. More of a zone, dare I say it, a "fuzzy-space". This is particularly so when one scans and follows the direction of movement. Somewhere a "twist" zone occurs (figure 2), be it somewhat non-specific but well defined.

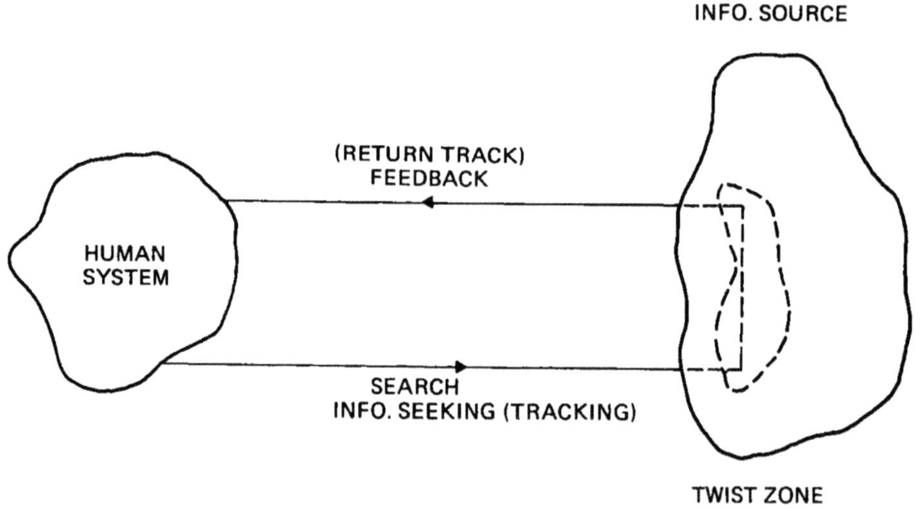

Figure 2

In both those impossible paintings the twist occurs somewhere between the "going-out" and "coming-back" cycle or loop. This may be taken as an example of confusion feedback where things are not what they seem. It is perhaps worthwhile to mention that the "going-out" and "coming-back" elements in this description produce in my mind associations of afferent and efferent nervous impulses in the transmission of information in our own nervous systems.

Turning now to what I call Cognitive Twists. I wish to refer here to two examples from topological mathematics,[8] which I believe to be relevant, and the two concepts from topology to which I wish to refer are the MOBIUS STRIP (figure 3) and the KLEIN BOTTLE (figure 4).

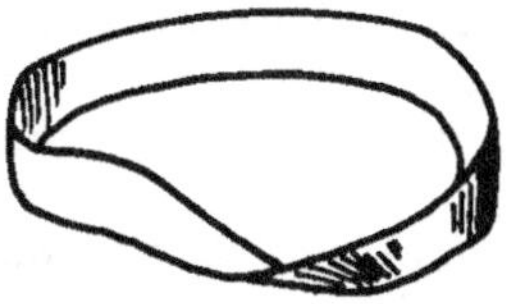

MOBIUS STRIP

Figure 3

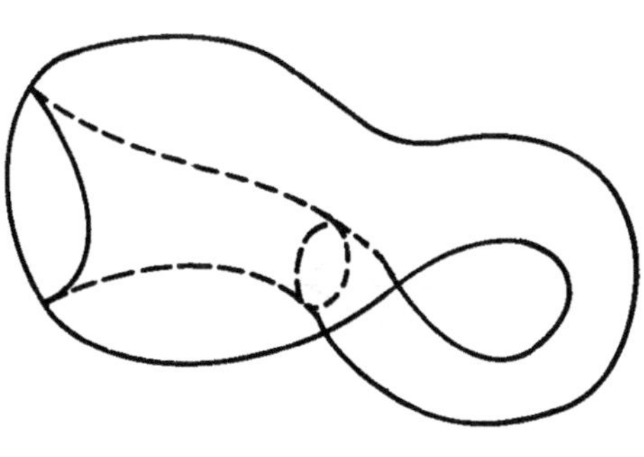

KLEIN BOTTLE

Figure 4

In the Mobius Strip we have the essential ingredients of a feedback system - a closed loop. This particular feedback system or closed loop is interesting in so much that there is again a twist or transition zone. (I shall use the word twist to describe an area of transitia, generally, ill defined.)

The twist of the Mobius Strip is best illustrated by the fact that an insect walking around the wall would cross-over from one side to the other with no particularly well defined change of state. Yet a transition had occurred.

Consider now the Klein Bottle. The Klein Bottle is an abstract topological construct of a bottle which it could loosely be said has no inside or outside. To my mind the more important aspect of this manifold would be revealed if we could observe the dynamic turning aspect. If this figure were to be displayed upon an electronic visual display it would represent better, perhaps the dynamic aspect of the "going-out" - "twist" - "coming-back" aspect I wish to convey and relate to the Escher pictures.

The rigours and finer points of Differential Topology apart, to my mind the concept of the Klein Bottle represents something of fundamental importance. That is, it is a construct that could stand on its own without the mathematics, but exactly how it should stand, in more detail, at the moment eludes me. Perhaps the Mobius Strip and Klein Bottle represent something of fundamental importance in the cognitive process.

It seems to me that the structure of the Klein Bottle could be gradually dissolved until we are left with the ring interface only, where the twist occurs (figure 5) highlighting the twisting dynamic transition area itself without the need to attempt the plotting of the loop itself. Similarly, we could dissolve away most of the Mobius Strip leaving the twist region only (figure 6).

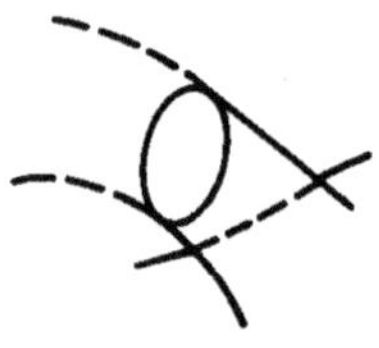

KLEIN BOTTLE
RING INTERFACE

Figure 5

MOBIUS STRIP
TWIST ZONE

Figure 6

Approaching now the nature of Behavioural twists or paradoxes, it does seem to me that in the Klein Bottle and Mobius Strip we have a bridge, a transition, be it very narrow, by which visual twists and behavioural paradoxes may be linked.

Alice[9,10] found herself in that strange world of paradox, of things not being what they seem or seeming what they were not. One of the more prominent generations of confusion seemed to be Humpty Dumpty who used words to mean what *he* chose them to mean, without an agreed consensus or clarification.

Perhaps it is not only lucky that Humpty Dumpty existed in a looking-glass land, but that he fell off the wall. Our luck has thus been enhanced because not only was he shattered, but also could not be assembled. On the other hand words do generate words; and it is the ways that words produce Paradox, mental confusion, conflicts and subsequent action or inaction (paralysis) to which I would like to direct the final part of this paper.

To begin with I would like to give a definition of paradox. Paradox may be defined as a "contradiction that follows correct deduction from consistent premises".[11] The nature of paradox is well presented in Watzlawick *et al* but it does seem necessary here to indicate that paradoxes appear to be contained within three classes. Watzlawick lists these as;

(1) Logico - mathematical paradoxes (Antimonies),

(2) Paradoxical definitions (Semantic Antimonies),

(3) Pragmatic paradoxes (Paradoxical injunctions and paradoxical predictions).

I would like to devote time to an explanation of the first two classes but Watzlawick does this very well. It could be possible that the Mobius Strip represents Class (1) and that the Klein Bottle represents Class (2). If this is so it is interesting to dwell upon the possibility of a topological concept representative of Class (3), that is to continually produce concepts and meta-concepts.

Pragmatic paradoxes involve, very much, human beings and actions and the essential ingredients given by Watzlawick are;

(1) A strong complementary relationship between the two people (Information Source and Information Seeker).

(2) Within the framework of this relationship an injunction is given which must be obeyed but must also be disobeyed.

(3) The person occupying the one-down position in this relationship is unable to step *outside* the frame of reference and thus dissolves the paradox by commenting on it, that is meta-communicating about it. A person caught in such a situation is in an untenable position.

(The implication for highly-structured, rigid, autocratic systems is apparent. While these may sometimes be necessary, they must in their function attempt to avoid these pragmatic paradoxes.)

To quote Watzlawick: "Thus while from a purely logical point of view these paradoxes are meaningless, from the point of view of their intersectional contexts, the phenomenon ceases to be merely a fascinating pursuit of the logician and philosopher of science but becomes a matter of stark practical importance for the sanity of the communicants be they individuals, families, societies or nations."

I am aware that my reference to "The Pragmatics of Human Communication" by Watzlawick, Bateson and Jackson is persistent, but this needs to be so. My quotations are extensive for to paraphrase would miss the twist contained in them.

In Chapter 6, Watzlawick gives eleven examples of Pragmatic Paradoxes. There is the example of the so-called mental retardent who can discuss the concept of mental retardedness and another example is the message, "Be Spontaneous".

There is again a good example of how ideologies are liable to become entangled in the dilemma of paradox - to quote again Watzlawick, "especially if their meta-physics consist of antimetaphysics". The example is from Koestler;[12]

> ... The party denied the free will of the individual and at the same time it exacted his willing self sacrifice. It denied his capacity to choose between two alternatives and at the same time it demanded he should

> choose the right one. It denied his power to choose between good and evil - and at the same time spoke accusingly of guilt and treachery ...

In a further example Watzlawick quotes from Proust[13] of the pragmatic paradox arising out of the frequent contradiction between socially approved behaviour and individual emotion.

Alexis in visiting his dying uncle is not allowed to show him his feelings of concern, and he feels he may appear unconcerned and unloving, which is quite to the contrary.

There is much more in Watzlawick *et al* under the heading of Paradoxical Communication. They go on to reveal the Double Bind Theory and its effects on mental health, Paradoxical Predictions and how sometimes it is better not to be "too sensitive" or "too intelligent" in dealing with them. From the point of view of games they go on to consider the Prisoners Dilemma, and how it comes about through the lack of trust between the individuals.

On the positive side it is pleasing to conclude on the note that paradox appears to have therapeutic effects (Chapter 7, Paradox in Psychotherapy) and that "many of the noblest pursuits and achievements of the human mind are intimately linked with his ability to experience paradox". However, it must always be remembered that we can be allowed to step out of the frame of reference and to be able to meta-communicate.

Returning to the beginnings of this final part on Pragmatic Paradoxes one can imagine the situation of the typist who *has* to put into one sentence the two statements that;

(1) Chicago is a populous city,

(2) "Chicago" is a trisyllabic.

(With apologies to Watzlawick *et al* p.195)

SUMMARY

It has been the intent of this paper to demonstrate a pointer to the way in which we may be led into paradox or confusion. The examples of visual and linguistic paradoxes have been used in an attempt to show that things are not always what they appear and to make the point that distortions or twists occur between the commencement of an information search and the final result.

It could be that somewhere between the initiation of the search for a particular kind of information and the ultimate information a "sleight-of-hand" or twist occurs, such that logical steps towards the information are side-stepped and the information seeker acquires "dummy information", or information upon which it is impossible to act. The pattern of his logical path may be traced and one could say *that* was the true path of "choice" and it is perhaps here again that I can be seen to disagree with the view of Dr. Smith. It does *appear* to be a maze-like problem-solving situation but the maze is a phantom or will-o'-the-wisp.

Electronics and automatic data retrieval systems are here and are increasingly finding their way into our society it seems.[14] However, at the moment, in an attempt to acquire information the seeker must still involve himself in a great deal of leg-work,[2] not only because of the diffuse nature of information but because often of the uncertainty of the type of question to be asked. To quote T.S. Eliot, [15] "Let us then not ask what is it, but let us go and make our visit".

Once in the library where the process of inquiry is frequently begun, and it is here that a twist may occur, the questions formulated become more vague or further removed from the original question in mind when we set out to gain information. Several articles on the psychological aspects of stress, aging, attention, short- and long-term memory could be cited.

There may be a circularity in our search or an oscillatory nature of inquiry set into motion - though we may not be aware of this. We may become lost in some fog of cognitive circularity or paradox much like a trip around a Mobius Strip, always aware of a dim nagging question.

> "In the room the women come and go talking of Michaelangelo.", T.S. Eliot.

To return again to the physical/biological aspect of gaining information. To acquire information, to get to the source, generally the library, requires physical effort. To then begin the cognitive aspect of information seeking or search it is perhaps better to have knowledge of the library system. It is necessary to communicate in a face-to-face manner with librarians and time pressures and work demands upon the librarians will impose a structuring effect upon this discourse.

The process of inquiry, of information seeking brings to the fore, to my mind, the dual nature of existence - body and mind. This well contained and integrated dualism may be very much highlighted by the process of information search, although it perhaps relates better to the observer/participator point of view previously mentioned. The participant, physically seeking information, and the observer, reflecting upon information. The psychological impacts on the physical/biological information seeker are highlighted elsewhere in this Institute.[16]

Finally it seems to me that in the process of seeking information we may often find ourselves in strange lands where paradox, yet valid, exists and just as Alice, Gulliver[17] and Orpheus survived their strange worlds so it seems, must the information seeker.

Perhaps the ultimate step of the information seeker and acquirer is to make the abundance of information and knowledge produce action and a better world, otherwise we may continually follow on circuits of paradox, twist and deflection. The step, is to go just that one step further, that next step, the leap which, it seems, eluded Gatsby;[18] "Gatsby believed in the green light, the orgastic future that year by year recedes before us. It eluded us then, but that's no matter - tomorrow we will run faster, stretch out our arms further ... And one fine morning -

So we beat on, boats against the current, borne back ceaselessly into the past."

Please note:

The points of view here expressed are those solely of the author and do not reflect those of JG Engineering Design Ltd.

References

1. Smith, Hugh T., "Human Decision Making in a Sequential Task", Proceedings of Seven Springs NATO Advanced Study Institute, 1972.

2. Donohue, J., "Research in Information Seeking - Its Place in the Teaching of Librarians", 1972 Institute.

3. Nevitt, B. (See these proceedings, Chapter 2).

4. Gregory, R.L., *The Intelligent Eye*, McGraw-Hill, New York, 1970.

5. Fairthorne, R.A. (Personal Communication).

6. Locher, J.L. (ed.), *The World of M.C. Escher*, Harry N. Abrams, New York, [1970].

7. Plates 258 and 230 of 6 above.

8. Kasner, E. and Newman, J., *Mathematics and Imagination*, Simon and Schuster, New York, 1940.

9. Dodgson, C.L. (i.e. Lewis Carroll), *Alice in Wonderland [etc.]*, Oxford University Press, London, 1971.

10. Dodgson, C.L. (i.e. Lewis Carroll), "Alice Through the Looking Glass" in the above edition.

11. Watzlawick, P., Beavin, J.H. and Jackson, D.D., *Pragmatics of Human Communication: A Study of International Patterns, Pathologies, and Paradoxes*, Faber and Faber, London, 1968.

12. Koestler, A., *Darkness at Noon*, Modern Library, New York, 1941.

13. Proust, M., *Les Plaisirs et Les Jours*, Gallimard, Paris, 1971.

14. Landau, R., (See these proceedings, Working Group 2 Report, and those of 1972 Institute).

15. Eliot, T.S., The Love Song of J. Alfred Prufock in *Collected Poems 1909-1962*, Harcourt, Brace, New York, 1963.

16. Klempner, I., Information technology and democratic governance, (See these proceedings, Chapter 5).

17. Swift, J., *Gulliver's Travels*, Oxford University Press, London, 1971.

18. Fitzgerald, F.S., *The Great Gatsby*, Scribner's, New York, 1953.

AN APPROACH TO INFORMATION AS A UTILITY

Tefko Saracevic

School of Library Science, Case Western Reserve University, Cleveland, Ohio, U.S.A.

THE NATURE OF INFORMATION PROBLEMS

Why should we attempt to consider information, or to be more precise, information systems, as a utility-type of enterprise? Generally for two reasons: first, the nature of information problems existing at present and anticipated in the future requires solutions that are considerably more user-oriented than we have now; and second, the precedent of successful operations of public utilities as user-oriented enterprises suggests that the concepts developed in connection with utilities may be applied to operations of some information systems. Let us first consider aspects of the information problems that invite a general reorientation in our thinking about information solutions.

The nature of the modern information problems can be perceived as a quantity/quality paradox: overabundance of information at the source ("Information explosion") and scarcity of relevant information for the user at the destination. To resolve the problems many types of information systems evolved with corresponding uses of information technology. However, from many user reactions it seems that the user's problem in getting relevant information has not been significantly alleviated and even worse, it seems that many solutions based in the new information technology resulted in new barriers, imposing new levels of complexity and difficulty for the user. Just from these points of view a reorientation seems in order.

And what about the future? Information was always vital to the functioning of social entities as well as individuals and it is becoming even more so, more directly.

The following future information problems and demands may be expected:

> The complexity of social structures and problems is increasing. As a result, the complexity of information requirements is increasing too; however, the ability of information systems to satisfy complex requirements is not increasing but rather even seems to be decreasing.
>
> The human social order is in a transition period from the industrial to a post-industrial society;[1] there is an increase (to the point of dominance) in types of work requiring higher education; knowledge is becoming the most important social force and knowers (professionals, technicians, managers etc.) the most important social group; as a result, communication and management of knowledge will be one of the most essential areas of the new post-industrial society, requiring new concepts for many types of information systems.
>
> More specifically, policy making and management related to environment may become a most important factor in the survival of our civilization; since rational policy making and management is based on information, we may expect an increase in demands for special kinds of new information systems related to policy making and management.
>
> In highly-developed societies everyday life is becoming more complicated and each of us, as an individual increasingly needs all kinds of information to function; furthermore, as we are becoming a people more and more assured of equality in goods and services we start seeking new services related to quality of life and new modes of participation in social processes... for all this we will need new types of information systems specifically oriented toward the general populace, not only toward the intellectual elite.

Obviously all of these problems and demands will not be resolved by approaching information systems as utilities. However, utilities may provide a conceptual framework, a general model, around which some information systems may be organized and some information problems may be alleviated. A number of proposals to this effect have already been made.[2] In summary, the recommendations for operating information systems as information utilities rests on two expectations: (1) on the level of psychological impact, the concept of a utility type of enterprise evokes a positive response because it is associated with the elusive but important idea of operating for public good - it is hoped that this will become the major orientation of information utilities; and (2) on the level of performance, the peculiar enterprises which are recognized as public utilities indeed were quite successful in dealing with problems of regulated supply of commodities or services with unique characteristics - it is hoped that information utilities will achieve a similar success in performance.

This paper is derived from a progress report of an investigation named Project INFUT (for INFormation UTility) which is attempting to uncover conditions under which a wide variety of information sources and systems could be considered as a utility.[3] The specifics of the investigation are limited to information systems within a university. There are, however, implications for other environments. The paper will concentrate only on two aspects investigated in Project INFUT: properties of public utilities and the implications for information systems. Other aspects of the project will be only mentioned.

SUMMARY OF THE PROJECT

Objectives

The objectives are to provide general answers to the following questions:

(1) What are the properties of public utilities that may be useful for the design and operation of information systems as a utility?

(2) How can we view the use of all information systems and sources in a university in a unified way that will allow for isolation of elements and interactions essential in a utility?

(3) What properties do and what properties do not characterize information sources and systems in a university as a utility?

(4) How to design an information utility?

Approach

The following approaches were taken in relation to each of the objectives:

(1) from a study of public utility literature and from interviews with people in public utilities we developed a conceptualization of generic properties of public utilities which we are able to use as a model for considering information systems as an information utility;

(2) from a study of: (a) various models and theories in communication and (b) the existing communications in the university, we developed a unified picture (a general model) of utilization of information in a university environment with an orientation toward the users;

(3) in a series of empirical studies we are attempting to obtain empirical evidence as to: (a) the validity of that unified picture and (b) the types of interactions that do or do not characterize the existing information sources and systems in a university as a utility.

(4) at the end of the project on the basis of all the studies we hope to suggest practical methodologies for study., design and development of information utilities.

Unified Picture

The University is a conglomerate of information systems, having the prime functions of teaching (communication of knowledge) and research (search for new knowledge and organization of knowledge). Elements of this conglomerate may be studied in isolation but, a unified view, (a common classification scheme, a general model) is necessary to study interactions in a consistent manner. A unified approach may moreover uncover sources and effects other than those commonly assumed. Our unified picture evolved from considering the Shannon-Weaver model of communication process,[4] combined with Wiener's cybernetics[5] and Goffman's epidemic theory.[6] However, we proceeded by placing the users as the central element in the conglomerate of systems, because users are the focus in establishment of a utility. We define four basic classes of interacting elements, each with subclasses:

Users (and potential users):
1. Students
2. Faculty
3. Staff and administration
4. Outside users

Formal Information Sources:
1. Curriculum
2. Library collections
3. AV/TV and media collections
4. Computer data

Informal Information Sources:
1. Personal collections
2. Invisible colleges

Management (policy making and allocation of resources)
1. At the level of each individual source
2. At the level of users
3. At the university wide level.

Graphically, we may represent these four classes and their interactions as follows:

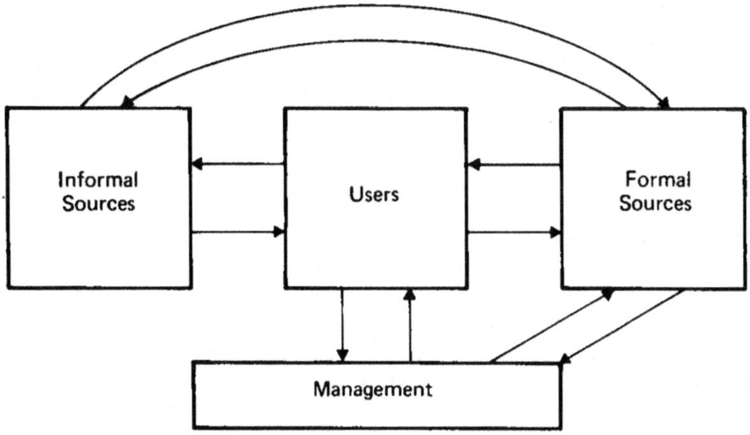

Empirical Studies

We assumed that the key for consideration of an information utility are the interactions between and within various elements (classes and subclasses) enumerated above. Thus, the empirical studies are oriented toward various stages of interactions: (1) inventory of factors in possible interactions within each element and, (2), interaction between elements from two, three, four different classes or subclasses. The following studies are underway, others are planned:

(1) Library use study: circulation: demands over time; demands by users, characteristics of non-users, un-met demands; demands by subjects in relation to curriculum.

(2) Informal sources: private holdings of faculty: extent, nature, investment, availability, relation to formal sources.

(3) Computer use study: facilities, nature of use and users, demands and un-met demands; relation to other formal sources.

(4) AV/TV use study: same factors as under (3) above.

(5) Faculty research study: pattern of relations between faculty on the basis of their publications, as opposed to departmental organization.

(6) Faculty teaching study: pattern of relations between faculty on the basis of the sharing of students.

(7) Curriculum study: pattern of relations between subjects

in the curriculum; demands on curriculum; access to curriculum.

(8) Management study: nature of regulations imposed; pattern of policy making; information on basis of which policy and allocations are made.

Organization of the Project

The very organization of the project may be in an anecdotal way quite illustrative of the problem attacked. The project is conducted under the auspices of the Complex Systems Institute, a newly-formed teaching and research unit that crosses departmental and subject lines. The aims of the Institute are both to study large complex phenomena and to come to an understanding of the political and moral factors in redesigning technology and the environment in such a way to satisfy human needs. Information and information systems clearly qualify as complex problems. It is also clear that the complexity of problems associated with information, their understanding and solution, cannot be sought and found within the scope of any one traditional source of activity, any one discipline. In Project INFUT these ideas served as a base for organization. The principal researchers came from three departments: Library and Information Science, Systems Engineering, and Operations Research; other researchers and advisors have come from Management, Computer Science and Sociology. Leaders of local utilities were interviewed as were key administrative personnel in the University: librarians, computer managers, and faculty on curriculum committees.

PROPERTIES OF UTILITY AND PUBLIC UTILITY

The word Utility refers to two related but still distinct concepts: (1) the state or quality of being useful and (2) the capacity to satisfy human wants or desires. Intuitively, we consider an enterprise as a utility when it exists primarily for the public good (whatever that elusive ethical concept may denote) - ideally, as individuals we would like to avail ourselves of the service provided by a utility when convenient to us and use the service for our purpose and in our own way relatively independent of the enterprise. The public utility as a distinct type of enterprise evolved over centuries as a result of the public desire for feasible and useful services (or commodities) that individuals organized in a society did not have a capacity to provide for themselves. This type of economic institution exhibited one distinctive aspect: the nature of the service was such that it required specific type of regulation.

In viewing the economic relations between producers/providers of services or commodities and consumers/users we may perceive a continuum from 'private' to 'public'. (This is not a consideration of ownership of an enterprise. At this point it is immaterial who is the owner.) Considering the end points as ideal, on the 'private' end we have the arrangements made directly between a provider and a user; interactions, effects and benefits are directly dependent upon any or all individuals involved. Moving away from the 'private' end of the continuum the relation becomes more complex and indirect, governed by forces of supply-demand, open market, planning or the like; the actions of some individuals may considerably affect the welfare of great number of others; the benefit an individual obtains from the service or commodity is dependent on the degree of his direct contribution to its provision. At the other end are the ideal 'public' conditions, where an additional and main governing force is public regulation, where benefit to the individual is relatively independent of his contribution. Between the two extremes are clusters of enterprises that relate in various ways sources and users of scarce services or commodities. Public utilities are that cluster of enterprises that moved or are moving more than other economic enterprises toward the 'public' end of the continuum.

As public utilities moved along the continuum they became characterized by a certain set of properties.[7] We may classify the generic properties of public utilities as follows: (1) *Monopoly*: a public utility enjoys a territorial monopoly because of necessity for large capital investments and public inconvenience of having direct competition; but competition is still possible as between utilities with general similar end uses (e.g., a user's choice between gas and electricity) and as between spending for a utility and any other spending (e.g., choice between buying an electric appliance or anything else). (2) *Social responsibility and licensing*: an enterprise becomes a public utility by a territorial license from some public body; the license carries the obligation to serve the entire encompassed population under reasonable conditions. (3) *Franchise*: along with license the public (i.e., public body) provides a public utility with some rights without which the service would be impossible (e.g., right of passage); therefore the public gives up something. (License is the charge and franchise is the means by which the charge can be carried out.) (4) *Economic regulation*: a third party regulates the price of service and is entrusted with prevention of abuses generally associated with monopolies. (5) *Service regulation*: third parties regulate the standards of service and the provision of facilities and ensure compliance with social responsibilities of a utility. (6) *Users and utilization*: the users of public utilities have certain regulated responsibility toward the utility, among others, in ways service is used; there is an effort in user education; in general, the service is brought as close as possible to the user, and used at the convenience of the user (at least ideally); utilities tend to have regular distribution of users: from a very few very large users to a large number of comparatively very small users.

Enterprises that desire or are forced to assume the role of a public utility have to assume to some degree the above properties. They may also have to develop methods and strategies necessary for fulfilling these properties. The many facets of public utilities are further illustrated by a number of fields in which the utilities are a subject of study, fields such as economics, sociology, jurisprudence, engineering, political science, ethics, management, etc. Throughout the world public utilities are subject to laws; the concerns of laws are remarkably similar regardless of political or economic system. Thus these general properties of public utilities exist across geographical boundaries.

Finally, one must discuss the basic nature of services provided by utilities. Public utilities provide services in these general areas: (1) provision of energy (electricity, gas, etc.); (2) provision of water; (3) transportation of people and goods (common carriers); (4) communication (telephone, radio transmission of signals, etc.); (5) provision of environmental services (sanitation, conservation, etc.). All these have one thing in common; they are the basic stuff of everyday living in an organized large and complex society. As individual human beings we need many other things to function and survive, but as integral members of modern and large social groups we also need services of utilities as an integral part of our life. The integration in everyday life of people in a society that has reached a certain level of development and complexity is the basic, common nature of services or commodities provided by public utilities. And this is the reason why concerns about utilities are fundamentally the same across national and political boundaries in societies that have reached that certain level. As the complexity and development of a society increases we can expect that utilities will play a larger role, that a number of enterprises previously not considered as utilities will assume properties of utilities, and that entirely new utilities will be developed dealing not only with functioning of life but more and more with quality of life in a society. Information utilities may be one of those.

INFORMATION AND INFORMATION SYSTEMS

Prior to considering properties of information systems as utilities we have to consider some general aspects of the underlying phenomena and related processes and systems, namely, information, communication and information systems. The nature of the underlying phenomena affects the specific characteristics of a utility.

Communication is a process (sequence of events) where something called information is transmitted from one object to another. The first object is usually called source, the second destination; there can be a dynamic exchange and feedback between them. The objects can be people, organisms, or machines. The concerns and problems in communication can be divided into different aspects or levels: technical, semantic and behavioral, or effectiveness. Relevance then can be defined as a measure of the effectiveness of the contact between a source and a destination in a communication process. Finally, systems that carry out communication processes are called information systems.

As a fundamental phenomenon we do not know what information is any more than we know what is energy, gravity, electricity, and so on. However, as we study the manifestations, behavior and effects of these fundamental phenomena and the properties of systems that deal with them, we can objectively study the manifestations, behavior and effects of information and properties of information systems. In investigation of any fundamental phenomena we start from intuitive understanding.

People have a strong intuitive understanding of 'information' They can and do use information with or without a formal understanding of information (e.g., Spinoza: "In order to know, there is no need to know that we know, much less to know that we know that we know"). It is with this everyday understanding of information that the vast majority of users approach information systems. They make requirements upon information systems within THEIR framework of understanding of information and not within the systems' framework. In this respect information utilities may differ from other types of utilities. We are suggesting that an information system in order to become a utility will have to be primarily oriented toward the intuitive understanding of information of its users. In general the intuitive understanding of information involves a variety of aspects, manifestations and relations, such as: (1) on the broadest level the understanding of information first and foremost involves the concept of knowledge - some content, something new, something previously unknown; (2) next, it involves objects - somebody or something who wants or needs it and somebody or something that contains, possesses or holds it; (3) a variety of possible processes - transmitting, providing, uncovering, observing, deducing, etc.; (4) a variety of possible forms - facts, data, messages, statements; (5) a variety of representations - language, symbols; (6) a variety of physical manifestations - signals, media; (7) and all the time and above all is implied the idea of some eventual use above and beyond information itself - for some purpose..to act, to decide, to enhance, or control other processes, to make money, to survive, to change or add to what we know, or even to enjoy or aesthetically appreciate.

In a formal vein the phenomenon of information is studied in many diverse disciplines and so confirms the complex ramifications and many manifestations associated with information. Intricately and integrally involved with information is the phenomenon of knowledge which is also investigated in many fields (e.g., Pears[8] relations between concepts of information and knowledge have been discussed at some length.[9] In dealing with the phenomena of information and knowledge we have to consider Russell's important point that a definition of knowledge (and information) involves certain inevitable vagueness and inexactitudes.[10] Consequently, we have to expect and adapt to certain inevitable vaguenesses and inexactitudes in any information system. Thus, in any information utility (with the possible exception of information utilities that exclusively deal with transmission of signals) we have to expect vaguenesses and inexactitudes in their service. Therefore, this may be one of the unique properties that an information utility does not share with other types of public utilities and may prevent evolution and design of information systems into utilities in a manner comparable to public utilities.

Information systems are an integrated assembly of many diverse elements, which may be classified and described in numerous ways. For consideration of utilities we may recognize that isolated elements of information systems may exhibit physical, biological, psychological and/or sociological properties. (However, we also have to recognize that the properties of the system as a whole, because of interactions, may be different than properties of its elements or slanted to some degree toward some elements):

Physical properties: involve among others: the physical nature of signals, symbols or patterns used in representing information in a system (electromagnetic, chemical, optical, etc.); the form and distribution of these signals, symbols or patterns;[11] the physical media in which they are recorded; interestingly, it is through these physical aspects that we associate information and energy.[12]

Biological properties involve: structure and processes of the brain and nervous system and effects of these on the structure and limitations of other man-built information properties such as redundancy, association ordering and quasimetric relations;[13] interestingly it is through the biological aspects that we associate knowledge and life - "life is a knowledge process".[14]

Psychological properties involve: states in individuals that enable us to perceive and refer to objects other than ourselves; ability of learning, i.e., ability to acquire, store, associate, modify, recall and apply knowledge; the property of compatibility between knowledge states and stores as a prerequisite for communication.

Sociological properties involve: the context of the social environment in creation, distribution and use; language; behavior of public knowledge: e.g., distribution of creators (Lotka's law) and of records (Bradford's law), epidemic interactions,[15] cumulation, obsolescence, etc.

Finally, we can now consider theoretically the properties of information systems as a whole. The elements in an information system interact to produce in the system as a whole the properties of: (1) redundancy of information; (2) association between elements of information (classification); (3) ordering of elements of information; (4) quasimetric relations between elements of information.

If these properties are not achieved we really do not have an information system, or we may have an information system that can not function effectively and cannot survive for any length of time.

In approaching information systems as utilities we will have to consider that the various combination of diverse elements, various properties and various interactions as found in various information systems will affect the degree to which we can institute these systems as information utilities. Some combinations may be more and some less suitable as a utility.

In summary, fundamental consideration of information systems as utilities are affected by:

(1) Intuitive understanding of information by its users.

(2) Inevitable vagueness and inexactitude associated with the phenomena of information and knowledge.

(3) Diverse combination of properties and interactions exhibited by elements in information systems.

TOWARD SPECIFICATION OF PROPERTIES OF INFORMATION UTILITIES

Information systems, such as libraries, have been a part of human civilization since it began. There is no civilization without formal information systems that cumulate, preserve and disseminate recorded knowledge. The great number of types of information systems that have evolved are due to a number of factors: (1) the environment required development of different systems that utilize different combinations of manifestations of information and assumed different combinations of properties; (2) the variety of uses of information and its purposes affected the development of different systems for different uses; (3) the development of various techniques and tools for use in information systems resulted in different systems for the same manifestations and uses; (4) a variety of economic constraints imposed upon systems forced development of systems to meet the constraints.

Eventually, we will have to investigate what types of information systems can be instituted as information utilities. In many countries some information systems have already fully assumed operations as a public utility. These systems are characterized by predominance of elements exhibiting physical properties and oriented toward technical aspects of communication: they transmit signals, e.g., telephone companies, common carriers of computer, TV, and other signals.

In addition, there are some other types of information systems that partially assumed operations as public utilities. These systems are characterized by predominance of elements exhibiting psychological properties; certain types of education systems, such as educational laboratories, are used as if they were utilities.

At present conditions in highly-developed countries may be such that there is a need for more information systems to assume operations as a utility, particularly the systems with a predominance of elements exhibiting sociological properties. To be more specific, certain classes of libraries, information retrieval systems, information analysis centers, public information systems, management, financial and business information systems, control and decision systems, transportation information systems, inventory systems, health information systems, government information systems etc. along with a certain class of education systems should move toward becoming an enterprise similar to a public utility. They should become an information utility.

Now let us turn to some preliminary observations from the various studies in Project INFUT in answer to the following question: to what extent are properties of public utilities applicable to information sources and systems in a university environment in general and in the specific environment provided by Case Western Reserve University in particular?

(1) Monopoly. Various information systems in the university do have monopoly on particular type and/or form of information for their users. In a university environment there are actually a number of information systems that eventually could be considered as information utilities, each relatively independent of the other, (in this respect there is a strong parallel to public utilities in a city). Direct competition of information systems and sources in a university is for funds not users. A great amount of redundancy of information is available from the relatively independent informal sources (private holdings, invisible colleges); these sources are not a utility, they supplement and/or replace at times the formal sources such as libraries and curricula (in this respect there is a strong parallel with the situation of transportation utilities). The basis of the distinction between formal and informal sources is the fact that policy for access for the former is determined by a management and the latter by the individual source itself.

(2) Social responsibility and licensing. Licensing of different information systems exists but is more of an informal nature. Some formal licenses are issued by the administration and/or faculty government bodies, e.g., approval of specific curricula, specification of the role and users of the computer complex, etc. In our particular situation of a private university the outside influences on licensing are minimal. Most licenses are due to long academic tradition. Interestingly, newer information systems and sources (lacking an understanding of their role as governed by tradition), are in constant transition and are formally licensed more than the older systems. The social responsibility of information systems is well understood by the professionals in the systems and by the users - however, as in public utilities these two understandings (by the enterprise and by the public) may not coincide, thus there is ground for tension.

(3) Franchise. For us this is the least understood property of an information system as a utility. Licensing is a charge to do something, franchise is giving something so that the license can be carried out (e.g., to a telephone company we give a right of passage over public or private land and more subtly and individually, we give up to some extent our right to privacy). But what can we give to (or give up for) information utilities. Some right to our minds? This property will have to be either conceptualized differently or simply considered not a part of an information utility.

(4) Economic regulation. In a private university there is no pricing of services by third independent parties as found in public utilities (in public universities there often is). However, there is a strong system of regulations in allocation of funds and resources by the management, i.e., the management assumes the role of an economic regulating agency. In general, there are no, and never will be, economic units as in public utilities suitable for pricing and thus direct economic regulation. Thus, we may expect that information utilities will never be directly economically regulated. For us, the economics of information is the second least understood property of an information utility. We may neglect or avoid the property and problem of franchise; however, there is no way to avoid the problem of economics.

(5) Service regulation. There is a minimum of direct service regulation (including prescription of standards) of information services by third parties as found in public utilities. (In public universities the attempts to regulate service by third parties are more prevalent and on the increase.) However, there is strong regulation of the condition of service as a part of any given information systems. The management and the professionals in the systems act as a service regulatory agency. As in public utilities the evaluation of quality (as opposed to conditions) of service is rather elusive and not formally pursued.

(6) Users and utilization. In distribution of use, information sources and systems in a university strongly parallel public utilities. However, the information systems are not really organized according to users' problems, needs and usage patterns, which results in user frustration, e.g., the curriculum is organized according to traditional departmental lines and the subjects and students more and more are not; libraries are organized independently of departmental or subject organization; computer facilities are organized around the computer, etc. One element is more forceful in university information systems and sources than in public utilities: utilization is regulated much more - when, why, what for, in what sequence, etc; there is an element of 'force' about use present in all education systems. Consequently, users do not feel the same type of responsibility as toward public utilities.

In summary, some information sources and system in our university have some hallmarks of a public utility, but it would be stretching it very much to say that any of them as presently organized is indeed a utility. To be operated as a utility among others there is a need for (1) a significant reorientation toward use and users; (2) a significant reorientation in the approach to cost and economics of information; (3) a much more specific structure of economic and service regulation; (4) instilling in users the feeling of responsibilities and relationships similar to those they have toward public utilities - and this may be the hardest to achieve.

CONCLUSIONS

Calling information systems information utilities will not make them into any type of a utility. Nothing short of a positive difference in the service to users can make them into a utility. Some general framework is needed within which this difference can be achieved. Basically, in this paper we speculated about the bare bones of a conceptual framework around which information systems can be operated as information utilities. However, much work remains to be done to complete the framework. We need further work on: (1) determining what types of information systems are and what types are not suitable to be operated as a utility; (2) clarifying what properties of public utilities are absolutely necessary for an information utility; (3) clarifying what properties are unique to an information utility; (4) determining what strategies are needed for achieving these properties.

References

1. Bell, D., *The Coming of Post-Industrial Society*, Basic Books, New York, 1973.

2. Slamecka, V., and Zunde, P., Science and information: some implications for the education of scientists, *The Proceedings of the International Conference on Training for Information Work*, F.I.D. publication 486, Rome, 1972, 357.

3. National Science Foundation Grant No. GN-36085 to Complex Systems Institute, Case Western Reserve University. William Goffman, principal investigator, Douglas Rothenberg, Tefko Saracevic, Paul Stephen, principals. Duration: Oct. 1972 - Oct. 1974.

4. Shannon, C.E. and Weaver, W., *The Mathematical Theory of Communication*, University of Illinois Press, Urbana, 1949.

5. Wiener, N., *Cybernetics: or Control and Communication in the Animal and the Machine*, 2nd ed., MIT Press, Cambridge, 1961.

6. Goffman, W. and Newill, V.A., Communication and epidemic processes, *Proceedings of the Royal Society, A*, 298, 316, 1967.

7. Kahn, A.E., *The Economics of Regulation: Principles and Institutions*, 2 vols., Wiley, New York, 1970, 1971.

8. Pears, D., *What is Knowledge?* Harper, New York, 1971.

9. The point was often made that 'information' can not be treated only in statistical terms and that for most practical purposes 'information' and 'knowledge' are so integrally related that they are nearly synonymous. (e.g., for a passionate argument of this point see essay by Lorenz in Note 15). We accept that 'knowledge' is a state, a notion, a concept - a phenomenon that can neither be measured nor precisely defined. Characteristics and manifestations of knowledge include: perception; search and creation; rationalization; organization, systematization, accumulation, and obsolescence; association, redundancy, feedback; language, representation and records; transmission, teaching and learning; uses and problem solving; environment, culture and society. We can observe that these manifestations can be grouped into two broad classes: those that refer to the 'holdings' in the minds of individuals and those that refer to communication from mind to mind. It is within the context of these latter communication-related manifestations of knowledge that the notion of information is brought in and related to knowledge. We may view 'knowledge' in a static and 'information' in a dynamic sense - knowledge may be compared to

potential energy and information to kinetic energy. In that sense, we may view knowledge as being in part a function of information. Knowledge refers to a person's range of information and information to items of knowledge; knowledge carries the connotation of a higher degree of systematization (this is evident in the often-used type of definition: "organized data is information, and organized information is knowledge". But then what is data?) However, it is also evident in another sense that information is a broader concept, involving in addition to knowledge, all kinds of signals and representations that affect our senses and biological processes. In any case, the variety of viewpoints is a reflection of a variety in reality, thus substantiating the argument that in approaching information systems as utilities we have to consider that the very nature of different manifestations or aspects of information may be the prime factor affecting the degree to which an information system can become a utility.

10. Russell, B., *Human Knowledge - Its Scope and Limits*, Simon and Schuster, New York, 1948.

11. For example, Shannon and Weaver, *op cit.*, and Zipf, G., *Human Behavior and the Principle of Least Effort; an Introduction to Human Ecology*, Addison-Wesley, Cambridge, 1949.

12. Tribus, M. and McIrvine, E.C., Energy and information, *Scientific American*, 225, 179, 1971.

13. Goffman, W. and Morris, T.G., Jr., Quasi-metric spaces and information systems, *Proceedings of the International Congress of General Systems and Cybernetics. (Oxford, England 1972)*, Gordon and Breach, London, (in press).

14. Pribram, K.H., ed., *On the Biology of Learning*, Harcourt, New York, 1969.

15. Goffman and Newill, *op cit.*

WORLDWIDE NETWORKS FOR INFORMATION, COMMUNICATIONS AND COMPUTERS

Kjell Samuelson

Stockholm University and Royal Institute of Technology,
Stockholm, Sweden.

INTRODUCTION

A number of global networks are in planning and are gradually coming into existence. Their main constituents are information stores or databases, computers, and telecommunications both nationally and internationally. Earlier studies have mainly been focused on optimization or simulation of traffic and loads under certain conditions. Few efforts have been made to distinguish the functions and properties that are significant for each of the different kinds of networks. The following work is an attempt at formalizing the main types of global networks in terms of structure, contents, distribution, and evolution. The nature of the situation is that of an open general system which might as well be treated as such, on the basis of past knowledge as applied to a world system. Also, the networking solution may become the loop-closing cure of something that is worse than information/message pollution, namely, "the ignorance explosion", which keeps people with ideas separated.[1]

GENERAL NATURE OF NETWORKS

The general nature of networks for information, computers and communications has been described by Borko[2] in terms of shape, interconnection, element type, and distribution. Some studies have been aimed at modelling the message transfer patterns,[3] while other work has been directed to structures, evolution, media, systems analysis, and international issues.[4] In earlier studies the network situation has generally been treated as a problem of transfer with less attention paid to media, storage, processing and various user communities. Today we may quite clearly distinguish the following networks: (1) Information Networks, (2) Computer Networks, (3) Communication Networks, (4) Relational Networks, (5) Network Hybrids. Their elements and interconnections, links and nodes are quite different, as are the structures and properties of the networks. The growth toward network hybrids is characterized by openness and the hybrids can be categorized as "constitutive" in the sense defined by von Bertalanffy.[5]

Information Networks

Today's information networks are usually built from nodes such as databases, library storage, or large files, that are continuously updated, extended, and maintained. Altogether the nodes function as a stack, divided into manageable, distributed portions that are organized and processed in a more efficient way (e.g., inverted files) than just a pile of knowledge. We recognize three basic elements (Fig. 1, 7, 8):

MB Masterbases. The original, centralized stores of full information divided into major content fields. Retrieval time is quite long and "cold" storage costs are moderately low.

DB Databases. The decentralized, retail stores that are tapped from masterbases, and mainly contain current information or surrogates. Retrieval time is swift and the more costly "hot" files are used for SDI.

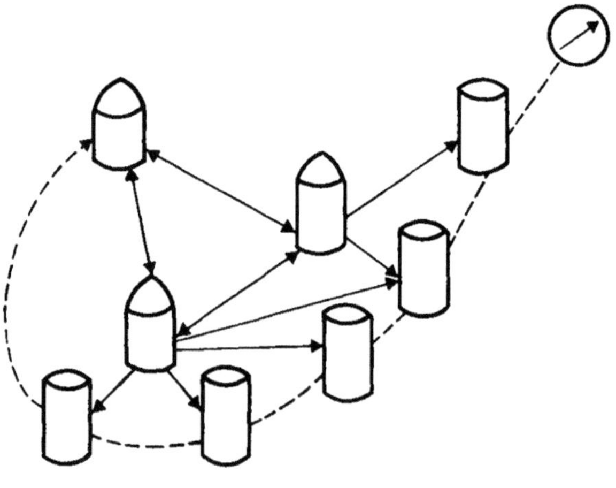

Figure 1 – Information Network

RS Relay-Switches. Specialized databases with locator files, may guide the information seeker to the most appropriate, likely and relevant database or masterbase. The very first entry may be facilitated through a manual, catalog-type referral-directory.

It is worth noting that in contrast to telecommunication practices, the RS is not used for traffic or transit switching to a known address, but for directory look-up in locator files, i.e., a specialized database that can tell where the address is, if it eventually exists.

In practice, the inter-database organization for MB and DB tends to become hierarchical because of the supply diffusion from the masterbases to the distributed retail databases. Often, the access-direction and search-routing run the opposite way, unless non-urgent but exhaustive and retrospective search is ordered primarily. The present situation for our information networks is that the processing of the DBs and MBs at the nodes is done through locally existing computers which are not interconnected and may in fact not be for quite some time, if ever.

Computer Networks

There exist a number of time-shared computers, that have been interlinked into networks (Fig. 2). Such computer networks have a number of local user terminals connected to each node computer. The size of the user population that can be reached is limited. The prime reasons for computer networking have been, increased reliability or fallback and shared software or CPU utilization. So far, only to a smaller extent has the interconnection contributed to the sharing of information, data, and files between users at remotely distant computer nodes. The computer networks are gradually improving with respect to switching or terminal and interface message processing. Also, new technologies or computer generations appear that lead to structural changes. Moreover, there ex-

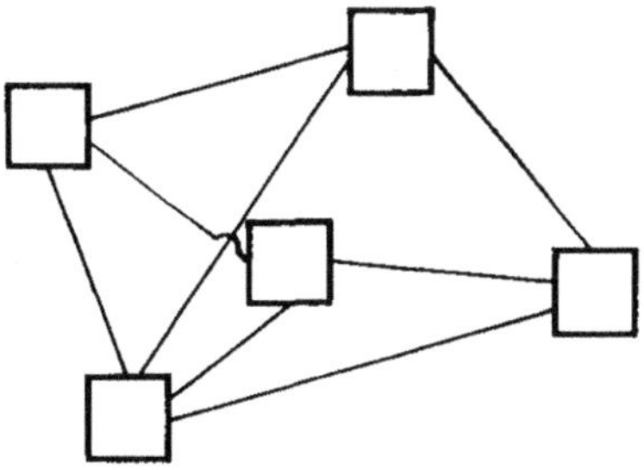

Figure 2 – Computer Network

ist computer networks that do not have retrieval packages and that contain no information or data of the kind handled by the information networks. These circumstances justify making a distinction.

Communication Networks

For many years a great variety of communication networks have existed (Fig. 3). The most well known ones are the widespread telecommunication networks represented by the common carriers, telephones, TELEX, TWX, TELPAK, etc., or mass communication, such as TV or CATV. Lately, a number of radiolinks, intercoms, and picturephones have come into existence. Service stations and outlet units are globally distributed among millions of organizations and people who have reasons for communicating. Communication networks in general do not have facilities for information storage, processing, and computation like the two preceding kinds of networks. Global communication networks are as widely distributed as can be.

Relational Networks

An exponentially growing number of relations exist between people and activities in this world. Although most of the relations cannot be formalized there are some that lend themselves to description in terms of relational networks (Fig. 4). The familiar types are citation indexes, sociograms, ethnical bonds, organization charts, at the macro-level and logical files, associative linkages, cognitive maps, thesaurus structures, classificatory ordering schedules, etc., at the micro-level. We shall mainly consider the macro-level logical networks. They are usually embedded in physical networks that can be formally described as cellular systems.[6] These are more general than networks since their parts may have more than two ends and can thus be connected to the remaining system through more than two node values. Communication often occurs along the relational paths. In a cellular system communication will be more frequent across the boundaries between neighbouring

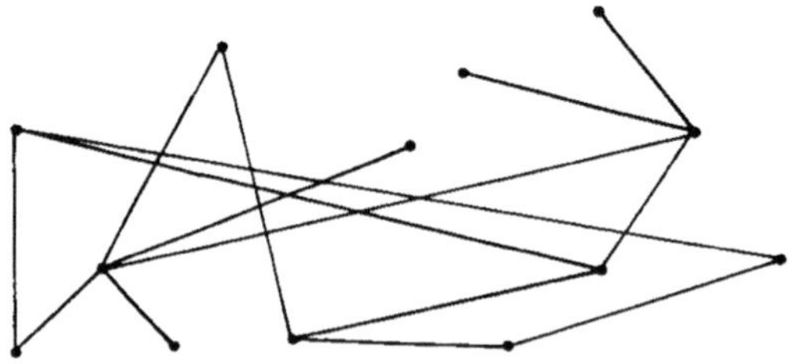

Figure 3 – Communication Network

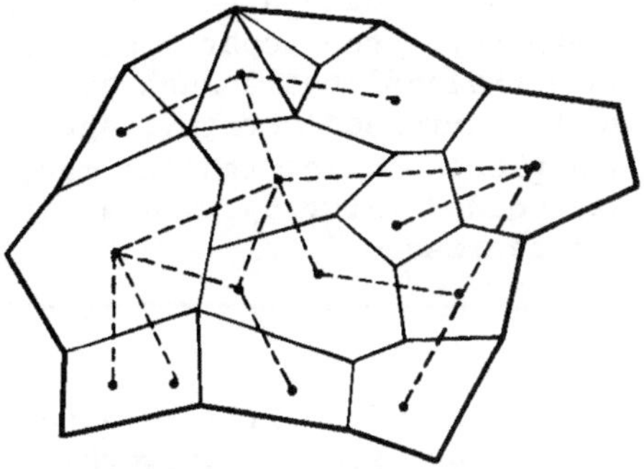

Figure 4 – Relational Network

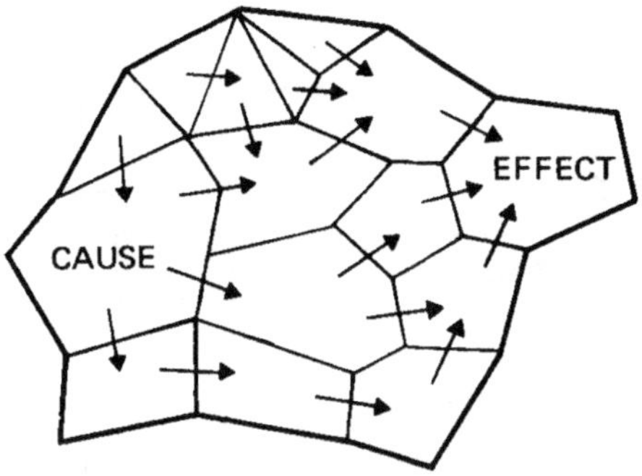

Figure 5 – Cellular System with Communications

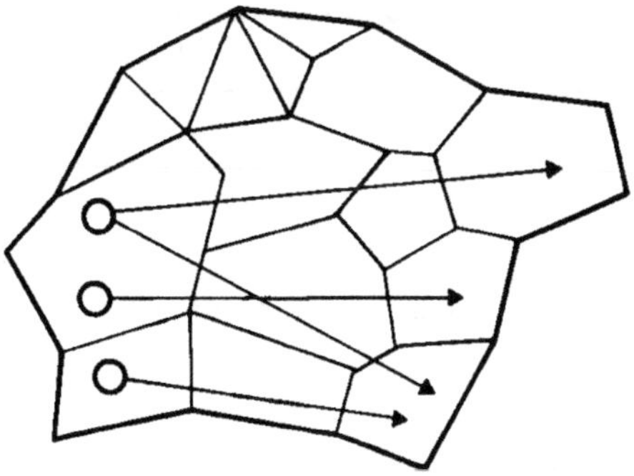

Figure 6 – Communication across Cellular System

cells (Fig. 5). For instance, only after long-distance physical telecommunication links have been superimposed will remote cells become connected. On a regional basis such extended interconnection may for a while lead to increased entropy, agglomeration, and congestion. However on a global scale with its many variety options those effects can be avoided (Fig. 6), and as in an open general system entropy may be decreased.[7]

Network Hybrids

Global networks in the future will be a conglomeration of many different networks. The four kinds of networks (information networks, computer networks, communication networks and relational networks) described above will become superimposed and result in network hybrids. They can be formalized in terms of a graph (Fig. 7) the flows and elements of which are indicated by a matrix (Fig. 8). Although these figures are selfdescriptive it is worth mentioning

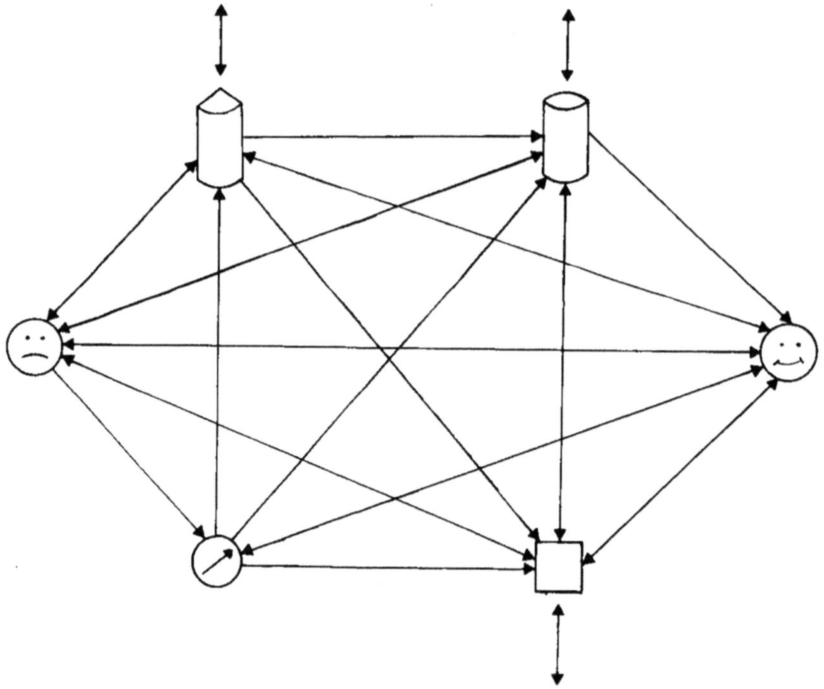

Figure 7 – Network Hybrid

IN ⟶	OUT: Finder					Receiver
Searcher, Seeker, (Finder)	0	1	1	1	1	1
Locator, Relay-switch, Referral-directory	0	0	1	1	1	1
Master-base	1	0	(1)	1	1	1
Data-base	1	0	0	(1)	1	1
Computer	1	0	1	1	(1)	1
Supplier, Sender (Receiver)	1	1	1	0	1	0

Figure 8 – Transition Matrix for Network Hybrid

that there exist big quantitative differences regarding the number of connections as well as the multitude of nodes, especially on a worldwide scale. Globally one may take into consideration that the switching and referral be divided among users' manual directories, national terrestrial referral-switches and a few international relay-satellites in orbit. The approximate magnitude of the elements in such global network hybrids is indicated in Fig. 9. The functional and quantitative characteristics are reflected in the overall network structural considerations. Probably the major benefit of network hybrids is that their characteristics are typ-

NODAL FUNCTION	ELEMENT	APPROXIMATE MAGNITUDE
Relay-satellite		h . 10
Referral-switch, ground locator		$i \cdot 10^2$
Masterbase		$j \cdot 10^3$
Database		$k \cdot 10^4$
Computer		$l \cdot 10^5$
Manual entry directory		$m \cdot 10^7$
Users at service stations/terminals		$n \cdot 10^8$

Figure 9 – Global Network Elements

ically those of "constitutive" systems in von Bertalanffy's sense.[8] The symbiosis and cross-fertilization between the four different network kinds lead to an amplification or reinforcement of their respective properties, to the effect that the advantages of the whole hybrid net are far more than those of the sum of its parts.

STRUCTURAL CONSIDERATIONS

The structural properties of networks can to some extent be described through graph theoretical methods, but they tend to be cumbersome for loop-counting when thousands of elements are interconnected. Each of the structural measures is sometimes particularly well suited for the representation of a specific kind of network. Only occasionally can the isolated networks be treated as having "summative" characteristics, to follow von Bertalanffy. With regard to the fact that the actual networks behave like "constitutive" systems having multi-purpose functional structures and dynamic interrelationships the following approach at formalizing complexity was made.[9] It appears that it is basically not the number of nodes N, that is most critical but rather the way in which they are interconnected, the number of such connections C, and the connection/node ratio. We therefore define as

(1) complexity: $\frac{C + 1}{N} \cdot (N - 1)$

(2) relative complexity: $\frac{C + 1}{N}$

(3) complexidense: $\frac{C + 1}{N \cdot \delta} \cdot (N - 1)$

The first two expressions describe different characteristics of the network structure. Harary[10] suggested that the above formalized complexity expression be divided by putting in the denominator the graph diameter δ, i.e., the longest track distance between two remote nodes or vertices. This reduces the complexity measure and yields higher precision but in a sense makes the causality less obvious for small systems and again makes shorthand calculation less rapid. The author therefore proposed that such a fraction as the third expression be used to differentiate between hard and loose coupling or, for example to compare two highly expanded flow production-lines, by expressing the inverse of a longest structural span or a sort of density. In that case the digraph flow directions should be known and the diameter should be exactly known in order not to introduce another source of error, while the earlier two formulas are less sensitive and can be used for rough estimates.

Other measures for judging network designs are accessibility and flexibility[11] or connectivity and cohesion.[12] These measures lend themselves best to computer networks and moderate-size communications. The complexity measures would serve information networks that may be so huge that they are imperceivable or unsurveyable. Flexibility of a network is defined as a quantity F, where $0 \leq F \leq 1$, N is the number of nodes and C is the number of connections, or

$$\text{(4) flexibility: } \frac{C - N}{N(N-2)}$$

The above expressions are shown (Fig. 10) for some simple and low-complexity structures. If the number of connections in a network is proportional to the number of nodes, then the flexibility varies inversely with the size of the network. For a fixed degree, the

	Connec- tions	Nodes	Dia- meter	Com- plexity	Relative complexity	Complexi- dense	Flexi- bilit
	C	N	δ	$\frac{C+1}{N}\cdot(N-1)$	$\frac{C+1}{N}$	$\frac{(C+1).(N-1)}{N.\delta}$	$\frac{C-N}{N(N-2)}$
	1	2	1	1	1	1	-
	3	4	3	3	1	1	-
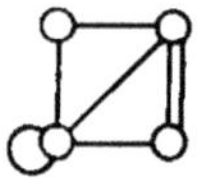	7	4	2	6	2	3	0.37
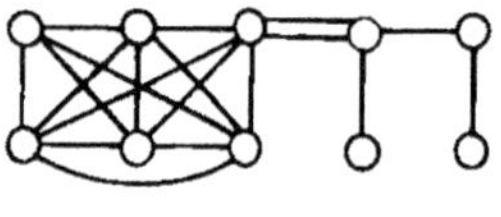	19	10	5	18	2	3.6	0.11

Figure 10 – Structural characteristics for simple and low-complexity systems network with fewer nodes has greater flexibility.

It is also desirable to have a measure of versatility, meaning that a variety of f files in the databases can be processed in different retrieval ways by using g program packages in the computers. A manifold of methods already exist for calculating reliability and availability that are all familiar.

Some design characteristics and crude features for the various network prototypes are indicated (Fig. 11) in line with the formal-

Design feature	Information Networks	Computer Networks	Communication Networks	Relational Networks
Evolution trend	Node dominated	Node dominated	Link dominated	Link dominated
Characteristic property	Few centralized nodes	Few distributed nodes	Many links	Many relations
Complexity	Low	Low	High	High
Relative complexity	High	Low	Low	High
Flexibility	Low	High	Low	Moderate
Node degree	Varying	Stable	Stable	Varying
Reliability	Low	High	High	Low
Accessibility	Low	High	High	Low
Structure	Hierarchical sometimes	Distributed	Distributed	Hierarchical sometimes
Density	Low	High	Low	High
Dependability	High	Low	Low	High
Efficiency	Low	High	High	Low
Use	Occasional	Occasional	Frequent	Frequent
Elements	Unique	Redundant	Redundant	Unique

Fig. 11

izations above. We hasten to say that this is by no means universally representative, but rather describes the conditions as they appear in many corners of the world.

In essence, one may conclude that it would be misleading to calculate structural measures from the mixed and synthesized network hybrids as they appear. The hybrids should rather be analyzed and decomposed into the main constituents. These may in turn be regarded as representing separate levels, scales and cones of resolution.[13] Thus, at the micro scale on the lowest plane, ideas are spread through interpersonal relations between individuals. At the next higher level different network communications take place, and so on up to the higher macro levels with international information flows.

GLOBAL NETWORKING AND DESIGN

From an overall design point of view it appears that a top down approach to implementation should be emphasized in order to avoid carving the iceberg stack of knowledge from the wrong end. Communications and service stations should be provided at an earlier stage than retrieval from databases or computerized processing. In reference to Fig. 9 the precedence relations for implementing the different kinds of networks and their elements should take place in an order that is shown in Fig. 12. These circumstances would give high priority to two-way, multipoint telesatellites for person-to-person, home-use communications, and mixed multimedia.

CONCLUSIONS

The globally-emerging network hybrids will no longer be limited by geographical boundaries or human groups in a cellular way but will comprise the whole planet.[14] The effect would, to quote Beer,[15] be a new means of managing modern complexity, as he refers to computers as part of a "nervous system" for the body politic and for control of the variety, i.e., the possible number of states of a system. The development of global network hybrids would make

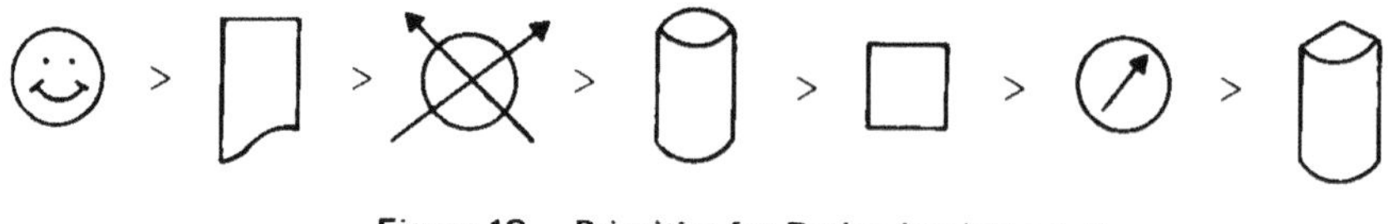

Figure 12 – Priorities for Design Implementation toward Global Network Hybrids

us advance to new and higher levels of abstraction.

The evolution of information networks and the orderly expansion of communications have been described in previous studies.[16] A few final remarks on the growth of network hybrids should be made. The evolution of information networks shows that the early structural stages of centralized mixed-star-trees or spiders become absorbed by interconnected short-cuts and mergers. Certain local efficiencies can always be achieved through nucleating activities. This will mainly be a matter of spatial economizing and is not intended for the sake of centralized control. In the long run the global networks for information will rather look like a handful of stars, with interconnected tips that again appear as smaller stellas. A similar constellation is for instance shown by the major airlines' route distribution. It is also estimated that the flows of information and materials transportation will mutually show push-pull effects for a while. After that period, information transfer may partially become a substitute to physical transport and commuting. The networks would not become fully connected or overly multi-connected, but rather be a mapping of socio-economical, geo-linguistic and ethnic-cultural patterns within a world system.

References

1. Lukasiewicz, J., The ignorance explosion: a contribution to the study of confrontation of man with the complexity of science-based society and environment, in *Transactions of the New York Academy of Sciences, Series II*, 34, 373, 1972.

2. Borko, H., The nature of networks, in *Global and Long-Distance Decision-Making, Environmental Issues and Network Potentials*, Samuelson, K. *et al.*, FID/TM at ASIS meeting in Denver, Colorado, 1971. (FID Publ. No. 493).

3. Nance, R.E., Korfhage, R.R. and Bhat, U.N., Information networks: definitions and message transfer models, in *Journal of the American Society for Information Science*, 23, 237, 1972.

4. Samuelson, K., Systems design concepts for automated international information networks, in *American Society for Information Science Annual Meeting, 32nd, San Francisco, October 1969. Proceedings*, 6, Britannica, Chicago.

Samuelson, K., World-wide information networks, in *Conference on Interlibrary Communications and Information Networks (CICIN), Airlie House, Warrenton, 28 September-20 October 1970*, American Library Association, Chicago, 1970.

Samuelson, K., International information transfer and network communication, in *Annual Review of Information Science and Technology*, Cuadra, C., ed., 6, Britannica, Chicago, 1971, Chap. 9.

Samuelson, K., Global and long-distance decision-making in *American Society for Information Science Annual Meeting, 34th, Denver, November 7-11, 1971, Proceedings*, 8, Britannica, Chicago.

Samuelson, K., Gezelius, R., Werner, H. and Johannesson, N.O., *Mixed Multimedia, Development Potentials for Picturephone, CATV and Teleprocessing*, FID/TM, Stockholm, Sweden, 1972.

5. von Bertalanffy, L., *General System Theory; Foundations, Development, Applications*, George Braziller, New York, 1968.

6. Langefors, B., *Theoretical Analysis of Information Systems*, 4th edn., Auerbach, Philadelphia, Pa., 1973.

Nador, P., Data administration and documentation associated with data base management systems, in *Systems, Cybernetics and Information Networks*, Samuelson, K. *et al.*, FID/TM, Stockholm, Sweden, 1972.

7. von Bertalanffy, L., *Robots, Men and Minds; Psychology in the Modern World*, George Braziller, New York, 1967.

8. von Bertalanffy, L., *General System Theory; Foundations, Development, Applications*, George Braziller, New York, 1968.

9. Samuelson, K., Management information systems and nervous systems - a comparison of design concepts, *Proceedings of IEEE Systems Science and Cybernetics Conference, October 22-24, 1969*, Philadelphia, Pa.

10. Harary, F., Personal communication.

11. Nance, R.E., Korfhage, R.R. and Bhat, U.N., Information networks: definitions and message transfer models, in *Journal of the American Society for Information Science*, 23, 237, 1972.

12. Wilkov, R.S., Analysis and design of reliable computer networks, in *IEEE Transactions on Communications*, 20, 1972.

13. Beer, S., *Management Science; the Business Use of Operations Research*, Doubleday & Co., Garden City, New York, 1968.

14. von Bertalanffy, L., *Robots, Men and Minds; Psychology in the Modern World*, George Braziller, New York, 1967.

15. Beer, S., Managing modern complexity, in *Management of Information and Knowledge; A Compilation of Papers Prepared for the 11th Meeting of the Panel on Science and Technology*, U.S. Congress, Science and Astronautics Committee of the House, U.S. Government Printing Office, Washington, D.C., 1970, 41.

16. Samuelson, K., International information transfer and network communication, in *Annual Review of Information Science and Technology*, 6, Cuadra, C., ed., Britannica, Chicago, 1971, Chap. 9.

CRITERIA FOR A VIABLE INFORMATION RETRIEVAL SYSTEM - ADAPTABILITY

James K. Mullin

University of Western Ontario, London, Canada.

INTRODUCTION

This paper deals with experience in constructing two commercial information retrieval systems and most particularly with current research in the implementation of an adaptive information retrieval system, the Western Automated Storage of Information System. The WASIS system uses an index utility function to learn from experience which indexes are most useful in pruning the set of possible records to the set of desired user responses. The system accepts suggestions from a data base administrator as to which fields of a record are to be considered as potentially indexable. The criteria of the values in the index attribute value pairs are also given by the data base administrator. The values may be ranges or more sophisticated functions than the conventional numeric or alphabetic $> < = \neq \leq \geq$. The actual development of a suggested index is determined dynamically by the system. An important feature is that the user does not know the name - or, indeed, the existence of any index. Thus requests to the system never depend on user knowledge of the current index structure as an IDS.[1] Because of this, the situation cannot arise that previous complex requests, which may later become infrequent, require the maintenance of indexes which are no longer economically justified.

SOME PROBLEMS IN DESIGNING RETRIEVAL SYSTEMS

There are a number of information retrieval systems available in the marketplace. MARK IV and GIS[2] are examples. All of these systems offer useful facilities for simpler file maintenance and retrieval than a customized programme. All of these systems also have some defects in that there are tasks which they cannot perform or cannot perform simply. In the ASIS exercise[3] Climenson points out good points and drawbacks of a number of earlier systems.

The author has been involved with two major non-academic information retrieval systems. Although large amounts of manpower, ten to twenty man years, were expended on these systems, neither could be termed a success. A major contributing factor in weakening these systems was a lack of adaptability. In trying to achieve generality, the systems became large and unwieldy. The original design goals - probably a good initial set of goals - were achieved, but the systems resembled dinosaurs and were not able to adapt to changing goals and usage. Some principal errors were:

Error 1: Predetermined record structure and record sizes. Surely this fault is and was obvious. However, one can be easily tempted - given a special-purpose agreed application to save time on generalization in order to get the system operational. Once the system approaches the operational phase, the changes come. One must avoid temptation.

Error 2: Predetermined "generalized" description of record structures and sizes. This is the same error cited above, but carried out at one level of indirection from the problem. The designers at this stage recognize error 1 and generalize the record structure. They then design a fixed non-general table to define the generalized record structure. All references to the data record are routed through the table. The problem with this approach is that references to this table may occur on a large number of programme modules - say thirty or forty. If the record description table proved inadequate for a class of future applications, then all the routines which make assumptions about the format of the table will need modification. In one application, it was necessary to process a large pre-existing class of files - COBOL produced files. The generalized file description tables proved inadequate to describe these files. There were then three alternatives:

(1) One could re-design the record description tables by increasing field sizes and adding the additional fields needed. This approach required the recoding of a large number of modules.

(2) One could "kludge" up the tables by using certain fields for a variety of purposes and point off to auxiliary tables. Certain fields would need to hold dummy entries to maintain proper functioning of certain existing modules. This approach lacks elegance and clarity.

(3) One could make all requests to table values through a set of communicating sub-modules. Only these sub-modules would know the actual table format. Then changing the table structure would require changes to only the communication sub-modules. Thus, the

description table format is hidden from all but a few communication routines. This is the suggested approach and that followed in the implementation of the system described in the next section of this paper. Unfortunately, this alternative is the most difficult approach after the fact. A-priori choice of this method would greatly simplify the problem of change and adaptability. The lesson to be learned is that one must plan for change ahead of the implementation.

Error 3: Determination by the implementation of how users will use the system. Implementors must attempt to make an efficient system. The implementors will make design decisions based on expected usage of the system. Update efficiency may be sacrificed for retrieval efficiency or the converse. Complex long records may be handled efficiently, but not eighty character images. Many of these decisions are unavoidable at the start, as there are no users. One of the more crucial issues is what shall be indexed in a record-oriented system. Often this decision is left to the user on file creation.

The problem is that at file creation, the data base administrator may not know the useful fields to index. Indeed, the "*useful*" fields may change with key. The TDMS system[4] indexes everything. Neither approach is likely to be optimal economically. The next section of this paper describes an attempt to avoid the third error.

AN ADAPTABLE INFORMATION RETRIEVAL SYSTEM

The purpose of this system is to serve as a test bed to investigate an adaptive retrieval system. First, we wish to create an easily modified system. Second, we wish to experiment with a system which adapts its indexing structure to its environment.

Indexes are often used in information retrieval systems to provide efficient operation. The indexes discussed in this paper are constructed as inverted files and are often termed secondary indexes. One problem is what shall be indexed, who says so, and when should a different set of indexes be used. Some systems permit indexing of a few fields selected at design time or file build time. The TDMS system[4] indexes everything. The system described in this paper, WASIS, adapts its index structure according to experience.

Decisions on which indexes to build and maintain are made on the basis of:

(1) General suggestions from the data base administration.

(2) Commands from the data base administration.

(3) Experience on which indexes were or would have been useful in previous operations.

One must note that indexes can be both a help and a hindrance and some measure of index utility is required. In retrieval, an index will be helpful if it reduces the number of file records to be read *and* it is frequently invoked by a request. In update, an index will be a hindrance if the indexed field is frequently modified. The system must change the data record and the affected indexes.

There is a special constraint imposed on a system which will dynamically notify its indexing structure. User requesting programmes must not make assumptions about the existence of an index or indeed know the name of the index. Thus previous user requests will continue to function independent of the existence of an index. Thus techniques such as those used in the IDS system, cannot be used where the name of an index chain appears in a user's request code. The system itself must determine the applicability of an index and use it properly.

There are four parts to the system:

(1) File creation.

(2) File update or change.

(3) File retrieval.

(4) Index instruction.

Except for one component, file creation is typical of many information retrieval systems such as UL/1.[5] Firstly, one describes the internal record structure giving:

(1) Names to items.

(2) Data types to items, alphabetic or numeric.

(3) The size of the fields.

(4) The multiplicity of the item.

Secondly, one must define a mapping between the external medium (usually cards or magnetic tapes) and the internal record structure. This is done by describing where the named items can be found on the external medium (e.g. salary-card 1, columns 10 to 15).

Thirdly, suggestions on what the system may consider as indexable are provided. This is the only really novel aspect of the file building process. The user provides a key suggestion section. The format of this section is:

```
KEYS
        field name : operator   value list
            .            .           .
            .            .           .
            .            .           .

END KEYS;
```

An example is

```
KEYS
LOCATION-CODE:  EQ    LONDON              [equals]
LOCATION-CODE:  EQ    TORONTO
SALARY       :  BT    10000. 20000.       [between]
SALARY       :  BT    15000. 25000.           "
COMMENT      :  CON   "GOOD"              [contains]
END KEYS     ;
```

The system has been instructed that the only indexes it may consider building are these five. *NO indexes are built at this time.* These are only the possible indexes that may be built if appropriate. The data base administrator may *suggest* a large number of possibilities giving the attribute or field name, the operator, and the value which the field must have. The data base administrator may add to or delete from this list of suggested indexes at any time under password control.

Retrieval is externally much like other systems. The UL/1 system[5] is a good example of a retrieval system. There are two phases in the WASIS system:

(1) A criterion is stated for selecting records.

(2) There is a mapping provided from the selected records to the output media.

A viable commercial system would have a third phase not currently present in this system:

(3) There is a provision for performing computations on the selected data. Most systems provide a means of computing totals and subtotals on selected fields and do output sorting. These features are planned extensions to this system.

The retrieved criteria for records is stated via AND-OR-NOT combination of operations on fields:

CITY EQ TORONTO

CITY EQ TORONTO AND((SALARY BT 10000 20000) OR COMMENT CONTAINS "GOOD"))

Such combinations in stating retrieved criteria are typical of many retrieval systems.

The mapping operation for output are also typical of most user-oriented retrieval systems. The name of the field is stated with the place on the printed page relative to the beginning of the record in question. A heading facility is provided. The data base administrator may provide a default output format for a file if he desires. The novel features of the retrieval system are in the internal use of indexes rather than in the external features. The internal index usage scheme is discussed later.

The system will operate using direct access by record key. This section is currently being implemented. No novel features are planned. Basically, the system will add, delete or change a data record *and* (a) discard an index if desirable, or (b) update an index.

The data base administrator can change the list of possible indices and supply default output formats. Most importantly, he can command that an index be built. The index is not immediately built by the system. It will be created at the first opportune time as described in the next section.

MAKING AND USING INDEXES

An index is actually created whenever it is established that it is desirable and when it is economical to do so. Consider for the moment the action of processing a group of *batched* retrieval requests without any indexes.

A record is read sequentially and the system checks each retrieval criterion to see if that record satisfies the retrieval criterion. If it does satisfy criterion k, the key [direct access address] of the record is added to output file k. On end of file, each output file associated with each of K criteria [$1 \leq k \leq K$] is used to access the records meeting criterion k and the output mapping is done. If the records of a request are to be sorted, the system will include the values of the sort/fields along with the selected keys. This permits the use of a simple sort algorithm. Refer now to figure 1.

Indexes which are to be created are processed in exactly the same manner as a batched request. The conditions for selecting the index keys [say, CITY EQ TORONTO], are logically equivalent to a

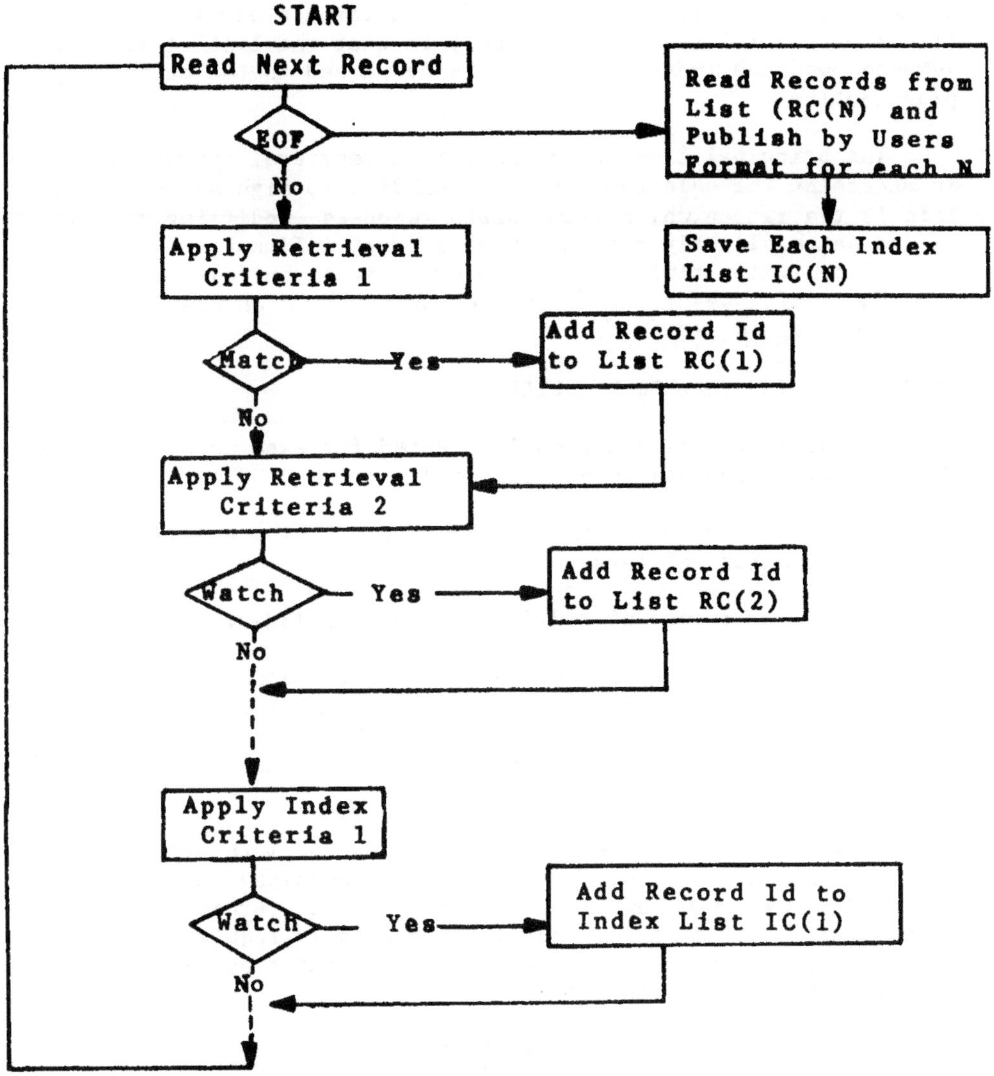

Figure 1. File Pass Operation of the System.

normal retrieval criterion. The index building process differs from a user retrieval criterion only in that the list of selected keys becomes the index instead of being passed to the output mapping routine.

The principal time delays in a file retrieval operation are in accessing the selected records from disk to high-speed memory. This is the reason why systems batch requests. Modifying the size of the batch slightly does not have a great influence on the retrieval selection timing. Thus the overhead in building indexes is low. *The WASIS system builds indexes only when a complete sequential file pass is required.*

SELECTING WHICH INDEXES TO BUILD

An index utility function is created for each possible index. This function examines three criteria:

(1) The frequency of *use* of the index attribute value pair in user requests for retrieval.

(2) The frequency of modification of the index.

(3) The cut factor of the index. This is defined as the number of entries in the index divided by the number of records in the file. Initially, this cut factor must be suggested by the data base administrator. Once the index is built, an exact value is known.

We are currently experimenting with appropriate utility functions and learning rates. The system has recognized and built indexes. This was, however, in a test environment rather than a real environment. Much more research in this area is required. The data base administrator may force an index to be built by setting parameters to the utility function.

RECOGNIZING THAT AN INDEX IS AVAILABLE

Besides creating appropriate indexes, the system must recognize when an index or combination of indexes is useful. Conditions are placed on records in the file by a logical triple: "name op value". Examples are "city eq London", "salary gt 3000" or "sex, eq male". City, salary and sex are assumed to be data items in the records. The available operators are the arithmetic: eq ne gt lt le ge. These apply to numeric and non-numeric items. Ct and nct are used to check whether a data item contains the value as a substring or not. Ex and nex are operators to check on the existence or non-existence of a value for a particular item in the record. One might state the criterion "city ex" or "salary nex", if one were trying to locate records with missing data. Logical triples may be combined via the 'and' and 'or' operators to form logical expressions such as:

city eq london and salary ge 4000;
city eq toronto or (salary ge 5000 and class eq 5)

Parentheses are used to change the normal precedence of the "and" and "or" operators. The retrieval criterion is parsed in a straight-forward way using precedence techniques. The output of the parse is a set of tables as shown in Figure 2. For each triple (name.op.value) the existence of an index is determined and the triple is changed to a 4-tuple where the fourth entry is the index number.

For within record tests, the use of these tables follows standard compiler techniques for logical expressions. The use of the tables for logical operations on indexes is of particular interest. There are some general rules for determining index utility which are best examined by considering the output of the parse as a tree. Only "and" or "or" are permitted as tree operators. Negation is treated by providing a set of complementary operators such as eq and ne or ex and nex. Check (v) indicates that an index exists for the triple.

(1) An index to an "or" operator is produced only if indexes exist for both inputs to the "or". If only one index exists, the other condition may be applied only by looking at each record in the file. If both indexes for input lines to the "or" exist, then the resulting output index is the union of the two input indexes.

(2) An index to an "and" operator is produced if one or both of the inputs are indexed. Even if just one of the inputs exists, the condition may still be used to limit the search. The output index is the intersection of the two input indexes if there are two. If only one input index exists, then the output index is the input index. If neither input index exists, then the output index is non-existent.

It is important not to do index operations (oring and anding) until it is determined that a useful index for the entire tree will be produced. In Figure 3, suppose that indexes exist for (class eq 5) and (salary gt 5000) but not for (city eq toronto). Then the output index will not exist since the left branch of the "or" box has no index. Doing the necessary disk calls to intersect the indexes into the "and" box, would be wasted effort, since a sequential file pass would be required to treat the "or" condition.

The algorithm must work from the leaves of the tree upward and generate, not execute, code for the index operations. As code for the index operations is found to be useless (due to missing

CITY EQ TORO AND SALARY BT 500 1000 AND CLASS EQ 5:

IETAB

1.	EQ CITY TORO B1
2.	BT SALARY 500/1000 B2
3.	EQ CLASS 5 B3

RCTA3

	OP				
-1	AND	1	-2	0	0
-2	AND	2	3	0	0

POINTER = -1

FINAL CODE X

```
INTERSECT B2, B3 TO B4
INTERSECT B4, B1 TO B5
```

Figure 2

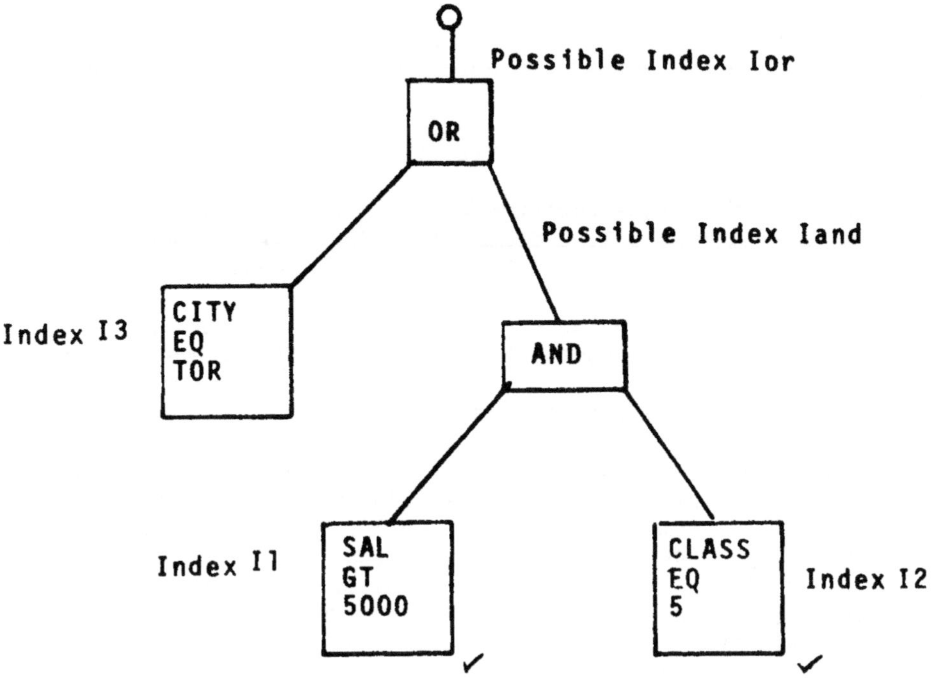

Figure 3

inputs to higher level "or" blocks), the generated code can be deleted before the operations are performed on the indexes. Again consider Figure 3. The code generated when I1 and I2 exist and I3 does not exist is:

Intersect I1, I2 to Iand.

When I3 does not exist, Ior becomes null and the above line is deleted.

If all three fields are indexed, the following code is generated:

Intersect I1, I2 to Iand.
Union Iand, I3 to Ior.
Ior is the output index.

The algorithm used by providing code to be interpreted to form the indexes, avoids any unnecessary calls to the disk for intersection or union of indexes.

PROBLEMS IN RECOGNIZING INDEX UTILITY

Three difficulties encountered are discussed in this section: indexes for multiple valued items, treating negation and the treatment of requests for items within a numeric range.

The problems encountered with multiple valued items are of interpretation of a request. When the user states a request as:

Author eq 'john doe'

the request is interpreted as any of the author fields are 'john doe'. Other interpretations are possible in different contexts. Some other interpretations could be: all the authors equal 'john doe', or most of the fields equal 'john doe', etc. After considering a class of special operators for multiple valued items, such as "any of", "all of", "n out of m of", it was decided to treat logical statements as "any of" as a default and to permit subscripting of the requests in cases where the "any of" interpretation does not apply.

Negation of an index operation such as "class ne 2" involves using the 'class eq 2' index and forming a new index which is a list of all records in the current universe less the index entries in 'class eq 2'. Hopefully, the universe of records will be reduced by previous conditions. If not, an index will not be used. Note that indexes are defined by triples: name, value list and operator. The usual operator chosen is equals. For certain requests, it might be advisable to modify the KEYS section of the record definition phase to permit declaration of other operators. This has already been partially done with the contains operator.

The use of value ranges in requests presents problems in recognizing that an index does exist for a range. Consider the request:

```
(A lt 1001) and (A gt 500) ;

     C1              C2
```

The two conditions are labeled C1 and C2. Assume that the KEYS section contained:

```
A:  0-500, 501-1000. 1001-1500. 1501-2000. 2001-2500 ;

      I1       I2         I3         I4         I5
```

Each of the ranges may or may not be indexed. Then C1 implies the union of I1 and I2 and a missing index for possible negative values. C1 will have no index. Even if there were no negative values, I1 might be a missing index. For similar reasons, an index cannot be built for C2. The problem is that I2 is precisely the index required, but by treating the conditions separately, it is impossible to recognize this.

Two solutions seem possible. One method would treat numeric lt, gt, le. ge. operators by building lists of indexes, i.e., C1's list would be I1, I2 and Ilow (a special non-existent index). C1 then implies the union of those indexes in its list. During an and operation, when index code is being generated, one could intersect index lists by name and possibly intersect out the missing indexes. In the example above, Ilow would be intersected out. This approach considerably complicates the programming involved. The method chosen is to ignore the use of indexes for the lt, gt, ge, le, operators and instead provide a range operator - between "bt". Here the two conditions may be conveniently stated together as:

A bt 500 1001;

This simplifies recognition of the compound condition and may in many instances be closer to the user's logic.

CONDITIONS ON THE ENVIRONMENT FOR SUCCESS

Since this system is intended to learn which indexes are useful and build these, it is necessary that patterns exist where certain items appear frequently in retrieval criteria. If all of the name value combinations appear with the same frequency in requests, then there will either be indexes for everything or the indexes will not often be available when they are needed. An ideal environment for this system is where a small subset of the name-value combinations appear very frequently. Such an environment could exist in a university class where students are probing a data base for facts on their current topic. Although the precise request forms are unknown, they would likely use a small number of keys. After the system adapted to these query types, operation might be considerably enhanced.

CONCLUSIONS

This paper has attempted to deal with means to avoid some pitfalls in the design of an adaptive information storage and retrieval system. Pride is the cardinal sin. An analog in systems design seems obvious. A-priori determination - the pride of deciding that the initial design is correct for all time - can lead to a fall. What can be done? Some initial determinations, formats and usage classes must be adopted to begin implementing a produce. Recent work by Parnas[6] offers some suggestions on creating adaptable systems. The main suggestion is to *hide a design decision in a small number of modules*. Program modules cannot be permitted to make assumptions about changeable parts of the system. This is good advice, but the problems are not solved. What are the changeable design decisions to be hidden? The initial design time will have to increase to isolate assumptions that the modules of a system are making. Several different modularizations will need to be compared and all assumptions on formats, orderings and access paths must be clearly stated. The intent is to clarify the adaption problem areas ahead of time in order to isolate future changes to a small number of locations.

The WASIS system has attempted to provide an adaptable test bed for an information retrieval system. It is implemented to accept change and further is designed to adapt its indexing structure to its environment with human guidance from the database administrator.

References

1. Honeywell Corporation, *Integrated Data Store Reference Manual*, Honeywell Information Systems File CPB-1565B, 1971.

2. Bryant, J.H. and Semple, P., GIS and file management, in *Proceedings of the 1966 Annual Conference, Association for Computing Machinery*.

3. Climenson, W.D., An exercise in computer file management, *Journal of the American Society for Information Science*, 21, 201, 1970.

4. Bleir, R.E. and Vorhaus, A.H., File organization in the SDC time-shared data management system (TDMS), in *Proceedings of the 1967 Annual Conference, Association for Computing Machinery*.

5. Olle, T.W., UL/1: A non-procedural language for retrieving information from data bases, *Proceedings IFIP Congress*, 1968.

6. Parnas, D.L., On the criteria to be used in decomposing systems into modules, in *Proceedings of the 1972 Annual Conference, Association for Computing Machinery*.

THE MEASURE OF A SUCCESSFUL INFORMATION STORAGE AND RETRIEVAL SYSTEM

Lee A. Friedman

Planning Research Corporation Information Science Company, San Antonio, Texas, U.S.A.

PROPOSITIONS OF SUCCESS

Formal ceremonials granting a merit badge of success to delivered information storage and retrieval (IS&R) systems[1] do not actually occur. Nonetheless, the principals involved will offer judgemental utterances expressive of positive or negative points of view about the system's utility. This is to say that any judgement of success, formal or otherwise, will depend on whose criteria of evaluation are used as the frame of reference - the system's engineer (or architect) or the system's user. Actually, the logical distinction between this two party value-system of reference is particularly marked, albeit some engineers would deem the subtle categorical differences as too arbitrary. Yet it is essential, at least for the sake of argument, to distinguish two propositions of success, the parochial and the global, as a prologue to the clarification of a three part system conceptualization problem - the subjects of this treatise.

The three parts involve: (1) defining a completive IS&R system and its purposes, (2) defining the system design concept and its applications, and (3) developing a system experimental test methodology to demonstrate the global proposition of success, i.e., to test and confirm the validity of (1) and (2) above (the all important feedback many system builders and theoreticians ignore). In the remainder of the introductory section, the two notions of system conceptualization are contrasted, followed by subsequent elaborations on completive system parameters, purposes, and postulates to explain parts (1) and (2) of the problem. The final section explains the proposed experimental test concept, criteria, and methods.

System Conceptualizations

Arguments citing the moot or imperceptible differences between the two propositions may be encountered at first, because proponents of both will boast of the same "holistic-system" idealism. Still, their respective notions about system conceptualization and construction do represent separate avenues of thought and action.

(1) The parochial proposition of success (Type I) is effected by a system design attitude, labelled as the mechanistic-analysis notion, which in practice *builds and combines separate components as a whole*. This approach views the IS&R system itself as being the determining factor, with the aim of optimizing product implementation and application. The result is a formalistic or conservative system that assumes the existence of determinate processes and outcomes.

(2) The global proposition of success (Type II) is effected by a system design attitude, labelled as the teleologic-synthesis notion, which in practice *distinguishes and builds an ensemble of functional components in relation to a whole*. This approach views user obligations and requirements as being prime determining factors, with the aim of optimizing the system's total functional performance as a user means-system. The result is an adaptive-predictive system that assumes the existence of uncertainties.

Properties of Types I and II Notions

Since the distinctions between Types I and II conceptualizations may still not be quickly apparent, the reader is cautioned against erroneously interpreting the two as either complementary notions or else defining one as a subset of the other. Perhaps a brief discussion depicting a few other attributes of both approaches will help to clarify what kinds of actions each does and does not represent. From a system engineer's point of view (and let us say that this would represent the most parochial, product-oriented view) the proposition of success is ascertained by a quantitative determination of whether the discrete operating components of the system (viz., hardware, software, humanware) can operate at a pre-specified level of performance. This approach reflects a certainly conservative, if not a *post hoc, ergo propter hoc*, design creed declaring that the delivered system will successfully serve whatever objectives are stated whenever component performances attain specified levels. This means, for example, that when components can achieve so many bits or updates per second, so much processing power, optimum acceptability of displayed messages, so much on-line activities accommodated, so much operator/equipment reliability, so many data elements stored and accessed, and so on, the users' requirements will be achieved or enhanced. The achieved system is perforce constructed of pseudo-determinate and formalized elements that would presumably "guarantee" or control all processes, solutions or outcomes, which therefore manifests many shortcomings when used in a basically indeterminate environment.

In this case, the system engineer's merit badge of success is also guaranteed. But yet there will be no assurances that the system will also achieve user and organizational purposes, verified either through design intentions or test methods. To be sure, the system engineers are not wholly at fault for adhering to this design logic. The client's original interpretation of system purpose may not have elicited anything more than a few parsimonious and discrete objectives, such as, "I want to be informed when X data elements arrive", or "We want this report composed of these data elements and displayed on-line", or "We want to update this file as soon as the data arrive", or "We don't want any unauthorized accesses".

In contrast, from a system user's[2] point of view (and let us say that it could also represent the view of a more sophisticated function-performance oriented engineer) the proposition of success is ascertained by more intangible methods of judging whether the delivered system's efficiency does actually serve as a viable means-system. The single user's interpretation of success will normally be based on unique experiences with the system - whether it accurately and completely answers his queries and enhances the achievement of his decision-making obligations. Now, it is this latter set of objectives that should become the criteria for judging system success and the keystone determinates of system purpose and subsequent system design concepts. The derivation of system purpose or benefit dispositions is the first step towards building an adaptive system within the framework of system synthesis conceptualizations, according to user expectations.

The remaining sections describe the specific attributes of the Type II notion, from the definition of system purposes through system concept definition and testing.

SYSTEM BENEFITS AND A DESIGN PARADIGM

At this point we turn to the first problem topic: completive and adaptive system benefits (or purposes), which system component ensembles must be designed to achieve. System benefits are exposed at two levels of abstraction. At the highest level, benefit statements, which resemble aphoristic goals emanating from corporate executive offices, define basic purposes. In their proper context such statements can be logically defined - an exercise that determines the second level benefits. Procedurally, dispositions of system benefits must be conceptually converted into criterial automata constructs, labelled as application and performance benefits, to govern both the properties of appropriate system capabilities and the operational measures to be used in testing degrees of system benefit accomplishment.

A few examples of these benefit classes are presented, for sake of clarity:

(1) *System Purpose*. This class would include statements of organization mission or purpose, like those below, presented as topic sentences in narratives elaborating the using organization's processes and "problems" requiring enhancement or melioration. The subject narrative would begin with, "The new system is intended to provide more effective resource management; improve assignment of resources to meet current requirements; improve ability to predict future requirements and resources; increase capabilities to perform reliable resource accounting and improve means of resource control". Subsequent systems analyses performed by system designers will result in the derivation of correlative system application and performance benefits. The narratives describing these latter two benefit classes would involve how the system is expected to enhance processes and meliorate problems.

(2) *System Application Benefits (as Criteria)*. This class would include such concepts as: task information *integrity*,[3] relevancy, certitude, closure, *input control*,[3] security, sufficiency, associativity, accuracy, utility, timeliness, etc. Hence, the aim here is to determine what combination of component capabilities (hardware, software and humanware) could achieve integrity and its associated benefits.

(3) *System Performance Benefits (as Criteria)*. This class would include such concepts as: information and/or system control, accessibility, speed, reliability, capacity, accuracy, user interface effectiveness, etc. Hence, the aim here is to determine what combination of component capabilities could achieve information or system control capabilities that would in turn optimize the accomplishment of application benefits.

Let us now view a simplified system design paradigm in Exhibit I which addresses the discernment and subsequent testing of all three benefit classes, and which leads to explanations of the second part of the problem. Note especially the three general hypotheses to be tested at Stage 4, as these are properties peculiar to the system synthesis notion and are central to the system design concept postulates and conditions described later. It is particularly important to recognize that tests of hypotheses 1 and 2 on system performance cannot be conducted via the Type I (parochial) approach, for whether effective component performance levels are achieved is strictly dependent on whether system benefits are accomplished. Figure 1, following the design paradigm, portrays the conceptual elements and temporal flow of the system design and

Stage 1 (t_1) Investigation and definition of user problems; requirements and obligations of current operation; organizational policies, goals, plans, configuration, etc.

(t_2) Specification and description of system benefits.

Stage 2 (t_3) Mapping of organization and potential user macro and micro processes; derivation of application and performance benefits and their operational definitions; development of system concept.

(t_4) Design of system operational requirements (hardware, software and humanware) in terms of application and performance benefit definitions. (At t_4 a link between pure and applied research and technology is established; for example, the selection of human information processing theories applicable to t_3 mapping for design of hardware or software surrogates.)[4]

Stage 3 (t_5) Implementing system design requirements - system development and operating phase.

Stage 4 (t_6) Conducting evaluation experimental program to test three subjective hypotheses - concurrently:

(1) The operational capabilities of system components will attain performance criteria and improved performance capabilities (the performance tests).

(2) The attained performance capabilities will accomplish the required application benefits (the application tests).

(3) The achieved performance and application benefits will improve user mission accomplishments - the system benefits (the mission tests).

Exhibit 1 - Design Paradigm

test scheme, as a baseline for locating the concepts discussed shortly.

Unique Variable Definitions

What the testing of these hypotheses will imply to the experimenter is that the system's operating capabilities, selected through design, will serve a dual role, as both independent and

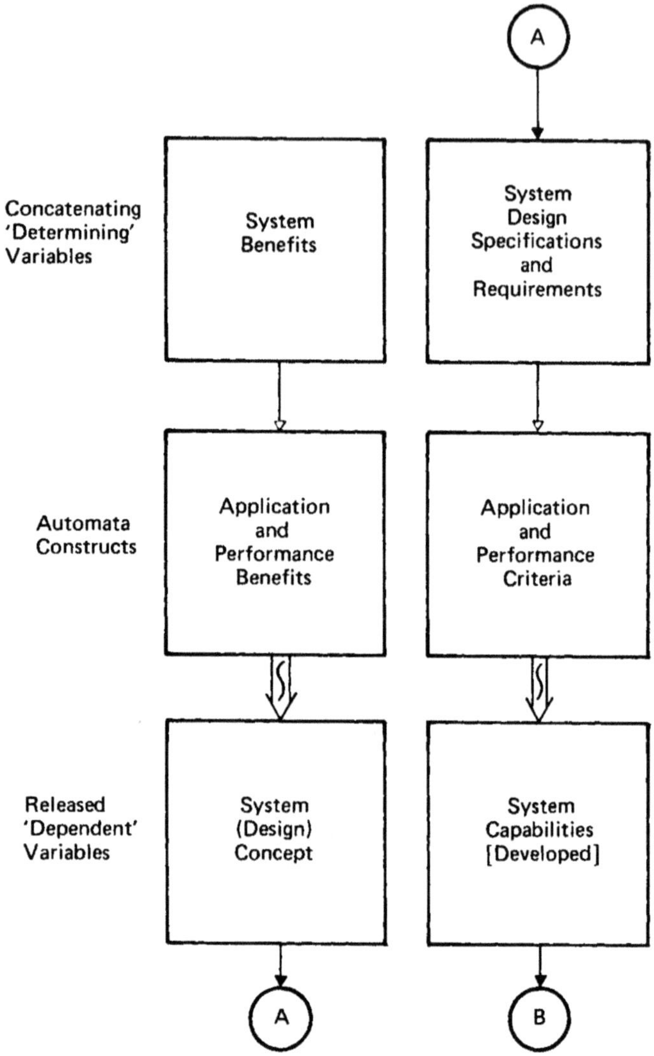

System Design Stages.

Figure 1 – Completive-Adaptive

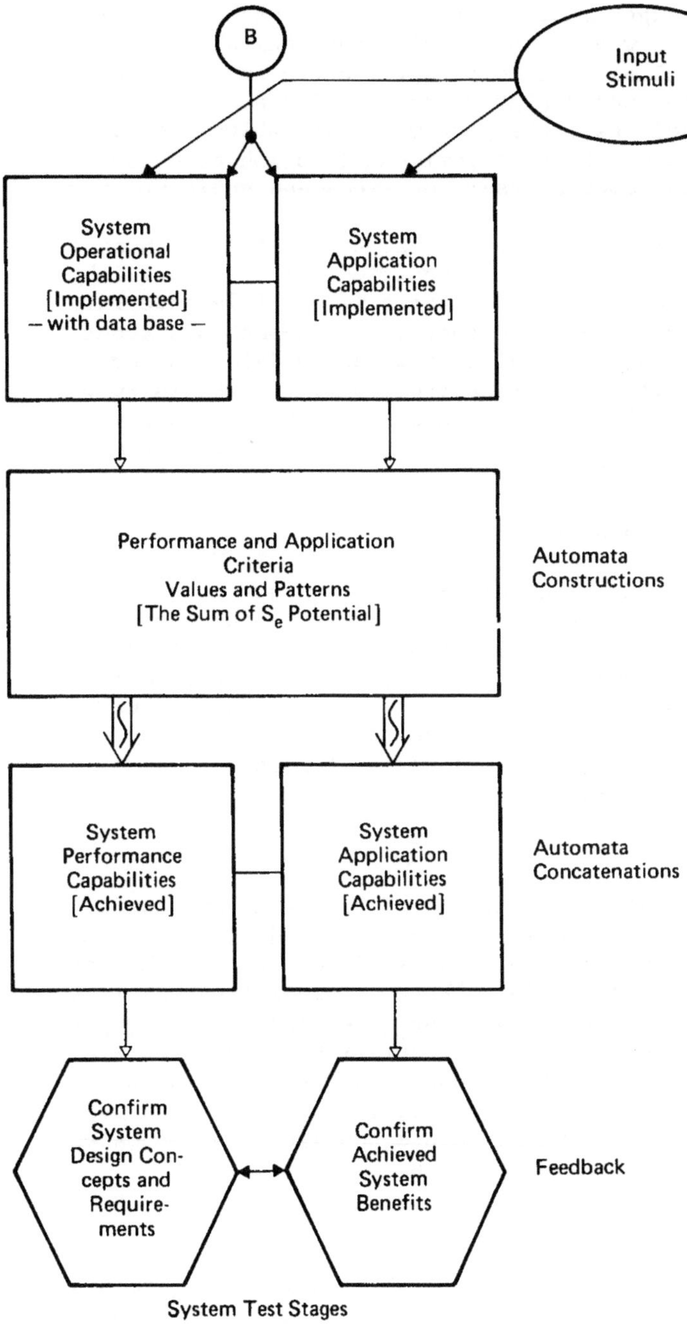

System Design and Test Concepts

dependent variables. In other words, the system's operating capabilities represent the experimental variables that we believe will induce certain benefits, when compared with the equivalent capabilities during the parallel test comparing the new and the old system. At the same time the designed capabilities are, during the parallel operational test, the controlled variables used to determine the degree to which the experimental variables effect positive or negative outcomes of user decisions and actions.

As a simple illustration, suppose that the system is found to achieve performance criteria, (e.g., specified workloads, reliability, etc.) as well as application criteria (i.e., file integrity). Now, while these test results would confirm our hypotheses about the designed capabilities, it is also critical to determine how well they aid in accomplishing the user's mission. Thus, in some instances the experimental tests may reveal that the achievement of certain application criteria may make little or no difference at all, or that the level of capability achieved may actually be an overkill in terms of cost or time, or level of effort, that would also degrade other user functions.

In sum, the system test, intended to confirm not only the validity of capabilities designed but also to determine their optimum degree of utility, is conducted in a parallel mode comparing both the new and old systems at the same time in an operational environment. In this manner, test results will provide a valuable *calibrated system model* identifying which capabilities should be retained, modified, or omitted, to balance system capabilities in terms of user operational requirements.

COMPLETIVE SYSTEM MODELS[5]

The essential parameters of the global or system synthesis approach, with reference to the second problem of defining the system design concept, can now be summarily explored, within the context of one functional system model. These parameterized baseline representations are important in that they are also indicative of system-state conditions governing the generation of specific system test hypotheses and the test methodology. Presented first, in Figure 2, are the prime functional parameters of an adaptive system as they would exist in a dynamic state configuration. Located here are two critical defining parameters, S_E and (f_N).

Dynamic System State Parameters

The S_E (system ensemble) is conceptually defined as a "mediator" consisting of system elements that react to an input stimulus (an event from E_I). S_E thus represents a complex of integrated functions organized into functional groups composed of (E, S, I)[6] elements or ensembles (e). Further, a unique combination of (E, S, I)

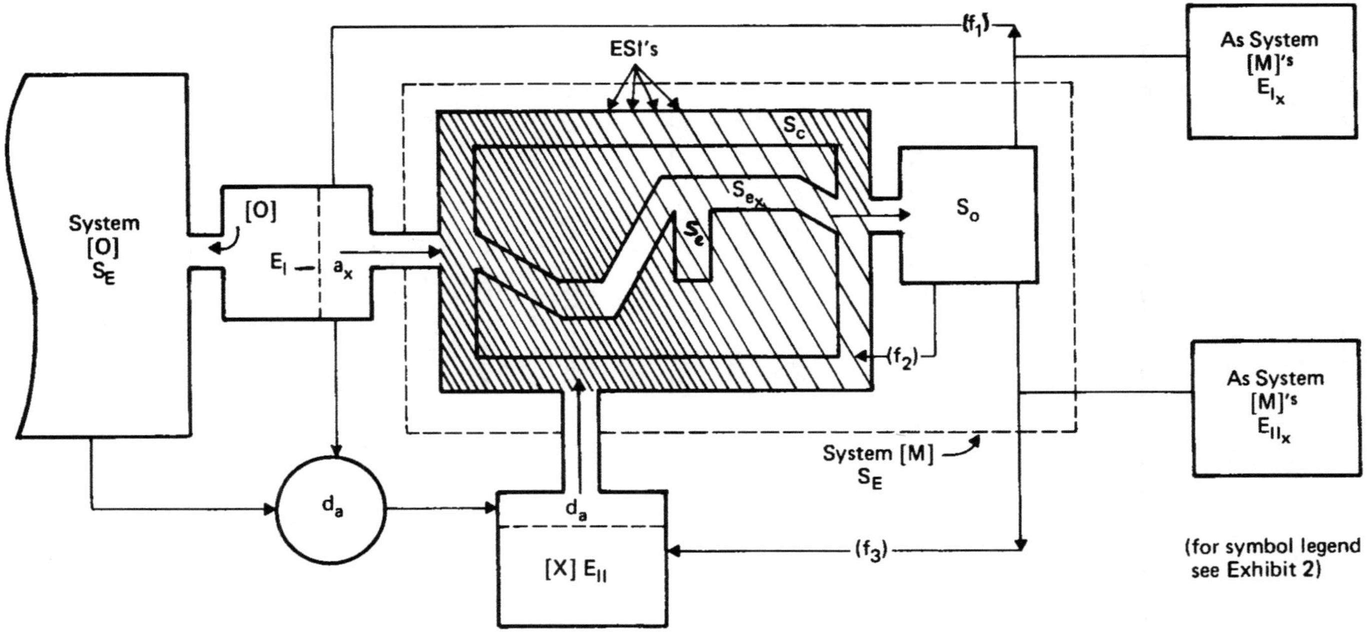

(for symbol legend see Exhibit 2)

Figure 2 — Functional Schema — Adaptive System Dynamic State

$(O)E_I$ = External Source of Input Stimuli (the a_x event)

$(X)E_{II}$ = External Source of Data on $(M)S_E$ and (M) Environment

$(M)E_I$ = (M) S_E Outcome or Conclusion to $E_{I_{a_x}}$ Recipients

$(M)E_{II}$ = (M) S_E Source of Data for External Data Source Control

$(M)S_E$ = Integrated System Ensembles (S_c, S_{e_i})

S_e = A Single Ensemble

S_c = The System Control Ensemble

(f_1) = Feedforward Information Transfer to $(O)S_E$

(f_2) = Feedback and Feedforward Information to $(M)S_E$

(f_3) = Feedback and Feedforward Information to $(X)E_{II}$

d_a = Data Transfer on $E_{II}a_x$ Events and E_{II}Environment State

Exhibit 2 - Legend of Symbols - Adaptive System Dynamic State

elements when reacting to a specific input element is labelled as S_{e_n}, with each element fulfilling a contributory function.

In terms of its referential properties an S_e is defined as a *concatenated*[7] ($\Rrightarrow$) (E, S, I) ensemble that is *released* (for action) upon the advent of a specific stimulus, and is part of a *contributory* chain: $(O)E_{I_a} \Rrightarrow (M)S_{e_a} \rightarrow S_o \Rrightarrow (O)E_{I_n}$. Therefore, an S_{e_n} is not to be considered as that combination of (E, S, I) elements determined or caused by a stimulus.

Synthesis Postulates

To identify the processes to which the terms "concatenated" and "released" refer, some further explanations of the synthesis conceptualization notion are required. First, there are three synthesis postulates (followed later by three defining conditions) which define the properties of S_E or S_e, viz., (1) that a discrete S_E is always part of a larger system, (2) that its S_e's are extended integrated functions of the larger system, and (3) that the discrete S_E is continually in an active state before, during, and after the advent of a specific stimulus - operating on related and non-related internal and external stimuli. It should be carefully noted that the third postulate is also the rudimentary condition for the time function, the important fourth element, to be described shortly.

To explain the meaning of postulates (1) and (2) let us again refer to Figure 2 and check the presumed boundaries of the S_E parameter. Note that the S_E parameter can be perceptually altered to represent alternative system fields, with floating boundaries. For example, (M)S_E could include all hardware, software and humanware (E, S, I)'s located within the structured boundaries of a specific organization's information system, including the system user. In this frame of reference the (O)S_E would become an external information system that requires (M)S_E to perform several (extended) parts of (O)S_E tasks. Now if the perceptual picture is shifted the (O)S_E could also symbolize an (M)S_E user (his mind/body system), in which case the E_{I_a} input stimulus would become the user's query statement to (M)S_E. Therefore, the released (M)S_e and S_c would be an integrated extension of that user's information processing and motor functions.

The meaning of postulate (3) might best be clarified in the form of a more mundane example, as follows: "While sitting at the dinner table I, (O)S_E, ask my neighbor, (M)S_E, to pass me the sugar bowl, (E_{I_a}). At that moment he is engaged in another conversation, (S_{eijlo}). He sights (E_{IIa}), the sugar bowl, picks it up with his right hand, (S_{eahl}) and hands it to me, (S_o)." This simple illustration points out that when any stimulus enters a system, certain S_e (E, S, I ensemble combinations) are released from among those actively operating on other stimuli. Of course, whether the S_{eahl} is released in the example above will depend on the valences allocated by (M)S_E to S_{eijlo} and to E_{I_a}. That is, if (M)S_E is deeply involved in an important conversation and the requestor and the request are of less importance, then the appropriate (M)S_e might not be released at all.

The activity of the second defining parameter (f_N) is the result of constants present in the S_o event. In essence, the (f_N) parameter, which constitutes a cybernetic constraint, effects the transfer of S_e generated feedback and/or feedforward information. As specifically pictured in Figure 2, the (f_1) flow would contain feedforward or predictive information intended to effect or control the state of (O)S_E; the (f_2) flow would contain both feedback and feedforward information - the former to include information for modification of S_e elements, and the latter to include information describing expected E_{II} data; and the (f_3) flow would also contain both forms of information - the feedback flow to describe the positive or negative features of E_{II_a} source data, and the feedforward to control future E_{II_a} data acquisition requirements.

Time Bound State Parameters

The inclusion of the fourth element, time, introduces conditions which now identify the S_E parameters as system variables. The defining conditions of the S_E or S_e parameters as presented below demonstrate the concept of S_e as system behavior, and reveal what S_E elements are to be tested and how they can be measured for contributory significance.

Three Principal Defining Conditions

Condition 1. A specific input stimulus (E_{I_x}) will elicit a different response by S_E, releasing different S_e behavioral patterns, as a function of postulate (3). This is to say that, in a dynamic adaptive system, if a single unique stimulus were to be entered into the system at subsequent time planes (t_1 or t_2 or t_3 ...) it would, of course, release different patterns of ensemble behavior, i.e., varied combinations of (E, S, I) activities, each time which will elicit different response patterns. Therefore it can be expected that between contiguous time planes each of the active (E, S, I) elements in each released S_e will manifest interplanar variations. (Although trend measurements over extended time might show, when E_I and E_{II} elements have determinate patterns, that the variations would converge into more discernable invariant patterns.)

Condition 2. Also implied by Condition 1 is that at any specific time plane certain intra-ensemble element variations can occur. Each element could vary in terms of its contributory significance toward an end result and in terms of its association (conjunctive, disjunctive, relational) with the other elements. The intra-ensemble elements of a particular S_e can thus vary, as a function of two situations, *viz.*, as a function of variable internal system states (e.g., the unique interpretation of a decision-maker, or equipment performance reliability existing at the specific time-plane), and/or the variable features of input stimuli (E_{I_x} or E_{II_x}).

Condition 3. This final condition defines the contributory significance of ensemble behavior. This condition states that any S_o (or specific system response) is an event which is the result of the average[8] behavior of the ensemble rather than by the exclusive behavior of any specific element within the ensemble. This condition also contains a cautionary sign for the architect and the experimenter. It says, in essence, that experimenters may very well have to rely on subjective rather than on objective (statistical) measures of significance in order to discern causal and relational events. In other words, experimenters may be able to measure the performance of a specific ensemble element (e.g., speed, reliability, number of products generated, product content quantity, etc.) but the contributory significance of any single element may not be effectively or validly identifiable through tests of statistical significance (metric or non-metric). What experimenters would eventually perceive are significant performance measures that are and are not really significant; and non-significant performance measures that are and are not significant. And, finally, the experimenter would perceive that certain effects were not due to element interactions, but rather due to their combined contributions as a whole.

SYSTEM TEST CONCEPTS AND METHODS

Parallelism

This section deals with the third and final part of the system conceptualization problem; specifically, defining the test program and methodology needed to verify the proposed system design concepts and properties, and their respective contributions toward accomplishing proposed system purposes and benefits. The generative theme of this test rationale was outlined earlier in the design paradigm (Exhibit I) and in Figure 1, which mapped the paradigm's transition states, forming the baseline model of the adaptive system test concept. Only a plan of execution is required to complete the test concept, which will be introduced by the scheme of parallel testing shown in Figure 3.

The concept of parallelism and its properties formalize the conduct of system tests under the constraints of real-time (actual) environmental conditions. The concept of parallelism further invokes the properties of two experimental test criteria: (1) that the newly developed system, $\boxed{N(M)S_E}$, as the experimental variable, is to be operated in parallel, for a limited time, with the current (old) system, $\boxed{C(M)S_E}$, as the control variable; and (2) that both $\boxed{N(M)S_E}$ and $\boxed{C(M)S_E}$ are to react in parallel to the same input stimuli.[9] This method of experimental testing, labelled the *Ada*ptive System *P*arallel *T*est and *E*valuation *D*emonstration (ADAPTED),[10] is built upon salient attributes adapted from several classic test design approaches, described shortly.

The Logic of Parallelism

Let me digress briefly to present arguments against the use of those alternative experimental test approaches usually suggested when the ADAPTED program is advocated. The suggestions are usually raised in question form, e.g., "Why can't the new system be tested in a non-parallel mode, using, say, *ex post facto* comparisons in lieu of parallel?" or, "Why can't the new system be tested under simulated conditions?" The answer to both questions is that neither is feasible, and only because two experimental conditions cannot be met: (1) that antecedent external environment patterns, viz., Figure 2's $(O)S_E$, and subsequent input stimuli (E_{I_a}) could not be replicated or controlled to determine whether $\boxed{N(M)S_E}$ could have done better than did $\boxed{C(M)S_E}$; and (2) that complete

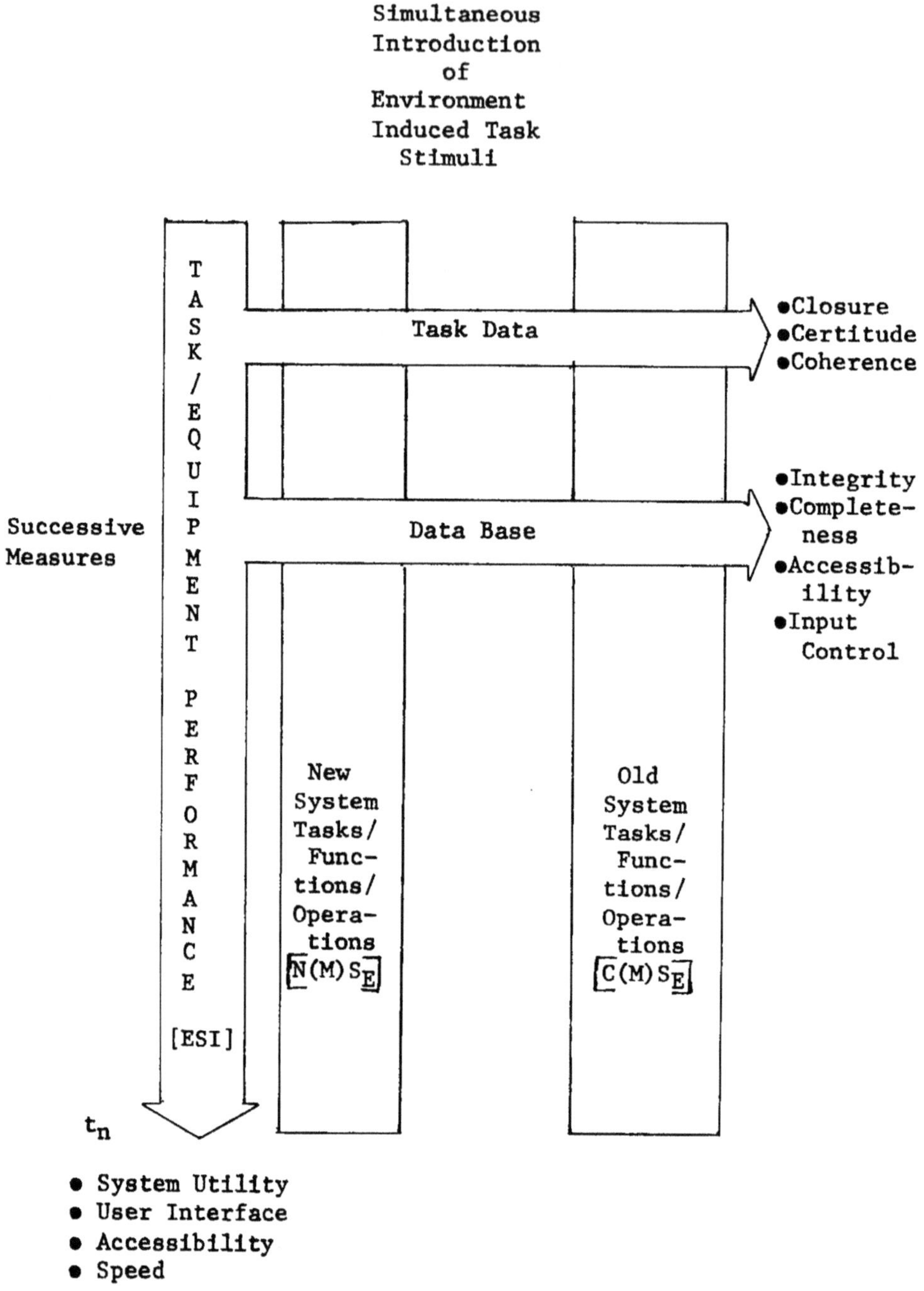

Figure 3 - Parallel Test Configuration Plan of Execution

data on $\boxed{C(M)S_E}$ past "parallel" operations (S_e's and S_o's) would be invalid for comparative purposes, even if available. To draw a conclusion, it can be said that the use of alternative test methods would vitiate ADAPTED test intentions simply because of the following facts, culled from the three postulates cited earlier: (1) the probability that any previous external environmental patterns and the intentions of subsequent input stimuli will ever again be replicated approaches zero; (2) consequently, any data available on $\boxed{C(M)S_E}$ past performance (in reaction to past stimuli) could not be used in a comparison with $N(M)S_E$ reactions to new input stimuli; and, (3) there would be no way of determining whether the $\boxed{C(M)S_E}$ would react any differently to the new input stimuli than could the $\boxed{N(M)S_E}$

Clearly, all that ADAPTED testing intends to do is to follow an experimental plan whereby the S_e reactions of the two groups are compared[11] on the same stimulus, to test the general hypothesis that $\boxed{N(M)S_E} > \boxed{C(M)S_E}$. By converting this general hypothesis into test statements covering the three hypotheses (noted earlier in the Design Paradigm, Stage 4) we will look for evidence that:

(1) $\boxed{N(M)S_E}$ component performance will be more efficient than comparative $\boxed{C(M)S_E}$ capabilities in accomplishing mission goals, and

(2) $\boxed{N(M)S_E}$ performance capabilities will more readily attain/sustain application benefits, and

(3) Improved $\boxed{N(M)S_E}$ application and performance benefits will produce greater or more positive effects (e.g., changing stabilizing environmental state, controlling external/internal perturbations, improving decision outcome reliability and validity, enhancing knowledge acquisition, etc.) than the $\boxed{C(M)S_E}$.

The following descriptive profile presents a typical ADAPTED chain of events demonstrating the conjunctive thread in hypotheses testing.

A Sample of Test Conditions

Comparative principals in both systems would perform the same task as initiated by an incoming task stimulus. Let us say that the task requires the generation of management reports that are to contain interpretations and evaluations of an event, a decision (recommendation) to alter or not alter organizational plans or policies or procedures, and the substantiating information supporting the evaluation and decision. Now, assuming we observe that within a specific time period the $\boxed{N(M)S_E}$ and $\boxed{C(M)S_E}$ managers produce approximately the same number of reports (maybe 150). And let us also note that several of the reports produced by $\boxed{N(M)S_E}$

managers included one or two more data items (or the same data items related differently) than the parallel $\boxed{C}(M)S_{\boxed{E}}$ reports. Finally, it is determined in the course of events that two of these $(N(M)S_E)$ reports have not only elicited exclamations of greater task closure and decision certitude by the recipients (report sources are masked) but have also caused a major change in certain organizational plans or policies. When such situations occur, and they usually will, the experimenter will be faced with several problematic conditions. First, he will note that no significant quantitative differences exist in the number of reports produced, in the time to generate the reports and even in the extent of report contents. Yet, the outcome of $\boxed{N}(M)S_{\boxed{E}}$ operations *did* produce a significant change in organizational programs and policies. Thus, the experimenter must find convincing answers to such cogent questions as: (1) does the occurrence of significant events verify the assumptions about the designed performance and application capabilities? (2) which ensemble behavior actually led to the creation of the two important reports? and (3) will such positive system-generated events ever be repeated in the future, based on the occurrence of only two special reports?

Criteria

Certain experimental test conditions, as implicit properties of the ADAPTED concept, are transformed into several classes of test criteria, and will serve two purposes: to effect the eventual selection and development of a test design plan (methodology) and also to govern the conduct of the test as well. These criteria constitute the formal rules defining the boundaries of, and procedures for, maximum control of test variables and attributes. It is therefore axiomatic that the degree to which variances from these criteria are either evidenced or permitted will determine the degree of test measurement validity and reliability. Four classes of defining criteria are noted:

(1) Test Subject Control

(a) Functional users of the new system, engaged as test subjects, must demonstrate a level of proficiency and experience equivalent to old system counterparts prior to test commencement.

(b) Test subjects must be operationally dedicated to designated new system positions (they might report administratively to old system managers, but test subjects will accomplish all tasks under the control of new system procedures).

(2) System Utilization Control

(a) Test data integrity and validity shall be guaranteed through the control or prevention of the exchange of information/products derived by either system; and new system products shall not be produced or issued in lieu or to satisfy old system obligations.[12]

(b) Test conductors/experimenters shall in no way interfere with, alter or manipulate the designated or designed procedures, processes, products or input stimuli of either the new or old systems.

(3) System Operations Control

(a) No improvements to component operations, necessitating factory or field change modifications, will be implemented during the course of the test program for the purpose of enhancing components' operational capabilities. This constraint pertains particularly to components officially accepted as being in compliance with designed capabilities and specifications, but whose designated capabilities are discovered during the course of the test as not being totally adequate or effective in support of actual system use and operations. (Such inefficiencies are to be recorded and subsequent recommendations for post-test improvements will be made.)[13]

(b) Component operations whose capabilities are determined during the test phase as not being in conformance with designed or specified capabilities, but which could be operable, will be corrected as required.

(4) Test Length

The length of the test period shall be planned so as to encompass the occurrence of salient events (E_{I_x}'s, E_{II_x}'s, and S_E's), that will enable the capture of necessary and sufficient data to confirm the test hypotheses. The criterion of test length includes a sufficiency of time requirement that is to ensure the closure of feedback loops. That is, sufficient time must be allowed to enable the acquisition of feedback information on the results of system generated events (S_o), to enable the measurement of effectiveness or significance of these events, and related S_e activities as well. The intention here is to ensure that all three hypotheses are tested. For instance, referring back to the sample scenario, only when feedback statements from the $[N(M)S_E]$ $[C(M)S_E]$ report recipients attests to $\boxed{N(M)S_E}$ report "improvements" and to their decided effect on the environmental state will the experimenter be able to test and confirm all three hypotheses.

Unfortunately, the cost of operating the two systems in parallel can be a prohibitive constraint on optimizing test length. In practice, test length plans would involve trade-offs between test priorities and the available test budget. Thus test design plans must realistically emphasize those critical tests which can be conducted within budgetary limitations. The appendix to this paper discusses alternative test procedure approaches that can reduce the scope of the test while precluding compromises in test objectives.

The ADAPTED Methodology

A problematic situation that could confront an adaptive system experimenter is that there is no single type of test design or experimental method that will readily accommodate the scope of ADAPTED program concepts, criteria and hypotheses. However, within the framework of the ADAPTED concept there are obvious traces of several experimental methods which can be adopted to suit our purposes. In short, the ADAPTED methodology will be constructed of principal attributes extracted from three discrete test design approaches and re-integrated or reconstituted into a new format. The three methods are generally called "controlled cross-sectional design", "controlled projected design", and "*ex post facto* design".[14] The definitions of three approaches, as discrete methods, are as follows:

(1) *Controlled Cross-Sectional Design.* This approach involves comparisons between two matched groups (experimental and control) at a single point in time. The objective is to *explain* events by identifying the basis of associations. While it does not establish the existence of a cause and effect relationship, it will result in operational indicators of possible cause and effect conditions and variables.

(2) *Controlled Projected Design.* This approach involves selected "before" and "after" comparisons between experimental and control groups, in which an attempt is made to measure what would be the effects of a new program at some future date, thus following the flow of events from a present date to some future date.

(3) *Ex Post Facto.* This approach involves the tracing of a present effect backwards to an assumed causal complex of factors existing at a prior date, using such recorded data as are available.

By now it should be easily understood why the fixed constraints and rationale of each one of these approaches are not independently applicable to the adaptive system test concept, and why only salient properties of the three approaches must perforce be extracted and reintegrated so that they can be applied concurrently to measure different facets of the same elements. For instance using Approach 2, the new IS&R system would represent the experimental program the effects of which would be compared with the old system technique. The collected measures of new system effects - improved integrity, accessibility, etc. - would enable a projection of the degree to which the new system would more effectively or reliably support the user's mission. These measures will also aid in confirming the new system's ability to achieve application benefits. Approach 3 would be used to "trace back" to those system performance events that "caused" the positive (or negative) effects measured in Approach 2. Approach 1 would be applied in cases requiring further explanation of the possible reasons for the occurrence of certain activities. (For example, what additional factors are related to the generating, by only one individual, of improved (new) system products?). Exhibit 3 presents an overview of the (reconstituted) ADAPTED methodology, including the three test approaches as they pertain to the test concept.

The final step in the suggested methodology, the development of detailed test design plans and procedures, becomes the first step for the experimenter. Such plans, which are beyond the scope of this essay, would include the identification of sub-hypotheses for each benefit, the associated system operational and performance measures, the method of data collection, sampling and analysis, test schedules, subjects and programs. Figure 4 presents a schematic overview of test procedure construction requirements.

CONCLUSION

The concluding remarks will address only the few positive and negative features of the ADAPTED approach - those which purchase definite advantages for designers, users and even scientists, and those which are problematic issues requiring eventual resolution.

Advantages Gained

(1) *Systems Model.* The results of the ADAPTED program will provide a verified and calibrated IS&R system model depicting the direct and indirect factors effecting IS&R system actions and objectives. The model, consisting of a comprehensive picture of the best-fit balance within and between S_e's, per achieved benefit, would be applied in the design of similar system configurations, and to provide the rationale for system capabilities, modifications or omissions.

(2) *System Synthesis.* The parochial system conceptualizations imposed by clients and system architects can be readily transformed into a system synthesis concept. This approach can lead to the identification of useful and meaningful system objectives and especially to a more valid definition of an IS&R system as a *system.*

(3) *Theory Testing.* The ADAPTED program offers a means to test relevant theories of information processing and retrieval and man/machine interfaces; the generation of new hypotheses and theories; and the acquisition of operational definitions of system concepts to aid in future system development and in research hypothesis testing.

Problem Issues

(1) *Costs.* The costs in time, energy and money to fulfill the ADAPTED program's aims can become a negative feature. The client has to be convinced that the advantages gained outweigh the direct

	Test
	Controlled Design ⟶ Ex Post Facto Elements
Concepts	System Determined: To trace and clarify our understanding of how some present causal factors may produce an effect at a future date; to verify predicted effect of system (user) activities on system performance; to determine underlying factors associated with system activities.
Hypotheses	That the designed operational capabilities will effectively achieve design criteria; e.g., input control, file integrity, security, etc.
Attributes	Match on similar new and old System Capabilities (procedures, manpower, data base) and user factors.
Variables: Independent	Client System Design Criteria and Operational Requirements
Dependent	System Operational Performance in achieving design criteria
Test Controls and/or Constant Conditions	• System Operational Capabilities and Configuration • Analytic techniques to measure system performance • Match on independent variables
Test Controls and/or Constant Constraints	• System Operational Capabilities and Configurations held as constants. (No changes or manipulation to determine varied effects)

Exhibit 3 (Part I) - The Properties

Type
Ex Post Facto Design → Controlled Design Elements
Mission Determined: Description of present situation as an effect of some previously acting causal factors, and an attempt to trace back over an interval of time to some assumed causal complex of factors to verify predicted effect of system performance on user activities; to determine underlying factors associated with mission activities.
That the system performance capabilities planned to achieve design criteria will effectively support user task obligations and requirements, and mission objectives.
Match on similar new and old System Techniques (procedures, data base use, user tasks) and user factors.
Systems Operational Performance Capabilities (New vs. Old System) User Performance and Mission products achieved
• New and Old System Operations held constant • Precision control through factor equation of analytic techniques • Match on independent variables
• No Randomization • No Manipulation of Input Activities • Standard Operations held constant • Indirect control of variables via symbolic manipulation, i.e., via analytic techniques of recorded activities

<u>of ADAPTED Methodology</u>

	Controlled Design → Ex Post Facto Elements
Evaluation	• Evaluate degree of design criteria achievement and effect of controlled variable (system capabilities) on mission accomplishment • Determine extent of system operability, maintainability and reliability, and effect on design criteria achievement
Analyses	System Performance: (a) Comparison of both systems' capabilities at a given time. (b) Analysis of the following system performance concepts: (examples) • Time - processing time, production time, ... • Completeness - data base completeness, ... • Accuracy - data base accuracy, ... • Frequency - system component utility, ... • Consumption/Cost/Resources • Operability/Reliability/Total System Performance (c) Analysis of those who have and have not been exposed to the new system and measuring the resultant effect, i.e., determining differences in operating techniques as they effect the achievement of the user's mission (e.g., file integrity techniques effecting task accomplishment levels)

<u>Exhibit 3 (Part II) - The Properties</u>

Ex Post Facto Design ⟶ Controlled Design Elements
• Evaluate effect of designed criteria achievement levels on degrees of mission accomplishment (i.e., evaluate changes caused by controlled variables) • Evaluate mission accomplishment differences
Mission Performance: (a) Comparison of both systems' capabilities over time. (b) Analysis of the following mission concepts: • Time - reaction time, time to complete task, ... • Completeness - report completeness, task completeness, ... • Accuracy - report accuracy, response output correctness • Frequency - amount achieved, magnitude achieved, ... • Judgements of user decisions • Consumption/Costs/Resources (c) Analysis of those who do and do not exhibit resulting effects and measures whether they had been exposed to a "cause", i.e., determining differences in mission accomplishment and trace back to system "causes", (e.g., level of task accomplishments as caused by file integrity operational capabilities)

<u>of ADAPTED Methodology</u>

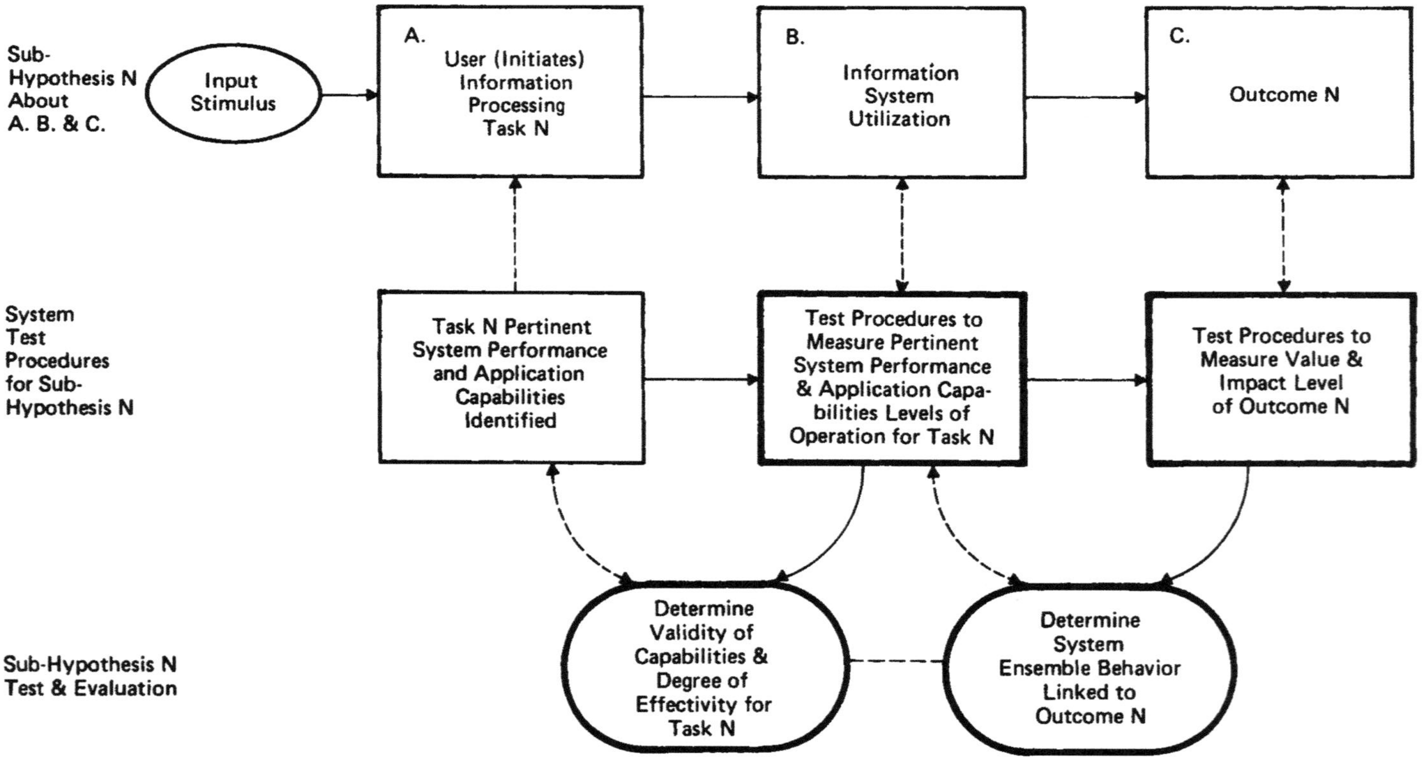

Figure 4 — Test Procedure Construction

costs to be expended. Such arguments, however, can only offer qualitative justifications of effectiveness since no quantitative values of ADAPTED program cost-effectiveness can be derived directly. And if these arguments fail to convince the client, then we can expect to implement greater trade-offs in test conduct. Trade-offs such as the use of simulation exercises and operational readiness testing will certainly reduce the knowledge to be gained on the system's operational effectiveness and utility in benefit achievement, to justify clients' investments. However, alternative solutions are available, as noted in the suggested test abridgement methods outlined in the appendix.

(2) *Statistical Uncertainties.* Any comprehensive discussion of the reasons for the general inapplicability of statistical tests of significance would be too extensive to cover in this paper. (Several remarks alluding to the uncertainties of statistical inference were presented in earlier sections.) To clarify at least part of the basis for these claims the following argument is offered. Essentially, the power of statistical tests of significance (metric or non-metric) cannot manifest the degree of sensitivity required to identify the ontological or absolute differences, at any level, among comparative $\boxed{N(M)S_E} > \boxed{C(M)S_E}$ or "interactive" $S_e(ES_I)$ measures. Hence, we may be able to observe that a released S_{e_n} did or did not effect some outcome, but we cannot determine which element had contributory significance (especially where $\boxed{N(M)S_E}$ element performance measures are *almost* identical to $\boxed{C(M)S_E}$ measures).

In truth, since theoretic distributions of probabilities of events (and stimuli) and of S_e activities are unknown, we cannot reliably derive the probability of event occurrences. This is to say, also, that for such states there is no statistically significant history of occurrences which could allow the inference of a probability of occurrences, and most important, the frequency and time span of occurrences. Such constraints also affect the utility of standard sampling techniques - to derive both statistical inferences and sample sizes. In short, sampling techniques cannot provide valid or reliable definitions of S_e events as samples, since each S_e event represents a single unlocatable sample from an unknown infinite universe. And this definition applies also to a class of S_e events that occur frequently over time in the generation of a specific management report.

Because of these constraints, ADAPTED program experimenters must rely on subjective evaluations of raw measurements to ascertain the contributory significance of system (S_e) behavior patterns (aided in part by graphic techniques). As the use of subjective interpretations could easily lead to untenable conclusions, we can expect experimenters to broadcast a need for a new *field* of quantitative evaluation techniques that addresses the complexities of dynamic, adaptive systems in motion. For these purposes the most promising approach could very well be the application of a simulation modeling technique to predict the system's behavior that would also depict its inadequacies under varied conditions. Such a model would be built in several steps, as follows:

1.

2.

n.

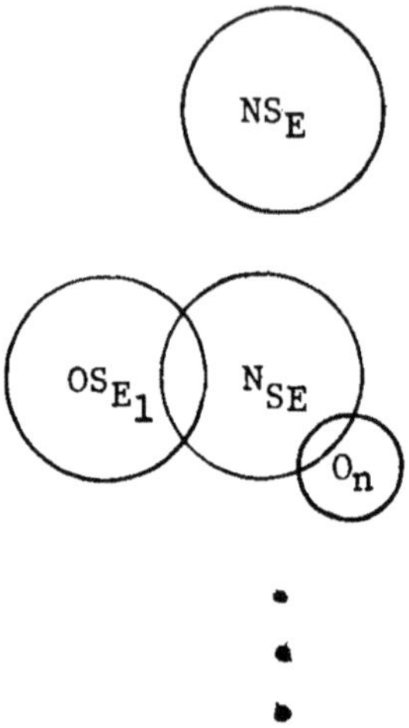

APPENDIX

SUGGESTED APPROACH FOR THE ABRIDGEMENT OF AN ADAPTED PROGRAM

Test Scope

There must be no abridgement of test programs intended to confirm all three hypotheses noted in Exhibit 1. Any degradation or omission of one of these hypotheses will certainly negate the intents and purposes of the ADAPTED program.

Test Depth

Ideally, to produce a reliable and complete calibrated system model all system users and all system uses should be subject to testing. Yet, some efficient abridgement of test depth and length can be effected without compromising system model development or ADAPTED program objectives. Of course, any degree of abridgement will require related degrees of relief from strict adherence to the defining criteria, noted in the text.

Essentially, the approach to abridgement control involves two related criterial steps: (1) function-importance scaling - defining the relative degree of importance of all system user functions in terms of actual and immediate levels of contribution toward accomplishing the organization's mission; (2) capability-importance scaling - defining the relative degree of each designed functional system capability in terms of expected levels of contribution toward the enhancement of user task performance.

The scaling analysis should provide the following kinds of information. The function-importance scaling effort should determine which organizational functions are allocated responsibilities for implementing the most direct and immediate tasks leading toward achievement of the organization's goals. For instance, a manufacturing firm's production line operation would have greater direct value than, say, the personnel or accounting functions. If necessary, further intra-function levels of importance should be determined. For example, production line control would be more important than inventory control; factory personnel recruiting would be deemed as more important than personnel records maintenance.

Next, the capability-importance scaling effort should determine which system operating capabilities have greater importance, ranked mainly in terms of selected function-importance categories. (That is, which operating capabilities - performance and application - are most important to the more important functional units?) For example, is data input control more important than, say, file integrity or security? And are there some attributes of data input control that are more essential than others?

After completing the scaling exercise, and some consensus is achieved, the experimenter can then determine the boundaries of the system test, using, as a baseline trade-off criteria, the known constraints, e.g., budget limits, test subject availability, required test data sample size, etc. The resultant decision should be to conduct tests only within the (selected) more important functions, for the more important tasks and collecting test data on only those operating capabilities designed to support those tasks. Moreover, not all tests need to involve completely matched-comparative (parallel) operations. To illustrate the latter point, briefly:

(1) Comparative-Match Tests: (New system test subjects and functions versus old system counterparts)

(a) Compare important functions on selected but matched tasks and collect universe sample data.

(b) Compare data base contents on important data elements using periodic but large samples.

(2) Comparative-Non-Match Tests:

(a) Compare task times, loads, and completeness with users using and not using the new system.

(b) Compare manpower loading and activities for those using and not using the new system.

By careful selection of the most critical functions and capabilities for testing, and by careful analysis of test results the viability, applicability and utility of the "untested" capabilities can be discerned through exceptions. That is, the experimenter can make certain positive assumptions about the untested capabilities as long as the major organizational functions are enhanced by the system. However, if certain degradations or dysfunctional conditions occur, then the experimenter must be able to determine whether the fault lies within the functions tested or is somehow contributed to by the untested capabilities, to some degree. In sum, test abridgement can be fruitful only as long as the experimenter maintains full control over and knowledge of the total system operation at any given time.

Finally, while the system developer's test objectives can be achieved through the suggested approach, certain of the shortcuts will have a more direct and negative impact on the achievement of user objectives. The main thrust of this impact will be on the user's ability to manage the post-test system. This is to say, that even if complete data on achieved performance levels were obtained, the possible effects of operational capabilities that were untested, within both tested and untested user functions, would be relatively unknown. Hence, in the event that certain user task performance degradations were to occur due to conditions contributed to by untested capabilities, system managers would be unable either to pinpoint the source or to determine the extent to which an errant capability required melioration or modification.

References

1. In this context, I define an IS&R system to mean: a large-scale, completive computer-based (man/machine) system, consisting of a complete and integrated data base and associated software serving as an integrated function among multiple (client) organizational functions and levels of responsibility, to optimize decision-maker's question-answering tasks, knowledge acquisitions and adaptive-predictive control of internal/external events. The category of completive IS&R systems does not include process controllers or simple function accounting systems. Rather, a completive IS&R system complements human information processing functions, as in an intelligent man/machine system. (See Friedman, L.A., The impact of technology on the future state of information technology enterprise, *AFIPS Conference Proceedings, 1973, National Computer Conference and Exposition*, AFIPS Press, Montvale, New Jersey, 42, 751, 1973.)

2. Churchman, C.W., *The Design of Inquiring Systems: Concepts of Systems and Organizations*, Basic Books, New York, 1971, 47.

3. *Integrity* as a criterion means prevention of lost or misfiled data. However, it must be made clear that the integrity criterion is an integrated element of other criteria, *viz.*, that the achievement of integrity capabilities is affected also by the correlative achievement of associated criteria such as relevancy, sufficiency, and closure. *Input control* means prevention of unauthorized or incomplete data from entering the system, and acquisition of complete data.

4. Friedman, L.A., Introducing new educational game principles: a case for educational and behavioral engineering, *Educational Theory*, 21, 59, 1971 (Reprinted in *Mental Health Digest*, 3, 1971).

5. See John, E.R., Switchboard versus statistical theories of learning and memory, *Science*, 177, 850, 1972, for analogous narrative models of neural information acquisition, processing and retrieval in animal behavior. (I have taken the liberty of borrowing several appropriate concepts and terms from Dr. John's treatise, the use of which here might not reflect his original intentions.) Note also that Dr. John's theme tends to address the difference between Types I and II approaches to the theoretical explanation of animal (brain) behavior.

6. (E, S, I) = *Energy*, (the processing or movement or transfer of information, people, equipment, materials, or the power to ...); *Structure*, (the hardware, software, humanware components); *Information*, (the data base contents, knowledge, experience, procedures, goals, rules, ...). A fourth element, time, is defined later. I am indebted to Dr. Shukei Aida, University of Electro-Communications, Chofu, Shi, Tokyo, Japan, whose coherent ESI model furnished the logical foundations of this discussion. See Aida, S., Informatics in eco-technology, to be published in proceedings of the *NATO Advanced Study Institute in Information Science*, Seven Springs, Pennsylvania, 1972.

7. A concatenated item or variable is a member of a class of items that can formally follow but are not caused by a related class of cogent antecedent items.

8. In this condition the concept "average" connotes a single value (subjectively and/or objectively derived) that summarizes or represents the general significance of a set of unequal events or values that are in a state of dynamic equilibrium.

9. However, during the test period the $C(M)S_E$ should be the only system responsible for officially supporting the organization's mission.

10. The system test methods described here were derived from actual

test programs designed and implemented by Planning Research Corporation/Information Sciences Company and the author in the development of advanced IS&R systems under USAF/RADC contract F30602-71-C-0157, and F30602-68-C-0108.

11. It is not essential that the new system's functions and capabilities match all of the old system (to eventually be replaced) in an ADAPTED program. Thus the new system could consist of only those critical functions or tasks requiring evaluations or improvements, and only a portion of the total data base included to service the incorporated functions or tasks. See the appendix for clarification.

12. To meet these criteria it might be necessary to locate new system test subject-users (and if applicable, their on-line consoles) in a remote and segregated area.

13. If such improvement were to be made during the test, this would invalidate many test procedures, requiring a halt in the test, the redesign of the testplan and starting the test anew. Greenwood, E., *Experimental Sociology: A Study in Method*, King's Crown Press, New York, 1945; and Chapin, F.S., *Experimental Design in Sociological Research*, Harper and Brothers, New York, 1955.

INFORMATION PROCESSING AND DISTRIBUTION UNDER (NEAR) MONOPOLISTIC CONDITIONS

Orrin E. Taulbee

University of Pittsburgh, Pittsburgh, Pennsylvania, U.S.A.

INTRODUCTION

Information scientists have found it extremely difficult to define precisely, either in a theoretical way or an operational way, the concept of information. Much progress has been made in this direction already - witness the large segment of literature devoted to this endeavor and the discussions in the Institute. Recognizing this difficulty in information science, others have and are continuing to investigate the "value of information" as an alternative to provide yet another stone in the foundation of the subject. Procedures and criteria for determining the value of information would be exceedingly valuable in establishing charges for value received, justifying the existence of an information system, developing budgets, and more generally, establishing accountability, credibility and economic viability in the field of information science.

The difficulty of placing a value on a chunk of information, for theoretical purposes, or so that it may be treated as a commodity item for pragmatic reasons, is not surprising when one permits his thoughts to dwell on the various aspects for a moment. A few comments will provide the basis for further individual exploration. Information that is exceedingly valuable to one person may, at the other extreme, have absolutely no value to another person. Even when two individuals agree that a chunk of information is valuable they may (and most likely will) not agree on its intrinsic value. These observations clearly result from the fact that differing uses will be made of the chunk of information by the individuals concerned.

I hasten to point out that there may be exceptions to the above statement. For it may be possible in some cases to establish a value for a chunk of information. This is true, for instance, for packaged information such as a book. The publisher sets a price; if an individual wants the book he must pay this price. An individual upon inspection of the book may decide that it is not worth that much to him and therefore does not purchase it. On the other hand, I'm sure that each of us has experienced the following situation after the purchase of a book. Upon use of the book we have observed that the book was worth many times what we paid for it; other individuals have observed that the book was disappointing and was not worth the purchase price. This leads me to remark that, though there was a purchase price established for the book, this price was in no way related to the intrinsic value of the book as an information package but was established as a basis for exchange in the marketplace. I would add that this price, established by the publisher, was based upon the cost of production, the competition, judgement of the market conditions and a reasonable profit. The conclusion, even in this case, is that there is an "intrinsic value" and a "market value" for the package. These two values will, in general, bear no relationship to each other.

Much the same situation exists when one considers the automobile or other product as an item of trade. The buyer of the automobile pays the purchase price and may in himself regard the intrinsic value as more than or less than this cost.

The previous discussion leads us to the point that more attention should be devoted to the value of information. Few additional comments will be made about the intrinsic value dependent upon the use made of the information. However, we shall adopt the position of the need for further exploration of the market value of information. In this connection we shall also view information as a service rather than a product. This provides the possibility of application of economic and marketing practices and theories to this nebulous concept of information.

Before proceeding further it is incumbent upon us to say something about an information system whose purpose is to provide this service. An information system consists of the staff and equipment along with procedures for collection, representation, organization, processing, retrieval and analysis of this ill-defined quantity called information. This rather broad scope of an information system permits the inclusion of such diverse systems as a command and control system, library, management information system, skills inventory system, parts distribution system or warranty repair system for an automobile manufacturer, control system for a manned space station, and the directory system for a given area code of the telephone system. A critical observation in this is that the system includes as an integral part a data bank of some kind which has been collected, organized and stored for search and retrieval purposes. In this concept of an information system no statement is made relative to the inclusion or exclusion of equipment of any kind - including the computer as part of the system.

This should be contrasted with what some have also referred to as an information system: an on-line system which permits the user access to storage space and contains the necessary tools for manipulation and processing of information which the user decides to store in his allotted (or purchased) space. A common example of this is the interactive computer system in which the user purchases access, through a terminal, to a computer. Note that the management of the computer facility provides nothing in the way of a data bank. I would characterize this system in contrast as an "information support system".

ECONOMICS OF INFORMATION SYSTEMS

Many information systems are supported on the basis of an administrative decision and appropriate budget. Others are supported on the basis of a grant from some outside agency (frequently government) on the belief of system management and the agency that it is necessary or good in the "public" interest. Here "public" will be taken to mean those who are users or potential users of the services provided by the system. For example, the public for a management information system established within a company is the collection of individual users (usually restricted to company employees or some subset of them). For a municipal library, the public consists of the residents of the municipality and others who are permitted to use it through established policy. In either case the individual privileged to use the system usually has unlimited use at no direct cost to him. Naturally, the user views the service as a free good, uses the services as frequently as he likes, and places an indeterminate value or no value on the service rendered or the information received.

System management and staff are as a rule very liberal about who is permitted to use the system except when some higher administrative authority dictates restrictions on use. The reason for this is patently obvious when one observes that these individuals justify the existence (and consequently their own importance) of the system on the basis of testimonials of satisfied users, number of patrons served, the number of searches conducted, the number of abstracts distributed, or some other statistics (counts) of utilization. These statistics of use give no indication of content of items and consequent intrinsic value of information delivered to the users. The statistics also do not indicate the use made of the information received from the system. Neither do they give any indication of the cost of making the service available on a per-time-of-use basis; neither do they indicate the cost of an item put into the data bank. I contend that we know no more about the economics of an information system than we do because administrative superiors have never asked that the system be justified or evaluated on the basis of budget (or actual expenses) versus income (or cost per search).

There is no reason to believe that every information system can or should be supported on the basis of income from sales of service. Certainly at a minimum we ought to experiment, perhaps for a system in a controlled environment, in determining costs of making services available, establishing charges to users, and comparing income against operating expenses. In this way we can gain much desired insight into the economic aspects of information systems.

It seems reasonably clear that selected information systems should be maintained and operated for the public good. It is perhaps equally clear that others should be operated on a self-sufficiency basis, i.e., supported only by income from sales of services; or else permitted to terminate their existence. No doubt it is a truism that, at least in the foreseeable future, there will exist more systems in the former category than the latter. There are others that should operate on a mixed basis, i.e., supported in part by income from sales of services and in part by an operational grant. Guidelines or criteria for effecting this categorization would be extremely helpful. For example, almost without exception, every library is sustained on the basis of public good. What rationale do we have for such decision other than precedent, perhaps resulting from the desire to "advance knowledge", and the avoidance of overhead resulting from developing costs and billing for charges?

What is even less clear are the conditions under which an information system may be converted from one supported in the public good to one which is self-sufficient. In case these conditions were clear there still remains the problem of the method of effecting the change.

MONOPOLISTIC CONDITIONS

Though it does not appear to be the case, most information systems operate in a (near) monopolistic environment. The reason for this is that either the information system is one of a kind in the sense that its data bank is unique, or systems duplicating all or a substantial part of a given system are relatively inaccessible due to location or other reasons to the users of the given system. Few information systems, with the exception of library systems, duplicate to any appreciable extent other systems. This is because of the expense, staff, and expanse of time necessary to develop an extensive information system.

A universal concern about an organization in a monopolistic marketing situation is that it may take advantage of the opportunity to establish rates and/or prices to its advantage and to the disadvantage of its customers. To prevent this it is common practice for government to appoint a regulatory agency to review and control to insure that the organization in a monopolistic situation is an absolutely essential item for its customers or potential ones.

It would appear then that "gouging the customer" by the management of an information system is not a real possibility at the present time since the practice is not to charge for the service and information derived from an information system is not generally deemed to be absolutely essential to a substantial percentage of the users of the service. Thus, even if charging for an information service was a common practice the prices could not be set too high, otherwise the customers will simply not use the service.

RESOURCE SHARING

The foregoing discussion leads inevitably to the question of resource sharing. This refers to the establishment of a relationship between two or more information systems so that each system does not attempt to be all things to each of its customers. But if a customer of one of the systems desires a service which is not available from the system which he customarily uses then the system management will refer him to another system which normally supplies that service. The obvious intent of resource sharing is to reduce the financial outlay for an information system operating in a particular setting. Up to the present time there has been little use in practice of resource sharing between information systems. There are several reasons for this:

(1) Concern for divulging confidential information
(2) Lack of knowledge on the part of the user or the staff of the information system which he ordinarily uses of what information is available in other information systems
(3) Inconvenience due to location or lack of knowledge on the part of the would-be user of the special techniques required to use another system
(4) Urgency, in the mind of the user, in getting access to the information
(5) Implications that system management has not done an adequate job
(6) Political considerations.

In spite of these reasons for not sharing resources there is one example which is widely known - the interlibrary loan. Many libraries have participated in this for many years. When a patron of the library came in looking for a particular document, book or other item which was not among the holdings of the particular library, the call went out to other participating libraries to ask that it be made available if it were contained in their collection.

How can the sharing of resources be effected? No doubt there are many and diverse methods for accomplishing this. One method which comes to mind immediately is that of appointing an individual on the staff of an information system who has the responsibility of being knowledgeable about other information systems and the services provided by them and who will additionally function as a coordinator/broker in interfacing between the user and other information systems.

The topic "sharing of resources" in relation to an information system immediately brings up the question of the definition of the word resources. What is meant by this term? It is not at all clear, for it means different things to different individuals. The first thing that is required is to clarify this concept. This should surely be done keeping in mind the type of capability and services that are available from well-known and widely-used information systems. Once this delineation is accomplished it would seem logical that the next step would be to catalog the resources that are available for sharing in many of those well-known and widely-used information systems. The catalog should consist of several sections which would be distributed accordingly. Among these would be at least: publicly-supported information systems for a broad segment of the public, governmental information systems, open (e.g., college and university libraries), and private (commercial) systems. It is to be expected that among the latter group only a limited number would be included because of the closed nature of the clientele who have the right of access and use. The exception to this would be those information systems that are profit-motivated.

A clear-cut case for resource sharing is where there are systems that are sufficiently different that it is advantageous to buy service from someone else's system rather than try to duplicate it within your own system. An obvious example of this is where there is a specialized, large data bank that is already existent in another system which would be too expensive to duplicate. Probably, the best example of all is where there is a need for information service within a given organization which does not have a viable information system to satisfy its needs. The admonition is clear in resource sharing: taking in each other's laundry is not the means of achieving effective resource sharing.

Governmental agencies have become concerned about the proliferation of information systems which they have sponsored; they wonder about the duplication which might exist among the systems, but also about the duplication of a proposed new system with an existing system. What control should be exercised by the agency? Just to give an indication of the complexity of the problem we might observe that the capability to be made available in a proposed new system could by modification be encompassed in an existing system. The catalog suggested above provides an important basis upon which to start discussion as to the types of control that should be exercised in the development of information systems.

PUBLIC UTILITY ASPECTS OF AN INFORMATION SYSTEM

The foregoing points made about information systems and others which have not been explicitly mentioned are displayed succinctly, in relationship to other commodities, in the accompanying Table. A few comments about the Table are in order. Its perspective is that of information systems in the aggregate. A particular information system may be in disagreement with the entries in the Table in one or more respects. No attempt has been made to be complete in the listing of the various facets of commodity production and distribution. Rather the facets shown are intended to be indicative. There are many others which can be included, e.g., training.

SUMMARY

By way of summary we have:

(1) Distinguished between intrinsic value and market value of information.

(2) Highlighted the difference between an information system and an information support system - the distinguishing characteristic being a management-provided data base in the former.

(3) Shown similarities and differences between an information system and a public utility. Caution should be exercised in viewing an information system as a public utility for information; an information system is a far more subtle concept than a public utility and the service it provides.

(4) Identified a number of avenues of investigation which appear to be worthy of exploration in advancing the state of the art in information systems.

(5) The concepts presented herein are in the nature of a preliminary report. Further investigation will be devoted to clarification and refinement of the ideas - particularly with regard to the Table.

	Automobile	Information Service	Electric Service	Taxi Service	Telephone Service
1. Competitive Market	Yes	No	No	Yes	No
2. Self-sufficient Activity (organ)	Yes	No	Yes	Yes	Yes
3. Indispensable	Yes	No, in most cases	Yes	Yes	Yes
4. Sell Item	Yes	Yes, for package	Yes	Yes	Yes
5. Resell some Item	No	Yes	No	Yes	Yes
6. Sell Tool for Access	No	Yes, in some cases, e.g., computer network	Yes	No	Yes, telephone instrument plus line
7. Identify Commodity Use	Yes	No, in most cases	Yes	Yes	Yes
8. Mechanism to Avoid Misuse	Yes registration	Yes, copyright	Yes, meter	Yes, meter	Yes, meter
9. Purchase Exclusive Ownership	Yes	No	Yes	Yes, while service is rendered	No
10. Regulatory Agency	No	Yes, in some countries	Yes	Yes	Yes
11. Self Service	-	Yes, in some cases	Yes	No	Yes
12. Marketing Packages	Yes, fleet sales	Yes, book	Yes, e.g. residential, industrial classes of service	Yes, 2 or more passengers cost basically same as 1	Yes, e.g., WATS Telpak
13. Antique Market (value not based on content or use)	Yes	Yes, for package (books)	No	No	No

Bibliography

Barzel, Y., The market for a semi-public good: the case of the American Economic Review, *American Economic Review*, 61, 665, 1971.

Choy, R.K.H., *The Production and Cost of a Free Urban Public Service: Resource Allocation in the Oakland Public Library Branch System*, by Center for Planning and Development Research, University of California, Berkeley, 1970.

Gruenberger, F., *Computers and Communications; Toward a Computer Utility*, Computer Communications Symposium, University of California, 1967, Prentice-Hall, Englewood Cliffs, New Jersey, 1968.

King, D.W. and Caldwell, N.W., *Study of the Cost Effectiveness of Retrospective Search Systems, A Report by Westet Research, Inc. to the American Psychological Association*, 1970.

King, D.W. and Neel, P.W., *Cost Effectiveness of On-line Retrieval Systems*, Westet Research Inc., Rockville, Maryland, 1971, ED 046 460.

Korfhage, R.R. and Delutis, T.G., *A Basis for Time and Cost Evaluation on Information Systems*, Purdue University, School of Industrial Engineering, 1969, PB 188 946.

Lancaster, F.W., The cost-effectiveness analysis of information retrieval and dissemination systems, *Journal of the American Society for Information Science*, 22, 12, 1971.

Lancaster, F.W. and Climenson, W.D., Evaluating the economic efficiency of a document retrieval system, *Journal of Documentation*, 24, 16, 1968.

Lamberton, D.M., ed., *Economics of Information and Knowledge*, Penguin, Harmondsworth, England, 1971.

Marshchak, J., Economics of information systems, *Journal of the American Statistical Association*, 66, 192, 1971.

Sackman, H. and Nie, N., eds., *The Information Utility and Social Choice; Papers Prepared for a Conference Sponsored jointly by the University of Chicago, Encyclopedia Britannica and the American Federation of Information Processing Societies.*

Sharpe, W.F., *The Economics of Computers*, Columbia University Press, New York, N.Y., 1969.

Veazie, W.H., Jr. and Connolly, J.F., *The Marketing of Information Analysis Center Products and Services*, Educational Resources Information Center, Clearing House for Library and Science, and American Society for Information Science, Washington, D.C., 1971.

Wolfe, J.N. *et al.*, *The Economics of Technical Information Systems: A Study in Cost Effectiveness*, 3 vols. and summary, University of Edinburgh, Edinburgh, 1971.

Chapter 4

USE OF INFORMATION

Use of Information

Editorial Summary

The distinction that was initially established by the Institute in differentiating information as a commody and information as a process serves to identify for review the papers present in this section. But a further word seems necessary - for no other reason than clarification. The use of information may be discussed in terms of two distinct but related contexts. First, information is discussed within a logistical context, that is, managerial, where the major aspects of the problem are such things as storage, access, and retrieval. The information in such cases is an object, a thing which serves as a mediating factor in meeting a specific need of the individual. The treatment of information by Brookes, Garvey, and Small are in this vein.

A second type of information has the character of a process. In this context, information represents a function, an operation on data which ultimately modifies the state of the user in his attempt to deal with the external world. The treatments by Triggs and Sackman are particularly concerned with this aspect. The logistical and functional concepts are clearly interrelated because the manner in which the individual formulates his need for data and proceeds to find it is determined by the way he shapes it "in his head". As to the role of an information system in all this, the information system is the environment (the technology-human mix) that facilitates the logistical-functional relationship. Lamont's discussion of an educational system is an example of this relationship. The discussion concerning the design and use of information systems is covered in other sections of this text.

In the logistical context we might ask: What kinds of information do scientists need? Can we predict these needs? Brittain's work in highlighting the difficulties of coming to grips with the needs of the user is now a classic. Brittain provides the Institute with a recapitulation of his perceptions. Numerous user studies have been conducted and, as Brittain claims, they are fraught with methodological difficulties. Perhaps more importantly, there is an absence of theory as to what constitutes a need for information and how this need can be measured.

Garvey has studied the socio-psychological nature of users extensively. In the particular study included in the Institute's program, Garvey relates his findings to scientists. Garvey claims that information resources required by scientists can be anticipated to a useful extent. Scientists, in general, proceed through research projects in an orderly manner which can be divided into stages. The information requirements are particular to each stage, both in types and quantities of information required. Information resources most often used at each stage will depend, also, on the type of research in which the users are engaged (e.g., whether physical or social science). The research might involve new subject matter or subjects that were previously studied, and the degree of experience of the individual scientist. Garvey has found that productive scientists are generally involved in several research projects at one time. Typically, the author of general articles starts his work 28 months before the article is published. It takes him about a year to complete his research and another 7 months to prepare the article. Consequently, to the extent that the rational structure underlying scientific effort can be described, information services can be designed to match the dynamic information needs of individual scientists.

Brookes' statement concerns the character of the data base which he refers to as knowledge. Data base (or knowledge) has a longevity dimension. Some examples of the "half life" of literature are: a few years for the literature of computer science; four and a half years for medical literature; and eight to sixteen years for mathematical literature. These are average and are simply illustrative of the rate of obsolescence. Brookes refers to Bradford's law which provides a mathematical model for measuring the degree of comprehensiveness of any computerized data or indexing service. Bradford's law may also be useful in measuring other aspects of social interaction. For example, the law has been useful in studying the borrowing of items from the library when the items are ranked, the borrowing of items from the library when the users are ranked, the publishing of books by publishers when the publishers are ranked, the indexing of papers input to a document system when the index terms are ranked, and the contributing to conference discussions when the contributors are ranked.

Brookes cites another finding that can be applied to the problem of structuring data bases to make them more effective as aids to the users' scientific work - Lotka's Law. Lotka noticed that in a cumulative index of abstracts, the number of contributors of a single paper was 4 times the number contributing two papers, nine times the number contributing three papers, and so on. Since the most significant contributions to any scientific field are very seldom made in the first papers in that field, Lotka's law can contribute, along with a number of other indicators, to a methodology for assigning significance coefficients to papers being considered for a bibliographic data base.

Griffith's presentation deals with the information commodity in yet another context. Griffith's interests center around the interrelationships that exist between different bodies of knowledge, or specialties. Small, his co-worker, regards co-citation to be a useful tool for mapping the specialties, and it has several advantages over other methodologies. Using the science citation index, one specialty can be compared with another. In one study, Small examined co-citation as a measure of the association between pairs of frequently cited documents. Co-citation is the number of times two documents have been cited together. This technique of using co-citations as a measure of associations frequently identifies relationships among papers which are not identified by techniques such as bibliographic coupling.

In the functional context, we might ask: How does man use information in decision-making and problem-solving? This question has been addressed by Triggs and Sackman. Triggs claims that vastly increased quality of data and information available to the decision maker may potentially improve the quality of his decision, but the processing demands placed on the decision-maker may offset the potential benefits. Decision-making aids may be useful in augmenting the performance of decision-makers under conditions of expanded information availability. Decision-flow diagrams, for example, may be useful. The decision-flow diagram may range from a simple indication of decision points in relation to an operational process, to a complex diagram indicating possible outcomes of each decision, probabilities of each of these outcomes and successful decision points related to each outcome. The decision-making process must be understood, however, before these decision-flow diagrams can be useful. Triggs discusses in some detail the "formalistic strategy" and "pragmatic strategy" in solving problems. The attempt is to identify formal algorithms that can be applied. However, Triggs quickly advises that attempts to impose a structure on an unstructured or fuzzy problem may distort the characteristics of the problem to the point where relevancy, and therefore usefulness, is lost. Decision-making in changing situations may not be adaptable to structuring at all. Key conditions of the environment change continuously, and it may be impossible to derive any structure that will adequately consider the environment at any specified point in time. A realistic approach might be to construct general procedures which would allow the analyst to quickly construct decision diagrams when needed for a practical situation.

Sackman's interest is the determination of how computers can aid man in processing data for problem-solving. Sackman's findings are particularly interesting in that they show that there is no difference in the way man proceeds to solve problems - with or without computers. Man achieves insight to a problem through the use of a computer when he begins to manipulate the data for a solution, or when an error occurs - which Sackman claims is rather typical. Sackman's studies allude to what can be called a "tunnel effect" where the solution of the problem is dictated solely by the software of the system. In general, computers are beneficial in providing the power for the manipulation of data after one determines what the nature of the problem is all about. Notwithstanding, working with computers will lead to their acceptance as a problem-solving aid.

In addition to decision-making and problem-solving, information technology has been applied to education. The Plato IV technology developed by the University of Illinois and discussed by Lamont is a significant attempt to demonstrate the educational effectiveness of computers in the classroom and to design a system that would be economical in competition with normal classroom instruction. Plato IV employs a plasma display panel instead of a cathode ray tube. The design characteristics of the system for instruction are nearly ideally suited for use in citizen interaction in community decision-making. The instruction capability facilitates bringing participants up to an equal level of understanding of a question or condition to be examined. Each viewer can then be asked to respond to a set of questions. In addition to responding to the questions, the citizen may also pose new questions or suggest new approaches or solutions which can then be discussed by other users of the system or by officials.

All in all, the evidence from the discussions on the use of information and information systems indicates that the ways individuals access information as a commodity can be readily appreciated. What is not appreciated is how individuals use information and related systems in decision-making, problem-solving, and communications in general.

INFORMATION NEEDS AND APPLICATION OF THE RESULTS OF USER STUDIES

J. Michael Brittain

University of Bath, Bath, England.

INTRODUCTION

It has been estimated[1] that there have been between 400 and 600 user studies in science and technology, the exact number depending upon the way in which user studies are defined. However, this figure may give an exaggerated indication of the amount of valuable activity in this field, because it has been estimated by the United States Federal Council for Science and Technology[2] that only 58 contributions from 450 publications identified in the field of user studies were based upon objective data. Since 1968 when the estimate of 400 - 600 studies was given, the literature has continued to grow. Lin and Garvey[3] noted, in reviewing the 1971 literature, that user studies continued to thrive in the United States and that beginning in the 1970s there was a good deal of activity in other nations; it has become a truly international phenomenon. Whatever the exact figure may be, there has been great activity in a relatively short space of time (the majority of studies have appeared since the Royal Society Conference, 1948), and it is reasonable to expect now that we should have a body of knowledge, verified and accepted, and capable of being applied, about the transfer of information, patterns of information seeking, and the needs and requirements of scientists and technologists for information data, documents, and information and library services. I suggest this is not the case.

In this paper I shall consider some of the reasons why this is so. In particular, I want to consider the objectives of user studies and the value of the results in the design and improvement of information services. I will not attempt to provide a review of the literature - this is done comprehensively each year in the first chapter of the *Annual Review of Information Science and Technology*; in addition a number of very thorough reviews have appeared during the last ten years. I expect these are known to the audience.

In considering the application of the results of user studies to the design of new information services, and the improvement of existing ones, I suggest it is of the utmost importance to consider the way in which the results of user studies accumulate to form a body of knowledge. I shall return to this question.

DEFINITIONS

The problem of defining information needs, demands, and requirements has plagued empirical studies. A satisfactory definition of need (as opposed to use or demand) has never evolved. The problem of definition has been dealt with very thoroughly elsewhere.[1] I will not go into detail here, although it is necessary to make a distinction between use, demand, need, and requirement. "Use" and "demand" are relatively easy to define operationally and therefore to measure. The demands that a user makes upon an information service can be recorded; likewise the uses made of services and documents can be recorded. The concept of "need" is more abstract. It can refer to one or any combination of the following: (1) needs stated by user; (2) needs that the user cannot state; (3) present needs - stated or unstated; and (4) future and potential needs - again these can be stated or unstated. The term "requirement" is often used as a general term, sometimes to indicate uses, demands, and needs and is difficult to define exactly.

I used to believe that the problem of measuring needs would be solved if the concept of need could be defined operationally. I no longer hold this belief and I am encouraged to hear at this meeting that an exact definition of the most abstract of concepts is not always a prerequisite to measurement. I will come back to this in the context of Cooper's suggestion[4] about defining concepts in terms of component statements.

METHODS

There are four main methods of obtaining data about information needs: questionnaire, personal interview, diary, and direct observation. The advantages and disadvantages of these methods have been well documented (e.g., by Parker and Paisley;[5] Borko;[6] and Bourne).[7] Many other methods have been tried, with varying degrees of success. More recently, with the introduction of on-line interactive retrieval services, it has been possible to record the man-machine interplay and to analyse these records. Also, queries put to information services have been analysed. However, when a broad picture of the information needs of a fairly large group of users is required, the most often used methods are the questionnaire and personal interview.

When data from user studies are required in the design of information systems, the limitations of existing methods are now very apparent. There has been a good deal of disappointment and criticism of the main methods but it is only recently, when attempts have been made to apply the results of user studies, that the limitations of the conventional methods have become apparent. In the past, user studies have been quite rightly criticised on methodological grounds: criticism has been made of sampling procedures, definitions of the variables studied and the absence in many cases of statistical tests of the significance of results. Now, the criticism is more fundamental. Doubt is expressed about application of the results in the design and improvement of information services. For example, the limited applicability of data in prediction has been noted recently in the UNISIST Report, which states:

> ... the information services of science are primarily intended for scientists, and it should be possible to collect from these all the data that are required - wishes, habits, needs, etc. - to determine an "ideal" information system ... Unfortunately, user studies, however thorough the exploration, however sound the methodology, do not warrant this optimistic approach.[8]

In discussing the relative merits of different types of user studies I suggest they should be divided according to the type of method used. This division corresponds, in a rough way, to the type of application possible. By this I mean that where very modest objectives exist (e.g., alterations in a lending system for documents based upon circulation figures), a sophisticated type of study is not required; a measurement of circulation may be sufficient. Methods used to establish information requirements include:

(1) Circulation data, demands made to an information system, etc.

(2) Analysis of queries made to an information system (e.g., content analysis, analysis of terms, relationship between terms)

(3) The feedback of data recorded in an on-line or mechanised information system to the system designer

(4) Analysis of user profiles (e.g., keywords, reports by information officers, changes of profile in the light of relevance judgements)

(5) Relevance judgements

(6) Data produced from interviews

(7) Data produced from questionnaires

(8) Other (e.g., diary keeping)

Studies which have used the first five methods have been a good deal less ambitious than those that have attempted to provide an overall picture of the information needs of many categories of users (mainly by questionnaire and interview methods). The results have been more successfully applied in the case of the less ambitious studies. The data obtained is often unambiguous, and can be fairly successfully applied to the re-design of library operations, dealing, for example, with loan periods, duplicate copies of monographs, etc.

Most criticism of user studies is directed at the wide-ranging studies, which have typically used interviews and questionnaires. I deal now with a few of these criticisms.

First, questions contained in questionnaires and interviews are often formulated by librarians and information scientists and may not truly reflect the problems the user encounters. For example, undue emphasis may be given to the use of bibliographical tools, either because the information scientist is not well acquainted with the way in which the user works, or because he imposes his own judgements about the way in which he thinks the user should work. I give an example from INFROSS (Investigation into Information Requirements of the Social Sciences).[9] Before we could design a questionnaire for use with social workers we had to interview them to find out about the details of their work. We conducted a pilot study using a questionnaire, but many of the questions were either irrelevant or were not understood by the social workers. As researchers and teachers, we had assumed that social workers were much more library- and bibliographically-oriented than in fact they were.

Second, data are usually gathered at one point in time, perhaps covering no more than 40 minutes of a user's working day. Interviews and questionnaire methods do not produce data that provide a longitudinal picture of information-seeking and gathering. Users often find it difficult to recall their habits or to name the most important sources of information, or forms of literature. We know that recognition scores are typically greater than recall scores. For example, if a user is presented with a list of articles relevant to his research he is much more likely to be able to recognise articles that he has seen or read (assuming of course that they are on the list) than to recall them from his own memory.

Third, surveys of information requirements by questionnaire deal mainly with the need for documents.

In the following sections I will consider approaches to the measurement of information needs that may avoid some of the difficulties listed above, but first I want to deal with the application of the results of user studies because the objectives of a user study should be made clear before designing a study and selecting measurement tools. The application of the results in the design of information systems should be one of the most important objectives of user studies.

APPLICATIONS

It is surprising how few studies have stated clearly their objectives. It has often been assumed that the major objective of gathering data about user requirements is for the design of information services and systems. In fact, very often lip-service only has been paid to this objective. It is also surprising to find that many information systems have been designed and developed in the absence of data about user requirements.

In the design of information systems, quantitative data are required about user needs and preferences, technical aspects of the hardware of information transfer, cost studies, and evaluation (including measures of performance and user feedback). The UNISIST Report emphasised the importance of user studies in supplying feedback from experimental services, but concluded that although

> ... much effort has been invested locally in user studies ... the outcome is not such as to provide systems analysts with data on user behaviour that could seriously pretend to a universal validity ...[10]

The universality of findings is of great importance when considering the application of the results. We also need to know whether or not the results of user studies accumulate to form a body of knowledge about information needs. At the present time I would say unequivocally that the results of user studies do not accumulate to form a body of knowledge, and very often results from one study are simply not comparable with results from another. This was vividly illustrated when an attempt was made to compare the results of INFROSS with the results of science user studies.[11,12] The objective was to find out the ways in which the information needs of social scientists were similar to, or differ from, those of scientists and technologists.

There are many problems involved in comparing the results of user studies with the objective of finding patterns which are repeatable from one study to another. Many studies contain factors that are not comparable. Other studies are comparable but of low validity, of poor quality, or otherwise inadequate. The following are some of the factors that can vary from one study to another:

(1) Population studied
(2) Environment
(3) Method
(4) Categories of questions asked
(5) Response rate
(6) Statistical techniques
(7) Presentation of results

In addition, there are other factors of a more general nature which make comparisons difficult. For example, a study may be purely exploratory, undertaken in order to formulate hypotheses which may be tested in subsequent studies. The results from exploratory studies are not amenable to much depth of analysis. On the other hand, some studies set out to test specific hypotheses, whilst others study a particular service with a view to improvement at a local level only. Even when the same survey method is used, major problems of comparison still arise because of incompatible categories and groupings of the data. There is usually considerable overlap between categories or omission of categories. It is essential for comparison that each survey states the categories used to describe the data.

Comparisons of the results of INFROSS and thirteen science user studies are given in Tables 1 and 2. The science studies were chosen on the grounds that they contained data on some of the areas covered by INFROSS, and data collection and presentation were in a form that made comparison possible. It turned out that the comparisons were very difficult to make, and this in spite of the fact that the best possible data for comparison were chosen. It is obvious that comparison of the results of studies selected at random from all user studies would be almost impossible.

Table 1*
MOST IMPORTANT INFORMATION SOURCE
(Percentage of respondents)

	STUDY			
SOURCE	ACSP# (1965)	Herner (1954)	Scott (1959)	INFROSS* (1971)
Monograph	-	12	4	17
Journal	38	11	73	17
Trade publication	22	16	-	-
Theses	-	4	-	3
Research report	1	15	-	8
Patent	1	1	-	-
Informal contact	31	-	-	8
Formal contact (conferences/meetings)	7	-	-	2
Other	-	41	23	45
Total	100	100	100	100

*Adapted from Skelton[12] #Advisory Council on Scientific Policy[11]

Table 2*
MOST EFFECTIVE RETRIEVAL METHODS
(Percentage of respondents)

SOURCE	STUDY							
	Scott (1959)	ACSP# (1965)	Herner (1954)	Fishenden (1959)	Herner (1959)	Martyn (1964)	Glass & Norwood (1959)	INFROSS (1971)
Citation in current reading matter	5	10	19	-	13	19	7	30
Abstract/index	4	40	15	23	12	13	6	18
Bibliography	-	-	14	3	6	4	2	12
Library Catalogue	18	10	8	12	9	4	-	8
Reviews	-	10	8	-	9	-	4	10
Personal	30	30	19	-	11	28	33	14
Chance	41	-	17	-	12	17	33	8
Other	2	-	-	62	28	14	15	-
Total	100	100	100	100	100	100	100	100

*Adapted from Skelton[12]

#Advisory Council on Scientific Policy

It would be impossible to use the type of data presented in Tables 1 and 2 for design purposes, other than as a general indication of trends. The differences between the social scientists and scientists (read across the rows in each table) is understandable. For a comparison of communication patterns in the physical and social sciences see Garvey, Lin, and Nelson.[13] No details were given about the group(s) of scientists; therefore it is impossible to know if Garvey et al. encountered difficulties in compiling a body of data about information-seeking patterns of scientists. It may be that they used the data on scientists gathered at Johns Hopkins Center for Research in Scientific Communication where heating engineers, scientists in mining, metallurgy and petroleum engineering, meterologists, geographers, sociologists, educationalists and other groups have been studied.[14]

The reason for the differences between the various studies of scientists is not clear. The differences may be due to different categorisations of users in each study, or different methods of measurement and analysis. From data of this type it is obvious that no general statements about the information needs of scientists or technologists can be made. Some of the problems of comparing the results of one study with another may be avoided if a detailed breakdown can be made of the data for each of the user groups included in each study. Comparisons of one study with another may then be more meaningful, and differences not so great. However, this is not always possible to do, either because (1) data are not reported in this way, or (2) sufficient details are not given about the user group to make comparisons possible (e.g., although two studies may both deal with physicists, one group may work in an academic environment and the other in industry). Very often the differences between groups of users included in a single study (e.g., chemists, physicists, biochemists) are smaller than the differences for a single group of users in different studies (e.g., chemists in studies x,y,z). This suggests that the differences between supposedly similar groups of scientists that have been reported by various studies are due either to different methods of measurement or different user groups. What is now required is a series of very tightly controlled studies - in some cases it may be possible to make use of existing data if it could be suitably re-analysed - in order to find the parameters of information requirements that are similar from one study to another and which designers could in future make use of.

An alternative approach is to undertake a study for each unique application. The user studies mentioned above were of relatively large groups of users, and were intended to be fairly comprehensive, covering many aspects of information need. Many studies have not had the objective of providing data that could be used, at least without further analysis, in the design of information services. When a study is confined to a particular group of users, a particular environment or information service, feedback data can be gathered with the objective of refining and improving the performance of the service. The data from this type of study is almost certainly going to be much easier to apply. However, such studies do not usually consider the question of generalising the findings to other information services or groups of users.

From this discussion two types of application emerge. The first type of application is the use of the results of studies to provide general guidelines, to ensure that the designer is "informed" in the sense that he knows the general limits within which the system must operate, but cannot use the data to determine the exact parameters of the system. The second type of application makes use of the numerical aspects of the data in design; for example, for the purpose of specifying exactly the parameters of a system. A few examples will make this distinction clear.

A good example of the first type of application can be found in Back[15] who considered the implications of the results of dissemination studies for the design of on-line reference retrieval systems. Back considered a number of studies and drew some general recommendations from them. For example, he made recommendations about the groups of users for on-line systems, the way in which such systems could compete with manual systems, and the type of material that should be included in the data base.

The second type of application involves the actual use of research findings from user and dissemination studies in the design of systems. I give as an example the creation of a machine-readable data base. It is conceivable that data are available about the preferences of users for abstracts, titles, titles plus abstracts, titles plus keywords, etc., and that data are available for each major group of users. If a machine-readable data base serves a large number of user groups, the retrieval operations and presentations will need to be programmed so that each group receives a different subset of the material. If data are available about the delay that is acceptable in receiving material it would be possible to design systems so that certain criteria were closely met. If data are available about the amount of information that each user group requires, the frequency with which they require it, and the amount that they require per unit time, it would be possible to design services so that they performed accordingly, irrespective of the total amount of information available. In most cases this type of data is not available.

We might ask how the information required by a systems designer would orient him in considering the data needed from a user study. A systems analyst, according to Katter,[16] might be interested in (1) methodologies especially calibrated and tuned to unique applications, (2) statements of information needs and uses with planning and design activities in mind, and (3) evenly-distributed sensitivities for the underlying social, psychological, technical, organizational and other factors that affect information needs and uses. The data required by a systems analyst could not usually be obtained by a single enquiry.

In discussing a systems approach to the design of information storage and retrieval systems Scheffler[17] maintains that a new philosophy (in tune with the stringent economic conditions of the 1970s) is needed in dealing with studies of user needs. Past studies have often been wasteful in terms of quantity of data gathered and lack of application. Scheffler argues that contact with the entire population of potential users on a one-to-one basis, deriving therefrom a summation of all needs, is not necessary. He reports that the University of Dayton Research Institute has had success in establishing user needs in small group sessions. A priority listing of needs should be established, ranked in order of importance. Users are in a much better position to state their information needs when they have experienced a prototype of the information system that is being designed; in other words, Scheffler argues the case for the continuous assessment (at least for the duration of the design and development work) of information needs in terms of user performance data and judgments made over a fairly long period.

As Katter[16] points out, it is usual to consider a redevelopment cycle in design activities; data about information need is only one part of the cycle, and may need to be up-dated and re-evaluated. In design, a redevelopment cycle consists of several kinds of activities. Katter suggests the following as a typical sequence: first, discover and state needs, derive and state system objectives; describe available resources and constraints. Then produce a system-operating concept; perform a functional phase and task analyses. From the results of these, produce detailed design specifications. On the basis of the specifications depict alternative systems. For each alternative, perform cost-effectiveness analyses for each element, component, and procedure. Apply trade-off analyses to cost-effectiveness data to select the best alternative. Produce detailed design specifications for the alternative selected and produce a preliminary operating version of the system. Perform shake-down and sensitivity testing and evaluation. Use the test results to modify the system and complete final acceptance testing. Produce, deploy, and install the system.

At many stages in such a redevelopment cycle user-generated data would be of value. It would be necessary to obtain user data about document needs, type and form of material and data required, timeliness of data, amount of material required, as well as user reaction to the various components of the system. Although user studies could be tuned to meet the needs of systems designers, this in itself would not ensure that the problems of measuring potential and unstated information needs would be overcome. I would like to consider this problem in some detail and suggest ways in which a solution might be found.

ALTERNATIVE APPROACHES TO THE MEASUREMENT OF INFORMATION NEEDS

Since one of the objectives of a user study should be to predict future information demands and requirements, it is important that studies are not confined to establishing current requirements only. The problem is a familiar one. For a number of reasons, users are very often not in a position to state their future information needs or the information that would make for progress in their discipline. Usage patterns and user-stated requirements are partly a function of existing services and the past experiences of users. However, it may not be the user's fault that information requirements cannot be established by the methods currently in use. A preoccupation with the inadequacies of the user is not the sort of attitude that the information scientist should take as he begins an enquiry about information needs and uses.[18] The burden of finding out about information needs should be placed with the information scientist and the librarian. Although I suggest that there is room for improvement in existing methods of enquiry (e.g., continuous assessment of information requirements in contrast to a single enquiry), I think we must look to different methods, and to some radical rethinking of the problem of information needs. Although users are very often in no position to state all their information needs, this is not to say that they are unable to state their current and future problems, activities, and types of work, including the decisions that they make and will have to make in the future. I suggest that users should be encouraged to formulate the objectives of their research or practice, produce an outline of their decisions and problems encountered over a fairly long period of time; then, in conjunction with information scientists, they would be in a much better position to formulate their future and potential information needs which would arise in connection with these problems and decisions. What then is the solution? A change in attitude on the part of information scientists to establishing needs? The development of new methods, or the refinement of existing ones? The redirection of interest to topics that users can make valuable judgements about? Or, what is perhaps the most radical suggestion, research into the ways in which data and information accumulate to form knowledge in science, including an assessment of future problems, developments, and decisions, in the light of which an information policy would be evolved? I will deal with some of these possibilities.

Refinement of Existing Methods

It is certain that the conventional methods of the questionnaire and the personal interview will continue to be used, in many contexts, to establish information requirements of scientists. However, the day when the general user enquiry, giving an aerial picture of a discipline or group of disciplines, was of value is perhaps past. It is now a case of refining methodologies, pinpointing specific questions, and gathering data that can be applied. From a methodological point of view, there are some refinements of conventional methods which are needed. These include the following:

(1) Questions should reflect user priorities and not those of the enquirer.

(2) Only questions which users are in a position to answer should be asked. General questions about future information needs should be avoided.

(3) Longitudinal studies are preferred to single studies: this has the advantage of establishing changing trends as well as giving the user greater opportunity to recall his activities. Users typically cannot give good recall in a single enquiry.

(4) Only significant results should be reported and this implies a rigorous application of statistical tests of significance.

(5) The objectives of the study should be clearly stated before the enquiry is made and an indication given of the way in which the results can be applied.

Below is an example of the way in which a user study might be conducted in the future. The study is concerned with the design of a new secondary service, let us say an abstracting service. It will be assumed that the value of a secondary service has already been established, and that the place of secondary services in the context of information requirements and all other information and library services available has been investigated. In part, previous user studies will be able to shed light upon these aspects. In fact, most user studies have stopped at this point: they have not gone on to consider the detailed parameters of a service. It is here that user studies have something to contribute in the future. The study would begin by describing and surveying the parameters of an abstracting service. These would include (1) field to be covered, (2) number of abstracting services in the field, (3) coverage of the primary literature, (4) overlap between existing services, (5) major user group(s) and subgroup(s). The following types of data would be required: (1) observation of users, (2) questioning of users, (3) classification of users (e.g., researchers, practitioners, administrators), (4) observation and questioning of intermediaries (e.g., information officers, library staff, staff of information centres, etc.), (5) the nature of subject expertise in the field in question and the value of information for research, laboratory practice, decision making, etc. In order to design a new abstracting service or redesign an existing one, it would be necessary, in addition to data gathered from users, to obtain data about the amount of material to be covered, the type of entry, etc. These aspects would of course involve some user-generated data but also other data (e.g., statistical parameters of the literature).

Methods from Other Fields

I have no suggestions to make about entirely new methods, although there are a number of methods that have been developed in other fields which should be tried in information science, in order to establish information needs. I would like to deal with four methods: the Delphi technique, techniques used in diffusion studies, a suggestion by Cooper[19] (which hardly amounts to a method) about the operational definition of information needs, and continuous assessment in the context of a mechanised or on-line information retrieval system.

First, I will deal with the Delphi technique. This was developed as a method of predicting and forecasting technological and economic requirements and has only recently been introduced into the field of information science by Reilly,[20] Borko,[21] Wennerberg,[22] Swedish Agency for Administrative Development,[23] Belzer, Isaac, Finkelstein, and Williams,[24] and Anderla,[25] although not in the context of measuring and estimating information needs. For example, Reilly dealt with the method as a potentially useful tool in information science research in general, and Borko used the technique to establish areas for research. A typical Delphi study might consist of the following stages.

(1) Opinions are obtained from a number of specialists about their expectations of what might happen in their field. In the case of information science, this might include judgements about information requirements, or tasks, or decisions requiring information. Participants in a Delphi study do not necessarily have to be experts or specialists. The technique has been tried with students and the public. Group participation (see, for example, Turoff[26]) is another variant of the technique.

(2) In the light of the loosely formulated opinions so obtained, a questionnaire is drawn up to obtain details about (a) when predicted events will occur, (b) probability that these events will occur in the period under study, and (c) degree of importance that should be attached to these happenings.

(3) A questionnaire is mailed to a group of individuals, usually experts in the area being studied, with fairly diverse backgrounds.

(4) Responses are scored and the means and standard deviations calculated for each rating scale.

(5) Each respondent is provided with a feedback of the results in the form of group norms, and respondents are requested to evaluate their previous preferences in the light of the feedback.

(6) Final analyses are made.

In short, the Delphi technique can be used to obtain a reasonably accurate group consensus for predicting the outcome of possible alternative courses of action, or future information requirements, et cetera.

Second, I will mention diffusion studies, which is a large field in itself, and continues to be very active. It is interesting to recall that in an early review of science user studies, Menzel[27] devoted a chapter to neglected approaches to the study of information needs. He mentioned a large number of approaches that have seemed both feasible and promising, although they have been largely neglected by information scientists. Menzel's comments on diffusion studies are illuminating. He points out that such studies take individual units of information (e.g., messages, ideas, data) and attempt to discover how they percolate through scientific and practitioner communities. Diffusion studies try to answer such questions as: "How long does it take for news of a given discovery to reach researchers in various fields, at various types of institutions, with various communication habits?"; "Do messages that fail to reach significant numbers have any special earmarks, for example, publication in obscure journals, lack of informative titles, special content?" "How many scientists are eventually reached by each kind of news, and what are the characteristics of scientists that are missed?".

Third, I will expand upon the suggestion made by Cooper[28] about the operational definition of concepts. Cooper suggests that the problem of measuring information need lies partly in the inexactitude of the term and the difficulty of defining it for measurement purposes. He suggests that a solution is to define information need as a set of component statements, where interdependencies may exist among the components. Cooper refers to such a set of statements as a statement tree; the topmost node of the tree has no special interpretation. We may equate this to the general term "information need". The nodes at various steps down from the origin are component statements which the user can formulate. The next level represents component statements which would be of interest if their parent nodes proved to be derivable. A third and fourth level, and so on, exist if all the previous levels are derivable. Most information queries stem from information needs which would require large trees for their representation. Cooper's suggestion is a more formal statement of what has often been regarded as the stumbling block in user studies: that is, users find it impossible to answer highly abstract questions about information need. In this sense, many studies of information needs have tried to define the concept at the level of the apex only. Others have contained a large number of questions, some of which may perhaps resemble Cooper's component statements, but the inter-relationship between questions has not been made explicit. In other words, no *tree* has been developed. The beginnings of a theoretical structure in user studies may be seen when component statements can be grouped together in a meaningful way and when the grouping does not vary at random from one study to another.

Fourth, I would like to refer again to the continuous assessment of user behaviour that is now possible with on-line interactive systems. Enquiries about information needs as such are not made, but from an analysis of user performance it is possible to derive statements about current information requirements. A good example of continuous assessment of user performance is seen in the LEADERMART information system at Lehigh University, described by Hillman.[29] Complete transaction files on system and user performance are created and maintained. Statistics can be produced about (1) delivery time, (2) selectivity, (3) accuracy of responses to demand bibliographies, (4) ease and rapidity with which a scientist can conduct a dialogue with a body of organized information in negotiating a request, (5) access speed of individual items, (6) browsing, and (7) interest profiles of users. On the basis of system and user performance feedback, appropriate modifications are made to each of the subsystems. In this sense, user-generated data are being used for design, in the continuous redevelopment cycle.

These methods can give valuable data; but the problem of measuring *potential* information needs will not be solved by the introduction of these methods. I suggest one last alternative approach, which is much more radical than any of the others. It is of course a counsel of perfection to demand of an information service that it provide users with things that they do not even know they want, but as Bernal has suggested:

> ... if our object is the advancement of science this is precisely its most important function, because those who know what they want can usually find it - even with difficulty, and after a certain time - but the unknown will never be found unless someone is attempting to provide the scientific worker with the possibility of discovering it.[30]

If information scientists and librarians are to provide scientists with information they do not know they want, I suggest this implies that the information profession must know something about scientific information, and the way in which it is disseminated, transferred and accumulated, which the average scientist does not know for himself, or at least has no inclination to find out about.

If information scientists can take on this role their horizons are immediately widened. They can then legitimately make recommendations about the relevance of information to specific scientific enquiries and problems, and introduce new information that the scientist may not see as relevant upon first acquaintance. The information scientist has usually been credited with superior bibliographical knowledge: that is, he is supposed to know more about documents and their organization and retrieval than the scientist himself. In the past, information scientists have not been attributed with any special understanding of the information and data as such contained in scientific publications. This has been deemed to be the province solely of the scientist. If the information scientist extends his range of activities to include some of those traditionally thought to be the province of the scientist, then the boundaries of information science are shifted. The information scientist may need to know more about the nature of subjects than he does at present, and make decisions on behalf of scientists. Obviously a change in attitude on the part of both scientists and information scientists will be required. The implications of such a change of role extend to the training of information specialists and the education of users.

At the present time it is difficult to say exactly what is the knowledge that the scientist does not possess but which the information scientist may (or could) possess, knowledge which would allow him to make predictions about the information needs of scientists or about current information which, though available, scientists, for one reason or another, do not know about and therefore do not request.

I would like to deal briefly with some of the characteristics of scientific information and knowledge that would be of use in establishing information needs and predicting them, and which would typically be the province of the information scientist (or in some cases the student of the science of science, the sociology of science, etc.). This does not mean that the scientist need remain ignorant of these aspects of his subject, but he would not be expected to research into them in order to gain a detailed understanding. Many of these aspects of scientific information are not particularly well understood at the moment, and in many cases have not been thoroughly researched. The beginnings are evident, but a good deal more work is required, obviously involving a co-operative effort between information scientists and scientists.

I will deal with three characteristics of scientific knowledge: (1) statistical characteristics of the literature, (2) the so-called "laws" of scientific knowledge, and (3) social and psychological factors involved in the communication of information.

(1) Statistical parameters of the literature include:

(a) Growth and size
(b) Scatter of literature across journal and subject disciplines
(c) Major divisions within subject areas (e.g., established by citation studies)
(d) Decay and obsolescence rates for literature
(e) Concentration of literature (e.g., established by citation studies, clustering studies, etc.)
(f) Currency of material (e.g., delay in publication, appearance in secondary journals)
(g) Characteristics of secondary publications (e.g., coverage of primary literature, currency, overlap, scatter, etc.)

(2) Characteristics, or "laws" of scientific knowledge
(a) Accumulation of data and information to form knowledge, building blocks, etc.
(b) Penalties for ignorance
(c) Substitutability of data and information
(d) Comprehensiveness, reliability of reviews
(e) Function of information and data in query formulation, hypothesis-testing, and theory construction

(3) Social and psychological aspects of scientific information and communication
(a) Relative importance of formal and informal communication
(b) Relationship between types of knowledge and types of channels (e.g., "know-how" information may best be transferred through informal channels)
(c) Evaluation of scientific work and literature
(d) As a stimulus to research and enquiry
(e) Information overload
(f) Processing of information by user (e.g., cognitive processes)
(g) Information and decision-making; information and action
(h) Productivity

Although there is now a substantial body of research in information science about some of the topics listed above, a good deal more is required before this type of data could be used in the design of information services.

Even at this stage, though, it is necessary to consider how such information could be applied. In some ways, application of this type of data, especially that relating to the prediction of future needs and demands, is easier than in the case of data that comes from the conventional user study. Where data are available about future demands, especially if it is of a statistical nature, the application is fairly obvious: it is more difficult to make sure that the indications for the future lead to the planning and development of suitable information services to meet these predicted needs. I give two examples of the way in which such data could be applied. First, consider the necessity of a scientist having to sample from the totality of information that is available to him. This has already been suggested by Bernal[31] as a partial solution to the problem of increasing amounts of information. A knowledge of the statistical parameters of the literature, especially data from (1)(e) above (i.e., concentration of literature across journals and/or subjects), would be useful in indicating the validity of a sampling procedure. For example, if there was a concentration of "key" articles in relatively few journal titles, sampling at random from a large number of journals would produce a large number of relatively unimportant papers and a small number of relatively important papers. Sampling would have to be stratified so that the important material was appropriately represented in the sample.

Second, I would like to refer again to the LEADERMART retrieval system in which a dialogue between the user and the machine is possible. The system is structured, drawing upon the parameters of the literature and subject matter in question, in such a way that the user is aided in his query formulation by the structure in the system. Describing the LEADERMART information retrieval system Hillman[32] states that retrieval

> ... takes the form of a dialogue between man and ... system, the intent of which is to arrive at a joint decision as to the nature, volume, pertinence, and utility of the information to be formatted, retrieved, displayed, and put to work.

The objective of the LEADERMART system is to provide a kind of road map, enabling users to trace important conceptual paths among topics either globally (i.e., system-established), or by local overlays (i.e., user-supplied connections). User-supplied connections refer to the input to the system by the user, and the system-established connections refer to what I have called the statistical parameters of the literature that can be built into the system, or in some cases perhaps established as new material is added to the system.

These alternative approaches to finding out about the information used and required by scientists, with the objective of providing data for the design of information systems, rely heavily upon indirect methods of assessment and prediction. And because of this these approaches are sometimes regarded with suspicion, especially by the library and information profession. This is because a degree of face validity is lost when users are not questioned direct. But these alternative approaches will no doubt gain in popularity in the near future, as a body of knowledge becomes available about the structure of communications and literature of science, and as all those responsible for the design and operations of information and library services come to realise that little more can be gained by asking the user direct about his information needs and requirements. If the information requirements of scientists change radically in the next decade then a return to the direct method of asking users may be of value, but if alternative methods of assessment are successful this occasion should not arise, because the changing requirements will be monitored and (hopefully) predicted to some extent.

CONCLUSIONS

(1) A large number of studies of the information requirements of scientists and technologists exist, but many different methods have been used. It is therefore difficult to compare results and make general statements about each aspect of information need.

(2) Very few results of user studies have been applied directly in the design of information systems. Many user studies have not been undertaken primarily to meet this objective, and where application of results has been the objective many studies have not produced data in a form that could be applied by systems analysts.

(3) There has been little concern about methodological issues. Many variables have been included but there has been little standardisation of variables or methods. Few statistical tests of the significance of findings have been applied with the result that perhaps only a very small proportion of the large amount of data that has appeared is worth reporting.

(4) Even if a more rigorous approach is taken to the measurement of information needs and requirements, the problem of assessing potential information needs would not be solved.

(5) Methods which (a) assist the user to formulate his information needs, (b) provide indirect assessment, and (c) are designed to gather predictions about future requirements, should be tried out in the future.

(6) Much more attention should be paid to the application of the results of user studies in the design of information services.

(7) The point of saturation may have been reached with respect to the collection of data about information requirements at a general level.

References

1. Line, M.B., The social scientist and his information needs, in Library Association - Reference, Special and Information Section, *Proceedings of the 16th Annual Conference, Durham, 1968*, London, 1968, 10.

2. United States Federal Council for Science and Technology, *Recommendations for National Document Handling Systems in Science and Technology*, U.S. Department of Commerce, Washington, D.C., 1965.

3. Lin, N. and Garvey, W.D., Information needs and uses, *Annual Review of Information Science and Technology*, 7, 5, 1972.

4. Cooper, W.S., A definition of relevance for information retrieval, *Information Storage and Retrieval*, 7, 19, 1971.

5. Parker, E.B. and Paisley, W.J., Research for psychologists at the interface of the scientists and his information system, *American Psychologist*, 21, 1061, 1966.

6. Borko, H., Determining user requirements for an information storage and retrieval system, in *Information Systems Workshop*, C.P. Bourne, ed., Spartan Books, Washington, D.C., 1962, 37

7. Bourne, C.P., A review of the methodology of information system design, in *Information Systems Workshop*, C.P. Bourne, ed., Spartan Books, Washington, D.C., 1962, 11.

8. UNESCO, *UNISIST - Study Report on the Feasibility of a World Science Information System by the United Nations Educational, Scientific and Cultural Organisation and the International Council of Scientific Unions*, Unesco, Paris, 1971, 25.

9. Line, M.B., *Investigation into Information Requirements of the Social Sciences: Information Requirements of Researchers*, (Research Report no. 1), Bath University Library, 1971.

10. UNESCO, *op. cit.*, 109.

11. Bath University, *Comparison of Results of Science User Studies With "Investigation Into Information Requirements of the Social Sciences"*, (Design of Information Systems in the Social Sciences, Working Paper no. 1, by B. Skelton), Bath University Library, 1971.

12. Skelton, B., Scientists and social scientists as information users: a comparison of results of science user studies with investigation into information requirements of the social sci-

ences, *Journal of Librarianship*, 5, 138, 1973.

13. Garvey, W.D., Lin, N. and Nelson, C.E., Communication in the physical and social sciences, *Science*, 170, 1166, 1970.

14. Lin, N., Garvey, W.D. and Nelson, C.E., *A Study of the Communication Structure of Science*, (Report no. 10), Center for Research in Scientific Communication, Johns Hopkins University, 1970.

15. Back, H.B., What information dissemination studies imply concerning the design of on-line reference retrieval systems, *Journal of the American Society for Information Science*, 23, 156, 1972.

16. Katter, R.V., Design and evaluation of information systems, *Annual Review of Information Science and Technology*, 4, 31, 1969.

17. Scheffler, F.L., A novel philosophy for the design of information storage and retrieval systems appropriate for the '70s, *Journal of the American Society for Information Science*, 24, 205, 1973.

18. Doyle, L.B., Perpetual user studies, *Datamation*, 12, 28, 1966.

19. Cooper, W.S., *op. cit.*

20. Reilly, K.D., Prospects for use of the Delphi method in information science research, in *Proceedings of the American Society for Information Science*, 7, North, J.E., ed., ASIS, Washington, D.C., 1970, 23.

21. Borko, H., Predicting research needs in librarianship and information science education, in *Proceedings of the American Society for Information Science*, 7, North, J.E., ed., ASIS, Washington, D.C., 1970, 27.

22. Wennerberg, U., Using the Delphi technique for planning the future of libraries, *Unesco Bulletin for Libraries*, 26, 242 and 252, 1972.

23. Swedish Agency for Administrative Development (SAFAD), *Report on a Delphi Study - Information, Documentation and Media*, SAFAD, Box 2106, 10131, Stockholm, 1971.

24. Belzer, J., Isaac, A., Finkelstein, E., and Williams, J., Curricula in information science: analysis and development, *Journal of the American Society for Information Science*, 22, 193, 1971.

25. Anderla, G., Technological forecasting and information handling, *Technology Ireland*, 4, 8, 1973.

26. Turoff, M., Delphi and its potential impact on information systems, in *Proceedings of the 1971 Joint Computer Conference*, American Federation of Information Processing Societies, 39, 317, AFIPS Press, Montvale, New Jersey, 1971.

27. Menzel, H., *Review of Studies in the Flow of Information Among Scientists*, 2 vols., Bureau of Applied Social Research, Columbia University, New York, N.Y., 1960.

28. Cooper, W.S., *op. cit.*

29. Hillman, D.J., The future of information provision, *IATUL Proceedings*, 5, 70, 1970.

30. Bernal, J.D., The supply of information to the scientist: some problems of the present day, *Journal of Documentation*, 13, 199, 1957.

31. Bernal, J.D., *op. cit.*

32. Hillman, D.J., *op. cit.*, 81.

Bibliography

Advisory Council on Scientific Policy, Survey of information needs of physicists and chemists, *Journal of Documentation*, 21, 83, 1965.

Borko, H., Determining user requirements for an information storage and retrieval system: a systems approach, in *Information Systems Workshop*, C.P. Bourne, ed., Spartan Books, Washington, D.C., 1962, 37.

Farradane, J., [Letter to the Editor], *Aslib Proceedings*, 10, 40, 1958.

Fishenden, R.M., Methods by which research workers find information, *Proceedings of the International Conference on Scientific Information, 1958*, National Academy of Science/National Research Council, 1, 163, 1959.

Garvey, W.D., The dynamic scientific information user, Paper presented to the *NATO Advanced Study Institute in Information Science*, Aberystwyth, 1973.

Glass, B. and Norwood, S.H., How scientists actually learn of work important to them, *Proceedings of the International Conference on Scientific Information 1958*, National Academy of Sciences/National Research Council, 1, 195, 1959.

Herner, S., Information gathering habits of workers in pure and applied science, *Industrial and Engineering Chemistry*, 46, 228, 1954.

Herner, S., The information gathering habits of American medical scientists, *Proceedings of the International Conference on Scientific Information 1958*, National Academy of Science/ National Research Council, 1, 277, 1959.

Lawson, H.B., Defense Documentation Centre reports user needs study results, *Defense Industry Bulletin*, 4, 9, 1968.

Lipetz, B.A., Information needs and uses, *Annual Review of Information Science and Technology*, 5, 3, 1970.

Lister, W.C., *Least Cost Decision Rules for the Selection of Library Material for Compact Storage*, Ph.D. Thesis, Purdue University, Lafayette, Indiana, 1967, (PB-174441).

Martyn, J., *Report of an Investigation on Literature Searching by Research Students*, Aslib, London, 1964.

Scott, C., The use of technical literature by industrial technologies, *Proceedings of the International Conference on Scientific Information 1958*, National Academy of Science/National Research Council, 1, 244, 1959.

Voos, H., *Organisational Communication: A Bibliography*, Rutgers University, New Brunswick, N.J., 1967, AD-657 647.

Wooster, H., *Policy Planning for Technical Information in Industry*, Air Force Office of Scientific Research, Directorate of Information Sciences, Arlington, Virginia, 1967, AFOSR-67-2588, AD-661 589.

STATISTICAL BIBLIOGRAPHY AND INFORMATION SCIENCE

B.C. Brookes

University College, London, England.

My interest in statistical bibliography arose from the fact that it offered a field of study which would most quickly yield testable hypotheses. If information science is to justify its explicit claim to be a science then it must be able to show that it has developed *theories*, i.e., well-corroborated hypotheses. Moreover, as a physicist by training, I agree with Lord Kelvin's dictum that we can hope to understand a phenomenon *scientifically* only to the extent that we can measure it.

A secondary consideration in selecting bibliography as a field of study was that any results obtained from the empirical analysis should have immediate application. It seemed to me important that any theories developed in the early stages of the new science should be seen to be useful. And a further consideration was that the research required could not readily command financial support to cover more than modest costs. Fortunately, the data relating to statistical bibliography of the kind I have in mind can be found in the back-runs of scientific periodicals available in any good science library. So the data are abundant; they are also publicly observable in the sense that anyone who might wish to challenge my results can do so. The techniques too are simple. Semi-logarithmic graph paper is the only aid required. Though we have developed computer programmes also, they are not absolutely necessary.

The *theoretical* objective of the work is to explore the implications of the empirical laws of obsolescence and scatter for the dissemination of knowledge. The *practical* objective is to structure bibliographic data-bases so that they can be more efficiently searched, more economically operated and more rationally designed. It seems to me that unless some programme of the kind I have in mind is undertaken, the large data-bases projected in the UNISIST proposals, for example, will collapse under the weight of the rubbish they accumulate. It seems to me absurd to continue to increase the sophistication of the computer technology applied to documentary systems without having a critical look at the material they handle and the use that scientists make of it.

THE PHENOMENA OF OBSOLESCENCE

Before the computer was applied to the literatures of science, the users of this literature, i.e., the scientists who also created them, had their own way of dealing with the problem of exponential growth. They masticated the new literature as it appeared, extracted from it whatever was useful and discarded the rest. This mastication is a continuing process in which new ideas and new empirical results are compared with those inherited from the recent past. The literature of a science is being continually revised and updated. In this process of revision, the new makes a dramatic appearance in the form of the new issues of the periodicals which fall heavily on our desks every day. But the discarded material is not rejected in any dramatic way; it is simply ignored and quietly assigned to the archives. So though the literature of science is, in general, *growing* exponentially, the literature in which scientists are interested is also being *discarded* exponentially. The volume of the data-base which scientists actually use increases as the *difference* between these two exponential rates.

This process of discarding has been disturbed by the operation of the computerized data-bases. In a recent user survey, of which a report by Sacha Presanis and myself will shortly be published by OSTI, one third of the users of a computerized data-base complained of the computer system's insistence on dredging up items which should have been discarded and forgotten. The users complain of the trouble and cost they were put to in tracing papers in obscure journals, a process which in Britain is easier than anywhere else in the world, only to find that they had wasted their time and money. Such effects are not avoided by putting the data system 'on-line'. The designers of data systems will have to pay more regard to the scientific use of the data they input and retain in their systems if they wish to continue in business.

It is important to establish the rate of obsolescence of the literatures of science. These rates can be surprisingly high. For example, at one stage of its development the literature of computer science had a half-life of about two years. Medical literature has a half-life of about four and a half years though this figure is an average which hides a range of rates for its component parts. The literature of mathematics has a longer half-life, of about twelve years overall, though this average too hides some intriguing anomalies. For example, *American* mathematical literature has a half-life of about eight years, *Russian* mathematical literature of about twelve years and *British* mathematical literature of about sixteen years. Why do these differences arise?

The measures of obsolescence I use are derived from citation data.[1] A sample of the literature, bibliographically well-defined, is taken from the periodicals which contribute to that literature. The sample is distributed over the base-year one wishes to measure from. The dates of publication of papers cited in this sample are then noted and cumulated as a distribution of frequencies of citation against data. The total number of citations is then decumulated year by year in succession to give a table which expresses the frequency of citation to each year or earlier. If these figures are then plotted on semi-logarithmic graph paper, a typical result shows that after a hesitant start (which represents the period of mastication of the new literature) the points fall on a straight line when the frequences are plotted against the logarithmic scale. Linearity implies that the rate of decay is exponential.

One of the implications of exponential decay is that, though the probability of citation of an ageing paper is always greater than zero, the negative exponential distribution converges to a limit. The total usage of the literature is therefore finite and measurable. If the rate of decay is known, the total usage of the literature by any one user can be estimated. As the rate of growth of a literature is equal to the rate of growth of the number of contributors to it, it is now possible to estimate the maximum usage that will be made of any bibliographical data-base.

Computerized data-bases were initially attractive to users because they presented a vision of a bright new world which contrasted very favourably with the musty archival runs of periodicals which cluttered the shelves of their libraries. But computerized data-bases too are now beginning to present to scientists some musty image of archival clutter. The bibliographic data imprinted magnetically on a computer tape is no more immune from the effects of obsolescence than are the printed journals on the shelves of the library.

THE PHENOMENA OF SCATTER

Our studies in London of the phenomena of bibliographical scatter are based on application of Bradford's law.[2] In the 1930's S.C. Bradford was Director of the Science Museum Library in London at a time when the library offered a valuable bibliographical service. Any would-be author could write to the Director asking for a comprehensive bibliography relating to any historical scientific topic. These compilations were done by competent bibliographers and resulted in a bibliography presented in a standardized format. The bibliography was organized for convenience of reference by the user.

All the journals cited were arranged in order of decreasing frequency of citation so that the list began with the richest sources offering many items and ended with a long thin tail of sources each offering only one item. Bradford insisted on checking all such bibliographies before he initialled them as worthy of the Science Museum Library's high reputation.

While doing this work he noted a statistical regularity. He formulated this regularity in two ways - by diagram and also in words. Unfortunately, though his diagram is clear, it does not quite conform with the verbal statement which is also ambiguous. Much effort has been expanded in clarifying the possible interpretations of these formulations. Contributions to this work have been made by Vickery, Cole, Leimkuhler, Goffman[3] Buckland, Fairthorne and, more recently, by Elizabeth Wilkinson[4] and myself;[5] more papers are about to be published. I believe that the uncertainties arising from Bradford's ambiguities have been resolved by Wilkinson's recent work.

If one considers Bradford's diagram it is possible to express the law of scatter in mathematical terms. The linear part of the graph, which has the rank of the periodical expressed on a logarithmic scale and which is the most important for practical purposes, has the form

$$R(n) = k \log_e n/s \qquad (1)$$

Here, R(n) is the number of items listed by the n richest sources, and k and s are constants for the particular case. I noted that a further clarification could be made. Bibliographies are necessarily finite so that the linearity cannot continue indefinitely. Where then does it end? If one assumes that the last item is a single contribution, then it is easy to show that k = N, the total number of sources. So one can then rewrite (1) as:

$$R(n) = N \log_e n/s \qquad (2)$$

It also follows that R(n) is the total number of items expected, as given by

$$R(N) = N \log_e N/s \qquad (3)$$

So, in order to discover how many items the complete bibliography contains, it is necessary to know only the numbers of items contributed by the 5, 6, 8...10 or so richest sources which are usually those most easily identified and most readily accessible. When the cumulative sums are plotted on semi-logarithmic paper, the points rise in a curve to some point at which the linearity begins. The slope, N, can be measured as soon as the linearity is established and, then, when N is known, the value of R(N) can be calculated.

We have added programs to the FAMULUS bibliographic package we have adopted at University College so that the values of N and R(N) are computed automatically for any bibliography compiled in the FAMULUS format.

We use the Bradford law in a variety of ways. For example, we measure the degree of comprehensiveness of any computerised data-base or indexing service we are interested in. The system under test is sampled and the data are plotted as I have described except that all the results from the system sample are included in the plot. The estimated values of N and R(N) can then be found and compared with the system values. If the system is not comprehensive then the linearity falls away in a 'droop' before it reaches the estimated end-point.

We do not, of course, expect all systems to be 100% comprehensive. It would be unrealistic to include in a data-base everything relevant that has been published because some fraction of it is of little use. But bibliographic systems ought to indicate clearly whether they claim to be comprehensive or not so that the consumer understands what he is getting for his money. It is perfectly reasonable to be selective, but if the compilers of the data-base are being selective then the basis of their selection should be made explicit.

The Bradford law is, we find, of wider application than we first thought. In fact, it pervades many aspects of social interaction and we have recently been trying to analyse the social mechanisms and wider social contexts in which the law arises. The great advantage of using bibliographic data for such analysis is that they are relatively free from the factors which may disturb other fields of social interaction. Let me mention some of the other contexts which have produced data conforming with the Bradford law:

(1) The borrowing of items from a library when the *items* are ranked.
(2) The borrowing of items from a library when the *users* are ranked.
(3) The publishing of books by publishers when the *publishers* are ranked.
(4) The indexing of papers input to a document system when the *index terms* are ranked.
(5) The contributing to conference discussions when the *contributors* are ranked.

All these systems consist of a set of resources to be grabbed and a population of grabbers. There are two interacting mechanisms: first, a free-for-all in which the population of grabbers can be ranked according to their success in grabbing and which can be represented by a simple hyperbolic law, and, secondly, a restraint which bears increasingly on the few who achieve a threshold of success. The effects of the restraint are indicated by the length of the run-up to the log-linearity; the longer the run-up, i.e., the more the graph deviates from the simple hyperbolic, the stronger is the restraint.

Elizabeth Wilkinson has recently been simulating the generation of Bradford distributions on the computer. She had to express the two mechanisms as a stochastic process in which the probability that the next event will occur in some specified way involves the recalculation of the new probability distribution created by each event. She found no serious difficulty in generating distribution which *looked* like Bradford distributions but which failed to meet the criterion, built into the program, that the slope of the log-linearity must be equal to N. Eventually she discovered that the key parameters of the conditions which express the two mechanisms had to be directly related in a precise way to get reproducible Bradford distributions. A report on this work will shortly be published by OSTI who funded the project and a paper is also being submitted for publication in one of the journals. Now that we know the answer from the computer simulations we hope to be able to solve our stochastic equations by analysis and then to clarify some of the uncertainties that remain.

The reason why so pervasive and exact a law of social interaction has not been perceived so clearly before is, I believe, that all who have worked on the problem, in bibliographic and other fields of social interaction, such as economics, have sought the *one single* mathematical formulation which would embrace the whole phenomenon. But any single mathematical expression can only be an approximation to a law which, it seems to me, requires *two* mathematical formulations for its exact expression, one for the run-up and one for the linearity, and which can also be suitably related. The Bradford law thus requires a new kind of formulation which has hitherto not been recognized in the social sciences. In mechanics, however, an analogous problem would be to determine the motion of a rocket projectile which is subject to two forces, first, of gravity which acts overall and secondly, of the propulsive force of a finite quantity of a propellant which acts only for a short time.

THE LAW OF AUTHOR PRODUCTIVITY

In 1926 Lotka[6] noticed that, in a cumulative index of *Chemical Abstracts*, the number of those who contributed a single item to the literature was four times as large as the number of those who contributed *two* papers, nine times as large as those who contributed *three* papers, and so on. We have tested this law on several large bibliographies and have found it to be generally valid.

Lotka's law has had only two practical applications, so far as I know. It indicates, for example, that as much as 61% of any bibliographic data-base is contributed by authors who have contributed only one paper. This result provides a useful check on some of the questionnaire results I have seen and is being applied to another field of activity with which I have been associated. It has been noticed that the most significant contributions to any scientific field are only very very rarely made in a *first* paper to that field.

So I am now considering this with a number of other indicators which together might enable an indexer to assign a significance coefficient to any paper being considered for input to a bibliographic data-base and which is derivable automatically and solely from data within the data-base.

In short, all the statistical activities I have outlined in this paper will be applied to the problem of structuring data-bases to make them more effective as aids to the scientific work of their users. Computer technology, it seems to me, is well ahead of our abilities to organize the items contained in the data-bases for effective application of all the computer can now do. Lotka's law is also of theoretical interest. Though it gives a good approximation to the data it is applied to, I believe that it can be formulated in a more meaningful but equally exact way. In effect it states that an author who has already contributed r papers to a subject finds it easier to publish his next paper the greater r becomes, and the probability relation is very precise. And recently Naranan has shown that Lotka's law is the inverted form of Bradford's law, though he made some approximations that I do not fully accept. But the implication of Naranan's result is that the empirical laws of Lotka and Bradford, like the two analogous laws of Zipf in linguistics, are manifestations of one single fundamental law of social interaction which can be expressed in two ways.

CONCLUDING COMMENTS

Though the empirical laws I have mentioned are expressible in very simple mathematical terms, the mathematical and statistical analysis of these laws can be surprisingly intractable. The Bradford distribution is a ranked array, not a conventional frequency distribution, and none of the distributions I have mentioned (other than the negative exponential) have finite variances or means. So the sampling techniques and significance tests that can be applied to most experimental statistical distributions do not apply. We are therefore beginning to believe that new kinds of statistical techniques are needed for our work and are hoping to attract some statisticians looking for research problems to come to our help.

The form of the Bradford law (1) intrigues me. It continually reminds me of the entropy formulae that frequently emerge from the application of Shannon theory and I feel that it should be interpretable as a measure of bibliographical entropy. But I cannot quite clinch the interpretation it seems to invite. I look hopefully to Professor Watanabe.

Another area of statistical bibliography that must be mentioned is Goffman's epidemiological model of dissemination, I am particularly interested in his evidence indicating that a threshold proportion of susceptibles to the new ideas must exist in a scientific population before a new science can hope to reach its phase of epidemic growth. It seems very relevant to our present problem. One of my next projects is to relate Goffman's epidemiological model of growth to my models of obsolescence and scatter.

Finally, I have to emphasize that the kind of work I have been describing does require a scientific discipline in application. From time to time I am sent papers which purport to show that one or other of the techniques I have described does not work. But all too often I quickly note that the data are poorly defined, that incompatible entities are conflated together, and that all kinds of scientific sins are committed. It is not a game that librarians play well. Yet I have to admit that it was a librarian - Jesse Shera - who first proposed it as a field for study.

NOTES

1. Brookes, B.C., Obsolescence of special library periodicals: sampling errors and utility contours, *Journal of the American Society for Information Science*, 21, 320, 1970.

2. Bradford, S.C., *Documentation*, Crosby Lockwood, London, 1948.

3. Goffman, W. and Morris, T.G., Bradford's law and library acquisitions, *Nature*, 226, 922, 1970.

4. Wilkinson, E.A., The ambiguity of Bradford's law. *Journal of Documentation*, 28, 122, 1972.

5. Brookes, B.C., Bradford's law and the bibliography of science, *Nature*, 224, 953, 1969.

6. Lotka, A.J., The frequency distribution of scientific productivity, *Journal of the Washington Academy of Science*, 16, 317, 1926.

MAPPING THE SCIENTIFIC LITERATURE

Belver C. Griffith

Graduate School of Library Science, Drexel University, Philadelphia, Pa., U.S.A.

INTRODUCTION

That science is a mosaic of specialties, and not a unified whole either socially or intellectually, is a frequently made assumption, but one which has only recently become the subject of systematic study. Most scientists have intuitive notions as to the subdivisions of their fields, but gaining an overall perspective on the scientific mosaic is impossible for any observer, however broadly trained. The continuing coverage of *Science Citation Index (SCI)* provides cross-disciplinary data appropriate to objective studies of the structure of the scientific literature. Pioneering studies by Garfield[1] and Price[2] utilized citation data to establish patterns of development within particular fields of science. Garfield conceptualized the citation relationships as directed lines connecting later documents with earlier ones. The totality of such relationships constitutes a "historical map", in that case, for genetics. Price's published study, a classic in conceptualizing the literature of science, presented citation data in the form of a matrix drawn from the literature of physics and depicting the development of a narrowly defined specialty. In a series of unpublished working papers, Price made several preliminary attempts to examine the structure of physics.[3]

These studies established early the feasibility of examining specialties in terms of citation data and have revealed the rich complexity of science when viewed at this level. Recently, Narin[4] has used data from the Institute for Scientific Information's *Journal Citation Index* to construct maps of science in terms of its journals. Emerging from this work is an overall view of the disciplinary structure of science, but not, however, of its specialties.

For our purposes the journal is too broad a unit of analysis to reveal the fine structure of science. For that reason we have focused our attention on the cited documents (articles or books) of science which, if they can serve as structural units, are clearly at a sub-specialty level. This concern with describing the specialties of science emerges from a series of recent studies, both sociologically and historically motivated, which have called attention to the importance of specialties as "building blocks" of science. Crawford's[5] paper described the efficiency of communication, structured about research centers, within a single specialty. Griffith and Mullins have argued that major changes in a variety of disciplines have been generated within small, socially coherent groups.[6] Crane's work on specialties in rural sociology and mathematics has led naturally to the idea of "mapping" the structure of specialties, and her recent book has directly argued this view.[7]

As basic units of analysis, we employed a sample of frequently cited documents (articles or books). The onset of rapid specialty growth appears to be accompanied by the emergence of key papers which are quickly and frequently cited. Goffman[8] and Jahn[9] have shown that the specialty is defined by a few critically important papers which appear early in the specialty's history. Small has observed similar patterns for the early development of nuclear physics in the 1930s. Certain discoveries such as the neutron and artificial disintegration of the nucleus were associated with key papers, which became highly cited very soon after they were published and became the focal points for later research.[10]

A recent study by Small examined co-citation as a measure of the association between pairs of frequently cited documents.[11] Co-citation is defined as the number of times two frequently cited documents have been cited together and provides a natural and quantitative way to group or cluster the cited documents. Small showed that co-citation identifies relationships among papers which are regarded as important by their authors but which are not identified by such techniques as bibliographic coupling or direct citation.

The strength of co-citation between two papers can be readily determined from a citation index. Each of the two papers is located in the index and their lists of citing papers are scanned. The number of *identical citing items* furnishes a quantitative measure of the strength of co-citation between the two papers. An identical citing item is a new document citing the earlier papers; therefore, co-citation is the degree to which two items of earlier literature are cited together by the later literature.

To be strongly co-cited, a large number of "new" authors must jointly cite the two earlier works. Therefore, co-citation is a relationship which is recognized and maintained by current researchers or, speaking generally, the degree of perceived relationship between papers by authors of a certain population of citing documents. Intrinsic to co-citation is this dependence on the population of citing documents; therefore, patterns of co-citation change over time. Bibliographic coupling, on the other hand, is static because it depends on citations contained only in the coupled documents. Co-citation changes as the interests and intellectual patterns of the field change and as new discoveries are made and introduced through the literature.

In the present paper, we report a first experiment to utilize co-citation as a measure of proximity between papers in mapping the scientific literature. We shall be concerned with the extraction of pairs of papers linked by co-citation, with the identification of specialties as groupings of closely-related papers, and with the display and interpretation of such groupings.

METHOD

Data Base

The source of data for the study was the magnetic tape version of the *SCI* for the first quarter of 1972. The general characteristics of this file have been described by Weinstock.[12] The file can be regarded as a fair representation of most major scientific literàtures.

A total of 1832 distinct cited documents were selected from the quarterly file on the basis of having been cited 10 or more times during the quarter. This high threshold immediately excluded specialties with very low citation rates such as those in mathematics or technology. The 1832 documents principally represented basic biomedical research, physics, and chemistry. Fragments of technology, statistics, computer science and psychology also appeared in the file. Presumably, a lower threshold will enable us to generate descriptions of these and other fields which have small literatures and which cite less frequently. The 1832 items represented only .21% of the 867,622 distinct cited items in the quarterly file. (A quarterly threshold of 10 is equivalent roughly to an annual threshold of 40 times cited.) The entire quarterly file contained 1,135,897 citations; however, 32,550 of them or 2.87% were to these 1,832 highly cited documents for an average of 17.8 citations to each selected item. In contrast, the quarterly file contains an average of 1.31 citations to each item.

Although we are dealing with a very small subset of the file, this sample of cited items receiving 32,550 citations was capable of retrieving 16,927 different source items (18% of the 93,782 source items in the quarterly *SCI*). This level of retrieval is, however, not unexpected if we compare its yield with what we could expect from selecting the same number of citations from the file at random. A binomial model of this process leads to an estimate that 25% of the source items would be selected with a 2.87% random sample of citations. A paper with n references has q^n probability of being missed, q = citations to selected papers/total citations. Therefore, it has $1-q^n$ probability of retrieval. The fact that we retrieved fewer source articles than predicted corresponds to the tendency of the sample of highly cited papers to appear together in the same source papers, that is, to be co-cited and in fact, the average number of citations per retrieved source article was 1.92. In contrast, the binomial model predicted an average of 1.45 citations per retrieved source article. This shows that the co-citation of highly cited documents is far more frequent than is predicted from a random distribution of citations throughout the literature.

Extraction of Pairs of Articles Linked by Co-Citation

The first step in the experiment was to extract, by computer, pairs of cited items linked by co-citation. The steps in this procedure follow:

(1) Using the first quarter 1972 *SCI* file, bibliographic information for both citing and cited documents was extracted to create a new file for the 1832 highly cited items.

(2) This file was sorted into a "source" file, i.e, the cited items were listed under citing items.

(3) For each citing paper, all pairs of cited papers were created. Within each pair, the two papers were ordered alphabetically so that pairs could appear in only one of their two possible arrangements. For a paper citing n highly cited papers, n (n-1)/2 pairs were thus created.

(4) The cited pairs were alphabetized across the entire file, which resulted in bringing identical pairs together.

(5) The number of identical pairs was counted; the count was printed by the first pair, and the remaining identical pairs were deleted to create a file in which each pair of cited articles was accompanied by its co-citation strength.

After the last step, each pair was duplicated with the item appearing lower in the alphabet now appearing first. Both the original and duplicate pair were then alphabetized so that all pairs involving any single item appeared together. After the co-citation frequency of the 1,832 selected items was determined, it was found that only 7,241 or 7.7% of the total source items in the quarter contributed to co-citation by virtue of citing two or more of the selected 1832 documents. More than 9000 documents of the 16,927 citing items retrieved, were retrieved by but a single cited item.

RESULTS

The Co-Citation File

A total of 36,316 co-citations were generated by the 7,241 source papers, creating 20,414 distinct pairs, i.e., pairs different in terms of one or both linked documents. The average co-citation strength was, therefore, total co-citation/number of distinct pairs or 1.78; however, all but 18 (<1%) of the original 1832 cited documents entered into at least one pair. Of the 1,677,196 pairs that might have been created from 1,832 items (i.e. (n) (n-1)/2) only 20,414 appeared. Thus, co-citation shows only .18% connectedness for this set of documents, evidence of the looseness of the overall structure.

The 20,414 co-cited document pairs vary in co-citation frequency from 1 to 81. This file, including all the pairs formed will be termed Level 1. To aid us in modeling structure, sub-files were created at three higher levels. Level 3 consists of all pairs co-cited three or more times, that is, after pairs co-cited fewer than three times have been dropped. The file at Level 3 consists of 3067 distinct pairs of documents. Level 3 pairs include 1310, or 70% of the cited documents initially selected, in contrast to 99% for Level 1, and at Level 3 there are 522 isolated cited documents. Table 1 summarizes the characteristics of the file of co-cited documents at Levels 1, 3, 6, and 10, in terms of number of pairs, number of cited items, and number of isolates. The average number of pairings in which a document appeared depends on the level being considered. At Level 1, there is an average of 22.5 entries per document, while at Level 3 the average is 4.7.

Table 1

Results of Extracting Co-Citation Pairs

	Number of distinct pairs	Number of documents paired	Number of isolated documents
Level 1	20,414	1,814	18
Level 3	3,067	1,310	522
Level 6	791	594	1,238
Level 10	213	193	1,639

In the next step we began tracing through the pairings at Levels 3, 6, and 10 to locate all pairs having one cited document in common with any of the pairs we had already. This was pursued by hand; the first author following through the chains generated around important papers in physics, the second author focusing on biomedical research. This task was then, fortunately, turned over to a computer program.

The File of Clustered Papers

The results of the clustering runs are most clearly presented using the terminology of graph theory. At a given co-citation level, the entire set of 1832 documents can be formally considered as a graph. After clustering, the graph will consist of a number of components, each component being a connected graph. The components will vary in the number of points (cited documents) they contain, and the number of lines (co-citations) connecting them. Since any two points can be connected by a co-citation strength anywhere from 1 to 81, some of the links are multiple lines, and the graph is, strictly speaking, a multigraph.[13] The number of isolated documents is simply the number of component graphs consisting only of one point.

Although for reasons of economy, there was no clustering done at Level 1, it was possible to gain some insight into the structure of this file. As mentioned previously, there were only 18 documents which did not pair at Level 1 and, therefore, 18 components of size one. A more surprising finding was that there was only *one* component of size two at Level 1. The distribution of cluster size for components consisting of more than two points is not known, but we can estimate that nearly *the entire set of 1812 documents* (1812 minus 20) is connected at Level 1 although its percentage connectedness was quite low (.18%). At Level 1, large clusters at higher levels, corresponding to nuclear physics, particle physics, and several areas in chemistry are connected by weak linkages to one another and to the large biomedical cluster which contains nearly half of the 1832 documents.

Table 2 displays the number of components of size n obtained for Levels 3, 6, and 10. At Level 3, the table shows that we have 522 isolates and 71 isolated pairs. The largest grouping consists of 801 cited documents. This single grouping consumes fully 44% of the cited documents in the original sample of 1832. At Level 6, there is, of course, a large increase in the isolates since now a document must have at least one link of strength 6 in order not to be isolated. At this level the biomedical grouping has been reduced to 200 documents, and at Level 10 to 72 documents. Not counting the isolates, the number of clusters, of all sizes, obtained for Levels 3, 6, and 10 were 115, 107, and 39 respectively. The trend toward a larger number of clusters with decreasing level, of course,

Table 2

The Outcome of Clustering:
Description of the Component Graphs

Component Size in Papers		Number of Components			Number Cited Items		
	Levels	3	6	10	3	6	10
1		522	1238	1639	522	1238	1639
2		71	60	21	142	120	42
3		13	17	3	39	51	9
4		9	11	6	36	44	24
5		7	5	3	35	25	15
6		5	2	4	30	12	24
7		1	1	1	7	7	7
8			3			24	
9		2	1		18	9	
10		1	2		10	20	
11							
12		1	1		12	12	
13							
14							
15		1			15		
16							
17			1			17	
18							
19							
20							
26			1			26	
27			1			27	
32		1			32		
41		1			41		
72				1			72
92		1			92		
200			1			200	
801		1			801		
	Total	637	1345	1678	1832	1832	1832

reverses at Level 1 where nearly everything connects.

Table 3

The Largest Clusters at Level 3

	Number of Documents	Number of Pairs	Percentage Connectedness	Average Date
Biomedicine	801	2205	69%	1963.3
Chemistry	92	291	7%	1962.5
Nuclear Physics	41	59	7%	1964.3
Particle Physics	32	99	20%	1968.1
Australia Antigen	15	70	66%	1969.0
Crystal Structure of Enzymes	12	30	45%	1968.0
Plate Tectonics	10	35	78%	1967.8
Virally Transformed Cells	9	25	69%	1964.2
Nuclear Magnetic Resonance	9	8	22%	1957.0
Neurophysiology of Vision	7	14	66%	1963.1

Table 3 lists the 10 largest groupings obtained at Level 3. Each of the groupings has been given a name which is intended to generally reflect the subject matter of the grouping. The Table also gives the number of documents in the grouping, the number of document pairs, the percentage connectedness (percentage of the possible pairings which are actually realized), and the average date of the documents in the grouping. The three largest groupings, biomedicine, chemistry, and nuclear physics all have identifiable subgroupings and this is reflected in their low percentage connectedness. The groupings are arranged in order of decreasing size, and as the cluster size decreases there is a clear sharpening in their subject focus, the smaller clusters dealing with highly specialized topics.

The largest grouping by far is the biomedical grouping, which consists of 801 cited documents. It is, however, a relatively loosely knit cluster consisting of many subgroupings as evidenced by its low percentage connectedness (.69%). At each level, the biomedical grouping is the largest. At Level 6, it has been reduced to 200 documents, and at Level 10 to 72 documents. Raising the co-citation threshold is only partially successful in breaking up this cluster into its subgroupings. At Level 6, we can identify 53 smaller groupings which broke off from the larger cluster, but a core of 200 documents remained connected at this level.

The idea behind breaking up large clusters by raising the threshold, is that links between clusters are weaker than links within clusters. This approach did not work as well for biomedicine as it did for other fields because of certain very highly cited biomedical papers which played a role in linking many sub-areas apparently unique to biomedicine. On examination, these were primarily methodology papers which were connecting together diverse specialties with strong linkages. These linkages were not eliminated by raising the threshold, a problem considered in more detail at a later point in the paper.

Clusters and Specialties

The question remains, however, as to whether clustering has provided groupings of the literature which correspond to identifiable subject matter specialties. Several errors are possible: Specialties could be linked together which are in fact not related to one another, or the clusters could be fragments of specialties which should emerge as single groupings. At the worst, the results could bear little or no relationship at all, to the specialties which would be identified by scientists.

Several limited validations of our technique were attempted. First, an inspection of the clusters offered *prima facie* evidence of validity. The first author immediately recognized the areas of particle physics and nuclear physics at Level 3; the second, the biomedical areas of reverse transcription (RNA on DNA) and Australia Antigen at Levels 6 and 10. Table 4 displays the reverse transcription linkages at Level 6. Of the 10 papers linked here, two were the reports in *Nature* of the codiscovery by Temin and Mitzutani and by Baltimore. An additional paper was the replication performed by Spiegelman and co-workers, and two other papers, Mitzutani *Nature* 228, p. 424, 1970 and Gallo *Nature* 228, p. 927, 1970, were among the 10 most cited papers within the first 9 months following the discovery.

With the exception of the papers which were clumped in the large biomedical cluster at each level, most groupings were readily recognized after a couple of paper titles were located and, generally, they linked to appropriate clusters when a lower level was

Table 4
Reverse Transcription Cluster at Level 6*

BALTIMORE D	NATURE	226120970	BALTIMORE D	P NAT ACAD SCI US	68150771	9
BALTIMORE D	NATURE	226120970	GALLO RC	NATURE	228 92770	7
BALTIMORE D	NATURE	226120970	MIZUTANI S	NATURE	228 42470	11
BALTIMORE D	NATURE	226120970	SCOLNICK E	P NAT ACAD SCI US	67178970	9
BALTIMORE D	NATURE	226120970	SPIEGELMAN S	NATURE	227 56370	17
BALTIMORE D	NATURE	226120970	SPIEGELMAN S	NATURE	227102970	12
BALTIMORE D	NATURE	226120970	TEMIN HM	NATURE	226121170	43
BALTIMORE D	P NAT ACAD SCI US	68150771	BALTIMORE D	NATURE	226120970	9
BALTIMORE D	P NAT ACAD SCI US	68150771	TEMIN HM	NATURE	226121170	9
GALLO RC	NATURE	228 92770	BALTIMORE D	NATURE	226120970	7
GALLO RC	NATURE	228 92770	TEMIN HM	NATURE	226121170	7
MIZUTANI S	NATURE	228 42470	BALTIMORE D	NATURE	226120970	11
MIZUTANI S	NATURE	228 42470	SPIEGELMAN S	NATURE	227 56370	6
MIZUTANI S	NATURE	228 42470	SPIEGELMAN S	NATURE	227102970	7
MIZUTANI S	NATURE	228 42470	TEMIN HM	NATURE	226121170	11
MOORE DH	NATURE	229 61171	SCHLOM J	NATURE	231 9771	10
SCHLOM J	NATURE	231 9771	MOORE DH	NATURE	229 61171	10
SCHLOM J	NATURE	231 9771	SPIEGELMAN S	NATURE	227 56370	6
SCOLNICK E	P NAT ACAD SCI US	67178970	BALTIMORE D	NATURE	226120970	9
SCOLNICK E	P NAT ACAD SCI US	67178970	TEMIN HM	NATURE	226121170	10
SPIEGELMAN S	NATURE	227 56370	BALTIMORE D	NATURE	226120970	17
SPIEGELMAN S	NATURE	227 56370	MIZUTANI S	NATURE	228 42470	6
SPIEGELMAN S	NATURE	227 56370	SCHLOM J	NATURE	231 9771	6
SPIEGELMAN S	NATURE	227 56370	SPIEGELMAN S	NATURE	227102970	7
SPIEGELMAN S	NATURE	227 56370	TEMIN HM	NATURE	226121170	15
SPIEGELMAN S	NATURE	227102970	BALTIMORE D	NATURE	226120970	12
SPIEGELMAN S	NATURE	227102970	MIZUTANI S	NATURE	228 42470	7
SPIEGELMAN S	NATURE	227102970	SPIEGELMAN S	NATURE	227 56370	7
SPIEGELMAN S	NATURE	227102970	TEMIN HM	NATURE	226121170	12
TEMIN HM	NATURE	226121170	BALTIMORE D	NATURE	226120970	43
TEMIN HM	NATURE	226121170	BALTIMORE D	P NAT ACAD SCI US	68150771	9
TEMIN HM	NATURE	226121170	GALLO RC	NATURE	228 92770	7
TEMIN HM	NATURE	226121170	MIZUTANI S	NATURE	228 42470	11
TEMIN HM	NATURE	226121170	SCOLNICK E	P NAT ACAD SCI US	67178970	10
TEMIN HM	NATURE	226121170	SPIEGELMAN S	NATURE	227 56370	15
TEMIN HM	NATURE	226121170	SPIEGELMAN S	NATURE	227102970	12

* A reproduction of the computer print-out, each line containing a document pair shown as abbreviated citations with the strength of co-citation. The document is indicated by first author, followed by journal, volume (3 digits), page (4 digits), and year (2 digits).

examined. Thus, reverse transcription tied into earlier work in molecular biology at a lower level, and the nuclear structure and particle physics clusters eventually linked together at logical points. However, as already noted, nearly all biomedical papers linked into an enormous graph at Level 3.

In certain areas in which papers are less frequently cited, the linked documents made sense even though few documents were sufficiently cited to be pulled into our file. Examples include: a small cluster of statistical texts, the basic Shannon papers in information theory, the four seminal papers in multi-dimensional scaling, a few controversies in experimental psychology and so on.

Second, groupings can be made more plausible by an examination of the patterns of word usage in the titles of the documents which cite documents in the clusters. If the vocabulary of the citing documents changes suddenly or discontinuously as we move through a particular network, this would indicate an incorrect linkage. Identical language patterns in the absence of linkages would suggest a failure to group together all the pairs of a single specialty.

Excellent illustrations of the relation of word analysis to the structure of clusters are two groupings in physics which emerge at Level 3, "nuclear physics" and "particle physics". Graphs of these two clusters at Level 3 are given in Figures 1 and 2 with co-citation strength coded as indicated in the figures. The "nuclear physics" grouping can be represented as a planar graph, that is, a graph in which no two connecting lines cross one another. (At Level 1, this graph ceases to be planar.) The significance of planarity is, first, that the structure is a loosely knit one and, second, that the number and strength of links can be interpreted broadly as distance. The average date of the cited documents in this grouping is roughly 1964, indicating relatively old material.

The graph of "particle physics", Figure 2, is in marked contrast. It was not possible to graphically represent this specialty due to a highly dense set of intersecting connections among the 22 documents in the upper part of the diagram, therefore, these were indicated by enclosing them in a large box. The cluster has a more simply structured tail which emanates from the paper by P.G.O. Freund.

Not only does "particle physics" differ from "nuclear physics" in terms of density of linkages (its connectedness was 20%) but also in terms of document age. The average date of the documents was 1968, four years younger than "nuclear physics". To conclude that a high density of linkages is associated with the recency of the specialty would however be premature, since many of the highly dense clusters we obtained were not as recent as "particle physics". In the case of "particle physics", however, the density of linkages

Figure 1
Graph of Nuclear Physics Cluster at Level 3

Each box stands for a cited document, and contains the name of the first author of the document and an abbreviated book title.

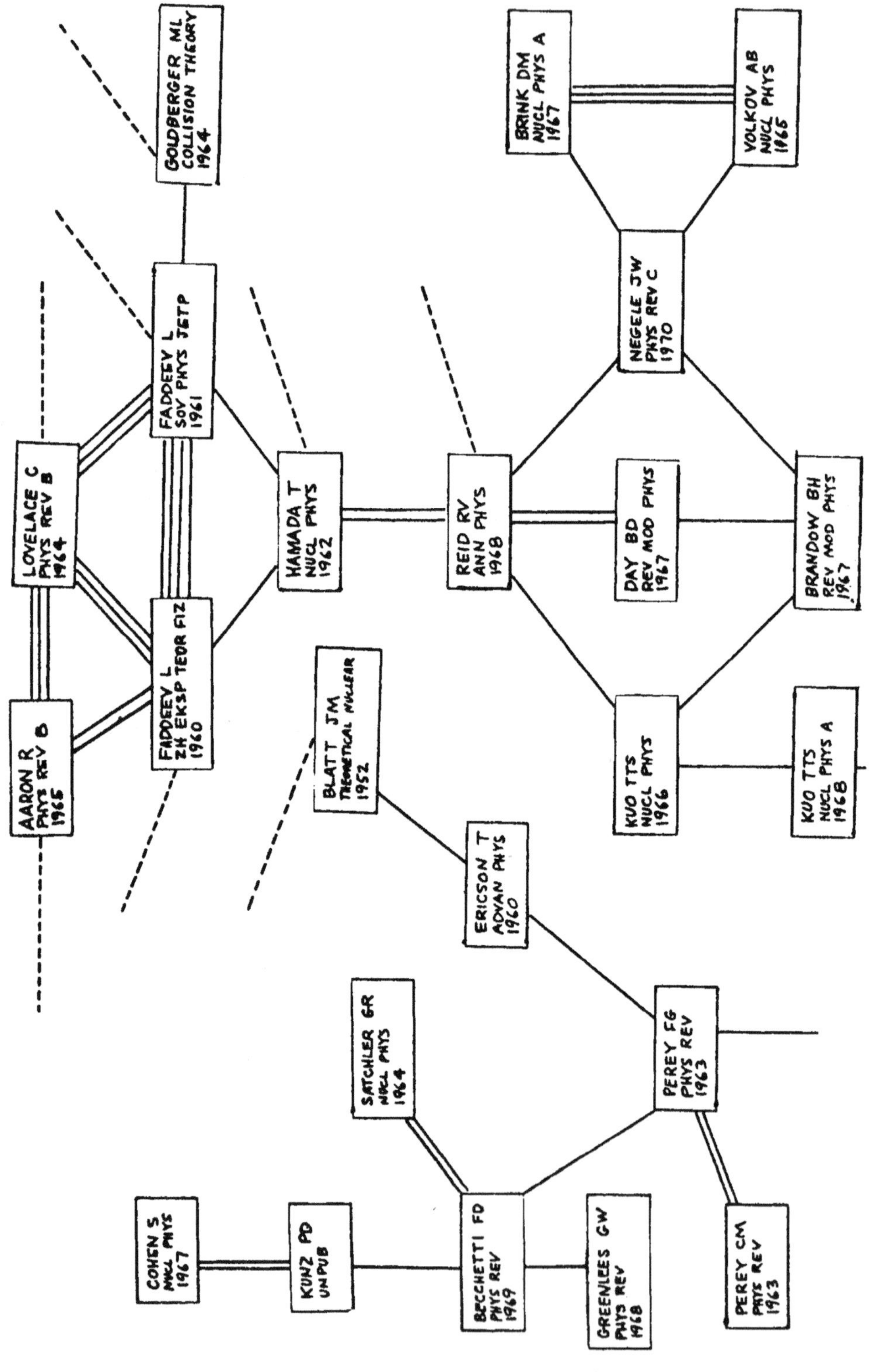

Co-citation Strength	Number of Lines
3	1
4 - 5	2
6 - 8	3
9 - 12	4

Dotted lines indicate points of lower-level connection to particle physics.

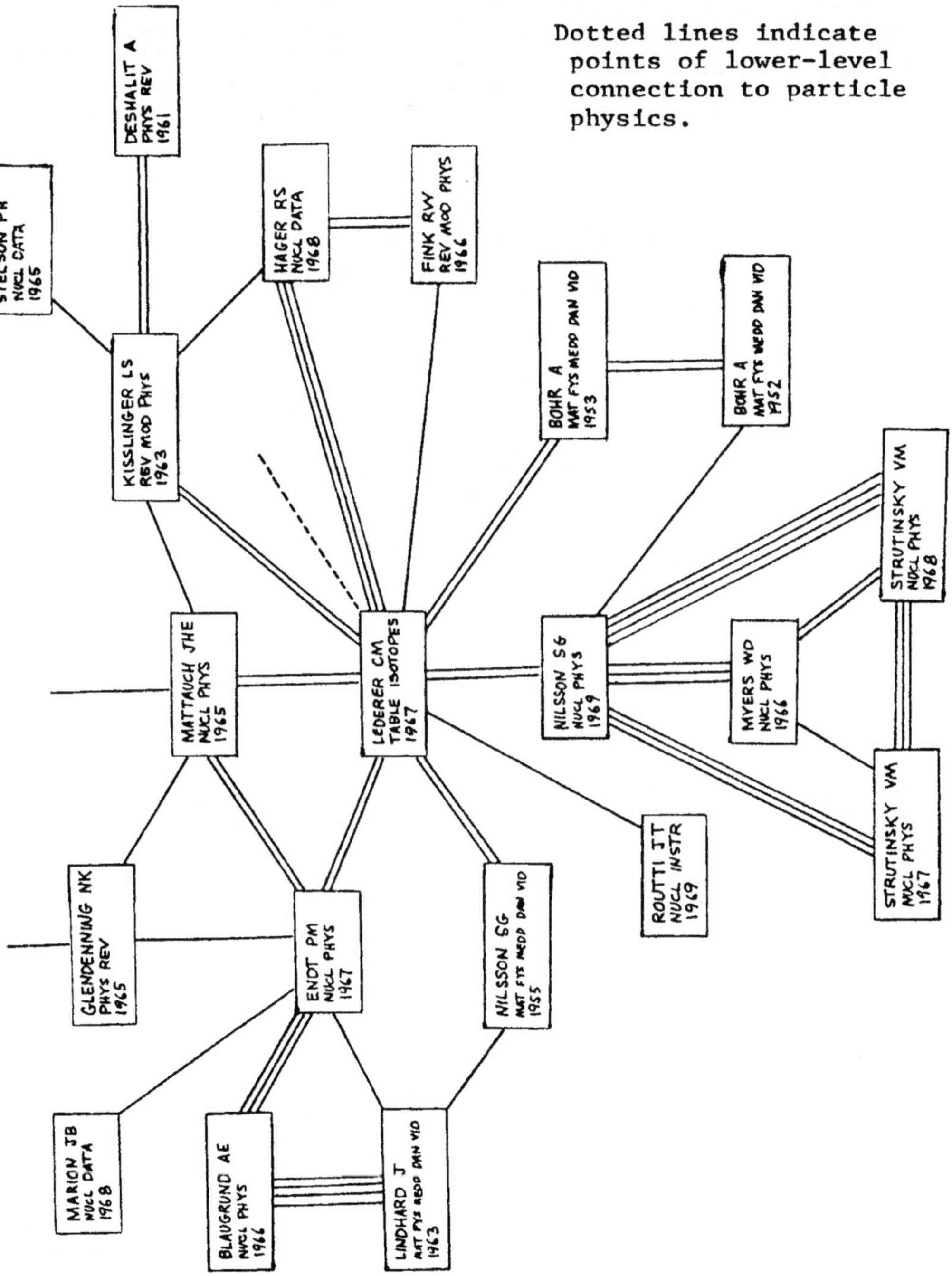

Figure 2
Graph of Particle Physics Cluster at Level 3

Documents and links are indicated as in Figure 1. Again, points of connection of particle physics to nuclear physics are indicated by dotted lines. The 21 documents enclosed in the large box were so strongly interconnected that depicting the pattern of linkages proved virtually impossible. The paper by Freund was linked to several of the documents in the large box, but was the only point of connection between that grouping and the lower part of the graph.

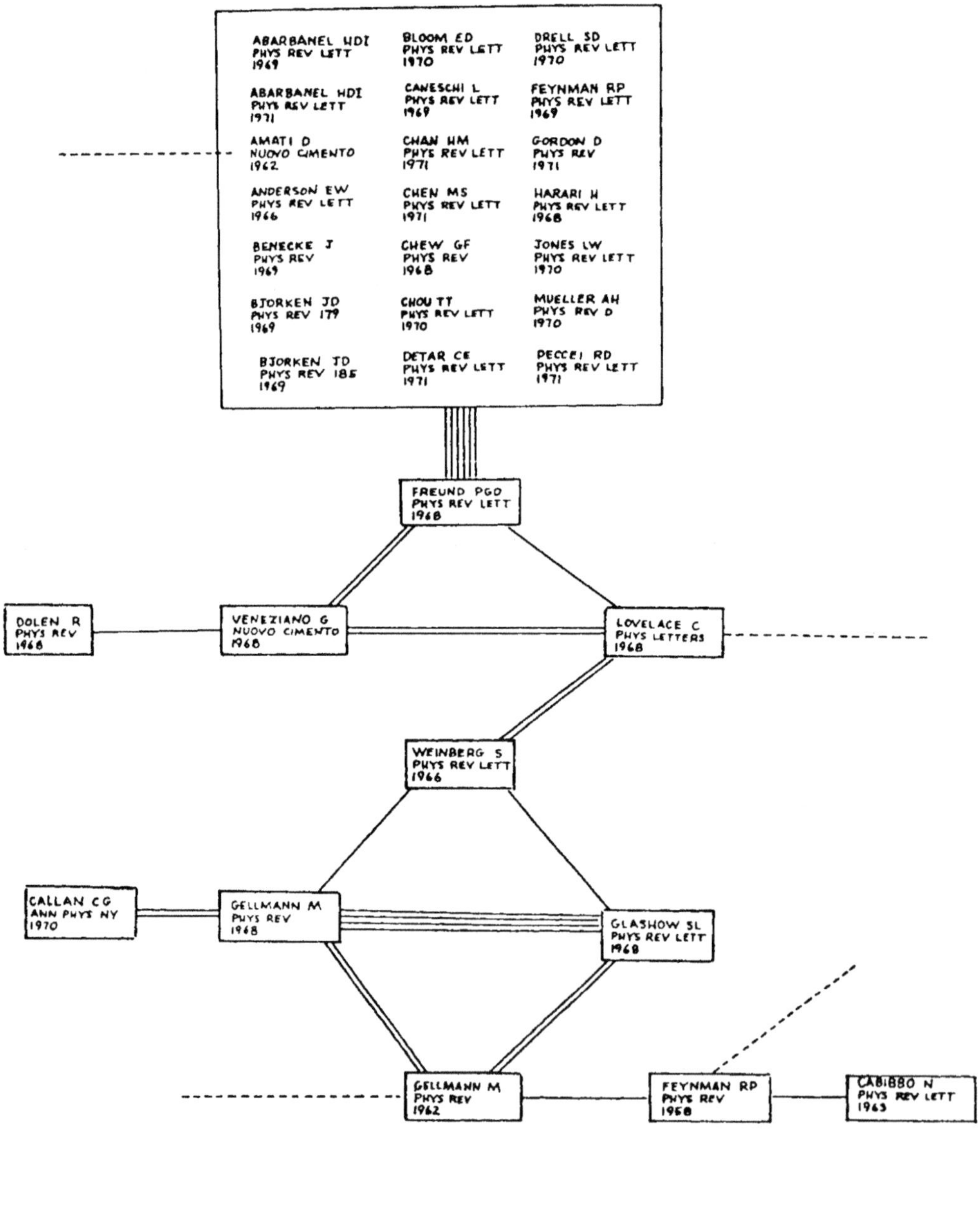

and the recency of the documents agree with our qualitative impression of this specialty as fast moving and highly problem-focused.

The groupings obtained at Level 3 for "nuclear physics" and "particle physics" were then investigated in terms of word usage of the citing documents. "Word frequency profiles" were created for the cited documents. For example, the 1965 paper by R. Aaron was cited by 14 documents in the first quarter of 1972, and in the titles of these 14 papers, the three most frequently appearing words, with the exception of "stop words" (see list of full-stop words in the *Permuterm Subject Index*), are "scattering" (5 times), "nucleon-deuteron" (3 times), and "polarization" (3 times). This word list and accompanying frequencies constitute a word frequency profile. Using first quarter 1972 data from the *Source Index* an analysis of the titles of all papers which cited the 1832 items in our sample resulted in a listing of words paired with cited documents. From this, word frequency profiles were created for the cited documents in both the "nuclear physics" and "particle physics" groupings. For each cited document, the four most frequently associated words have been selected from the listing, and arranged in Figures 3 and 4 in the same configuration as the document groupings.

By examining the word frequency profiles for each of the cited documents, it is possible to observe how the subject matter changes as we move through the network. For example, in the "nuclear physics" grouping, the subcluster around the L. Faddeev paper is concerned with "scattering" and as we move from this to the adjacent subgrouping around the B.D. Day paper the word "matter" is prominent, which in this context implies theories of nuclear matter. Going through the documents by J.H.E. Mattauch, C.M. Lederer, and then to S.G. Nilsson (1969) we pass from "reactions" to "fission". Starting at the bottom of the "particle physics" diagram, around the papers of M. Gell-Mann, work is concerned with current algebras and broken chiral symmetry (SU_3). Moving up through the network, we encounter applications of the Veneziano model. As we enter the highly dense cluster, the vocabulary shifts to a concern with inclusive reactions (introduced by Feynman in 1969) and the "deep" or internal structure of elementary particles. The vocabulary of these networks seems to behave in a way which is consistent with our informal labels of "nuclear physics" and "particle physics". There is no evidence of discontinuities in the nomenclature which would indicate the joining of unrelated specialties.

As already mentioned, linkages between the two groupings were found at levels lower than 3. At Level 1, the groupings are joined by 14 pairings of strength 1. The dotted lines on Figures 1 and 2 show the points of connection between the clusters. As is evident, most of the connections originate from the "scattering" subcluster of "nuclear physics" and tie in to the loosely structured tail of the "particle physics" cluster. This linkage proved of interest

Figure 3
Word Frequency Profiles for Documents in Nuclear Physics Cluster

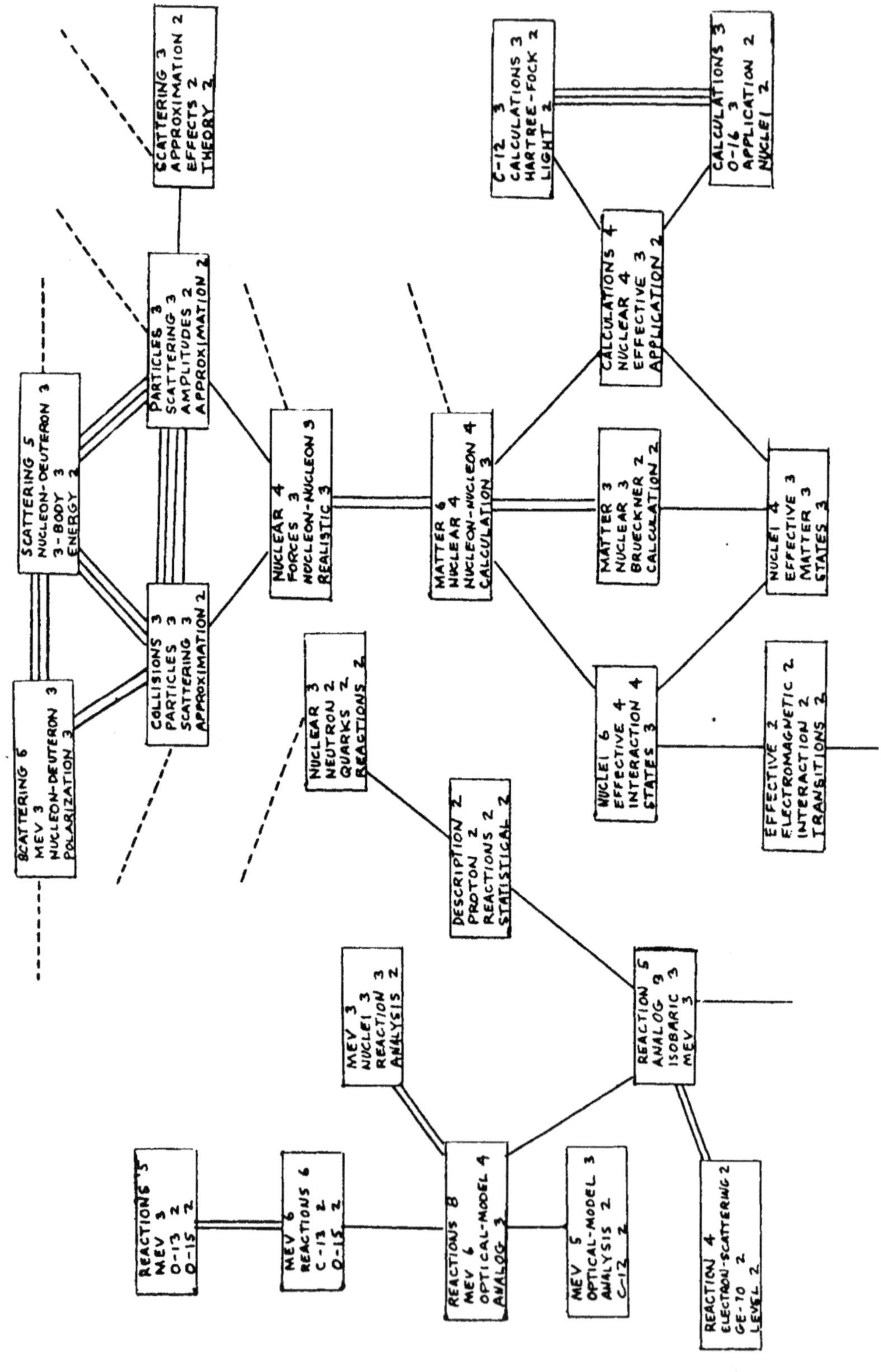

The four most frequently associated title words for each of the cited documents have been indicated, each word followed by its frequency.

Figure 4
Word Frequency Profiles for Documents in Particle Physics Cluster

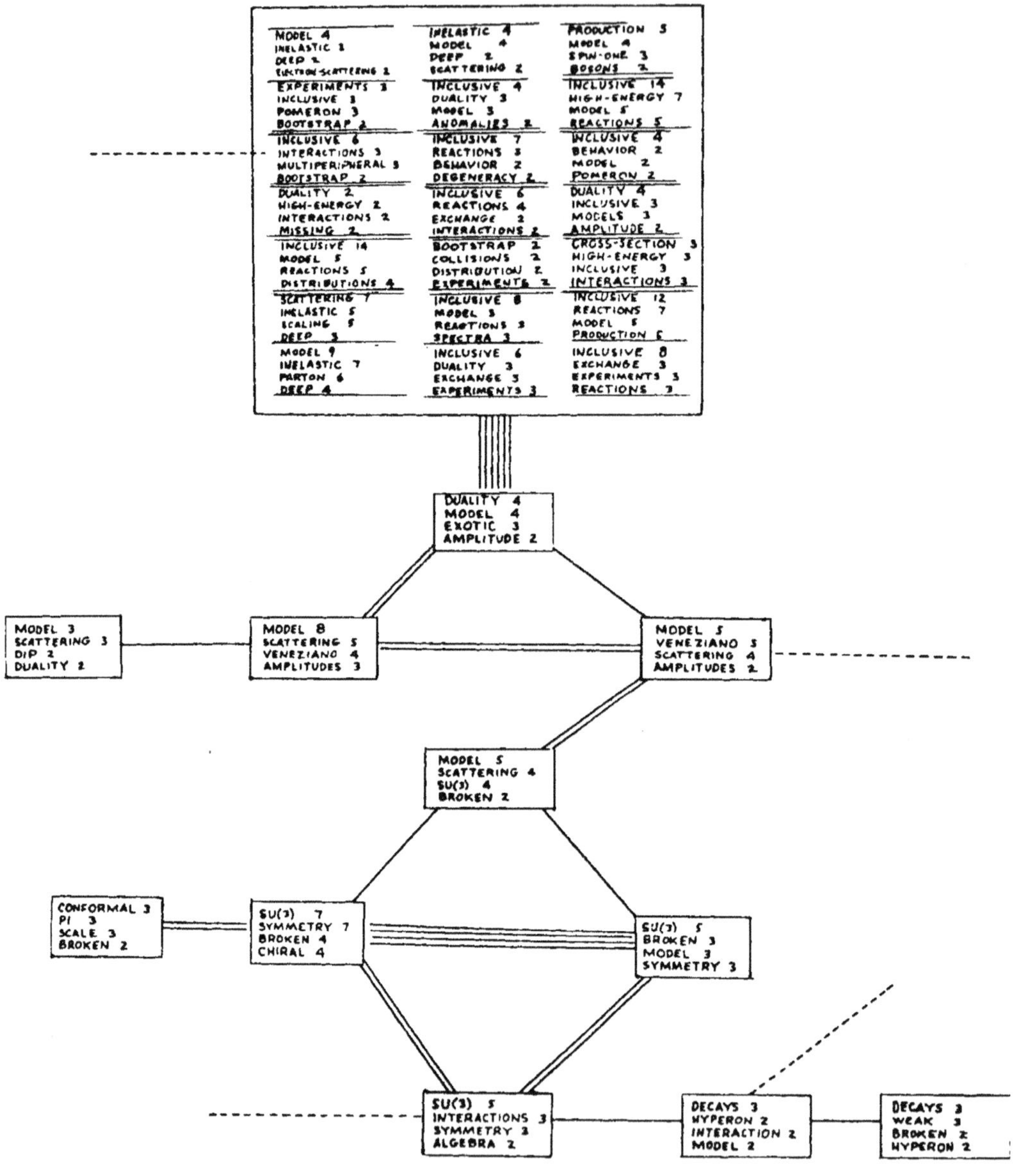

and was intellectually valid. The author of one of the linking documents, a review article on what are called "final-state interactions", states: "The main subject of this review article is a chapter of physics which is recognized as common ground for nuclear physics and what is currently known as 'particle physics'."[14] Thus, the author of a citing paper placed his work at exactly the point relative to these areas of physics as we found his paper serving as a link within our graph.

The Biomedical Cluster

Earlier the unusual characteristic of the cluster within the biomedical literature was mentioned. Now that we have examined some typical clusters within natural sciences, we would like to return and discuss some of the ways in which the biomedical literature proved to be unusual. Table 2, which displays the distribution of component sizes within the complete graph of the 1832 papers selected for our experiment, shows the abnormal size of the cluster of the biomedical literature at each level. The first insight into the nature of this cluster was gained by locating the most highly co-cited paper at Level 10 and then the thirty papers which were most frequently co-cited with that paper.

In any reasonably fair sample taken from *SCI*, one paper, that by O.H. Lowry on the measurement of protein, emerges as the most cited, and this paper was the center of the cluster and had the highest co-citation strength, 81, found in the entire file. The other papers highly co-cited with the Lowry paper appear in Table 5. Of these papers, only two failed to have titles giving clear indication of being method papers and all of them were among the most

---

Table 5

Titles of Biochemistry Methods Papers

1. Andrews, P., *Biochemical Journal* 91, 222, 1964, "Estimation of the Molecular Weights of Proteins by Sephadex Gel-Filtration".

2. Andres, P., *Biochemical Journal* 96, 595, 1965, "The Gel-Filtration Behaviour of Proteins Related to Their Molecular Weights Over a Wide Range".

3. Bartlett, G.R., *Journal of Biological Chemistry* 234, 466, 1959, "Phosphorus Assay in Column Chromatography".

4. Bray, G.A., *Analytical Biochemistry* 1, 279, 1960, "A Simple Efficient Liquid Scintillator for Counting Aqueous Solutions in a Liquid Scintillation Counter".

5. Burton, K., *Biochemical Journal* 62, 315, 1956, "A Study of the Conditions and Mechanism of the Diphenylamine Reaction for the Colorimetric Estimation of Deoxyribonucleic Acid".

6. Chen, P.S., *Analytical Chemistry* 281, 1756, 1956, "Microdetermination of Phosphorus".

7. Davis, B.J., *Annals of the New York Academy of Sciences* 121, 404, 1964, "Disc Electrophoresis-II. Method and Application to Human Serum Proteins".

8. Dubois, M., *Analytical Chemistry* 28, 350, 1956, "Colorimetric Method for Determination of Sugars and Related Substances".

9. Ellman, G.L., *Biochemical Pharmacology* 7, 88, 1961, "A New and Rapid Colorimetric Determination of Acetylcholinesterase Activity".

A. Fiske, C.H., *Journal of Biological Chemistry* 66, 375, 1925, "The Colorimetric Determination of Phosphorus".

B. Folch, J., *Journal of Biological Chemistry* 226, 497, 1957, "A Simple Method for the Isolation and Purification of Total Lipides from Animal Tissues".

C. Gornall, A.G., *Journal of Biological Chemistry* 177, 751, 1949, "Determination of Serum Proteins by Means of the Biuret Reaction".

D. Kernovsky, M.J., *Journal of Cell Biology* 27A, 137, 1965, "A Formaldehyde-Glutaraldehyde Fixative of High Osmolality for Use in Electron Microscopy".

E. Koshland, D.E., *Biochemistry* 5, 365, 1966, "Comparison of Experimental Binding Data and Theoretical Models in Proteins Containing Subunits".

F. Lineweaver, H., *Journal of the American Chemical Society* 56, 658, 1934, "The Determination of Enzyme Dissociation Constants".

G. Lowry, O.H., *Journal of Biological Chemistry* 193, 265, 1951, "Protein Measurement with the Folin Phenol Reagent".

H. Luft, J.H., *Journal of Biophysical and Biochemical Cystology* 9, 409, 1961, "Improvements in Epoxy Resin Embedding Methods".

I. Marmur, J., *Journal of Molecular Biology* 3, 208, 1961, "A Procedure for the Isolating of Deoxyribonucleic Acid from Micro-organisms".

J. Martin, R.G., *Journal of Biological Chemistry* 236, 1372, 1961, "A Method for Determining the Sedimentation Behavior of Enzymes: Application to Protein Mixtures".

K. Monod, J., *Journal of Molecular Biology* 12, 88, 1965, "On Nature of Allosteric Transitions - A Plausible Model".

L. Ornstein, L., *Annals of the New York Academy of Sciences* 121, 321, 1964, "Disc Electrophoresis-I. Background and Theory".

M. Reynolds, E.S., *Journal of Cell Biology* 17, 208, 1963, "Use of Lead Citrate at High ph as an Electron-Opaque Stain in Electron Microscopy".

N. Sabatini, D.D., *Journal of Cell Biology* 17, 19, 1963, "Cytochemistry and Electron Microscopy-Preservation of Cellular Ultrastructure and Enzymatic Activity by Aldehyde Fixation".

O. Shapiro, A.L., *Biochemical and Biophysical Research Communications* 28, 815, 1967, "Molecular Weight Estimation of Polypeptide Chains by Electrophoresis in SDS-Polyacrylamide Gela".

P. Spackman, D.H., *Analytical Chemistry* 30, 1190, 1958, "Automatic Recording Apparatus for Use in the Chromatography of Amino Acids".

Q. Venable, J.H., *Journal of Cell Biology* 25, 407, 1965, "A Simplified Lead Citrate Stain for Use in Electron Microscopy".

R. Vesterburg, O., *Acta Chemica Scandinavica* 20, 820, 1966, "Isoelectric Fractionation, Analysis and Characterization of Ampholytes in Natural pH Gradients-IV. Further Studies on the Resolving Power in Connection with Separation of Myoglobins".

S. Warburg, O., *Biochemische zeitschrift* 310, 384, 1941, "Isolation and Crystallization of Enolase".

T. Watson, M.L., *Journal of Biophysical and Biochemical Cytology* 4, 475, 1958, "Staining of Tissue Sections for Electron Microscopy with Heavy Metals".

U. Weber, K., *Journal of Biological Chemistry* 244, 4406, 1969, "Reliability of Molecular Weight Determinations by Dodecyl Sulfate - Polyacrylamide Gel Electrophoresis".

V. Yphantis, D.A., *Biochemistry* 3, 297, 1964, "Equilibrium Ultra-centrifugation of Dilute Solutions".

highly cited papers in the file. Whenever a biomedical cluster existing in one level was traced to a lower level, one or more of these papers would connect directly into this cluster, most connections being effected by the Lowry paper alone. Accordingly, we then attempted to determine what would happen if one or several of these papers were removed.

At first, these attempts to break up the biomedical cluster had little success. By removing seven papers, including the Lowry paper at Level 3, we managed to reduce the biomedical cluster by about a third in terms of the total number of pairs. Slightly more success was achieved by removing only the Lowry paper at Level 6; the biomedical cluster broke into 14 different clusters with merely the removal of that single paper. The operation that proved most successful was to remove 29 "method" papers at Level 3. When the papers were again clustered at Level 3, the effects of even this removal were comparatively minor. However, it was observed that most of the remaining linkages tying the biomedical cluster together were at Level 3, and this literature was then re-clustered at Level 4. This combination of strategies, raising the clustering threshold and removing the leading method papers, proved successful. The biomedical cluster broke up into 60 sub-clusters, the largest of which consisted of only 66 cited documents.

Our conclusions are, first that the biomedical literature presents a structure that is quite different from the physical sciences in these data and second, that some comparatively simple strategies can be employed to break up clusters which are pulled together by only one type (i.e., methodological) of paper. In the next step in this research we have planned to apply an algorithm which will relate the component clusters in the biomedical literature, one to another, and which will precisely describe the role of the method papers in tying together this entire literature.

DISCUSSION

Although our intent has been to reveal specialty structures using co-citation, our clustering methodology did not assume the existence of specialties at all. We could not anticipate at the beginning of our experiment whether any or all the documents selected by being cited ten or more times would be grouped. The specialty hypothesis suggested something between complete unity or complete isolation for the documents. What we found was a wide range of cluster sizes, all the way from the huge, though loosely connected, biomedical grouping to isolated documents. The number and structure of the groupings were also dependent on the Level at which the clustering was conducted. Large groupings would in some cases be fragmented by raising the threshold level, suggesting that in many instances links within clusters were stronger than links between clusters, and demonstrating in such cases the correctness of our intuitive impression of specialties as centers of scientific activity.

The role of certain highly-cited papers in holding together large chunks of biomedicine gradually became clear, and the obvious step of deleting these papers allowed us to separate the large chunks from one another. (A parallel pattern, although less pervasive, was observed in the nuclear physics cluster.) The special role played by these papers in connecting specialties suggests experimentation to develop a less arbitrary methodology for dealing with them. One would involve a normalization of the co-citation measure achieved by dividing the co-citation frequency by some function of the individual citation frequencies of each of the co-cited documents. Thus the measure of similarity between two documents would be adjusted by each document's potential to be linked. Such normalization would automatically compensate for the broad impact of the method papers, and it may furnish a more conceptually valid view of structure. Evidence from the pattern of word usage, and our knowledge, albeit limited, of the history of recent science, suggests that many of the groupings of documents obtained do correspond to the outlines of scientific specialties. Each specialty is represented in terms of the frequently-cited documents, which together with the authors of those documents, provide a kind of profile of that specialty. Most of the authors so identified are among the "important" individuals in their respective specialties, and such eminence is nearly always accompanied by a high citation rate, although high citation frequency, in itself, is not sufficient for eminence. Of course, caution should be exercized in associating authors of highly-cited documents with intellectual leaders in the specialties. In the first place, we have indicated in the figures only the first author of each highly cited document, and the first author is often times not the most eminent. Secondly, some documents become highly cited because they contain descriptions of important methodologies or frequently-used data. Authors of these works are usually not considered in quite the same light as authors of conceptually important work.

The nuclear and particle physics groupings illustrate both the advantages and difficulties inherent in graphical representation. When the graph is planar, as is the case for the nuclear group, it is relatively simple to draw it. All of the linkages can be drawn as straight lines and their strengths indicated. When the group is highly interrelated, as it is for particle physics, graphical representation is practically impossible and we resorted to merely lumping all the documents together, allowing no indication of internal structure. Needless to say, the idea of graphing the 801 papers in the biomedical cluster boggles the imagination. Since a great many of the groupings are internally complex, the advantages of a spatial representation, such as multidimensional scaling, become apparent.

The mapping of specialties, to show both their internal structure and their relationship to one another, is a task of prime importance for the understanding of science and, we hesitantly venture, for the understanding of scientific knowledge. Co-citation offers a solution to the mapping problem which has several advantages over other methodologies, as for example, those based on questionnaires or studies of isolated fields. When utilizing citation data, we avail ourselves of a comprehensive, objective, and internally consistent data base, the *Science Citation Index*. One specialty can be compared to another, and in principle, at least, one period in science can be compared to another, by utilizing consecutive annual files.

The investigation utilized only one quarter of an annual *SCI* file, and only a 3% sample of that quarterly file. Yet the groupings obtained from this very small file revealed, for the three major fields of the natural sciences, a structure of specialties which was consistent with intuitive notions. It also revealed many previously unknown, or at least unanticipated, relationships. These relationships, such as the linkage between nuclear and particle physics, are known to the specialists themselves and were of course, established by their citation patterns. This is the crux of the co-citation method: the observed relationships are nothing more than those which have been established by the collective efforts of the community of publishing scientists. In effect, our task is to depict them in ways that shed light on the structure of science.

References

1. Sher, I., Garfield, E. and Tropie, R., *The Use of Citation Data in Writing the History of Science*, Institute for Scientific Information, Philadelphia, 1964.

2. De Solla Price, D.J., Networks of scientific papers, *Science*, 149, 510, 1965.

3. De Solla Price, D.J., unpublished studies, 1964-1967.

4. Narin, F., Carpenter, M. and Berlt, N.C., Interrelationships of scientific journals, *Journal of the American Society for Information Science*, 23, 323, 1972; and Carpenter, M.P. and Narin, F., Clustering of scientific journals (in press).

5. Crawford, S., Informal communication among scientists in sleep research, *Journal of the American Society for Information Science*, 22, 301, 1971.

6. Griffith, B.C. and Mullins, N.C., Coherent social groups in scientific change, *Science*, 177, 959, 1972.

7. Crane, D., *Invisible Colleges: Diffusion of Knowledge in Scientific Communities*, University of Chicago Press, Chicago, 1972.

8. Goffman, W., Mathematical approach to the spread of scientific ideas - the history of mast cell research, *Nature*, 212, 449, 1966.

9. Jahn, M., *Changes with Growth of the Scientific Literatures of Two Biomedical Specialties* (unpublished Master's thesis, Drexel University, 1972).

10. Small, H., *Nuclear Physics in The Physical Review, 1927-1940* (unpublished report of a study at the American Institute of Physics).

11. Small, H., Cocitation in the scientific literature: a new measure of the relationship between two documents (in press).

12. Weinstock, M., Citation indexes, in *Encyclopedia of Library and Information Science*, 5, Marcel Dekker, New York, 1971, 16.

13. Harary, F., *Graph Theory*, Addison-Wesley, Reading, Mass., 1971, 17.

14. Slobodrian, R.J., Final-state interactions, *Reports on Progress in Physics*, 34, 175, 1971.

Acknowledgments: The authors, Belver C. Griffith, Graduate School of Library Science, Drexel University, and Henry G. Small, Institute for Scientific Information, wish to acknowledge that the research for this paper was supported by the authors' institutions and by 1 R01 LM911-01 to Drexel University. They also wish to acknowledge the able assistance of Mr. Edward Cline of the Institute for Scientific Information for the systems design of the clustering algorithm, and Ms. Judy Stonehill of Drexel University for programming same.

THE DYNAMIC SCIENTIFIC-INFORMATION USER

William D. Garvey

Johns Hopkins University, Baltimore, Maryland, U.S.A.

INTRODUCTION

In this paper we shall explore the concept, "information user". Recently, our research has been directed at examining some of the variations in scientific-information-user behavior in order to detect and describe the underlying themes. Our assumption is that by understanding and organizing these dynamic processes, we should be able to arrive at a more efficient matching of the output of information services with the user's information needs. We shall attempt to describe some of the dynamics of the information user by presenting our findings relative to two variations in the scientific-communication behavior of scientists. First, we shall discuss "intra-individual variations"; changes which occur within individual scientists as their scientific work progresses. Second, we shall discuss "inter-individual variation"; differences between individual scientists (traditionally referred to as "individual differences").

THE SCIENTIFIC-INFORMATION USER

Before discussing our findings, we should like to explain what we mean by the "scientific-information-user". In our studies, we have concentrated on a limited stream of activity in the overall process of communication of scientific information - the flow of scientific information to and from active scientific researchers. Within this framework, it has become quite clear to us that the major users of scientific information are also the major producers and evaluators of scientific information. We shall limit our discussion to this group of scientific-information users. There are, of course, other streams of activities which parallel and interact with the flow of information in science. However, the activities of technology and utilization, while at various times interactive with science, have developed for the most part independently and are qualitatively different in their nature.[1] The scientific communication behavior which we shall be examining in this paper will be mainly that of scientists actively productive of scientific information. The term "use" has various meanings in information science, i.e., scientists "use" scientific-information media in various ways. Taking the scientific information published in journal articles as an example, we find that the scientific community "uses" such information in several general ways.

The attainment of professional goals. There seems to be little question that scientists use journals as a means of gaining and maintaining visibility among their scientific peers. In the "academic world" great emphasis is placed on the publication of journal articles when selecting and promoting (or firing) faculty. Also, since scientific careers are enhanced by identification with scientific discoveries, scientists use journals to identify publicly the priority of their discoveries.

The establishment of a public body of scientific knowledge. Scientists may use journals purely as archival repository for their scientific works. The main purpose here, from the producer's point of view, is to add his findings to other recently-generated scientific information so that it can become integrated into scientific knowledge. The scientist, as a consumer, uses this body of material to acquaint himself with the accumulated knowledge that exists whenever he embarks on *new* research or inquiry. (This material is also used as the basis for instruction of future scientists.)

The facilitation of day-to-day scientific work. There are some uses made of scientific-information media which are not so much to acquire "needed" information as to facilitate the routine tasks required of a working scientist. The nature (and importance) of these "uses" may be best demonstrated by listing a few of them.

(1) A scientist often acquires and uses information in a journal article long before the article is published. (He somehow has gained the information through informal media.) When this scientist "uses" the published article he is simply examining it to determine if the "quality control" of the journal's editors and referees has resulted in any changes in the information he *previously* used in his work.

(2) When a scientist says he uses preprints of journal articles, he frequently means that he distributes them to gain critical feedback about his manuscript before submitting it to the journal-editorial process.

(3) Scientists often use their requests for reprints of articles as indirect ways of disseminating information about their own work. For example, a request to the author for a copy of his article may be accompanied by a few of the requestor's own papers, which he believes the author of the article will find especially relevant to his work. Here it seems that the scientist is using a request to gain access to an informal communication network in order that henceforth he might be included in the exchanges of this more or less exclusive network.

(4) Scientists use particular journals to publish their works for a variety of reasons other than disseminating information about their scientific works *to the most appropriate audiences*. For example, (a) the journal may be where the scientist has always published (he understands the journal's editorial policies and style and finds it easy to conform to them, (b) certain journal characteristics are attractive (no page charges, free reprints, relaxed refereeing, quick publication, etc.), (c) high prestige (e.g., a pure rather than applied science journal simply because the former is more prestigious), (d) the manuscript may have been solicited by an editor, and (e) the article may have been rejected by more appropriate journals.

We have mentioned three general categories of scientific-information-use behavior because we wanted to acknowledge their existence (and importance) to the scientific enterprise and distinguish them from the kind of information-use with which we shall be concerned in this paper.

<u>The advancement of the research front</u>. Here we are concerned with the scientist's use of information to advance the frontiers of scientific "research". Scientists have a considerable amount of knowledge of their own at any one time, and so need not start each new stage of scientific work, naively; but in the progress from stage to stage, situations often require more information. Information is needed to facilitate progress in the scientific work. Scientists seek this information; and *use* a variety of sources to acquire it. This investigator is the kind of scientific-information-user we shall examine in this paper.

SOURCE OF DATA

Since 1966, we have been following on a regular basis the scientific information exchange activities of several thousand scientists who are associated with nine scientific disciplines.[2] These disciplines include physical and social sciences and the members involve themselves in a broad spectrum of basic and applied science activities. We initiated our investigation with studies of the annual national meetings of each of the professional societies associated with the nine disciplines. A year later, we followed up the journal publication of presentations made at these meetings. In subsequent years, we have returned on a regular basis to the authors of these presentations and articles as well as to other authors of articles which were published in the same core of journals. In the most recent study in this series[3] we resurveyed the authors of articles (published in 1968-1969) about which we have had already collected survey data to determine their continuing scientific-information productivity and their current information needs. This study was a questionnaire survey and the overall response to this survey from 2030 scientists, represented 80% of those surveyed.[4] These are the data which we shall report today.

STAGES OF RESEARCH IN SCIENTIFIC WORK

One of our initial interests in the study under discussion was to find out if there was any reliable progressive order in the stages of research being conducted by our population of scientists. The question we wished to answer was: If we know the stage of research in which a scientist is currently involved, can we assume which stages he has completed and which stages he has yet to complete? The data shown in Figure 1 shed some light on this question. In this graph we show the percentage of scientists who were involved in the various stages of their *current* research (on the abscissa) which they had either completed (solid line) or were actively performing (broken line) at the time of the survey. The solid line shows a significant decrease in completion of activities and (the dotted line) a concomitant progressive increase in scientific work. This sequence of stages is much like that we were taught to follow in graduate school. But the main thing is that the ordering exists, and somehow we should be able to take advantage of it.

It is also clear from this graph that the percentages associated with the broken curve (present stage of current scientific work) add up to more than 100% and, therefore, our individual scientists must have been involved in more than one stage of scientific work at a time. We wanted to know if there was any discernable organized pattern. To answer this question we examined *for each scientist's most recently completed stage*, his other *current* involvement in stages prior to and after the completed stage. These data were next appropriately adjusted and combined, and then organized for graphic display in the following way. For example taking the ordering of stages A through K shown here as our basis, if a scientist had *just completed* Stage E (calibration, pretesting, etc.) we determined what activities he was *currently involved* in. If he was currently involved in Stage D (Preliminary experimentation, field trials or mockups) we considered him being currently involved in a stage of scientific work which was normally one stage prior to Stage (E) he had just completed. Likewise if he were involved in Stage H (Collection of data) we considered him being involved in a stage of scientific work which was normally four stages later

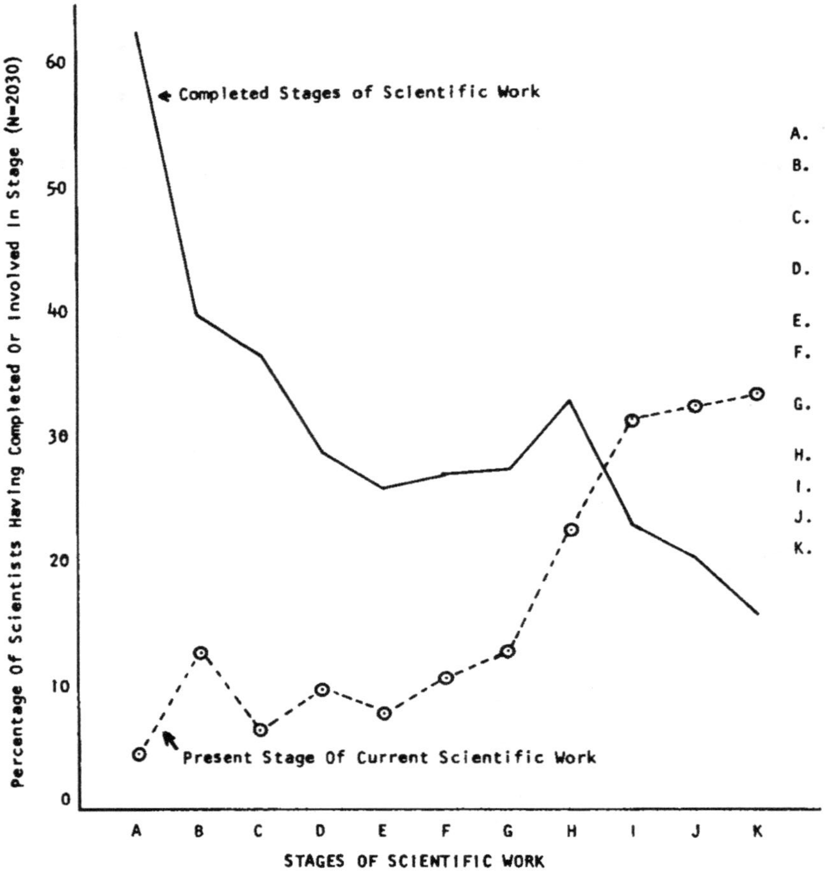

STAGES OF SCIENTIFIC WORK

A. Preliminary planning (general)
B. Specific planning: Theoretical/conceptual
C. Preparation of written research proposal
D. Preliminary experimentation/field trials or mockups
E. Calibration, pretesting, etc.
F. Design and development of equipment/apparatus
G. Formulation of experimentation/study design
H. Collection of data
I. Analysis of data
J. Interpretation of results
K. Preparation of report of work

Figure 1: Stages of research which scientists had completed or in which they were currently involved

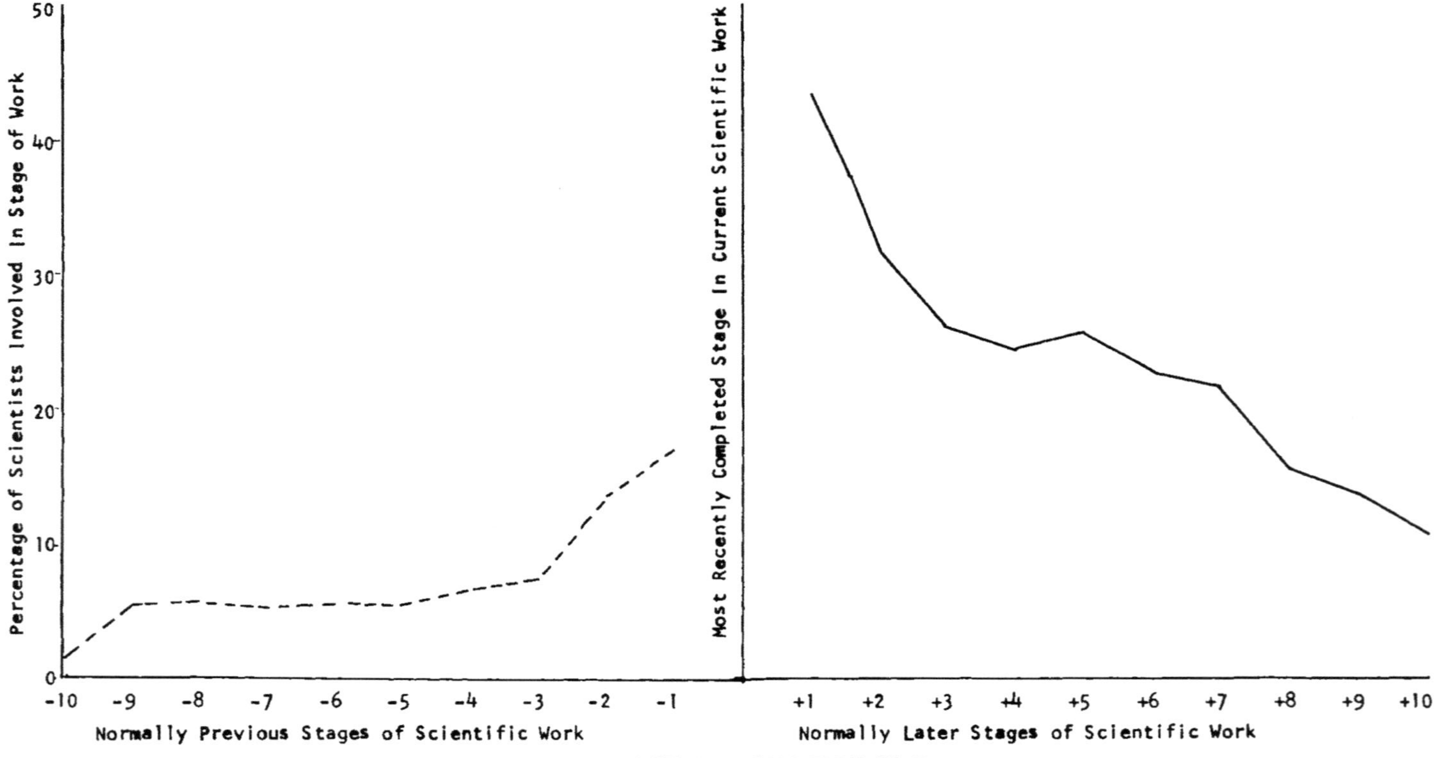

Figure 2: Current Stages of Scientific Work in Relation to the Most Recently Completed Stage of That Work

[a]The percentages are based on the number of scientists having completed each stage of scientific work.

than the one which he had just completed. The data resulting from this analysis are shown in Figure 2.

It is apparent from this graph that the sequential ordering displayed in the previous figure is similarly maintained for simultaneous activities. That is, at any given stage of his scientific work the scientist may be involved in more than one stage of scientific activity but these stages are typically clustered in a "forwardly progressing" direction, which suggests that scientists advance through their scientific work in orderly sequences.

VARIATION IN INDIVIDUAL SCIENTISTS' INFORMATION NEEDS AND USE OF SOURCES AT DIFFERENT STAGES OF RESEARCH

We have been assuming that the typical scientist has need for information (beyond his own knowledge) at each of these stages. To test this assumption we asked each respondent to indicate, for his most recently completed scientific activity, whether or not he needed any scientific or technical information to advance to the next stage of his work. Figure 3 presents these data for each of the eleven stages of scientific work we have been discussing. There are differences in the extent of information needs at different stages of scientific work, but they are not great (though some of them are statistically significant). The important thing to be learned from this graph is that at every stage in the process of scientific work the vast majority of scientists involved therein needed and sought information beyond their own knowledge.

Variation in information need. An examination of the various stages of scientific work and particularly the specific activities which went into them suggests that different kinds of mental processing are going on at different stages of scientific work. Consequently, we might expect that different kinds of information may be needed at the different stages. Table 1 shows the nature of information needed at each stage by the majority of scientists involved in that stage of scientific work. During the early stages of any research endeavor the scientist especially needs information to aid in the perception of his problem and in formulation of procedures appropriate to his inquiry; also he seeks to place his own work in proper context with other current or recently completed work.

In the intermediate stages his information needs become more specific, for example, details of techniques and methods. In the final stages his needs shift to the general body of scientific knowledge as he seeks to fully interpret his data and integrate his findings into the current state of scientific knowledge.

The pattern of information needs relative to stages of scientific work shown in this table seems to match our model of the

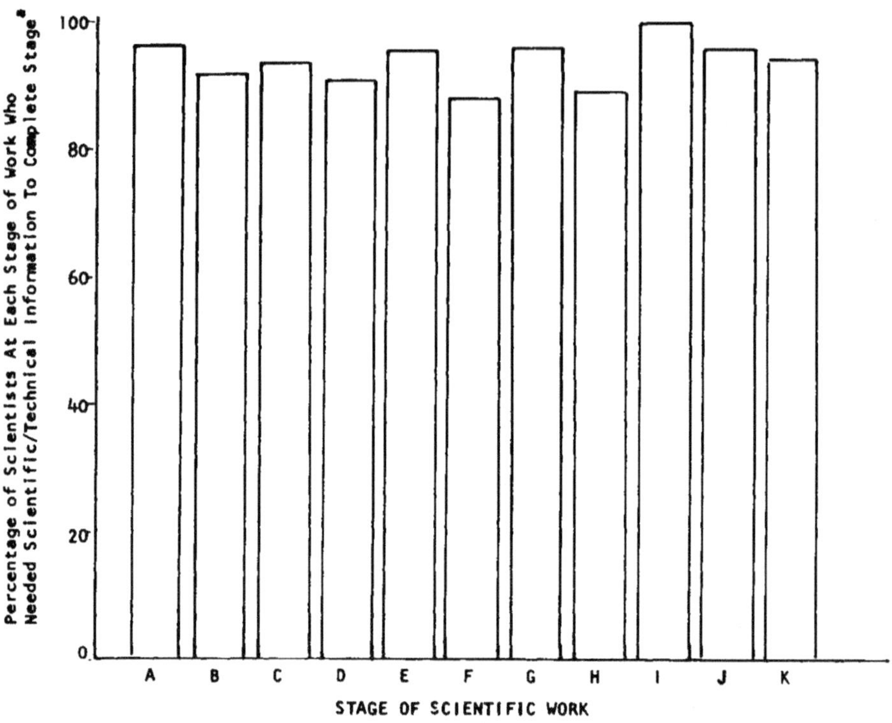

A. Preliminary planning (general)
B. Specific planning: Theoretical/conceptual
C. Preparation of written research proposal
D. Preliminary experimentation/field trials or mockups
E. Calibration, pretesting, etc.
F. Design and delevopment of equipment/study design
G. Formulation of experimentation/study design
H. Collection of data
I. Analysis of data
J. Interpretation of results
K. Preparation of report of work

[a]The percentages are based on the number of scientists involved at each stage.

Figure 3: Percentage of scientists at each stage of research who needed scientific/technical information to complete stage

Table 1: Nature of Information Needed By The Majority of Scientists Involved In Each Stage of Scientific Work

NATURE OF INFORMATION NEEDED	A. Preliminary planning (general)	B. Specific planning: Theoretical/conceptual	C. Preparation of written research proposal	D. Preliminary Experimentation/field trials or mockups	E. Calibration, pretesting, etc.	F. Design and development of equipment/apparatus	G. Formulation of experimentation/study design	H. Collection of data	I. Analysis of data	J. Interpretation of results	K. Preparation of report of work
To aid in perception or definition of problem	X	X	X	X							
To formulate a scientific or technical solution	X	X									
To place work in proper context with similar work already completed	X	X	X	X	X		X	X	X	X	X
To relate work to ongoing work in area	X	X	X	X				X	X	X	X
To select a design/strategy for data collection					X		X				
To select a data-gathering technique						X	X				
To design equipment or apparatus					X	X					
To choose a data analysis technique					X				X		
To enable full interpretation of collected data				X				X	X	X	X
To integrate findings into current state of knowledge in area								X	X	X	X

conduct of "normal" science. But the important thing for information technology is that these needs do vary as the stages of scientific work change - information provided at the initial stages of research may be irrelevant at that time (and even mask needed information) but could be relevant at a later stage. It is also worth noting that there is one variety of scientific information which the scientist is *constantly* seeking - namely, information which he needs to place his work in context with similar work recently completed or contemporary with his own. Here the scientist is seeking to maintain his work at the cutting edge of science, and, therefore, he constantly seeks new, viable information throughout all stages of his scientific work so his own work will not be *passé* by the time it is completed.

Variation in sources used. Next we would like to explore the sources which the scientists *successfully used* to obtain scientific or technical information needed for the most *recently completed activity* in their scientific work. (First, it should be mentioned that almost all of the scientists who needed information at a particular stage did eventually satisfy their needs; however, they often used a variety of sources to do so.) Figure 4 shows the major sources which provided the needed information. Of the two main sources one is informal (local colleagues and students) and the other is formal (journals). These two categories also interact; colleagues may alert an investigator to a pertinent journal article, and a journal article may identify a colleague who can aid an investigator. At this point the picture becomes even more complex, for further analyses of the data revealed a highly significant interaction between stages of scientific work, nature of scientific information needed, and source of the needed information. Without going into all the details of our analysis we should like to describe some of the major features.

We have shown earlier (Table 1) how the nature of the scientific/technical information needed varies as a function of the stages of scientific work. Our data also indicate that some sources are more useful at some stages than at others. For example, nonlocal colleagues are most useful at the stage of interpretation of results; technical reports, at the stage of calibration and pretesting, etc.; preprints, at the stage of preliminary experimentation/field trials or mockups. A more important question, however, is whether or not some sources are more useful in providing certain *kinds* of information than other sources. We then compared two information sources (journals and local colleagues) to see if one was more frequently used than another for obtaining particular kinds of information necessary to "advance" to the next stage of scientific work. It is quite clear from these data (Figure 5) that each of these two major media provides different kinds of information. They also seem to complement each other perfectly - for the kinds of information which journals provide most adequately, colleagues

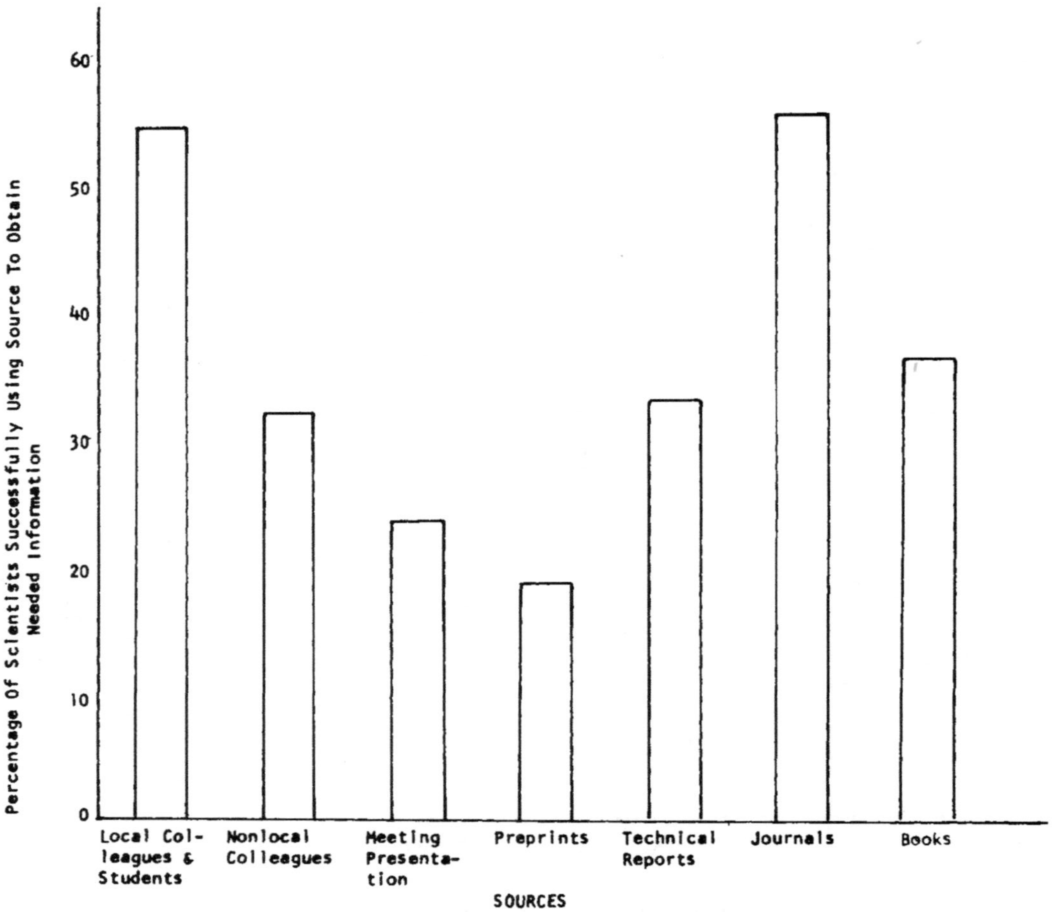

Figure 4: Source of <u>obtained</u> information needed for most recently completed activity

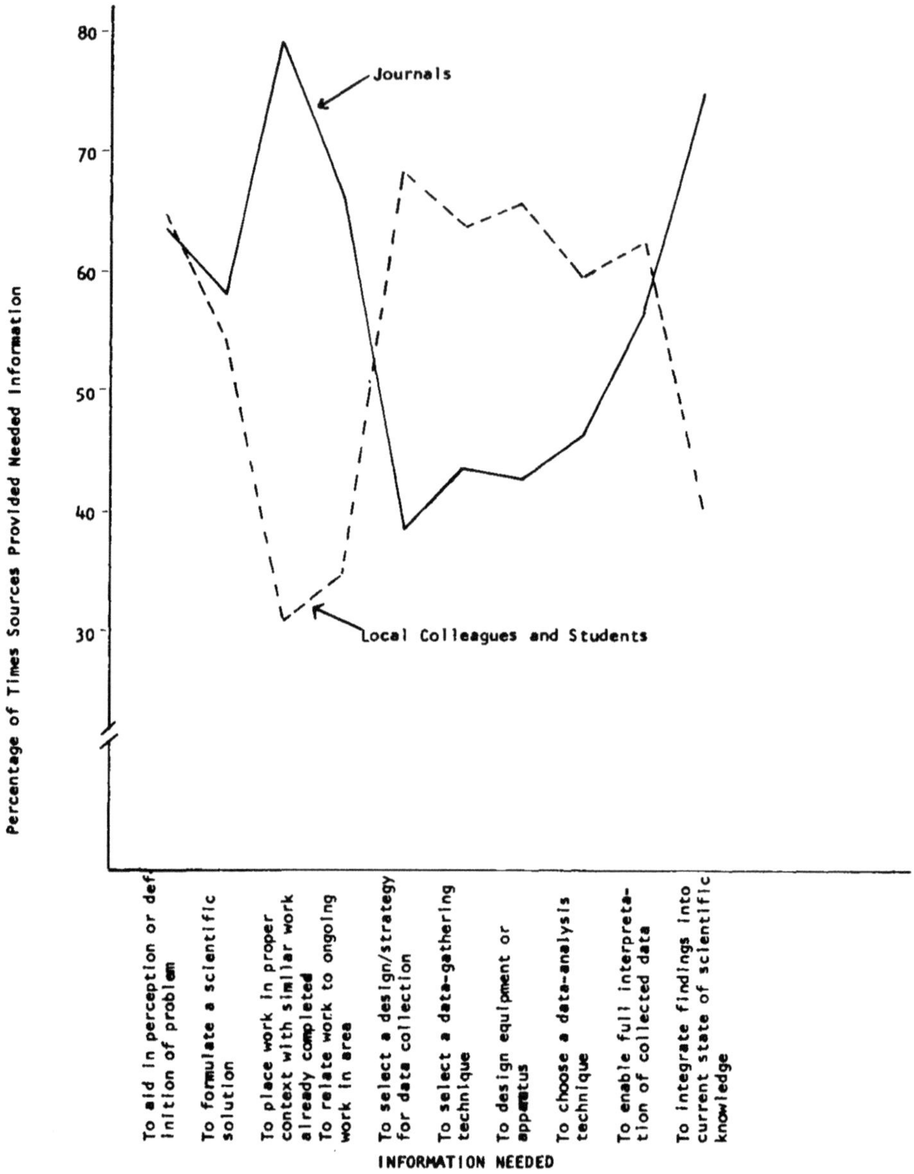

Figure 5: Uses of two major sources to obtain needed scientific/technical information

[a] The percentages are based on the number of times scientists reported needing X type of information.

are least helpful and vice versa. A few examples will illustrate how these two sources are differently used. Journals are most useful in providing information needed (1) to place a scientist's work in proper context with similar work already completed, and (2) to integrate his findings into current scientific knowledge; these two are the kinds of information least frequently obtained from local colleagues and students. On the other hand, local colleagues and students are superior providers of information needed to select a design for data collection, to design equipment or apparatus or to choose data-gathering technique.

Each of the other information sources has its peaks and valleys when plotted on a graph similar to the one shown in Figure 5. Books seem to play a dual role in providing needed information: on the one hand they are effective in providing (general) information needed to formulate a scientific solution and on the other (specific) information needed to choose a data analysis technique. Preprints and meeting presentations are most effective in providing information needed to relate a scientist's work to *ongoing* work in his area. Technical reports, although yielding less useful information in general than journals or local colleagues and students, displayed a rather flat spectrum of kinds of information useful to the scientists. Relative to the other kinds of information they provide, they appear to function best in yielding information needed to design equipment and to select data-gathering techniques.

VARIATION IN INFORMATION UTILIZATION DUE TO SIMULTANEOUS INVOLVEMENT IN MORE THAN ONE RESEARCH WORK

Up to this point we have tried to show that scientists conduct their work in orderly stages; they are usually involved in several stages at once and have more information needs at some of these stages than at others, but these stages cluster in an orderly fashion. If this were all there were to the scientific communication process, information technology should not have too great difficulty providing scientists with needed information.

We would now like to discuss some of our data which disclose the complexity of this situation. To begin with, most of the scientists in our sample were highly productive - the majority publishing two or three articles per year. (You should recall that our population of scientists were selected as the *producers* of scientific information, i.e., they were already authors of meeting presentations and journal articles.) If a scientist's publication of three articles per year was a matter of starting research and completing a manuscript every four months the issue would be simple. Our data, however, suggest that the patterns of productive scientists' work are much more complex. Figure 6 illustrates the character of this complexity. The column of boxes on the left shows the average times major events take place in the production of a jour-

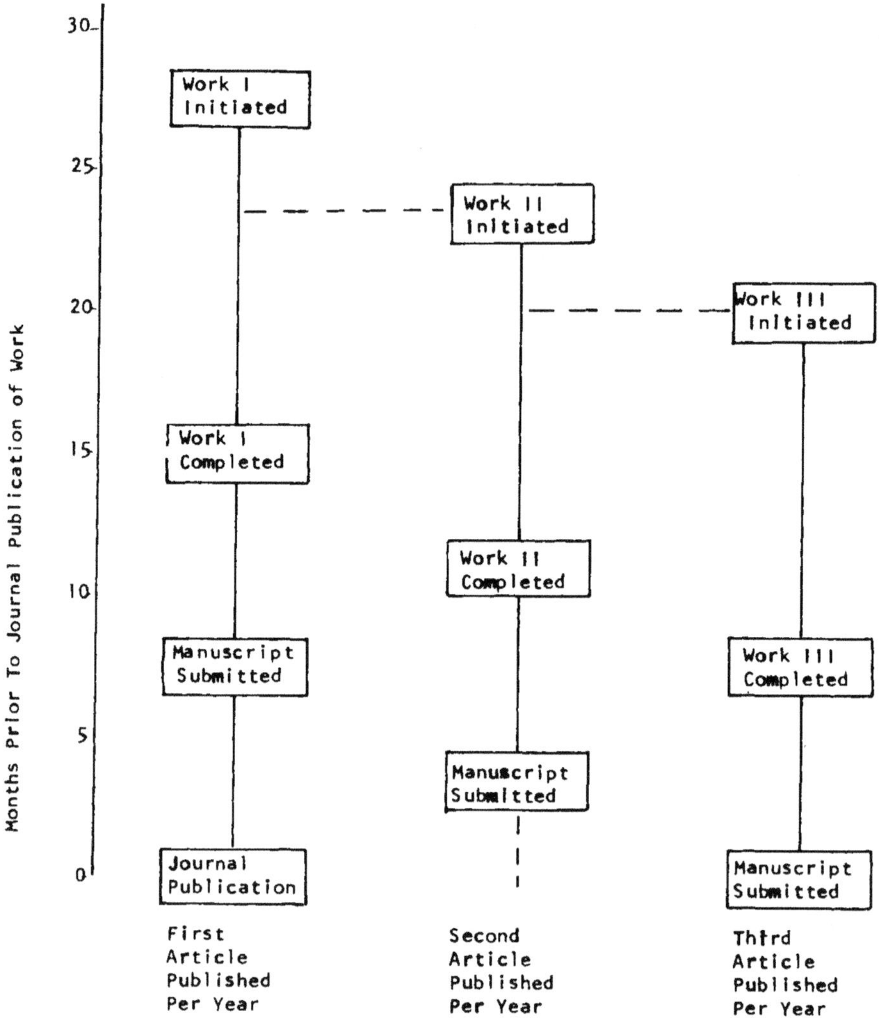

Figure 6: Average sequence of events in the scientificworks of scientists' production of these articles per year

nal article. Typically, the author of a journal article starts his work 28 months before the article is published. It takes him about a year to complete the work and another seven months to get it into shape so that he can submit a manuscript reporting it to a journal. The boxes to the right show the time schedule of the works of those authors who produce at least three articles per year. This diagram is an average, and so the individual scientist's patterning of simultaneous activities is still more complex. The main point, even in this simplified scheme, is that the productive scientist is generally engaged in the conduct of more than one scientific "work" at any one time. Consequently, at any one time he may have different information needs relative to several different activities.

INTERINDIVIDUAL VARIATIONS IN INFORMATION NEEDED AND SOURCES USED

We have been discussing the typical scientist in our sample of 2030 scientists - the typical information needs and the typical sources used to satisfy these needs. Science information technology tends to be developed to serve this typical (ideal) information user. We shall next examine how the component groups of scientists in our sample differ from this typical scientist.

Physical versus social scientists. First, we divided our sample into social and physical scientists. The nature of the information needs of these two groups differ, as shown in Figure 7. Some of these differences are not unexpected - physical scientists more frequently need information to formulate technical solutions and to design equipment or apparatus, but there are other differences here which we should note. Social scientists seem to be in greater need for information to select a design or strategy for data collection and to choose a data-analysis technique. Also, although they seem to need more information to aid in the perception of their problem they tend to need less to interpret their data and integrate their findings into the current state of knowledge.

Figure 8 shows the sources which provided the information these two groups of scientists need for their work. Here we see that some media are more useful to social scientists and others are more useful to physical scientists.

Applied versus Basic Science. Some of our scientists were engaged in basic science and some applied - others were doing some of both. Figure 9 shows the different information needs of scientists involved in these two kinds of scientific work. To begin with it would appear that applied and basic scientists have similar information needs for the more mechanical aspects of scientific work; for example, to formulate a scientific or technical solution, to select a design strategy for data collection, to select a data gathering technique and to design equipment or apparatus. Basic

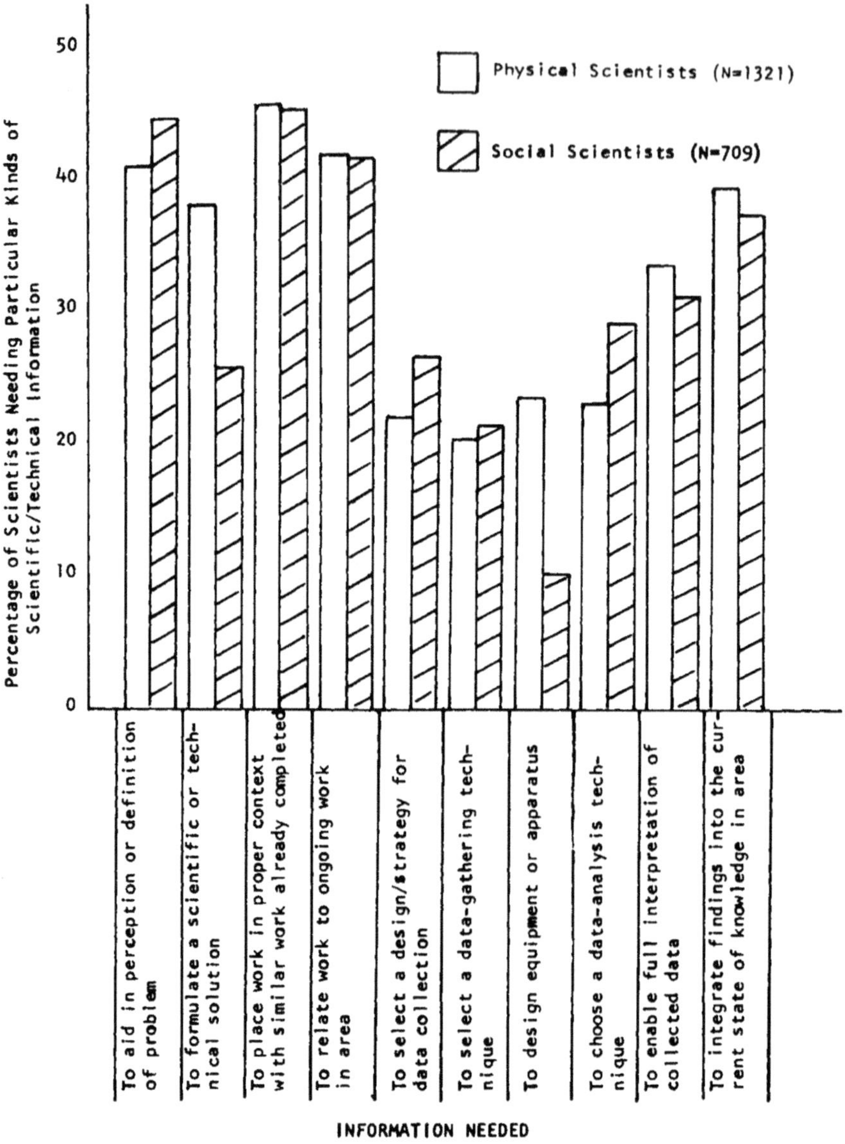

Figure 7: Comparison of the information needs of physical scientists with those of social scientists

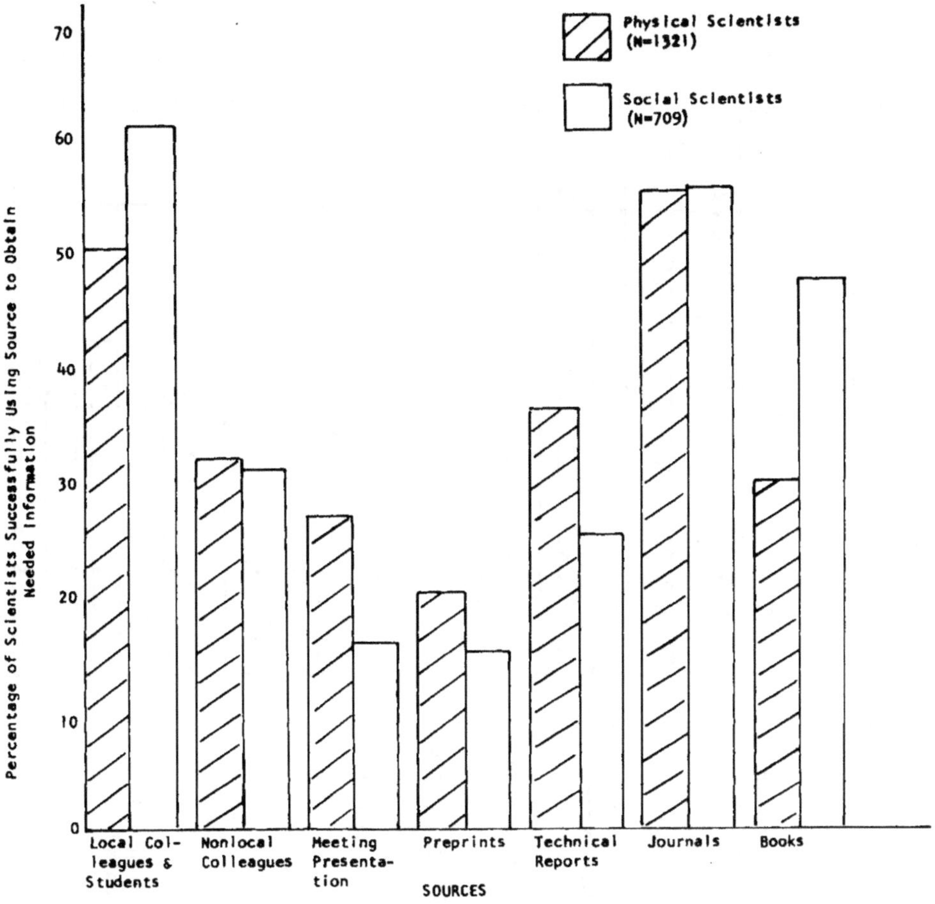

Figure 8: Sources of obtained information needed for most recently completed activity

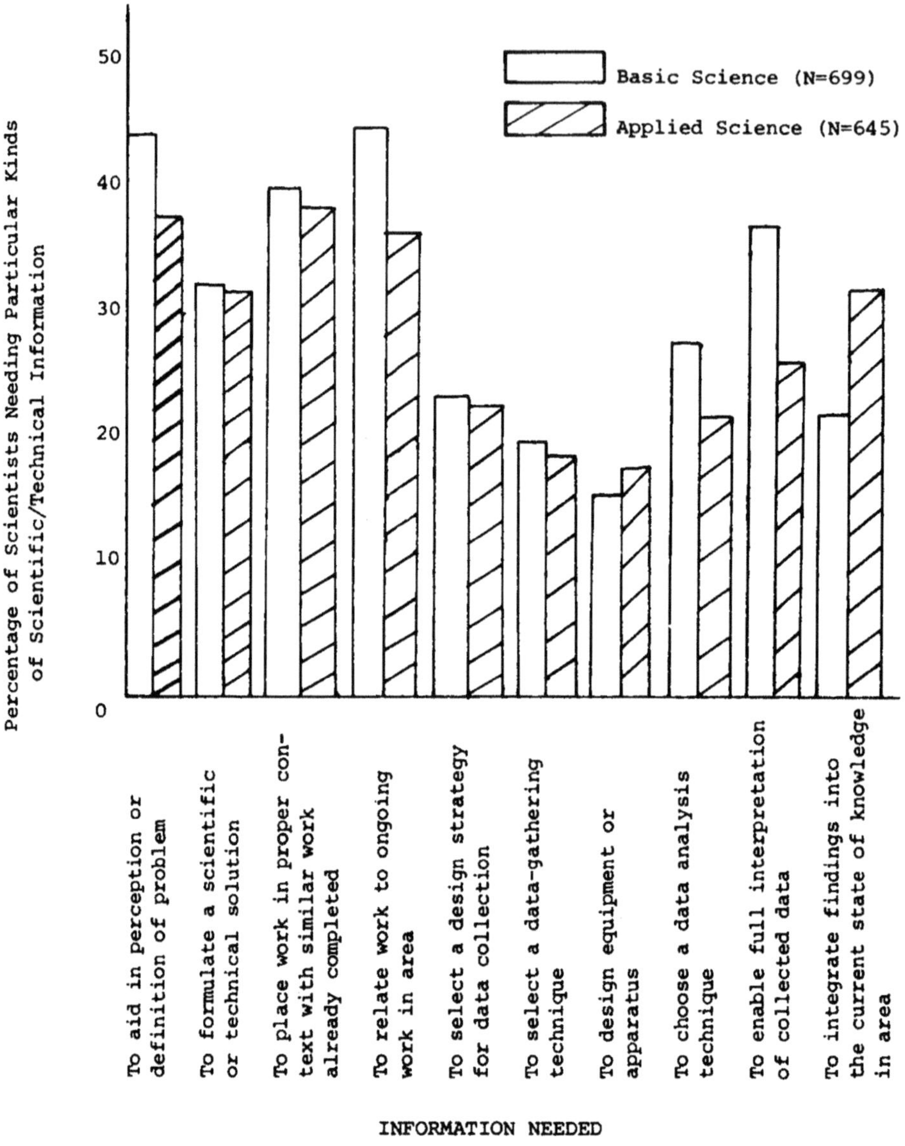

Figure 9: Comparison of the information needs of scientists conducting basic work with those conducting applied work.

scientists appear to have greater information needs to aid in the perception or definition of problems, to enable full interpretation of collected data and to relate work to ongoing work in area. The only aspect for which the applied scientists display greater information needs (beyond their present knowledge) was to integrate their findings into the current state of scientific knowledge in their fields.

Figure 10 compares the usefulness of various information sources in satisfying information needs of scientists involved in basic work and those involved in applied work. It should be noted that in general these media provide more useful information to the basic scientist. The only exception is the technical report which yields more needed information to the applied scientist. Apparently, the greater need of the applied scientist for information to enable him to integrate his findings into the current state of knowledge in his field is *not* being fulfilled by journals and books, which are the most pertinent sources for the basic scientist. Currently, the applied scientists rely more heavily on technical reports for this information.

New subject-matter area versus same subject-matter area. Many scientists change the subject-matter areas of their scientific work relatively frequently - a third of our sample had changed the areas of their scientific work within 18 months after the previous survey. Figure 11 shows a comparison of the information needs of those scientists who remained in the same subject-matter area with those who changed their areas. In general, those who changed areas tended to have more frequent need for scientific/technical information.

The sources used to obtain information necessary to conduct the scientific work of these two groups are shown in Figure 12. With the exception of two sources, nonlocal colleagues and preprints, those scientists working in a new subject-matter area used all sources more often to satisfy their information needs. It is worth noting that journals seem especially useful in satisfying information needs of scientists working in new subject-matter areas.

Less experienced versus more experienced scientists. We divided our sample of scientists into three equal groups according to when they received their highest degree. The most experienced group received their highest degrees between 1916 and 1958 (N = 654); the least experienced, between 1965 and 1969 (N = 681). Figure 13 shows a comparison of the information needs of these two groups. As might be expected, in most instances the information needs of the least experienced scientists are greater than those of the most experienced scientists. In Figure 14, we show the sources which provided these two groups of scientists with the information they needed. One source, local colleagues and students, appear to be

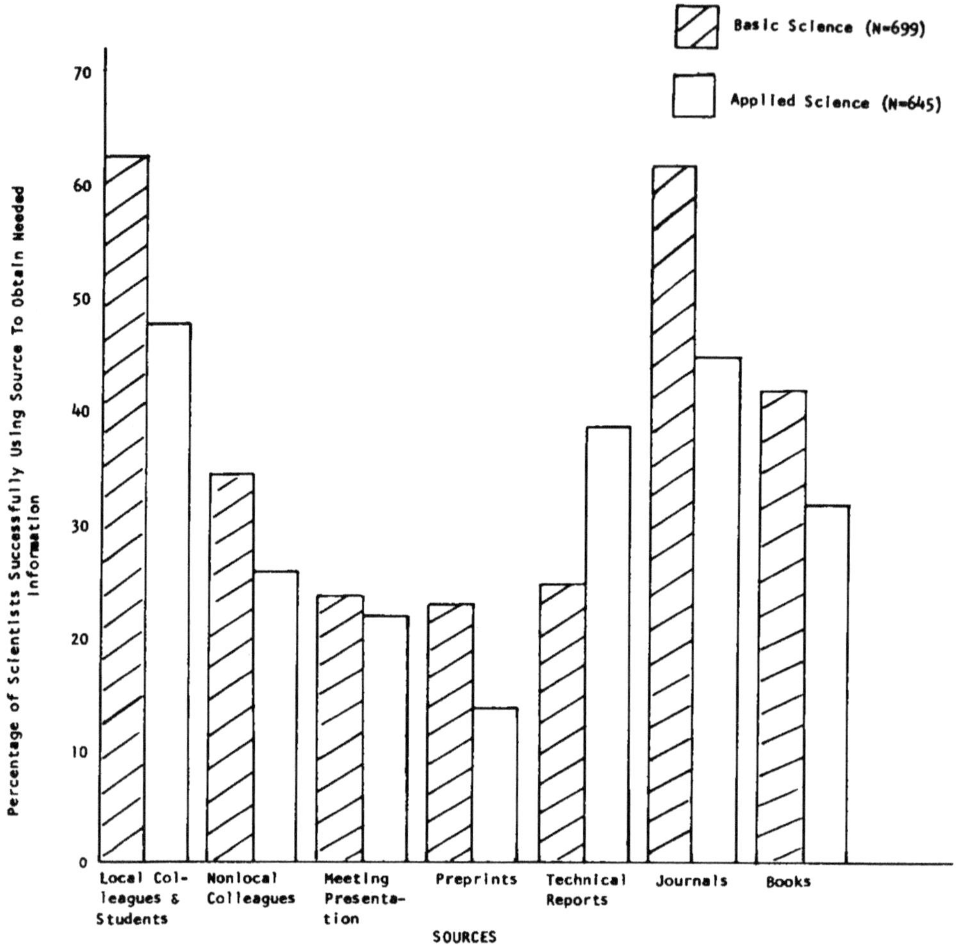

Figure 10: Source of obtained information needed for most recently completed activity

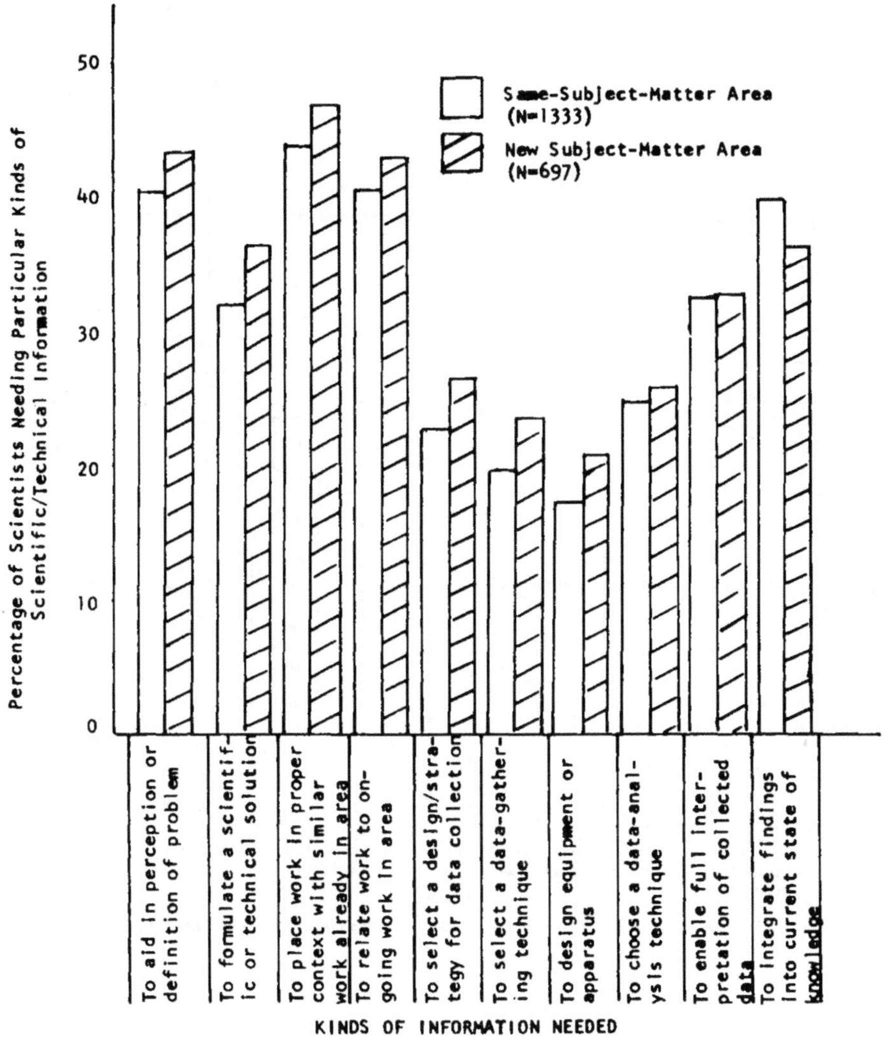

Figure 11: Comparison of the information needs of scientists who recently changed subject-matter area of their scientific work with those who had not

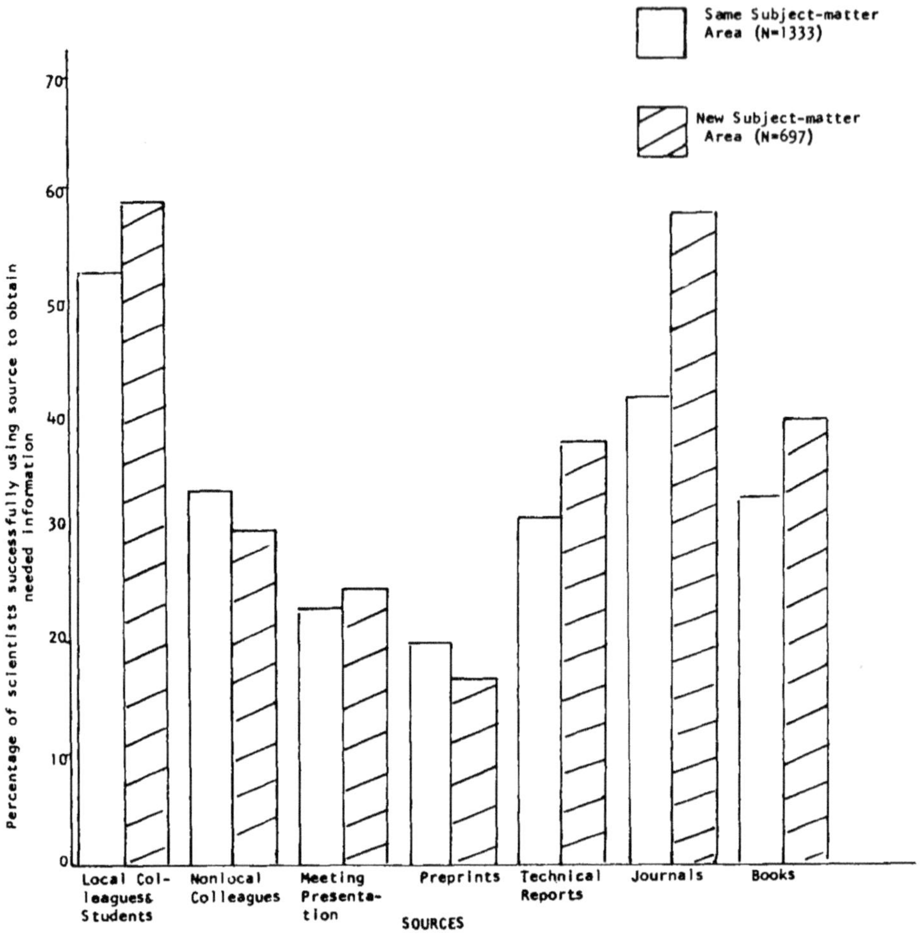

Figure 12: Source of obtained Information Needed For Most Recently Completed Activity

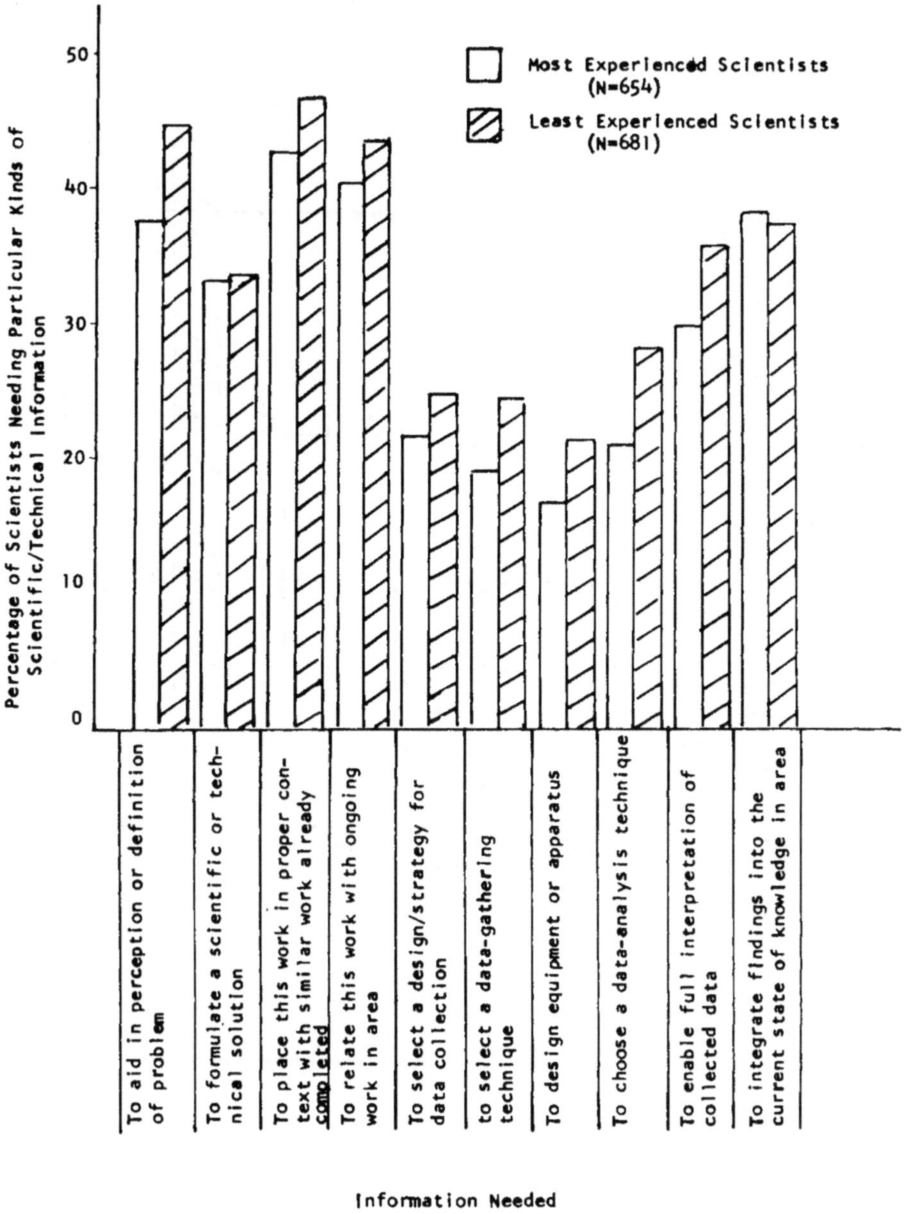

Figure 13: Comparison of the information needs of "most experienced" with "least experienced" scientists

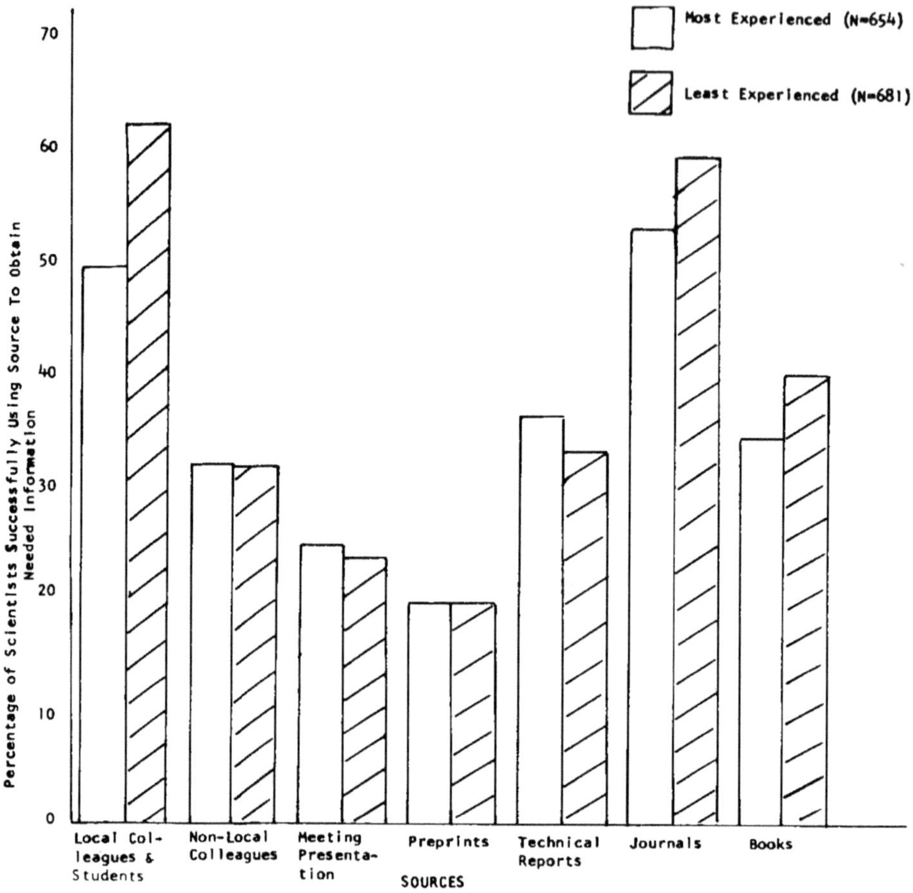

Figure 14: Source of obtained information needed for most recently completed activity

especially useful to the least experienced scientists. Journals and books also provide more needed information to this group.

In trying to show the different information needs and uses of media of the different kinds of scientists involved in the production of science, we have presented only the simplest picture. The interaction among these variables is widespread. Let us take a couple of examples to demonstrate what we mean.

The data in Figure 14 demonstrated that the least experienced scientists acquire useful information from some sources more frequently than the most experienced scientists. The graphs in Figure 15 show the percentage of scientists acquiring needed information from each of the sources as a function of the nature of the information needed. It may be seen that the relative effectiveness of any single source to the least- or most-experienced scientist is often determined by the nature of the information needed by the scientist.

Also it is worth knowing that some of these variables interact only with certain others. The graphs in Figure 16 illustrate this point. You may recall that some sources were more effective in providing needed information to scientists working in a new subject-matter area. The data in this figure indicate that it depends on whether the new area scientist is involved in basic or applied science. If he is engaged in basic science, then the local colleagues and students, technical reports, journals and books are more useful in providing him with information he needs to conduct work in his new area. On the other hand, if a scientist is working in applied science it seems to make little difference in terms of the usefulness of the various sources, whether he is working in a new or the same subject-matter area.

SUMMARY AND DISCUSSION

The purpose of our current study of the scientific information user is to examine the inherent variation in user behavior and to detect and describe patterns of use that may help "information specialists" to provide improved information services to scientists. To facilitate our task, we concentrated on one aspect of scientific-information-user behavior - that having to do with research-front scientists' needs for the additional information required to advance from stage to stage of their investigations.[5]

Our first step was to determine if there was any general order in the process of conducting research. We found that the typical scientist appears to progress from stage to stage of his work in a rational sequence. If you know what stage he has just completed, you should be able to forecast the current and future stages of his research, as well as to identify what other stages he has

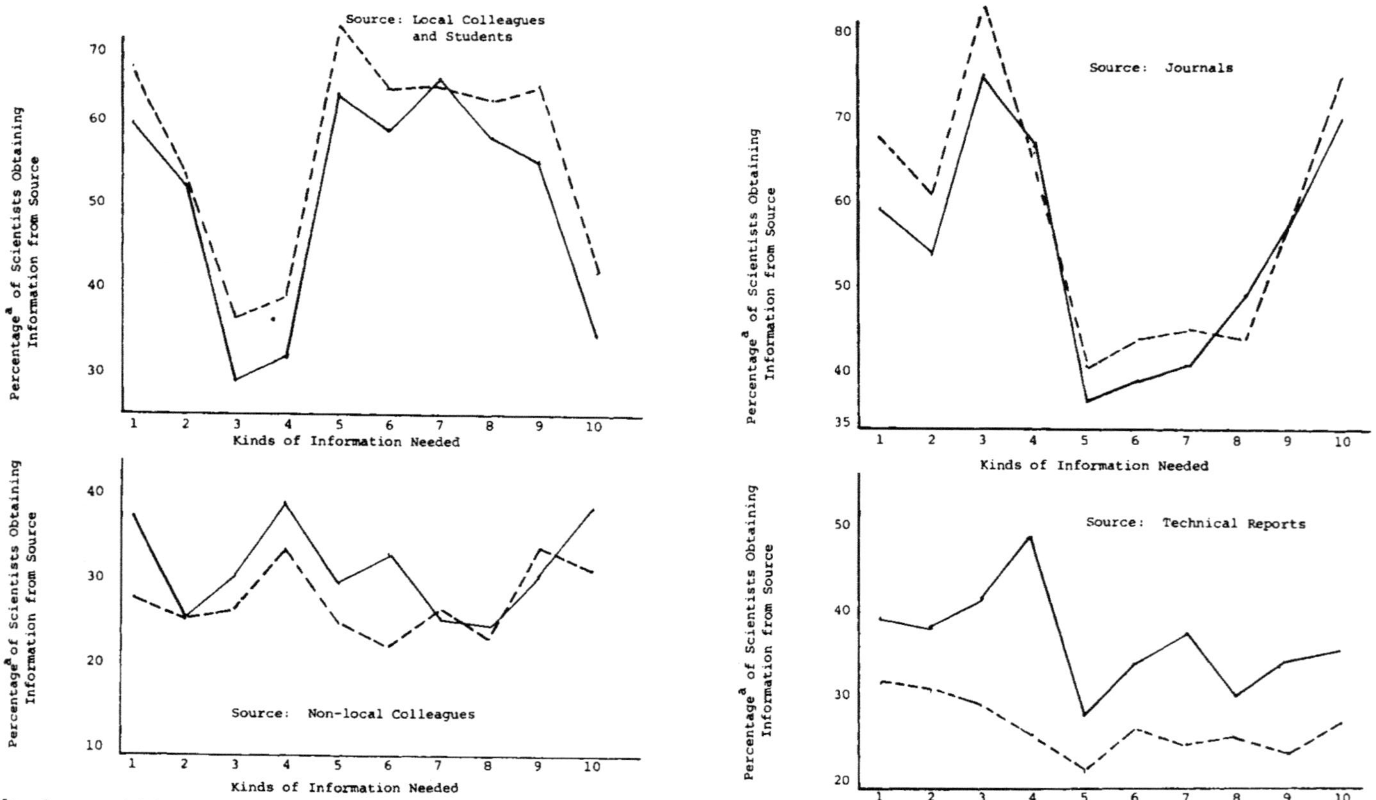

Figure 15a: Sources yielding useful scientific or Technical information as a function of the kinds of information needed

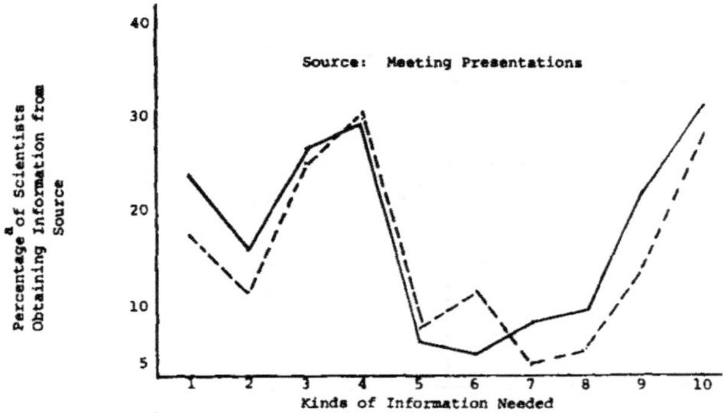

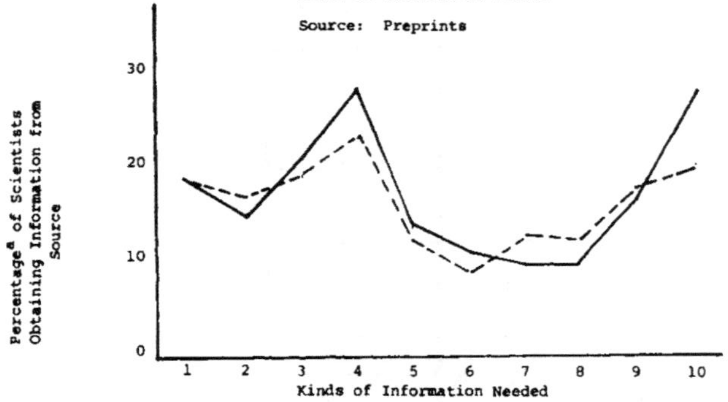

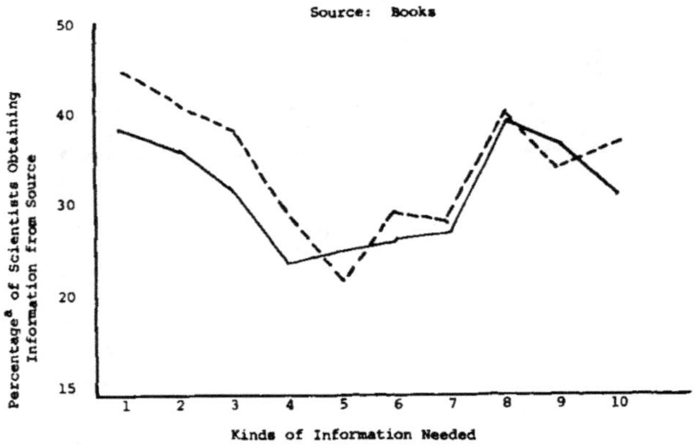

——— "Most Experienced" Scientists
----- "Least Experience" Scientists

Kinds of Information Needed

1 To aid in perception or definition of your problem
2 To formulate a scientific or technical solution
3 To place your work in proper context with similar work already completed
4 To relate your work to ongoing work in area
5 To select a design/strategy for data collection
6 To select a data gathering technique
7 To design equipment or apparatus
8 To choose a data analysis technique
9 To enable full interpretation of collected data
10 To integrate your findings into the current state of knowledge in area

Figure 15b: Sources yielding useful scientific or technical information as a function of the kinds of information needed

[a]The percentages are based on the number of times scientists reported needing particular kind of scientific information.

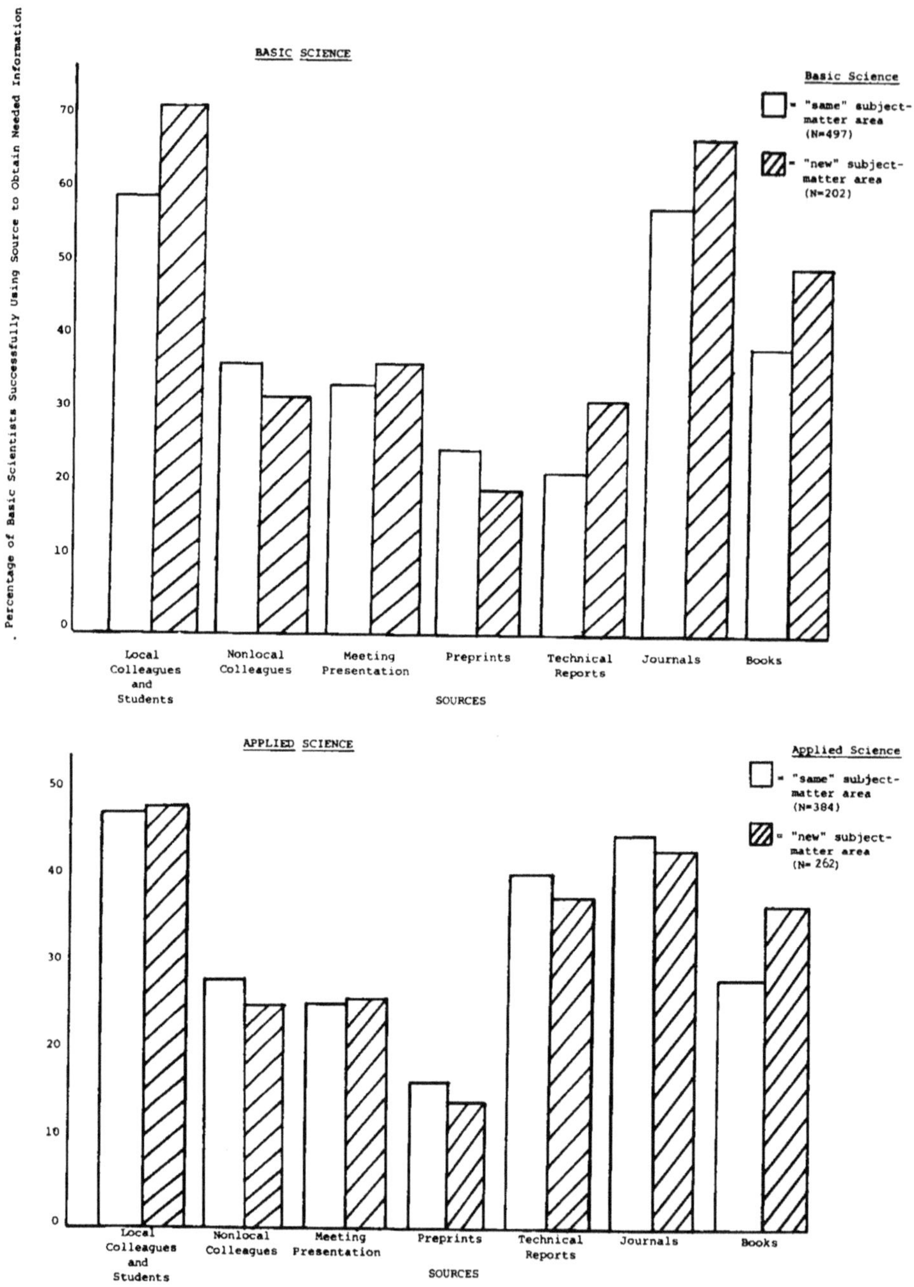

FIGURE 16: Source of Obtained Information Needed for Most Recently Completed Activity.

successfully completed.

An examination of the scientific activities involved in these various stages indicated that *behavior* changes as scientists' research progresses. This immediately raised the question of intra-individual variation in scientists' evolving information needs. Our data demonstrated that there were different information needs associated with different stages of scientific work. Thus, it seems clear that it is important that the information service be informed of the stage of research for which information is being sought.

The information needed was usually acquired by the scientists in our sample and they were able to proceed to a new stage of their work. The information sources used to fulfill these needs, however, were found to vary according to the nature of the need. For example, journals were most effective in yielding certain kinds of information and local colleagues most effective for others. These data suggest that information services may be more effective if they concentrate their efforts on certain sources (and ignore others), depending on the stage of the scientists' work and the particular information need he may have at that stage.

The behavior of productive scientists is more complex than we have just described. The scientists we studied were more productive than average, and typically were involved in more than one stage of research on the *same* research project and in other stages of one or more "different" research projects. Scientists' being simultaneously involved in more than one stage of a "single" research project does not appear, from our data, to raise too great a problem since the simultaneously conducted stages on the same project seem to cluster in rationally ordered sequences.

Scientists' working simultaneously on several "different" research projects could be a more serious problem for information services. Recently, we have been examining the continuity in scientists' research projects and we find that the typical scientist seems to lace together his work in a systematic way. The great productivity of some of these scientists implies that their scientific behavior must have considerably more rational organization than moving from one research project to another by fits and starts.

We have also been examining inter-individual differences among the 2030 scientists in our sample. Not only does an individual's scientific behavior, information needs, and media use change as he progresses from stage to stage of his scientific work but such needs and behavior vary among scientists according to a number of factors. Some of those we have examined are: (1) physical versus social scientists, (2) basic versus applied scientists, (3) experienced versus inexperienced scientists, and (4) scientists' working in the same subject-matter area versus scientists working in a new subject-matter area.

All of these were found to produce significant variation in the information needs of scientists and in the sources they used to satisfy these needs. Also we found that most of these variables tend to interact with each other. Physical and social scientists do (or don't) need different kinds of information at different stages of their work depending on whether or not they are: (1) involved in basic or applied research, (2) have had much or little experience in scientific work, and (3) involved in the same or new subject-matter area of scientific work.

This picture of the scientist's production and use of scientific information is indeed complex. We believe, however, that within the overall complexity of this picture, the behavior of the *productive* scientist is highly efficient. To the extent that we can describe and explain the rational structure underlying this efficiency, hopefully we shall be able to shape information services to match the dynamic information needs of individual scientists.

Notes

Note: Any difference between stages, groups, sources, etc., stated in this paper have been tested by appropriate statistical techniques and found to be significant at the .05 level or less.

1. Marquis, D.G. and Allen, T.J., Communication patterns in applied technology, *American Psychologist*, 21, 1052, 1966.

2. The nine disciplines studied were represented by American Geophysical Union, Optical Society of America, American Meteorological Society, Association of American Geographers, American Institute of Mining, Metallurgy and Petroleum Engineers, American Society of Heating, Refrigerating, and Air-Conditioning Engineers, American Sociological Association, American Educational Research Association, and American Institute of Aeronautics and Astronautics.

3. Garvey, W.D. and Tomita, K., Continuity of productivity by scientists in the years 1968-1971, *Science Studies*, 2, 379, 1972.

   We were especially interested in learning if they were currently working in the same subject matter area as their previous articles. The results of the first study of this particular series suggest that apparently every two years about half of the authors of journal articles move into "new" subject-matter areas, frequently returning to previous areas.

4. For detailed descriptions of these surveys, questionnaires used and general findings see: Garvey, W.D., Lin, N., Nelson, C.E. and Tomita, K., Description of a machine-readable data bank on communication behavior of scientists and technologists, *Selected Documents in Psychology*, 2, 3, 1972. This document can be obtained from the American Psychological Association, 1202 16th Street N.W., Washington, D.C., 20036, U.S.A.

5. This aspect of scientific-information-user behavior proved to be amenable to study and seems to be clearly significant from the information specialists' point of view. We wish to state, however, that other "uses" made of scientific information (mentioned in the first part of this paper) are in our opinion of comparable importance to the scientific enterprises. Moreover, there is evidence that they are not independent of each other. In fact, the history of "information services" suggests that some failures were due to a disregard of the importance to science of *all* the kinds of "uses".

Note: The research reported in this paper was supported by a grant (NSF GN-514) from the Office of Science Information Service of the National Science Foundation.

DECISION AIDS AND PROTOCOL ANALYSIS IN HUMAN PROBLEM SOLVING

Thomas J. Triggs

Monash University, Australia.

INTRODUCTION

To date, the most common command-and-control system application is that of information storage and retrieval. However, there has been a strong realization for about the last 10 years that the human information manager requires decision aids that go well beyond the data storage and retrieval stage, and actually intervene in the human judgement and decision-making process. Modern techniques of data collection across a wide variety of fields have increased the processing demands on the decision-maker; and although increased data may often act as an aid to the human, the problem of interpreting and integrating these data may, in fact, have been made more complex.

Significant attention has been devoted to the problem of how decision-making aids might increase the efficacy of decision-making, and augment the performance of decision-makers in command-and-control systems. Just 15 years ago, decision theory, although reasonably well formulated, had experienced little practical application. At that time, decision theory, as characterized by applications of Bayes' theory, had not been mated with systematic attempts to analyze systems of interest. During the last ten years, however, there has been an increasing wave of interest in the application of systems analysis to complex systems involving the human as a decision maker. Sometimes this has involved rigorous applications of elements of decision theory, but in addition it has also been used to permit a general understanding of a complex process involving both chance and decision-making under uncertainty. In this latter case, the concern is usually with understanding the structure and uncertainties of the environment within which the human must operate, while in the former the analyst is also much concerned with specifying preferences and subjective probabilities as well.

Frequently, the results of the analysis of the environment are stated in the form of a decision flow-diagram or decision tree. These diagrams may be in relatively unembellished form; by only indicating at which stages of the process the human decision maker should act, and what his choices are at that point; by showing where events are not in the hands of the decision-maker, and what the results of chance might be. In a more detailed analysis, the diagram may have probabilities of occurrence associated with outcomes at the various stages of the process out of the control of the analyst. The decision flow-diagram is then a format for indicating the order and sequence of analysis in a decision problem. In more rigorous decision analysis, the branches of the decision tree may contain probabilities and costs.

Such diagrams can also be deterministic where the process is sufficiently well understood and lacking in uncertainty, and is simply a convenient means of directing the handling of information in making assessments, with little sophisticated human judgement in the process (such as Kleinmuntz's technique for the automatic handling of MMPIs).[1]

Such diagrams, whatever their form, have been found to be useful ways of spatially representing a sequential decision process under uncertainty, and a number of different areas from basic research to business applications have made use of them. Researchers in these various areas have emphasized different aspects of the decision-making process. To some extent the nature of the application dictates the emphasis on one aspect of the decision and a de-emphasis of other aspects. For example, acceptance of the profit nature of business applications simplifies greatly the task of imposing a preference ordering on possible decision outcomes; whereas in the military situation, this preference ordering of possible decision outcomes must involve a complex weighting of many factors - including lives, territory, and material.

There is evidence that decision aids can improve the ability of humans to make decisions, for example, the work on the so-called Bayesian approach. This technique allows the use of incoming relevant data to impact on the probabilities of a set of mutually exclusive hypotheses. As has been frequently pointed out, this approach furnishes a "normative" model based on probability theory, and in particular on Bayes' theorem. It provides a mechanism for suggesting how a human should act as a decision-maker in the face of incoming probabilistic data. Discussions of Bayes' theorem have received great emphasis in the literature,[2] especially the role that such a theorem can play in augmenting human decision making in simulated systems.[3] As Parsons[4] has pointed out, the simulations at the Ohio State command-and-control facility have shown the potentials of Bayesian processing, as the results suggest that "the quality of decisions was improved 10-15 percent". However, despite these encouraging results, such procedures remain very little used. Perhaps the success in these simulations was achieved because the set of possible operations was quite restricted despite the relatively complex experimental environment, so that in Edwards' words "the subjects can ... accumulate relevant frequencies linking data with hypotheses".[5] Many real operational command-and-control systems may not lend themselves to such a structured approach.

In what follows, I will discuss some dimensions to decision making in systems and the human limitations in this process. Some approaches to understanding the decision problem and the cognitive operations involved will be described. This will be followed by a formulation of what decision flow diagram aids can do for the analyst.

RELEVANT DIMENSIONS OF DECISION MAKING

Approaches to the Study of Decision Making

One can distinguish (at least conceptually) two rather different approaches to research on decision-making and problem-solving. The distinction to be made here is not the one that is frequently made between normative and descriptive models. It concerns not so much the nature of the model one is attempting to develop but rather the strategy one employs in developing it.

Formalistic Strategy. On the one hand, there is what we will call the formalistic approach. The strategy in this case is to attempt to conceptualize the decision-making or problem-solving process in abstract terms and in a way that would be generally valid across different problem areas. The approach is exemplified by the work of Luce and Raiffa,[6] and by that of Schlaifer.[7] The approach is illustrated well in some recent work by Gorry[8] in an interactive, Bayesian computer program. The problem in medical diagnosis is to infer the state of a system (disease of a patient) from a set of system attributes (symptoms). Any one attribute typically may be characterized by different attributes, each with a particular probability. Furthermore, attributes cluster together in different coherent patterns: for example, one attribute may be more or less likely, given a particular state, and depending on whether another attribute has been identified. The non-independence of attributes in such cases means that any Bayesian procedure must take into account the interrelationships between attributes as the information about attributes is collected. The task of the diagnostician is to determine the system state, and Gorry's program operates interactively to aid him in this process. The program performs sequential diagnoses, obtaining additional information from tests that it selects for the user to perform. In evaluating tests the program considers both the costs of tests and the seriousness of possible misdiagnosis. Additional tests are costly in terms of discomfort or danger to the patient, the time of skilled people, and money. This is balanced against the value of the additional information and how much its acquisition will reduce the expected cost of misdiagnosis. Hence, the program functions both as a Bayesian aggregator of probabilities and as a complex data base that has stored knowledge about probabilities, costs, and their interdependencies.

There are two aspects of Gorry's approach that are important considerations in trying to develop a useful computerized system. First, the application is limited to a very well defined problem area, such as renal failure, where the cost of misdiagnosis is very high (often life or death of the patient). In such a case, there is a body of procedural knowledge that resides only in a few experts who are concentrated in a very few places. Not infrequently a critical situation develops because a patient in an outlying location is not initially diagnosed correctly. The expert body of knowledge needs to be more widely distributed at least for the purposes of preliminary diagnosis. It is this kind of selected subproblem where the potential of a computerized Bayesian approach is high.

The other aspect of Gorry's approach that should be emphasized is his attempt to build a probabilistic information structure that represents the knowledge of the best experts in the area. This can be done by questioning experts, or even by having experts themselves try to represent the structure of information within the necessary format. The important aspect is not the particular methodology used, but the necessity of modelling the best human processors if an effective system is to be developed. The Bayesian approach must be adapted to the particular decision structure involved in the problem at hand.

The problem with this approach is that relatively few nontrivial, real-world problems are amenable to the type of analysis that the approach requires. That is to say, seldom can one define all possible states that a complex system can assume. Nor is it usually possible to specify the probabilistic dependencies between observations and system states. Neither is it possible to identify in an unambiguous way the costs associated either with the acquisition of information or the numerous ways in which faulty diagnoses can be made. The great advantage of the approach is its tractability. If one can in fact satisfy the requirements, which is to say, if one can define the problem space in all its particulars, one does have an optimal algorithm (Bayes theorem) to apply to the problem of extracting information from data and using it to arrive at a reasonable terminal decision. What one must do in order to apply the technique, however, is to be very selective with respect to the problem on which he is to use it: the analyst must seek out problems which lend themselves to this type of structuring. Unfortunately, this appears to exclude a large proportion of the decision-making situations that people face in the real world. However, the possibility always remains that this can be done in part - a problem can be structured at many levels. If done at a sufficiently general level, the structure perhaps can be accomplished in a way that is helpful to the investigator.

Pragmatic Strategy. Another approach to research on decision-making is to begin with a problem as it presents itself in the real world, and to attempt, by whatever empirical méthods are available, to discover within, or impose upon, the situation a structure which will reduce the problem to one of manageable proportions. A characteristic of many real-world decision-making situations is their apparent lack of structure, if not their extreme fuzziness. It is often necessary to resort to relatively "brute-force" techniques in order to tease out whatever structure is inherent in a situation, or alternatively to discover what structure can be imposed on the situation without changing its character to the point of discarding the problem of interest. The possibility of qualitatively changing a decision-making situation by forcing it to fit a prescribed model is a very real one. The most obvious danger here is that of over-simplification. The temptation is very great to make a complex problem tractable by forcing it into a conceptual representation with which one knows how to cope. The price of tractability, however, often can be relevance. What one must eliminate or ignore in order to force the fit may well be the very aspects of the problem that gave it its distinctive character.

For example, there are problems in complex industrial and chemical plant operations where humans perform in a very skilled fashion as information managers in the absence of a careful (or indeed, any) description of how the work is performed. Furthermore, traditional task analysis yields very little understanding of how the operator is performing, and he himself has great difficulty in defining his job directly. There are several examples in the literature where "brute-force" analyses have been performed on operators of complex plants in order to establish their modus operandi: examples include Beishon,[9] Bainbridge, *et al.*,[10] and Dutton and Starbuck.[11] Several different techniques for obtaining data have been used, but central among them is the taking of verbal protocols. Verbal protocol analysis is one, but not the only, technique for extracting information that is pertinent to his internal processing.

Human behaviors that have been analyzed using such a "brute-force" pragmatic approach range from the simpler problems of personality analysis using the MMPI,[12] to more complex analyses of cryptoarithmetic, logical thinking, and chess, Newell and Simon.[13] For example, Kleinmuntz[12] was able to develop a decision-flow diagram indicating how an operator handled information solely on the basis of listening to protocols repeatedly and coding the sequential decision rules. Attempts also have been made to develop fully automatic protocol analysis software such as PAS-1.[14] There exist several levels of sophistication in how such data can be analyzed and the complexity of the process must be matched to the problem at hand.

A Comparison of the Formalistic and Pragmatic Approaches. It is difficult at this stage to make a value judgement concerning the relative merits of the formalistic versus empirical approaches. Indeed, probably both have their place. The greatest need perhaps is for more determined efforts to use both in concert. The formalistic approach begins with theory and general representations and attempts to move in the direction of applications. The empirical or pragmatic approach begins with real problems and attempts to move in the direction of formalization. One would hope that there would be a common ground where these two approaches would come together in developing structures which characterize decision situations. Unfortunately, there is little evidence in the literature that this has happened. One of the reasons perhaps is that both positions have advocates who tend to play the role of apologists for their respective positions, which precludes the possibility of a fruitful combination of both of them.

Structured Approach to Decision Making. In one sense, the operational decision-maker will never be called on to make decisions in a completely unstructured situation. He will always have some notion of the characteristics of the situation, and some of the variables which influence the situation. However, in a complex situation, this may not really help in working out rationally what his alternatives are, or in choosing the appropriate action. The decision problem is still effectively unstructured.

Until really quite recently, all complex decision problems fell under this rubric. But, since the advent of decision analysis, the decision-maker has the alternative of systematically focusing on all elements of his problem, and the interactions between them. Such analysis requires that the decision analyst make explicit his attitudes towards the various consequences of his decisions, and what uncertainties he perceives in the situation. When made public and communicated, these assumptions can be argued about. The various possible alternatives are first made explicit, and then argued about. As Raiffa says "... By decomposing the problem into its basic parts, decision advisors can quickly focus on those issues on which they have fundamental disagreements".[15] By structuring the problem and recording this structure in a decision flow-diagram form, the problem can be constantly revisited with no loss of memory or detail. It can be refined when new information comes in, so that it can be used as a predictor for the future, because it supplies some answer to "what if" types of questions.

Several arguments can be made against the procedure of structuring decision problems. First, they are both time-consuming and expensive. In more dynamic situations, the time for action may pass before the analysis is completed. Second, the use of a structured format may, in fact, inhibit the richness of the decision-maker's cognitive operations. The requirement for the analyst to record specifically the steps involved in his mental processes as he moves towards a decision may alter the way in which the details would be usually handled, and might not detail how the analyst was unconsciously combining processes and making inferences from data. However, in balance, the case for structuring problems seems to be convincing, however, and can now be regarded as well-established.

Decision-Making in Changing Situations

Characterizing some decision situations as static may be a perfectly valid assumption. However, very few decision situations of operational interest are truly non-changing. Even in slowly changing business environments, aspects of the problems are constantly being modified. Nevertheless, although analysis of some problems in this area may take considerable periods of time, the situation may not have changed significantly enough to render invalid the characteristics of the underlying problem space. The analysis still allows the analyst to make reasonable decisions, to ask "what if" questions, and to use the analysis to direct him in seeking out relevant information.

On the other hand, rapidly changing problem environments may be less amenable to formulating a specific decision analysis. A requirement to construct a decision flow-diagram for each problem situation as it emerges would clearly be unreasonable in such situations. Examples of such situations would be many military resource allocation problems and flight management problems.

An alternative approach might be to have a family of decision flow diagrams conceived and developed prior to the actual occurrence of the particular situation. The implication here is that the possible environments that could be encountered in operational circumstances are definable in advance, and that their number is such that a development of the type could be kept to manageable proportions. An example of where this could be a feasible task is in an industrial plant that makes several product lines. Individual decision analyses could be carried out for each product. Another is the Dartmouth MEDEX program which has developed a wide class of diagrams to be used by paramedics in medical diagnosis and treatment.[16]

However, in many operational situations, particularly those which are extremely ill-structured (such as many military situations) such definition of a restricted set of environments might not be meaningful. In fact, a taxonomy of such situations might have to be restricted to very general classes of environments and decision-making requirements[17] which might not then be useful for many specific situations.

More realistic for highly variable dynamic environments would be the establishment of general procedures which would either be useful to the analyst in directing him in quickly constructing decision diagrams for particular situations, or general procedural diagrams (or logical flow-charts) which could direct him in the handling and processing of information necessary in making a decision. Rather than the procedure dealing with the actual decision process, this alternative suggests the development of a normative or canonical technique for handling data, rather than dealing with the actual possible characteristics of the data base itself. In general, there are more applied examples of the specific flow-diagram form constructed by a skilled worker and involving a very particular problem[18] than the more generic type of diagram.[19]

The Components of the Decision Making Process

Up until now, we have discussed decision making in a global way as though it represented a single, albeit complex, activity. The decision-maker, in fact, is often faced by a number of separable problems. Decision flow-diagrams, decision trees or logical flow-charts will be useful to a greater or less extent depending on which of these problems is being attempted by the analyst.

The decision-maker's task can be conceptualized as having the following components:

(1) The need to generate hypotheses about the current state of the system the analyst is concerned with.

(2) The requirement to develop initial probabilities (or some equivalent of human conception) for the various hypotheses that are being considered.

(3) The task of gathering information to both modify the possible hypotheses or to upgrade the assigned probabilities.

(4) The assignment of preferences to the decision outcomes, which includes not only the actual states of the system, but also what the decision selected by the analyst was whether he was correct or not.

(5) The establishment of criteria which will govern the way in which decisions are made, and finally

(6) The operational selection of an action.

The decision-maker himself does not invariably get involved in all of these components explicitly. In particular, the importance of probabilities and probability assessment in actual system applications is not at all clear. Although, following the details of Decision Analysis as laid out by investigators,[20] all of these components would be an integral part. In fact, these components are representative of those classes of activities that should be found in any complete decision-making procedure. However, all of these activities are not necessarily carried out by a single person and, in fact, in many cases would not be. For example, information gathering may be carried out by several specialists. Furthermore, the emphasis on the various components will change depending on the nature of the decision task and its complexity.

Also, these component activities in decision-making do not simply constitute a well-organized, serial, open-loop activity in which one first gathers information, then develops a set of hypotheses, and so on until the final action is selected, thus completing the cycle. These components may overlap considerably in time, and have to be frequently reiterated.

Logical diagrams can be used as a way of representing the products of these various tasks in some operational sequential order. The diagram can represent a division of the problem space, i.e., a large-scale decision requirement can be broken down into a series of sub-decisions, to each of which the above series of decision components may be appropriate. The activities of generating hypotheses, information gathering, probability estimation, and identification of action alternatives are involved in establishing the decision structure as manifested by setting up a diagrammatic representation. Those remaining activities that can be subsumed under the rubric of "action selection" are more involved with the usage of a specific decision tree once it is actually constructed in order to reach a terminal decision.

The more recent emphasis on decision analysis and its emphasis on structuring the problem space, and appropriately representing it before carrying through the decision selection process throws more emphasis onto the hypothesis generation process as opposed to hypothesis evaluation. The first is an inductive, creative process. The second is deductive and analytic. The generative process involves the determination of possible states of the system of interest and what courses of action are open to the decision-maker. The evaluation process is concerned with establishing the probabilities of the alternatives identified.

Unfortunately, in many theoretical treatments of diagnosis, the process of hypothesis generation is essentially ignored. Hypotheses are taken as given, and the emphasis is placed on the process of evaluation. For example, Bayes' rule, which has received considerable attention from investigators of decision-making, presupposes a completely-developed hypothesis space. If the true state of the system is not included in the set of hypotheses with which the decision-maker is working, the Bayesian procedure will not provide it. This comment does not refer to the problem of assigning prior probabilities, but to that of identifying what the possibilities are. Bayes' rule is no help in this regard. This limitation should not be minimized, inasmuch as for many diagnostic tasks the really difficult problem is not to test for the correct hypothesis, but to generate it in the first place. One of the most challenging questions confronting researchers who wish to develop techniques to facilitate decision-making, or thought processes in general, is that of finding ways to facilitate the generation of reasonable hypotheses.

Once a reasonable structure of the environment has been established and has been represented in an interpretable form, the analyst is able to ask "what if" questions which can aid in the generation of plausible hypotheses regarding the state of the system of interest. No rules-of-thumb appear to exist for carrying out this procedure for hypothesis generation. It depends greatly on the specifics of the environment, and the skill and background of the analyst. Suitable structuring of the decision problem with its attendant characteristic of "Divide and Conquer"[21] can be an additional aid to the analyst to indicate what additional information is needed or would be useful, although it would normally provide no indication of possible sources for this information. Information seeking is, of course, an extremely general type of behavior, and often occurs throughout the whole decision-making process. Information gathering may be a large requirement in order to discover what action alternatives are open to the decision maker, and in all but completely static activities will be a continuing activity. The objectives of information seeking, and the type of information that is required, will differ for different phases of decision making, but almost always will be an important component. If a decision aid can help focus on where information is deficient, then it could be of significant help to an analyst.

Alternative Types of Decision Aids

As was implied before, Bayesian decision-makers for operational problems can be considered to have set the pace in how human judgements can be aided in information management systems. Bayes' theorem was viewed as a normative model against which the human was assessed to see how his judgements departed from the best that could be expected. Sometimes the experiments have been carried out for real-life problem situations, such as target identification;[22] the dispatching of aircraft on military missions;[23] determination of aggressor intent and threat evaluation.[24] One very general assumption that seems to underlie this work is that humans can be regarded as good perceivers of the situation, but bad aggregators of probabilities. Most frequently, in this type of research, single-stage decision processes were studied, allowing incoming data to affect directly the contingent probabilities associated with the validity of the possible hypothesis, without intervening stages.

Another approach used in the study of human decision processes and judgements is that referred to as the "regression" approach to human judgement. Slovic and Lichtenstein[25] have recently contrasted the Bayesian and Regression approaches to human decision-making as follows. In the regression technique, the human's integration of information is described in terms of correlational statistics, the aim being to model the human operator's method of combining and weighting information by developing mathematical equations which describe his combinatorial processes. The approach borrows heavily from the psychology of Brunswik, who analyzed the manner in which one learns the characteristics of his operational situation using data he acquires from his environment. In their consideration of these two fundamentally different types of judgemental models, Slovic and Lichtenstein indicated that adherents to each approach communicated little with each other, and tended to work with different experimental paradigms and problems. Importantly, the Regression approach de-emphasizes that the human's aggregation process is faulty and concentrates on the unreliability of human judgements. This raises the suggestion that focussing on other ways of aiding the human might be fruitful. In fact, the evidence to date has been taken by Slovic and Lichtenstein to indicate that subjects are processing information in ways fundamentally different from the Bayesian and Regression models. They favor an approach which studies the molecular analyses of heuristic strategies employed by human operators, as indicated by experimental measures such as verbal protocols, introspective reports, and the like. Ironically, of the two approaches described above, the recording of insights of the human on how he utilizes information has been investigated only within the Regression paradigm. The approach favored by Slovic and Lichtenstein is in part incorporated in the general approach of such investigators as Newell and Simon.[26]

Rather than treating the decision-maker as a faulty aggregator, the decision-maker can be viewed as needing help in perceiving the consequences of his actions correctly. For instance, Newell and Simon[26] have regarded the human as an impaired processor in the sense that he tends to misperceive the situation and the results of his actions. Smith[27] has taken this approach in developing computer aids to assist the task of resource allocation in an industrial task.

Development of Predictive Models

One way to develop a predictive model in human decision making is to go through a process of "divide and conquer", as Raiffa[28] refers to it. By decomposing the complex problems confronting the human in a command-and-control system into a set of simpler problems, the various parameters, subdecisions, and utilities acting on the situation can be made explicit as was discussed previously. Once this decomposition has occurred (resulting in a decision tree or decision flow-diagram), "what if" questions can be asked and the analyst can systematically examine the consequences of his possible actions. Bayesian handling of data may be involved in these simpler problems, but heuristics and other possible fast-time models may constitute some of the other subprocesses.

As Raiffa says, "Decision analysis requires the explicit articulation of a thought process." I have already discussed some of the methodology involved in gaining access to this thought process, and the ways this can be applied to produce the required decision analysis. It appears that in some problem areas, such as business analysis, this process tends to be somewhat informal, perhaps because the various alternatives can be made quite explicit, and hypothesis generation is less of a problem. However, in more complex situations, such as military command-and-control systems, it is unlikely that casual procedures for explicating thought streams will be satisfactory because of the complexity of the data bases and decisions involved.

I will now discuss the use of verbal protocol analysis as a methodology for explicating thought streams.

PROTOCOL ANALYSIS AND HUMAN COGNITIVE PROCESSES

The understanding of cognitive functioning while the human is carrying out various types of tasks has been of significant interest to experimental psychologists for over a hundred years. Although one can find this common thread of interest running through work in human performance during this time, the motivation for such interest and the methodology used in such study has changed over the years.

Thus we have the early proponents of the introspectionist "Wurzburg School" who judged that it was a worthwhile and scientifically legitimate practice to obtain data using introspective experimentation. Then, around the turn of the century, two developments - improved scientific measuring devices (e.g., chronometric devices) and strong behavioral criticism of introspection - tended to downplay the use of introspection and its further methodological growth.

Although the behaviorists (perhaps to this day) believe that introspection was defeated, deGroot[29] maintains that it was primarily a social victory. As he succinctly puts it, introspection "just went out of fashion". Johnson[30] has pointed out that the work of the Wurtzburg introspectionists was never refuted as a means of understanding cognitive functioning nor was it laid aside as unimportant; it was simply discontinued as an observational tool. Psychologists developed feelings of discomfort about using introspective data and tended to use such information only occasionally, and then, poorly. Even more recent investigation[31] tends not to be very explicit about how problem-solving verbal protocols were actually used. Earlier work on introspection may have hastened its demise by focusing on the *content* of the verbal protocol rather than on the *processes* involved in cognitive functioning that, through analysis, could be recovered from the verbal stream.

Following the demise of introspection, the fields of cognitive research and applied experimental psychology yielded other techniques that have been used to understand at least at some level the cognitive functioning of the human. Very frequently objective response data are analyzed, and the results used to infer what types of mechanisms are being called into play when the human is processing information. These analyses may be modelled at many different levels, ranging from detailed physiological processes to higher-level cognition, depending on the aims of the researcher. On occasion, these data are backed up by reference to introspective observations by experimental subjects. Such supplementary protocol data is likely to be useful only if the modelling is aimed at higher-level cognitive functioning.

Recently, there has been a considerable growth of interest in modelling those cognitive processes that are more accessible via introspectional data than via quantitative behavioral measurements. While there is by no means universal agreement about the efficacy of analyzing such verbal information (see, for example, Edwards' less favorable comment),[32] this approach appears to have come of age in modelling human decision-making processes, problem-solving, etc.

In what follows, I will describe some of the tasks that have been subjected to verbal protocol analysis, and the procedures used for gathering and analyzing such data.

Verbal Protocol Analysis as a Methodology

A number of researchers[33] have based their cognitive models on the verbal protocols of experimental subjects. Basic to this use in such model-building is that the protocol is used to uncover the process or processes being brought to bear on the task rather than the content per se of the protocol. The verbal data can be used at different levels. For example, Bruner, Goodnow, and Austin[34] used the verbal data only to recognize what hypotheses the subjects tested in a concept learning task. Having defined these hypotheses, they made judgements about the strategies and processes that the subjects used in making decisions. On the other hand, Feldman uses the thought stream of their subjects in detail in a binary-choice experiment and developed his understanding of the cognitive function of the subject from the very explicit statements that the subjects made. Feldman, having built a process model, had it formulated in a detailed enough fashion that he could simulate a subject's behavior to a good level of agreement. This would not have been possible in the more general conceptualization of Bruner, *et al.*

When using the term verbal protocol analysis, we mean to refer to the detailed consideration of verbal streams elicited from the subject, where such a stream constitutes a complete account of the subject's information processing during the task. Frequently, several other terms have been grouped together under the same rubric: introspection, thought-stream analysis, thinking aloud, and systematic retrospection. While these terms may refer to somewhat different techniques in gathering such subjective data, the same general criticisms are used against all. For instance, the behaviorist generally criticized the use of introspection because it was a reflection of a hypothetical process, not directly related to the actual behavior of the subject. The emphasis in the work of the early introspectionists was on the content which led to coding distinctions that just were not consistent or verifiable. In more modern times, verbal protocols are interpreted more as a reflection of underlying processes, and they are regarded as reflections of actual behavior (albeit internal) rather than the view that these mental events actually exist.

Nevertheless, the use of subject protocols is still criticized for various reasons. Bruner, *et al.* argue that verbal reports provide insufficient data for analyzing internal processes. It has been suggested that a person's ability to verbalize provides an account too far removed from fundamental internal operations to provide support for theoretical concepts. It has been emphasized by Miller, Galanter and Pribram[35] that introspections can be somewhat unreliable, and show considerable variability between subjects. Subjects have been often found to forget to verbalize at important moments.[36] However, this is less of a problem in well-trained subjects. Subjects have been found to be inaccurate occasionally in the ways that their internal mental processes are described.[37] It has also been pointed out that verbalization can affect the perception of the task and how the task is carried out. Thought processes may, in fact, be inhibited by the requirement to place it in public view. The requirement to verbalize might, in fact, impose a structure or organization and make the thought stream sound more sequential and organized than it actually is. Some researchers[38] go even further in their criticism of such techniques, and suggest that a person's ability to verbalize may provide an account too distant from fundamental internal operations, under some circumstances, to provide highly satisfactory support for theory.

However, despite these criticisms the methodology of protocol analysis (and related methodologies) can be regarded as firmly established. Investigators have noted that the verbal stream analysis can provide good data on internal processing.[39] Reitman[37] finds that despite lapses, subject's protocols are most usually appropriate to the task at hand. Miller *et al.*[40] imply that it is easier to simulate verbal reports than the fine-grained structure of the decisions apparent in behavior. Furthermore, protocols have been found to be highly predictive of actual human behavior.[41] In most situations, the taking of protocols has not been found to distort the task performance. For example, Reitman[37] found that the data collected by means of protocols in a memory task did not differ significantly from data collected without protocols and Newell and Simon[42] cite some similar evidence.

To sum up, despite the pessimistic statements on verbal protocols and some pessimism over the progress of cognitive theorists to model human problem solving, there are at least several impressive simulations (albeit not many) based in part on thinking-about techniques that have reproduced, in computer programs, the processes involved.

Observations on the Analysis of Introspective Reports

As Frijda[43] points out, highly detailed analysis of introspective reports has seldom been practiced in the recent past. Introspection gathered in a systematic way is possible to accomplish, but is frequently pursued in a somewhat ad hoc, ill-defined fashion. Frequently, the details of the introspective technique, the manner in which the introspection was solicited, and even any description of the analysis procedure is omitted.[44]

deGroot[45] identifies the principal tasks involved in protocol analysis (which can also be applied essentially to analysis of other introspective reports) in order to obtain systematic description of cognitive phenomena. These tasks are: the translation of statements into behaviors, the reconstruction of chronology, the coding of responses into subordinate processing mechanisms, and the development of an abstract statement of the sequence of events so identified. Such coding of protocols and interpretation are considered by deGroot to be of crucial importance, and the tasks of analysis are usually found to be very time-consuming. It should be noted that despite the fact that deGroot means such a series of coding tasks to apply to "thinking aloud" techniques, he himself relies on a variety of introspective techniques. Laughery and Gregg[46] carried out their very detailed protocol analysis in similar fashion to these stages of deGroot, and indeed this sequence of steps underlies the more modern attempts at "automating" protocol analysis as discussed by Waterman and Newell.[47] This attempt to develop fully automatic protocol analysis software is referred to as Protocol Analysis System I (PAS-I). Fine-grained, accurate introspective analysis is of crucial importance whenever, for instance, a machine-generated protocol, the output of a modelling process developed out of protocol analysis, is intended for comparison with human verbal output. For the program-generated protocol to provide a good representation of the human protocol, the coding rules must be carefully defined, and the hierarchy of processes carefully stated. Laughery and Gregg's and Feldman's work[48] provide isolated examples of program construction on the basis of protocol analysis, and both of these studies carried through very detailed and painstaking analyses of the verbal protocols. Frijda[49] gives an explicit description of the steps and procedures involved in the simulation of human information storage and retrieval, and shows how the procedure must be iterated in order to discover deficiencies and to sharpen up the process description.

Given the existence of a model that generates protocols, how can the actual human protocol be matched with the protocol output from the model? Frijda[49] suggests that no formal methodology has yet been developed in this area. The comparisons are essentially subjective and anecdotal. To quote Frijda, "Common sense impressions of similarity seems the only basis for judgement." However, on occasion, there is such strong correspondence of the substance of what human subjects do and what the model generates, that a "lining up" between the events of the two protocols constitutes a form of Turing's test[50] and can be utilized as an evaluation technique. An example of strong correspondence between the human and model generated outputs is contained in the recognition of sequential patterns of Simon and Kotovsky.[51] In order to avoid the problem of modelling idiosyncrasies and non-relevant parts of the human protocol, a subjective selection process by the analyst is necessary and carries the attendant risk of pruning essential ingredients of the protocol. Frijda suggests that when a model protocol is being applied to a new situation to check its adequacy then the simple statements in the new protocol should be scored in terms of the processes whose output they are intended to represent. Scanning the new verbal protocol and classifying it according to the processes or subroutines would allow comparison with the subroutine sequence of the model, and even allow one to calculate an index of fit based on the sequence of processes and their frequency of use.

Newell[52] gives a description of the general procedure for carrying out protocol analysis in terms similar to that used by deGroot, but in considerably more detail. In Newell's conception, the initial stage in protocol analysis is to code their verbal information into operationally relevant phrases. The next task, then, in the process of extracting information pertinent to the human's internal processing is to construct a Problem Space, which consists of the states of knowledge about the situation. Both the prior conditions and what final information is needed must be understood here. Also, the operations which can apply to a particular state of knowledge must also be defined. It is then necessary to develop a Problem Behavior Graph which is a representation of the possible paths along which the human can move towards a final solution. It is the final form of this diagram which represents a decision flow-diagram of interest in explaining complex behavior.

If the components of this diagram are not fine-grained enough, the essential processes might be carried within the framework of one of these components, and since the diagram would not then explicate the important processing steps, it would have to be termed *superficial*. If the analysis, on the other hand, is too fine-grained then it might not reveal the level of higher organization or control necessary to carry out the process. In such a case, Newell and Simon have used the term *disaggregated* to describe a situation. Finally, these authors used the term *epiphenomenal* to describe the situation where the obtained diagram is an artifact of the truly operative cognitive processes.

The formal procedure utilized in PAS-I is similar to that used by Newell and Simon. The verbal stream is coded into approximate linguistic form and is organized into segmental topics. Then based on the constraints of the problem, these segments are converted into semantic elements (of what is known and the operators that are relevant). These, in turn, can be transformed into operator groups, consisting of an operator, and the information it uses (its input) and the new information it produces (its output). Finally, the program (governed by the constraints of the problem space) yields the nodes of the decision flow-diagram and the linkages between nodes.

To use the detailed procedures developed by Newell and Simon and their co-workers (as they point out) entails very significant amounts of time and programming effort. It is also clear from the work of Baylor[53] in his analysis of mental imagery which required a long period of detailed work based on seven minutes of verbal protocol. On the other hand, Kleinmuntz[54] was able to develop a useable decision flow-diagram by a paper-and-pencil technique purely on the basis of listening to protocols repeatedly and coding the sequential decision rules. This latter experience shows that, although time-consuming, the analysis of verbal protocols can remain a tractable task, at least for some problems.

There continues to be attempts to extend and refine the methods of protocol analysis, such as developing of measures to supplement the protocol (e.g., eye movements, Winikoff)[55] and deGroot refers to the possibility of linguistic analysis of subject's statements. Newell and Simon and their co-workers continue to develop programming methods to assist in protocol analysis. However, as Frijda[56] points out, a very large part of the analysis work is performed prior to the state when programming can really help. This is understandable in Frijda's view since process description at a fairly high cognitive level is what is psychologically relevant.

Newell and Simon argue with this view that the significant contribution can be made in the early stages by the human interpreter. They consider that when psychology uses human analysts as measuring instruments, multiple scoring is required to develop measures of reliability. They feel that in the long run the solution to questions of validity and reliability lies in mechanizing the procedure - in getting the human out of the procedure of selecting processes and developing the understanding of the inferences. However, they indicate that such endeavors are, indeed, at an early stage.

SUMMARY.

This paper has considered the role of the decision-maker in command-and-control systems, and the possible role of decision flow-diagrams in aiding the human. Two fundamentally different approaches to research on decision-making are delineated, along with other dimensions of the decision-making process.

The role of protocol analysis in developing models of human cognitive processes is then discussed, considering some of the more recent literature on the analysis of verbal introspective reports. It is fair to say that no single standardized methodology has developed for the obtaining and analysis of verbal protocols. This may be in part because of the difficulty and time-consuming character of such analyses. However, it is clear that the classic method for studying human information processing for higher level tasks has become verbal protocol analysis. The techniques used for comparing the verbal protocols and the output of models representing the human cognitive processes involved in a problem-solving or decision-making task are still at a very rudimentary stage, where most investigators rely only on an anecdotal description of the protocol and the model output.

References

1. Kleinmuntz, B., ed., *Formal Representation of Human Judgement*, Third Symposium on Cognition, Carnegie Institute of Technology, 1967, Wiley, N.Y., 1968.

2. For example: Edwards, W.D., Dynamic decision theory and probabilistic information processing, *Human Factors*, 4, 59, 1962 and Edwards, W.D., Phillips, L.D., Hays, W.L. and Goodman, B.C., Probabilistic information processing systems: design and evaluation, *IEEE Transactions on System Science and Cybernetics*, SSC-4, 248, 1968.

3. Howell, W.C., *Some Principles for the Design of Decision Systems: A Review of Six Years of Research on a Command-Control System Simulation*, Report TR-67-136, Wright-Patterson Air Force Base, Aerospace Medical Research Laboratories, Ohio, 1967, and Edwards, W.D., Phillips, L.D., Hays, W.L. and Goodman, B.C., Probabilistic information processing systems: design and evaluation, *IEEE Transactions on System Science and Cybernetics*, SSC-4, 248, 1968.

4. Parsons, H.M., *Man-Machine Systems Experiments*, Johns Hopkins, Baltimore, 1972, 402.

5. Edwards, W.D., Dynamic decision theory and probabilistic information processing, *Human Factors*, 4, 59, 1962.

6. Raiffa, H., *Decision Analysis; Introductory Lectures on Choices Under Uncertainty*, Addison-Wesley, Reading, Mass., 1968.

7. Schlaifer, R.O., *Analysis of Decisions Under Uncertainty*, McGraw-Hill, New York, 1969.

8. Gorry, G.A., Modelling the diagnostic process, *Journal of Medical Education*, 45, 293, 1970.

9. Beishon, R.J., Problems of task description in process control, *Ergonomics*, 10, 177, 1967.

10. Bainbridge, L., Beishon, J., Hemming, J.H. and Splaine, M., A study of real-time decision making using a plant simulator, *Operational Research Quarterly, Special Conference Issue*, 19, 91, 1968.

11. Dutton, J.M., and Starbuck, W.H., Finding Charlie's run-time estimator, in *Computer Simulation of Human Behavior*, Dutton, J.M. and Starbuck, W.H., eds., Wiley, New York, 1971.

12. Kleinmuntz, B., *op cit.*

13. Newell, A. and Simon, H.A., *Human Problem Solving*, Prentice Hall, Englewood Cliffs, N.J., 1972.

14. Waterman, D.A. and Newell, A., *Protocol Analysis as a Task for Artificial Intelligence*, Paper presented at the Second International Joint Conference on Artificial Intelligence, British Computer Society, London, September 1-3, 1971.

15. Raiffa, H., *op cit.*

16. Sox, H.C., Sox, C.H. and Tompkins, R.K., The training of physician assistants: the use of a clinical algorithm system for patient care, audit of performance and education, *New England Journal of Medicine*, submitted for publication, 1973.

17. Sidorsky, R.C., Houseman, J.F. and Ferguson, D.E., *Behavioral Operational Aspects of Tactual Decision Making in AAW and ASW*, Report NAVTRADEVCEN 1329-1, General Dynamics Corporation, Electric Boat Division, New London, Conn., 1964, and Vaughan, W.S. and Mavor, A.S., Behavioral characteristics of men in the performance of some decision-making task components, *Ergonomics*, 15, 267, 1972.

18. Tuddenham, W.J., The use of logical flow charts as an aid in teaching roentgen diagnosis, *American Journal of Roentgenology*, 102, 797, 1968, and Rubel, R.A., *Decision Analysis and Medical Diagnosis and Treatment*, unpublished doctoral dissertation, Harvard School of Business Administration, 1967.

19. Gorry, G.A., Modelling the diagnostic process, *Journal of Medical Education*, 45, 293, 1970, and Horabin, I., Improving human performance, *A Research Quarterly*, 1, 28, 1972.

20. Examples are: Howard, R.A., The foundations of decision analysis, *IEEE Transactions on Systems Science and Cybernetics*, SSC-4, 211, 1968, and Raiffa, H., *Decision Analysis*, Addison-Wesley, Reading, Mass., 1968.

21. Raiffa, H., *op cit.*

22. Kaplan, R.J. and Newman, J.R., *A Study in Probabilistic Information Processing (PIP)*, Report TM 1150/000/00, System Development Corporation, Santa Monica, California, 1963.

23. Miller, L., Kaplan, R.J. and Edwards, W.D., JUDGE: A value-based tactical command system, *Organizational Behavior and Human Performance*, 2, 239, 1967.

24. Schum, D.A., Goldstein, I.L. and Southard, J.F., Research on a simulated Bayesian information-processing system, *IEEE Trans-*

actions on Human Factors in Electronics, HFE 7, 37, 1966.

Southard, J.F., Schum, D.A. and Briggs, G.E., *An Application of Bayes Theorem as a Hypothesis Selection and in a Complex Information-processing System*, Report AMRL-TDR-64-51, Wright-Patterson Air Force Base, Aerospace Medical Research Laboratories, Ohio, 1964.

Southard, J.F., Schum, D.A. and Briggs, G.E., *Subject Control Over a Bayesian Hypothesis Selection and in a Complex Information Processing System*, Report AMRL-TR-64095, Wright-Patterson Air Force Base, Aerospace Medical Research Laboratories, Ohio, 1964.

25. Slovic, P. and Lichtenstein, S., Comparison of Bayesian and regression approaches to the study of information processing in judgement, *Organizational Behavior and Human Performance*, 6, 649, 1971.

26. Newell, A. and Simon, H.A., *Human Problem Solving*, Prentice Hall, Englewood Cliffs, N.J., 1972.

27. Smith, H., *Human Decision-making in a Sequential Task*, paper delivered at Advanced Study Institute, NATO, Aberystwyth, 1973.

28. Raiffa, H., *op cit.*

29. de Groot, A.D., *Thought and Choice in Chess*, Mouton, The Hague, 1965.

30. Johnson, D.M., *The Psychology of Thought and Judgement*, Harper, New York, 1955.

31. For instance: Duncker, K., On problem solving, *Psychological Monographs*, 58, No. 5, 1945.

32. Hayes, J., Strategies in judgemental research, in *Formal Representation of Human Judgement*, Kleinmuntz, B., ed., Wiley, New York, 1968, 251.

33. For example: Feldman, J., Simulation of behavior in the binary choice experiment, *Proceedings, West Joint Computer Conference*, 19, 133, 1961.

34. Bruner, J.S., Goodnow, J.J. and Austin, G.A., *A Study of Thinking*, Wiley, New York, 1958.

35. Miller, G.A., Galanter, E. and Pribram, K.H., *Plans and the Structure of Behavior*, Henry Holt, New York, 1960.

36. Newell, A. and Simon, H.A., *Human Problem Solving*, Prentice Hall, Englewood Cliffs, N.J., 1972.

37. Reitman, W., What does it take to remember? in *Models of Human Memory*, Norman, D.A., ed., Academic Press, New York, 1970.

38. For instance: Dennett, D.C., Machine traces and protocol statements, *Behavioral Science*, 13, 155, 1968.

39. de Groot, A.D., *op cit.*

40. Miller, G.A., Galanter, E. and Pribram, K.H., *Plans and the Structure of Behavior*, Henry Holt, New York, 1960.

41. Winikoff, A.W., *Eye Movements as an Aid to Protocol Analysis of Problem Solving Behavior*, unpublished doctoral dissertation, Carnegie-Mellon University, 1967, and Feldman, J., Simulation of behavior in the binary choice experiment, *Proceedings, West Joint Computer Conference*, 19, 133, 1961.

42. Newell, A. and Simon, H.A., *Human Problem Solving*, Prentice Hall, Englewood Cliffs, N.J., 1972.

43. Frijda, N.H., Problems of computer simulation, *Behavioral Science*, 12, 59, 1967.

44. Bainbridge, L., The nature of the mental model of process control, *Proceedings of the IEEE International Meeting on Man-Machine Systems*, Cambridge, England, 1969.

45. de Groot, A.D., *op cit.*

46. Laughery, K.R. and Gregg, L.W., Simulation of human problem-solving behavior, *Psychometrika*, 27, 265, 1962.

47. Waterman, D.A. and Newell, A., *Protocol Analysis as a Task of Artificial Intelligence*, paper presented at the Second International Joint Conference on Artificial Intelligence, London, September 1-3, 1971.

48. Feldman, J., Simulation of behavior in the binary choice experiment, *Proceedings, West Joint Computer Conference*, 19, 133, 1961.

49. Frijda, N.H., Problems of computer simulation, *Behavioral Science*, 12, 59, 1967.

50. Newell, A. and Simon, H.A., *The Simulation of Human Thought*, RAND Report P 1734, Santa Monica, California, 1959, and Hayes, J., Strategies in judgemental research, in *Formal Representa-*

tion of Human Judgement, Kleinmuntz, B., ed., Wiley, New York, 1968, 251.

51. Simon, H. and Kotovsky, K., Human acquisition of concepts for sequential patterns, *Psychological Review*, 70, 534, 1963.

52. Newell, A. and Simon, H.A., *Human Problem Solving*, Prentice Hall, Englewood Cliffs, N.J., 1972.

53. Baylor, G.W., A treatise on the mind's eye, *Institute de Psychologie*, University of Montreal, July 1971.

54. Kleinmuntz, B., The processing of clinical information by man and machine, in *Formal Representation of Human Judgement*, Kleinmuntz, B., ed., Wiley, New York, 1968.

55. Winikoff, A.W., *Eye movements as an Aid to Protocol Analysis of Problem Solving Behavior*, unpublished doctoral dissertation, Carnegie-Mellon University, 1967.

56. Frijda, N.H., Problems of computer simulation, *Behavioral Science*, 12, 59, 1967.

REAL WORLD PROBLEM SOLVING WITH AND WITHOUT COMPUTERS*

Harold Sackman

Rand Corporation, Santa Monica, California, U.S.A.

The study that I'll present today is a pilot study of a National Science Foundation grant on the topic of the psychology of man-computer problem-solving. This study is the initial effort that we've made in connection with this National Science Foundation work. I can only just summarize the highlights for you so that they will be available in the proceedings. The full details have been published as a Rand report. Any of you who are interested in all the fine details, how we went about it, what precise data we had, and what statistical tests we used and so on - an empirical study - please write to me at the Rand Corporation or let me know here.

Most of the extensive problem-solving literature over the past decades has been concerned with paper and pencil testing or with artificial makeshift problems such as short puzzles, and games conducted in a single laboratory session. This study is aimed at real problems in the natural working environment of the individual or the group; it is a contribution to growing trends in real-world experimentation. Since computer usage is spreading rapidly throughout society toward its ultimate form as mass information utilities, and since we know so little about real-world man-computer behaviors, it behooves us to investigate the use of computers in solving real-world problems, to help anticipate and plan more intelligently for this massive social transition.

A leading motivation for this approach is cross-fertilization of the large existing behavioral literature on human problem-solving with the small but steadily growing findings on man-computer problem-solving. Both domains are expected to benefit from joint studies. The basic approach to accomplish this objective is system-

* *See p. 551.*

atic comparison of problem-solving behavior with (versus without) computers - throughout all stages of problem-solving. This pilot study accordingly involves a development of an applied model of man-computer problem-solving to describe why, when, and how people use computers to help solve significant problems related to their work. The model is aimed at characteristic stages of individual and group problem-solving utilizing real-world problems as the basic empirical unit of study. In line with these considerations, the four basic objectives of this exploratory study were as follows:

(1) systematically compare problem-solving with and without computers for real-world problems;

(2) determine special characteristics of man-computer problem-solving including significant trends for increasing social use;

(3) test the utility and validity of an initial 9-stage model of problem-solving stemming from behavioral systems and planning theory;

(4) to recommend directions to follow-on research.

Let us look now at the problem-solving model that we came up with.

A critical examination of the extensive behavioral literature on problem-solving and a review of the more recent, but limited, literature on man-computer problem-solving has led to an applied model. The initial model was developed in part from classical stages of generalized problem-solving. The most widely-studied and controversial paradigm has been Wallace's 4-fold analysis of problem-solving that you probably remember from your undergraduate psychology days: preparation, incubation, illumination, and verification. In this study the earlier classical stages were updated with newer inputs from systems and planning theory. The result is a revised paradigm. Here are the nine stages which are positive. I can just go over them very briefly; I can't really go into too much rationalization. This is basically an initial effort in this direction.

First, the emergence of the problem, which consists of developments leading up to definitive recognition of the problem and initial assessment of problem goals. This stage marks the initiation of the new problem and its link with related problems often neglected in classical stages.

The second stage is competing problem approaches. This refers to competing exploratory alternatives in early formulation and growth solution of the problem before any significant commitment of resources is made. This stage permits an early test of originality, in conception and approach, to distinguish between authentic problems and routine tasks. One of the difficulties you get in working in this area is, problem-solving could be, if you are very liberal in your interpretation, coextensive with human behavior broadly considered, and a problem needs a problematic aspect to it, a requirement for something novel, different, unusual in working toward a solution.

The third stage is the proposed problem plan. This refers to statement of the problem and a general approach to its solution in the framework of a proposed working plan including estimates of required time and resources. This stage highlights the integrative and evaluative functions of planning and early problem-solving.

The fourth stage is consensus and commitment, which refers to review and modification of the problem proposal with peers, management, funding agency, whatever the social context might be, leading to acceptance of a negotiated plan with commitment of resources and/or personal commitment of time and effort. This stage distinguishes between potential problems that are aborted and selected problems that are actually pursued.

The fifth stage is development of problem methodology, which refers to detailed exploration, specification, and working definition of the problem, including design and development of detailed methodology to solve the problem. This corresponds to classical stages incorporating a working or operational definition and methodology for the problem.

The sixth stage is solution-testing, which refers to the application of selective methodology with iterative tests and evaluations leading to an accepted solution. This corresponds to classical stages concerned with discovery of acceptable solutions.

The seventh stage is consolidation and refinement, which refers to final revisions, elaboration, and rationalization of accepted solutions. This corresponds to the classical verification stage.

The eighth stage is communication of results, which refers to demonstrations, written or verbal reports, to others on problem findings. This stage points up the extensive follow-on communication activities typically omitted from classical stages.

The final stage is feedback and evaluation, which refers to impact on others in the problem community, including feedback on dissemination and acceptance, or controversy generated from the study. This stage stresses social reaction and social continuity with respect to the problem-solving activity and the link to new problems - "follow-on" problems. Although the model is not viewed as a rigid chronological sequence, empirical correlations between hypothesized sequences and actual sequences are expected to be high or positive. It is an iterative model with overlapping categories marked by recurrent test and evaluation of successive stages and it does not necessarily preclude looping and parallel operations.

The model I've just described builds upon classical problem-solving stages and reflects current trends toward more directed, formal, planning in early stages of problem-solving, toward increased participatory planning prior to commitment of resources, toward wider dissemination of results and more extensive feedback on solutions from the problem community. The increasing use of computers in problem-solving and the concomitant competition for limited computer-related resources has presumably quickened these kinds of trends. Let us now examine the methodology that was worked out to provide an empirical test of these concepts.

A questionnaire was developed to meet pilot study objectives. The form required a group briefing followed by self-administration involving roughly 1/2 to 3/4 hr. for each problem. The items include background data on the respondent and his problem, open-end descriptions of problem stages and computer usage, objective items evaluating the model and the use of computers, and additional items describing problem insight and special difficulties. One group consisted of eleven UCLA graduate students from the School of Business Administration. The other included eight Rand professionals mostly in the Information Sciences and Mathematics Departments, where I come from. Each individual served as his own control by reporting a non-computer problem and a computer problem. Generally speaking the Rand respondents were more computer-sophisticated than the UCLA sample.

How was "problem" defined? Reported problems fulfilled the following general criteria, for real-world problems: (1) that they are completed problems, (2) that they are necessary and important for the individual's position or job, (3) that they are relatively recent problems (for more accurate and more complete recall), (4) that they do not represent a splinter nor fragment of a larger problem, and, (5) that the progress of the task that they reported as their problem required some originality in approach and execution; it did not constitute a routine execution of the test that was well understood. No restrictions were placed on the time that the individual spent on the problem, nor on the size of the team of which the individual was a part. The individual subject had the option of taking a short period of time or a very long period of time. These characteristics collectively constitute the operational definition of job-related real-world problems for the study.

The data permitted two basic types of tests: (1) computer versus non-computer problems matched by individual, (2) Rand versus UCLA students.

The first type of test allowed checking for similarities and differences in problem-solving patterns between computer and non-computer problems. The second permitted exploration of similarities and differences between two different samples - graduate students versus Rand professionals. Another set of measures related to a matrix of the nine problem stages I mentioned to you previously, versus 10 types of problem resources.

Let me just briefly indicate how resources were classified: there were four categories in resources referring to people:

the individual;
the team;
management;
other people outside of the above three.

There were 3 categories referring to computer services:

(1) on-line computer services;
(2) off-line computer services;
(3) computer-produced aids.

And there were 3 general categories, rounding it out to make ten altogether. These included other machine aids such as audio-visual aids, documents and references, and other manual aids (chalk, blackboard, pencil and paper etc.). The matrix of 9 stages versus 10 types of problem resources was filled in with a standardized importance-rating scale for the object problem, which permits systematic analysis of variants with differences between problem stages, problem resources, and interaction effects.

Many useful leads can be pursued in such analyses. For example, in which stage in problem solving are computers used most often and least often? Are documents or computers consistently rated as more important by subjects for solving problems involving computers? Still another, are managers rated as more important or less important in helping to solve problems involving computers as compared to problems without computers. Other measures refer to evaluations of the 9-stage model. Shift in attitudes towards computers as a consequence of the problem experienced, and the estimated impact of computers in contributing to problem insight are further measures. Methodology was eclectic. This is essentially a retrospective, case history technique. It included objective and open-end items, quantitative and qualitative analyses, intra-individual or ideographic measures versus full sample type of measures, frequency hypothesis testing and exploratory hypothesis testing, liberal use of participant opinion and evaluation. Each of these contributed checks and balances to the overall results.

Findings are grouped under three major objectives of the study:

(1) general characteristics of real-world problem-solving;
(2) special characteristics of man-computer problem-solving;
(3) critique of the stages of problem-solving.

Many of the quantitative findings were statistically significant. These were based on small sample statistics, for this particular pilot sample. However, and this needs to be emphasized, no claims are made that the pilot sample is necessarily representative of Rand professionals, nor UCLA graduate students, let alone computer users, broadly considered. That is why I keep reiterating that this is a pilot study. Results are strictly exploratory and suggestive, rather than definitive. Any credence accorded to them must rest on more comprehensive, follow-on studies.

Let's look at the first piece - portrait of real-world problem-solving, common to both computer and non-computer problems. First let me give you some notion of some base-line statistics. The average problem duration for the sample we studied was 4.1 months, ranging from a fraction of a month to 18 months. Thirty percent of the total sample were individual problems; 70% were team problems, an average of 2 1/2 people per problem, with a maximum team size at 7 people. 81% of all problems involved a written report. Another 13% required demonstrations, or an oral report. Only 6% did not involve any type of report at all. Problem content was extremely diversified, ranging from mathematics to English problems, from medical modeling to selling a company, from off-line statistical analysis to on-line simulation.

One of the key findings was that similar stages were discovered for both computer and non-computer problem-solving. We did not find any significant differences as far as the characteristic stages that people went through in solving their problems. Paired comparison tests of each individual's computer and non-computer problem revealed consistent and statistically significant similarities in problem-solving stages, proportion of time spent on each stage, and comparative utilization patterns of human documentation and machine resources. This generalized functional similarity probably holds because individuals direct and optimize a solution process to satisfy human needs and expectations at all problem stages, whether or not computers are used.

The second finding is the primacy of the individual in problem-solving, regardless of the type of problem. The individual was consistently rated on a comparative rating scale I mentioned earlier as *the* most important resource in getting work done, and in meeting problem objectives throughout virtually all stages of problem-solving, for group as well as individual problems, and for computer as well as non-computer problems. With reference to a standardized scale, the individual was rated as more important than the team, management, others, and more important than documentation and machine resources. Reported major problem insight was primarily an individual phenomenon. In these data the individual emerges as the prime mover in problem-solving.

Another finding refers to the relatively stable problem-solving roles that seemed to emerge. The team was rated second in importance to the individual. Management was third, and other people were last, in overall average importance ratings. The team received moderately important ratings throughout the entire problem-solving cycle. Management was rated relatively higher for initiating problems, committing resources and in evaluating final results. "Others" received highest ratings for final evaluation. The division of problem-solving labor seems to involve basic work and creativity from individual effort; interactive support, feedback, and specialization of roles, stemming from the team; authorization and resources derived from management; and evaluation coming from both management and a wider problem community.

Another key finding refers to the systematic effects correlated with the size of the problem and the team that worked on it, and the amount of resources associated with the problem.

A pivotal finding emerged from substantive changes in the structure and function of real-world problem-solving, along the continuum from smaller to larger problems. The smaller problem involved shorter periods of time, the small teams or just single individuals, incorporating limited resources, often appearing as a fragment or sub-problem of an imbedding problem. The larger problem tends to involve more people, bigger budgets, more expensive resources, including computer resources, and more typically, a self-contained project. The pilot sample did not test the limits in either direction. Results show the above grouping of variables to be moderately positively correlated with each other. Larger problems tended to exhibit more differentiated problem stages. Problem stages are a function of problem size, a more formal operating structure, and a more specialized division of labor. In addition, larger problems reflected greater involvement from management, more importance according to the team, more diversified use of resources, and higher proportions of problem insight attributed to group discussions. Other implications of this continuum are discussed later under findings for computer problems and problem stages.

Another finding refers to the interrelationships demonstrated between documentation and creativity. External documentation was consistently found in all types of problems to provide the main source of creative suggestions in early problem-solving. In early planning stages documentation was typically rated second only to individuals, more important than the team. This suggests a strong parallel with scientific endeavor where the continuity of collective findings, the state-of-the-art, and the associated literature search sets the tone and pattern for subsequent scientific inquiry. The completed problem, in turn, seems to provide fresh documented input for new problem mutations, not only in the scientific domain, but also in this particular real-world problem community.

Another finding was something called the "push-pull" theory of insight. The varieties of insight and conditions associated with insight were quite diversified but reflected some notable trends. One mentioned earlier is that major problem insight tended to be individualized. A second is that insight was more often associated with problem method and organization of resources than with the inherent nature or structure of the problem. (These are recorded instances of insight on the part of the subjects.) The third set of observations suggests the push-pull theory of insight. Most subjects reported insight occurred either during active deliberate exploration and testing of data, which we call "pushing" the problem, or as a reaction to a problem anomaly; for example, errors, difficulties, that had to be resolved, which we call being "pulled" by the problem. Pushing is planned and expected; pulling is unplanned, disruptive, and unexpected. These characteristics may provide an experimental basis for real-time, behavioral testing of errors and mistakes. A very preliminary theory is posited for optimal balance between pushing the problem and being pulled by it to achieve insight.

Finally, reported recommendations for improved problem-solving. Participants cited variations on the theme of poor or inadequate planning as the most frequent shortcoming in real-world problem-solving. Human problems and data difficulties were second and third in order of frequency for all problems. More effective planning was the major reported recommendation to improve problem-solving, followed by better selection of individuals and teams. That completes the common findings for both types of problem.

Now let's look at the findings for man-computer problem-solving for computer-related problems. The findings summarized should be interpreted as distinguishing features of computer-aided problem-solving, over and above the common features of real-world problem-solving previously discussed. The composite of both summaries constitutes the portrait of man-computer problem-solving derived from this study.

The first finding we cite is a new class of "unthinkable" problems. Computer-aided problems generally represent "unthinkable, manual problems in quantitative and qualitative respects". Quantitatively, they typically incorporate computational loads that can only be handled by computer assistance, such as large scale statistical processing. Qualitatively, and this is in large part a function of user experience, they involve methodology, possible only with computers, such as complex simulation, on-line videographics, and heuristic modeling. Taken together, computer problems effectively constitute either a quantitative mutation of manual problems, or new qualitative off-shoots stemming from the capabilities and properties of computers, software, and computer-accessible files, which in turn are continually undergoing rapid technological change.

A second finding refers to the role of computers as information workhorses. What do we mean by that? Most computer problems encountered in this study involved relatively unimaginative brute force computation - the quantitative variation of computer problems just mentioned. The more advanced applications were fewer in number and were performed by more experienced and more sophisticated computer users.

The third finding is referred to as "bottom-up" man-computer problem-solving. What does that mean? Results consistently reveal a "bottom-up" syndrome of computer usage in problem-solving. This refers to the virtually exclusive use of computers in later, rather than earlier, stages of problem-solving. This is quite striking. Computers were rarely used in early planning stages, regardless of users' sophistication. The infiltration pattern of computers in real-world problem-solving is from the bottom, up, appearing primarily in later stages of problems after most key decisions have been made. The predominance of brute force computation, coupled with the bottom-up syndrome, supports the characterization of computers as information workhorses for most current applications.

A fourth finding refers to computerization and problem formalization. All other things equal, when matched by individuals as in this study, computer problems tend to be larger, longer, more expensive, with bigger teams, than non-computer problems. As mentioned earlier, larger problems manifest more differentiated stages of problem-solving and more formal working structure. Since real-world computer problems tend to be larger problems, they also tend to be more formalized in inception, planning, implementation, and termination.

The fifth finding is interesting. We call it the "management avoidance" syndrome. Since computer problems tend to be larger and more formal, in the structure of the present investigation, management would normally be expected to play a more important role. The pilot data showed the opposite - management was rated lower in importance for computer problems and in practically all stages of problem-solving, as compared to non-computer problems. This finding is provisionally interpreted as a "management avoidance" syndrome, stemming ostensibly from the extra added cost in time and effort, and the risk of technical embarrassment for managers to get on top of more difficult and complex computer problems. The pilot data are meagre and need considerable follow-on work to substantiate this avoidance interpretation.

An interesting finding was the proselytizing impact of computers. The majority of computer users reported an increase in positive attitudes toward computers as a result of their problem experience. We asked them, "was your attitude more favorable, less favorable, or the same, after you finished your computer experience with this particular problem?" Only one out of nineteen participants reported more negative attitudes toward computers after his problem was completed. In spite of the usual complaints over poor service, program bugs, hardware problems, the great majority found computers indispensable for successful, timely, and cost-effective solutions, and vital for problem insight. Several spontaneously indicated that they enjoyed working with computers. Sheer increase in computer experience seems, according to these data, to contribute to winning friends and influencing people to join the computer bandwagon. This result, however, may be heavily biased by the positive computer orientation of the pilot participants. Tests of validity generalization are needed in other disciplines to establish the strength and direction of this effect, if it does in fact exist.

The next finding refers to computers and problem insight. Most participants rated computers as *very* important in contributing to major problem insight. Insight was primarily individualized as with non-computer problems. However, it occurred more often with the computer-related activities, for example, at the terminal or poring over computer printouts, than it did at the individual's desk. Most reported insight for non-computer problems occurred at the individual's desk. Exploratory computer runs, such as sensitivity analyses and unexpected computer-related anomalies (weird printouts for example), seemed to enhance the previously-mentioned push-pull process for generating insight in the problem situation.

I am sure most of you are familiar with the next finding, which we call procrustean computer problem-solving. In some cases, virtually total dependence of the problem-solver on the computer system and its software, led to marginal or unsatisfactory problem solutions, with essentially no way out. Rather than try to move the problem over to a different computer, or change of software, or both, to get at better solutions, many did not have the time, budget, inclination, or skills to retrofit computer services to the problem. The result was that the problem had to be fitted to the computer, sometimes with devastating procrustean consequences. Boehm has noted this procrustean effect with aerospace engineers, where management deliberately made computer systems less accessible to discourage stereotyped computer solutions and stimulate more creative thinking. A hypothesized long-term psychological effect is computer tunnel vision where human solution space is progressively compacted into the block universe of available software.

There were some recommendations for improvement in man-computer problem-solving. These referred primarily to later stages in problem-solving and were mostly of the garden variety type - more reliable systems, easier languages, cheaper software, more forgiving software, more powerful compilers, more extensive user training, and more accessible consultants and computer specialists. There were virtually no recommendations for improved problem-solving in early planning stages, except for the "burnt child" reaction - that pilot runs of software with preliminary problem data be instituted as early as possible before plans are cast in concrete. This early warning policy applies equally well to non-computer problems, stemming from established practice in scientific method and perhaps to show business: rehearse first.

Let's look now at the third area of our results - stages of problem-solving. What findings did we get here? Very generally we got a tentative validation with this small sample of the initial model. The key validation procedure in this study was simply evaluative ratings on how well the reported problems fitted into the proposed 9-stage model of problem-solving. Ratings of good or better occurred for 62% of all problems, with an associated standard error of plus or minus 11% for the pilot sample. Your 95% confidence belt would be 22% on either side of that so you can see that a small sample doesn't really tell you very much. The median number of recommended changes to the model was one, typically to decrease the number of stages. Thus, the pilot results tentatively validate the proposed model for real-world problem-solving. Qualifications though, are required. The first qualification is with respect to problem stages as a function of problem size. The earlier summary of systematic effects of problem size exerted a major impact on stages of problem-solving. Very small problems exhibited fewest recorded stages. Medium-sized problems fitted well into the proposed nine-step model and very large problems not recorded in this study would probably reflect more differentiated formal structure than that proposed by this model. One problem shoe apparently can't be stretched to fit neatly into all problems. The pilot evidence indicated that the proposed scheme fitted computer problems significantly better than non-computer problems and generally fitted the larger problems better than the smallest problems in terms of better ratings with fewer recommended changes. Almost 80% of all computer problems were rated as good or better against the proposed model. Thus, the results for man-computer problem-solving tend to be encouraging.

Now what are some recommendations that come out from these results? The fundamental initial congruence between non-computer and computer aided problem-solving demonstrated in this study indicates the value of free cross-fertilization between the small but growing scientific literature on man-computer problem-solving and the established behavioral literature on problem-solving and its allied areas. An example of overlap is the use of computers to enhance human exploratory behaviors and to expose key problem errors, both of which are vital precursers for significant problem insight, for example, the "push-pull" concept of insight previously noted. Behavioral scientists are needed to become more deeply involved in man-computer problem-solving. Those versed in Computer and Information Sciences should become more familiar with related methods and findings of Behavioral Science. The marriage is long overdue. Results of this study suggest several practical leads for more effective social policies and problem-solving strategies. The early-planning stages in real-world problem-solving seem to be haphazard and almost chaotic. Institutionalization of more formal and more objective planning procedures was indicated by the respondents. The requirement for a formal written plan and the institution of more comprehensive and objective reviewing procedures would seem to be a key step in the right direction. A more refined requirement would incorporate systematic pilot studies in early plans particularly for proposed use of computer techniques and services. Early computer runs on representative or even benchmark problem tasks would provide sorely-needed early warning on "swing" decisions to use or not to use computers and on difficulties likely to be encountered after a full-scale commitment to a particular approach is made.

The entire area of initial planning and early warning pilot testing is still largely shrouded in a veil of mystery. The management avoidance syndrome with computer problems, if valid, hints at the possibilities of an altered role for management, anticipated in part by Toffler's concepts of "adhocracy" in Future Shock. With increasing complexity of problem-solving tools and techniques, perhaps it is too much to ask managers to keep on top of accelerated technological advances to evaluate the technical effectiveness of problem solutions. A new management role with a new division of evaluative labor might be required where a specially designed panel or group of technical experts, or perhaps informed peers, provide balanced and independent evaluation of problem-solving, with the manager acting as a neutral catalyst to organize and synthesize such expert evaluations into the larger institutional setting. Such procedures invite new difficulties, though, and are not at all free from abuse. In any case it was quite clear that the last evaluation stage in problem-solving was typically a casual affair with virtually no balanced systematic feedback for any of the principals. The suggested shift in roles is from relatively autocratic, management-only, evaluation toward participatory evaluation of problem-solving effectiveness with pluralistic checks and balances.

The above leads are only suggestive and tentative. Their purpose is to hint at the potential social value of more comprehensive investigation of real-world problem-solving. Follow-on studies are recommended featuring improved versions of problem-solving models, more rigorous operational definitions of problems and more systematic and representative samples of real-world problems and problem-solvers. Push-pull theory of insight, management avoidance syndrome, and computer tunnel vision effect merit further investigation. Longitudinal in vivo, as opposed to retrospective, studies of real-world man-computer problem-solving are especially needed for independent behavioral testing and as an empirical corrective to retrospective studies. Results of such real-world studies are likely to lead to more effective modes of problem-solving and to significant advances in the humanistic design and development of computer systems.

*The presentation included in these proceedings is the transcription of the tape recording of the talk given in the Institute. The technical report on which it is based, and in which the data and findings are more inclusively recorded are contained in Rand published Technical Reports.

COMPUTER-BASED COMMUNICATIONS MEDIA AND CITIZEN PARTICIPATION

Valarie C. Lamont

University of Illinois, Urbana, Illinois, U.S.A.

INTRODUCTION

Since the early 1960's a movement has been growing in the United States calling for increased citizen involvement in planning and decision-making at the local level. As a result of these demands, old forms of participation have been infused with new life or discarded where no longer relevant. At the same time, new forms of participation are being explored. Special interest groups are continuously being created; radio talk shows and television documentaries have increased; experiments with two-way cable television and telephone-computer couplings are underway.

This presentation describes research being conducted at the University of Illinois, Urbana-Champaign, utilizing computer-based communications media for purposes of citizen participation in social planning and decision-making.

TECHNOLOGY

The technology which I refer to throughout this discussion is the PLATO IV teaching computer system. The PLATO IV system consists of a Control Data Corporation 6400 computer and, at the present time, over 300 terminals. Over the next 2-3 years, this computer will access 4000 terminals, all of which can operate simultaneously. When the system is completed, it will have more simultaneously-operating terminals than any other multiple-access computer system.

A terminal consists of a typewriter-like keyboard by which the user sends messages to the computer and a television screen which displays computer-generated graphic information and computer-selected photographic slides to the user.

PLATO IV terminals are now distributed throughout the state of Illinois and at remote locations in other states, including California, Colorado, Iowa, Indiana, New Jersey, Maryland, Massachusetts, Canada, and Paris, France.

The projected plan is that Illinois will be the central location for the surrounding midwestern states, while duplicate PLATO systems will be developed for other geographic regions. The second PLATO system will be built for Lowry Air Force Base in Colorado. California, Arizona, and Florida will probably be the sites for the next systems. Internationally, Heidelberg, Germany is presently discussing the possibility of acquiring between 500 and 800 PLATO IV terminals.

The PLATO system has been developed primarily as an educational device. To briefly review its history, initial work on this system began over 13 years ago when a group of engineers at the University of Illinois posed the question: How can computers be used in education? This question led to the formation of a University committee to discuss this coupling and finally to the establishing of what is now referred to as the PLATO Project. The primary concerns of this project over the years have been (1) to demonstrate the educational effectiveness of computers in the classroom, and (2) to design a system which would be economically competitive with normal classroom instruction.

The technology itself, the computer system, has gone through 4 "generations". PLATO I was simply a single station with a keyset and a cathode-ray tube (CRT) for simultaneous display of computer-generated characters and photographic slides. PLATO II consisted of the same equipment, but with 2 stations operating simultaneously. PLATO III marked a major advance in the project: the system consisted of over 70 PLATO III CRT terminals, 20 of which could operate simultaneously from a CDC 1604 computer. During the time of PLATO III, advances were being made in the technology itself, in the programming language, and in the educational programs being developed on the system. The extensive use of PLATO III in the classroom at the elementary, junior high, high school, and college levels demonstrated that the computer was educationally effective in the classroom. The second major concern, economic competitiveness with normal classroom instruction, was overcome with PLATO IV. The distinguishing feature of the PLATO IV system is the plasma display panel, which has replaced the cathode ray tube. Basically, the plasma display panel consists of 2 pieces of glass with gaseous elements sealed in between the panels. The elements are "zapped" with an electric charge, and the panel will stay lit until turned off. This implies that the terminals have inherent memory. Terminals are connected to the computer by low grade telephone wires. It is expected that the display panel will eventually reduce costs of computer-aided instruction to between \$.25 and \$.50 per student contact hour.

RESEARCH IN CITIZEN PARTICIPATION

Combining the interactive capabilities of this system and its widespread distribution as an educational device, we have come to regard the PLATO IV system as a potential new mass communications technology with feedback capabilities. As such, the main focus of our research with the PLATO system has been to investigate its potential as a communications medium in the community to involve citizens in social planning and decision-making. To this end, our main task has been to develop a variety of programs which hopefully demonstrate the different ways PLATO or PLATO-like systems can address general community or specific interest group needs. The following programs describe the range of uses we see for the PLATO system.

Boneyard Creek

This program demonstrates how the PLATO system can be used to discuss local issues. The topic presented concerns the environment, specifically, the future of a stream which runs through Champaign-Urbana, Illinois. The program is divided into the following major segments:

(1) History.

(2) Past and present policies.

(3) Alternatives to present policy.

(4) What interest groups are doing about the issue.

(5) Questions to determine which alternatives are preferred by the participants.

(6) Presentation of the preferences of others in the community.

(7) Questionnaire to determine participant reaction to the program.

The Boneyard Creek program was the first "experiment" conducted to demonstrate the feasibility of using the teaching computer as a medium for involving people in community planning. The program was presented to a non-random sample of the population, including government leaders, civic leaders, and interested citizens. While the results of this experiment have been written up in another paper which is available, I would like to briefly list what the data from this experiment suggests:

(1) People from the community were willing to come to work through the program.

(2) Participants found the medium useful for presenting concise and relevant information.

(3) Presentation of information in this way may be particularly useful for those who have not yet formed an opinion on an issue.

(4) Participants found particular advantage in being able to make comments and respond to questions.

(5) Participants indicated that they would like to see more issues presented in this way.

After the development of this program, our interests turned to different program formats and similar issue programs were not developed. However, recently we have been working more extensively with interest groups in the community, and, as a result, we are now developing a set of programs on the mental health issue in Champaign-Urbana. These programs will reflect particular needs of the people involved in mental health - specifically, community education on mental health and training para-professionals to become familiar with what services are available for people with mental health problems.

Delphi Exploration

"Delphi" is a computer-based exploration of alternative futures. Participants have the opportunity to "create" their ideal future by investing positively or negatively in issues and developments. They may then see secondary effects which their investments have on other issues and developments. The following format describes the program:

(1) Decision-making framework

(a) The explorer selects a development, whose probability of occurring by the year 2000 he would like to change, from a list of 35 social and technological developments. Examples of developments are "air cushion vehicle", "genetic manipulation", "cashless society".

(b) The explorer makes an "investment" in the development which is an indication of his estimate of the development's desirability.

(c) The computer calculates the secondary effects of the investment according to estimates of secondary effects made by the programmer and a panel of "experts".

(d) For each development the old probability (before investment) and the new probability (after investment) are displayed.

(e) The above sequence of operations is then repeated as the explorer selects another development to invest in.

(2) Information units
(a) A short descriptive statement.
(b) Background paragraph.
(c) An associated probability of occurrence by the year 2000.

(3) Mathematical model
(a) Matrix of interactions or secondary effects.
(b) Algorithms which change probabilities.

Data collected from the "Delphi Exploration" indicated that this program was particularly useful for "perspective stretching". Participants not only gained experience in thinking about the future, but they became acquainted with thinking in terms of interrelationship among items.

Discuss

This is a conferencing program similar to Murray Turoff's "Discussion" program. The program allows participants to communicate between computer terminals by storing messages in the computer memory which are then available to all other users. The program proceeds in the following way:

(1) A list of participants, their locations and time of last entry into the program are shown.

(2) The participant selects one of the following options:
(a) Review messages in sequential order.
(b) Review messages by keywords.
(c) Delete an item.
(d) Add an item.

(3) Unstructured conferencing can take place over a period of weeks.

The format used in this program can be used first to facilitate dialogue among people in the same community. That is, people would have an opportunity to discuss with others their ideas, fears and concerns about their community and to find out other people's views. Secondly, this program can be used by people who are geographically separated to exchange information on their community organizations, strategies, comments on citizen participation efforts, etc.

The program is presently being used for informal discussion among people in Champaign-Urbana; Evanston, Illinois; Washington, D.C.; and Indiana University. We are currently planning to use the

"Discuss" program for conferencing at the city, state and national levels:

(1) At the city level, we would like to invite citizens to use the PLATO IV terminals scattered around Champaign-Urbana for an informal discussion on the future of Champaign-Urbana.

(2) At the state level, we would like to have people at the remote sites - Chicago, Springfield, Normal and Evanston - participate in a discussion on plans for the Bicentennial Celebration in the State of Illinois.

(3) At the national level, we would like to have people at the remote sites - Indiana, Iowa, Canada, Washington, D.C., New Jersey, Massachusetts, Maryland, California, Colorado - use the PLATO IV terminals for an informal discussion about their research and citizen participation experiments.

Cities

This is an information retrieval program on the PLATO computer system which provides data on the Champaign-Urbana area. The strategy employed in this program is the branching sequence. Participants obtain information about their community by making a series of selections beginning at the general and going toward the specific. The program is divided into the following major areas:

(1) Organizations.
(a) Government.
(b) Business.
(c) Labor.
(d) Civic Groups.

(2) People.

(3) Maps.

(4) Statistics.

This program can provide basic information which interest groups should find useful in developing policy and strategy. In addition, the program can also be used by citizens to increase their "community fluency".

Perceptions

This is a questionnaire program on a variety of issues relevant to Champaign-Urbana, Illinois. Participants use the program in the following way:

(1) The participant selects an issue: drugs, finances, education, religion, politics, race relations, University of Illinois - Community relations.

(2) The participant is asked a series of questions on that issue.

(3) The participant then views aggregate responses of others who have worked through the program.

(4) The participant also views the results of a sample survey using the same questions.

This program is useful to citizens who would like to have some idea of how others in the community feel about certain issues, as well as a priority list of problem areas. This type of program could serve as a catalyst for the formation of interest groups as people begin to realize that others are concerned about the same issue. Equally important, people have the opportunity to compare their perceptions of reality with others in the community and perhaps with a national sample. This may lead to rethinking one's position on an issue or seeking out more information on an issue.

In addition to demonstrating alternative uses of the PLATO system, these programs also demonstrate different types of information exchanges which may be facilitated by computer-based communications media. For example, consider the following simplification of a community dialogue model developed at the Charles F. Kettering Foundation:

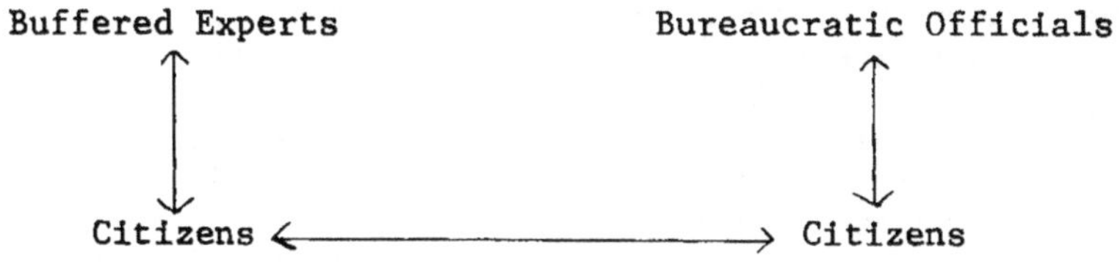

Community Dialogue Model

Diagram 1.

The 3 communication links depicted are between (1) citizens and experts, (2) citizens and officials, and (3) citizens and citizens. The PLATO program just described could be used not only to provide information generated by experts and officials to the populace, but also to provide information generated by the populace to experts and officials. Hopefully, one consequence of this information exchange would be decisions which better reflect needs and demands of the citizenry. In addition to this type of information exchange, computer-based communications media could also be used for dialoging between citizens. While this particular information exchange is not often strongly emphasized, in today's society there seem to be indications that people are becoming more interested in finding out how others in the community think about issues, exchanging information with others on issues, and selecting others with similar interests to form interest groups on issues. The conferencing programs on PLATO seem to be particularly suited to facilitate this type of communication.

RESEARCH IMPLICATIONS

In summary, the programs described suggest possible community uses of PLATO in the areas of community education and discussion of issues, information retrieval, and conferencing. However, PLATO-like systems will probably not have a significant impact on general community decision-making processes for another 15-20 years. The more immediate impact will probably be in the area of community education for specific interest groups.

Thus, while we cannot predict specific societal changes as a result of the introduction of this technology into the community, we can suggest the following areas which may be affected by PLATO-like systems:

(1) Increased citizen involvement in social planning and decision-making. In addition to cable television, video cassettes and other computer systems, computer-based communications media will provide an alternative mode for entry into the political system. Hopefully, the technologies available will lead to more interest in using the technologies for purposes of citizen participation. Further research and experimentation should focus on making these technologies accessible and easy to use by the community.

(2) If citizens have more opportunities for input into the political system, it is not unreasonable to assume that certain types of decisions, at the local level, will probably be made exclusively by citizens. Administrators today complain not only about the number of decisions they must make but also about the types of decisions they must execute. Sharing some of this burden with the citizenry will leave more time for decision-makers to engage more fully in medium- and long-range planning activities.

(3) Hopefully, with more citizen input into the decision-making process, the quality of decisions will be improved. Of course, "quality" is a nebulous concept and criteria must be specified in

order to be able to measure quality. Generally speaking, however, people tend to be satisfied simply from knowing that they participated or had some input into decisions which affect them. By having more input from the community, experts and decision-makers will hopefully formulate policies which more accurately reflect the needs and demands of the people they serve.

Chapter 5

INFORMATION TECHNOLOGY AND ITS IMPACT ON MAN AND SOCIETY

Information Technology and its Impact on Man and Society

Editorial Summary

The rapid development of technology, in particular communication and information technology, with its significant impact on man and society is reflected in the concern of information scientists expressed in much of the literature, directly or indirectly. Some papers concentrate upon the often undesirable side effects of technology as the major topic of discussion.

A readily-perceived acknowledgment of the impact of information technology on man and society is technology's direct influence on the "quality of life". Man is genuinely concerned that his mode of living - the nature, quality and quantity of resources available to him - is gradually eroding. Kochen argues that information systems can decisively influence the quality of life by providing man with a greater awareness of the options open to him when confronted by events and problems. Information systems allow the manipulation of data concerning man's world, which results in different representations of the world, and consequently provide the opportunity to revise and use the data accordingly. These representations enable man to formulate long-range plans under varying conditions. Information systems as such enable man to minimize effects, shape conditions, and in general, increase the exercise of wisdom.

The "quality of life" that Kochen talks about is particularly important to those countries that we consider "under-developed". Maria Lima discusses the development of man as it appears through the eyes of the anthropologist to bring into focus the need to emphasize the human condition in Information Science. Lima alludes to the importance of communications in human development and hints that the most crucial question facing information scientists is the matter of information transfer. The collective presentations of Lima and of Drs. Pablo Lima and Maria Rosa De Alvin Costa provide a challenging, human view of the important considerations facing information scientists regarding social issues. Pablo Lima makes an eloquent plea against the suppression of the human spirit through the influence of materialism on the culture. Unfortunately, Pablo Lima does not extend the argument sufficiently for us to realize productive measures that can be generated to balance the influences that he details with an air of pessimism. Dr. Alvin-Costa, on the other hand, draws on the educational problems of children in Portugal and implies that information science could do something about them. The problems with the education of children seem to be inherited because of past indifference of the population to the importance of education in the development of human life in Portugal. Alvin-Costa does not provide us with an awareness of how prevailing information technology can ameliorate such conditions. Her concerns have been chiefly towards establishing a political base. Only the beginning of the paper by Pablo Lima has been included, and the paper by Alvin-Costa has not been included in these proceedings, mainly because the editors thought that much of the content would receive better emphasis through summarization.

Salancik, a psychologist, deals with the behavior of people in a state of uncertainty and speculates on the possible impact of information technology for social control. While refusing to make value judgments as to whether this impact is good or bad, he is concerned that information scientists should be aware of the influence that such systems may have on social behavior and the value of life. He regards such systems as possibly increasing public exposure and reinforcing subsequent pressure to conform.

Klempner also has views on the subject. Information technology, effectively used by government, provides additional power to those influencing the social, economic and military systems. The relationship between information technology and democratic governance is of particular interest to Klempner. He raises questions of a fundamental nature, aimed at showing the potential danger of having effective information technologies misused by technocrats.

Charles Meadow and Robert Wood present a critical analysis of the failings of information systems, with regard to their time value for human life and society. They show by using some examples what has been, and often still is, the origin of new information systems: new technology, rather than need. In addition, technology-oriented systems designers tend to give too little heed to the end user and the operator. Also, at least in the past, they had very little concern for the global implications of their creations because they ignored the possible interactions with the environment in which the system, once introduced, has to work. Meadow and Wood see some hope in more recent trends, which make designers and decision-makers more aware of future effects on man and society, in part because of the development of more systematic methods of technology assessment.

On a philosophical base, De Grolier attempts to provide an understanding of the complexity inherent in the information-society interrelationship. He asks the question, how can we best discover the impact of information technology on individuals and institutions? Based on examples and an analysis of conventional methods applied to his investigation (Delphi method, case studies, cross-cultural comparative studies, and historical studies), he answers that there is no best method, but that much can be gained from interdisciplinary studies. The body of his paper presents an overview of the interactions between information technologies and social developments. He uses this to develop six hypotheses, generalized observations related to this topic. He claims that modern information technology has the potential means for reversing the effect of mass media to return to interpersonal, communication with direct feedback. He recognizes a potential for equality rather than polarizing differences in power, culture, and economic strength by proper application of information technologies. He points to the paradox of simultaneously having too much and not enough information: mental pollution by modern society paired with an inability to obtain the right information when it is needed. His treatise concludes with his concern for professional ethics and the role of the information scientist in the decision-making process, an area in which information science has had little, if any, impact.

Finally, what is reflected in these papers, which are addressed directly to the impact of information technology on man, can be sensed as a biasing factor in many other aspects in the Institute. In particular, the critical analysis of information science and its present course, as discussed by Rosenberg and Nevitt, falls into this category.

INFORMATION AND THE QUALITY OF LIFE

Manfred Kochen

University of Michigan, Ann Arbor, Michigan, U.S.A.

Maintaining or improving quality of life is a paramount issue today. Investigators of urban problems are concerned about deteriorating quality of life in inner cities. While technological advances have, in the past, contributed to the quality of life, they may in the future detract rather than contribute. The information technologies are no exception. But need this be so?

Many public policies are - or should be - aimed at preserving or increasing the quality of life for those who are to be governed by these policies. Information science and technology can play at least two roles in shaping enlightened public policies. First, it can contribute ideas as ingredients for such policies. Secondly, its techniques can be applied to help shape and enforce appropriate public policies.

We plan, in this paper, to clarify those two roles. We hope to show what contributions towards the quality of life have, can, and should come from the information sciences. To start, we need a better explication of the "quality of life" concept. This includes some appreciation of efforts to find an index.

TOWARDS ASSESSING THE QUALITY OF LIFE

Comparison of Some Proposed Measures

In Table 1 we summarize and compare four sets[1] of general indicators for the quality of life that have been proposed.

The first row covers variables such as mean life span, infant mortality, incidence of morbidity of diseases. The general heading

Table 1

Some Sets of Indicators for the Quality of Life

I. Lasswell & Kaplan	II. Maslow	III. Foa	IV. Chapin
Welfare	Survival	Services	Safety, health
Wealth	Security	Money	Economy
Affection	Belongingness	Love	
Enlightenment		Information	
Skill			
Rectitude			
Power	Materialism (prestige)	Goods	
Respect (Deference)		Status	
	Self-Actualization		
			Convenience
			Amenities

"welfare" seems a bit broader than "survival" but narrower than "services". It certainly includes safety and health. The 8 indicators of Lasswell and Kaplan were used to compare the quality of life in different countries. On a national scale, a country's concern with survival might be measured by what fraction of its GNP it spends on defense and health, though this would overlap somewhat with security. Wealth on a national scale is measured by GNP or GNPC (per capita). On an individual scale it is total economic worth or net assets of a person. This relates to financial rather than physical security. Psychological security is related to both of these. Affection and Belongingness are really different psychological dimensions. But the need to give and to receive affection, like the need to include or be included, or the need to dominate or be dominated are personality variables. They fit together in that sense. If a person or a country rates high in each of these indicators, he or it is very likely to enjoy a high quality of life. If he rates very low on at least one of them, he quite probably might wish to improve the quality of his life, if he were asked in such terms. Nonetheless, all these indicators together do not seem to capture the essence of the "quality of life" concept. It may well be that a universal conceptualization or a single index is unattainable.

Desiderata for an Index

Consider the following set of desirable properties we propose for a good index of the quality of life. Let IQL(P,t) be such an index for a living object P at time t (years) after birth.

(1) IQL(P,t) is a partial ordering of the states in which P can be. By a state of P we mean the values of: (a) environmental variables that affect P's behavior and (b) internal variables that determine how P "feels", behaves and what P may be conscious of. Thus, P may be in a state of dreaming a pleasant dream. Some states are preferred by P to other states and P is indifferent among certain states. That is how IQL(P,t) is interpreted. If at time t, P prefers state s to state s', then we can write (s IQL(P,t) s'), reading IQL(P,t) like the symbol $\succeq$.

(2) Let P be a set of living objects, P_1, P_2, P_3,... . Suppose all the $P_i \in P$ are characterized by the same set of states. Denote by P* the largest subset of P such that $IQL(P_i,t) = IQL(P_j,t)$ for all P_i, $P_j \in P$ and all t. If these stringent requirements are relaxed, P* can be more inclusive. Given two orderings IQL and IQL', the one which has the above property for the larger group P* is preferred. This is essentially an assumption of homogeneity to deal with the problem of aggregating individual's preferences. This is, of course, a large and difficult unsolved problem in all the social sciences.

(3) If states s and s' differ only in that s provides P* with more options among equally valuable resources needed to recognize and cope with problems, then (s IQL(P*,t) s'). For example, if P* is a community faced with problems of protection against burglary, and s is a state corresponding to a greater variety of burglary prevention-devices available in that community than state s', then IQL is such that s is preferred to s'.

(4) If states s and s' differ only in that s has more resources available to P that P can either expend directly to solve problems or exchange for what he needs, then (s IQL(P,t) s'). Clearly, if P is in a state s of greater wealth - available cash - IQL should be such as to reflect his preference of s over s'.

(5) If s and s' differ only in that s is a state in which P is more *aware* of the gap between what is and what could be, and if he is also more aware that he can fill the gap and how, then (s IQL (P,t) s').

(6) If s and s' differ only in that s enables P to exert more influence, power and thus control his fate, then (s IQL(P,t) s').

(7) If s and s' differ only in that state s P has a greater probability of survival, then (s IQL(P,t) s').

The role of time t in IQL(P,t) is best summed up in the following statement (paraphrased) from Lotka.[2] The goal of evolution is perfect adjustment of feelings and actions towards increasing *consciousness* of an essential unity with the universe. This guarantees survival at higher quality. The living object whose will is so adjusted is Fortune's favorite. His will is Nature's law.

Essence of the Concept

We will not discuss the critiques and implications of these assumptions, because there is more to the concept of "quality of life" than can be captured by an operationally defined index or a mathematical object (partial ordering) with the properties specified above. We would like to apply this notion to all living units in nature. These units include plants, lower animals and men. Only men, however, have discovered the "quality of life" concept and become concerned with it. For the first time in evolutionary history, a part of living nature has become aware of being able to shape its environment so as to affect the quality of its life and that of future generations of its species.

If a useful single index to assess the quality of life can be found, then this may be our awareness of our options, actual and potential, present and future, and actions based on this awareness. We view such awareness as a *representation* by P (the actor) of the world and himself which refers to P's ability to revise and use the representation itself. *The quality of this representation is reflected in P's use of it to make long-range plans as if P had infinite life at the same time that P uses it to act as if life would terminate immediately.*

Uses and Abuses

If we had a good measure for IQL(P,t), what should it be used for? An obvious answer is to change such conditions as would increase it.(Here and in the rest of this section we speak of the index as if it were a number along a scale rather than a 2-place predicate. Under certain conditions, the 2-place predicate can be reduced to a single ordinal scale.) Suppose the designer or planner of a new town understood and could control some of the conditions causing low IQL for certain, selected individuals, e.g., fostering crime, insecurity, unemployment, affecting particular people. First he may try to prevent such conditions from arising. This requires considerable understanding and wisdom. But the cost per unit of increase in IQL is relatively low. Secondly, he may eliminate or reduce the unfavorable conditions if they have already arisen. The cost of this is a bit greater per unit increase in IQL. Finally, he may minimize *the effects* or consequences of these unfavorable conditions (e.g., minimize crime but not its cause). This is most costly per unit increase in IQL. For certain conditions, IQL can no longer be raised at any cost. It also requires far less wisdom than preventing the causal conditions from arising in the first place. This use should be viewed primarily as experiments towards a science aimed at improving IQL. It is not intended as massive reform or revolutionary social change.

A second important use for a good IQL is that of *shaping* conditions that make us more *aware* of the quality of our lives. The designer or planner of a new community could change or shape P's perception of IQL(P,t). We might have $IQL(P,t_2) \geq IQL(P,t_1)$ for $t_2 \geq t_1$ solely by a change in P's representation.

It is equally important to ask what a good IOL should *not* be used for. One such possible use is deciding when a life is not worth preserving. Morison[3] has argued that an aging person's IQL begins to decrease after certain golden years. During that period the cost of maintaining that life also increases. Kass[4] criticized this view of a continuous transition from birth to death. He objected to decisions which affect lives on the basis of complex cost-benefit analysis. Similar objections also apply to decisions about the very young and the unborn. Severely retarded infants can attain high IQL only at very high cost, and for some, it may not be feasible to raise or even maintain it at tolerable levels at all. Of course, parents carrying defective genes should be informed of the risks, so that they can make their decisions with full awareness of the risks. All decisions involving the preservation of a life almost always involve religious, moral, emotional and other very human considerations that no calculation based solely on the optimization of an IQL should attempt to by-pass.

The second potential use to which an IQL should *not* be put is social engineering. It seems desirable to create conditions underlying the growth of a community which would seem to maximize the IQL for most of its members. Such predictions are usually based on assumptions and values of a designer/planner. The scientific foundations for community planning are not sufficiently developed to enable anyone to make valid predictions on major issues, not even to assess the validity of existing predictions. Today's solutions contain seeds for tomorrow's problems. Conditions necessary and sufficient for raising the IQL for "enough" members in a community are very complex, variable and very difficult to understand. In sum, well-intentioned social reforms or revolutionary social changes have, in some instances, succeeded in raising the quality of life primarily through clever political and organizational moves rather than deliberately planned, scientifically based steps.

Possibly IQL(P,t) could take the form of a questionnaire that elicits from P at time t (1) his long-range plans or dreams, and (2) a record of his most recent acts. From these responses would be computed a score that satisfies as many as possible of the various properties and serve the two uses listed here.

INFORMATION PROBLEMS IN IMPROVING THE QUALITY OF LIFE

Let us assume that the key determinants of the quality of life are awareness of problem-situations and the number of options and wisdom to choose. Suppose that a person's long-range plans reveal his awareness of a great variety of tasks, or opportunities with which he would like to cope as well as awareness of constraints and other aspects of life quality. Suppose also that his day-to-day actions are consistent with these plans and reflect the wisdom of his decisions. We would then say that the quality of his life is high.

One of the most common causes for low IQL(P,t) would be, according to the above, failure by P at time t to recognize, select and cope with tasks and opportunities facing him. Failure to select or cope may be due to not utilizing the options among problem-solving resources that are available to P. This can be minimized by providing P with an information service of the kind provided by information and referral centers, with an aid towards increased awareness. P can be helped to avoid failing to recognize tasks or opportunities by an information service that teaches "wisdom".

The problems to be dealt with by information systems aimed at helping a person improve the quality of his life can thus be divided into two main classes: (1) increasing his awareness of his problem-situations and options, and (2) spreading and increasing wisdom.

Understanding and Awareness

Let us analyze these in turn. Let us define a *problem-situation* for a person P as follows. It is a state, together with a set of actions, from which P can choose one which leads to a consequent state that P would prefer greatly to his current state. Alternatively, there may also be an action that P could choose which leads to a consequent state such that P greatly prefers his present state to it. Among P's options for action may be both such extremes. He should avoid the latter and select the former type of action. But he must *know* which actions, if taken when P is in a certain state, will lead to various consequent states.

This is exemplified by many common fairy tales. The hero faces a precarious choice. If he chooses right, he wins the maiden and lives happily ever after. But woe to him if he chooses wrong. As in real life, the hero is not told which is the right and which the wrong choice.

People may be acting upon hypotheses of the form: "If I am in a state s and I choose a then I will next find myself in state s' which I greatly prefer (or the opposite) to state s". To the extent that such hypotheses are both useful and confirmed, they represent what P *knows*. An hypothesis is useful for P if he can base his choice of actions on it and this leads to state-changes he favors. This use of "knows" is consistent with that in the standard analysis of knowledge.[5] There, P is said to know a proposition p (an hypothesis) if (1) is true (same); (2) P is sure of p (corresponds to being useful); and (3) P has a right to be sure (corresponds to confirmation).

Knowledge of what will happen if an action is taken when a certain condition holds is quite different from awareness of the condition as a problem-situation. Ability by P to recognize a state as an opportunity or a task for P is prerequisite for improving P's quality of life. The greater the variety of states that P can correctly recognize as problem-situations, the greater his potential for improving the quality of his life.

To be aware of a problem-situation, P must be aware of his own part in it. He must be conscious. We propose to explicate this by requiring that P must be able to form and use hypotheses about his own ability to form hypotheses. P must know that he knows. He must also know that he doesn't know certain things that he needs to know. He must be able to ask questions that reflect his understanding of some of the relations among and between what he knows, and what he does not know.

An "information system" can help P improve his awareness of problem-situations by providing new perspectives, changing orientation, triggering shifts of representation, modifying attitudes. That is the connection with information science. To be sure, what is needed is not a conventional information system that provides access to documents or recorded data. It is one that reorganizes and repackages knowledge, and it educates rather than transfers or communicates. We are only now beginning to understand the desired characteristics of this needed kind of information system.

Wisdom

Let us proceed to the second challenge for information scientists: aiding the dissemination and building up of "wisdom". Wisdom goes beyond awareness and understanding much as understanding transcends knowledge. Person P is wise if he not only *knows* what would happen if he acted in certain ways and if he not only *understood* the limits to his knowledge and how what he knew was interrelated, but if he chooses what understanding to use when.

Conventional wisdom encompasses the 3 c's: caution, compassion, curiosity. If our understanding or knowledge is limited, we are wise to be kind and cautious. "Human existence depends on compassion and curiosity" said the physicist V. Weisskopf. "Curiosity without compassion is inhuman; compassion without curiosity is ineffectual."[6] Quality of life depends on the presence of both human and effectual elements.

Such conventional wisdom as can be effectively passed on from one generation to the next must be spread. Novel kinds of information systems can help this process. They must do far more than mere transfer of knowledge or even imparting of understanding and awareness. They must modify behavior. Again we are only now beginning to understand the characteristics such needed information systems should have.

But is there enough wisdom to spread? Keynes said in *The Economic Consequences of Peace*[7] shortly after World War I that those who govern do not even know what is to be known (we would replace "known" by "understood and wisely applied"); that they cannot or will not foresee the likely consequences of their decisions; that they are unaccustomed to competent thought, ignorant of what knowledge is, and that they do not understand that it matters. There is vastly more knowledge today for those who govern to be ignorant of. If Keynes meant to suggest that it was merely a question of bringing those who govern into closer rapport with what is known, he was probably only half-right. We tend to assume that by now our vast treasury of knowledge has grown divergently from the growth of the problems we are becoming aware of. We need to redirect the growth of knowledge towards greater relevance. We need even more to accelerate and direct the growth of understanding, of the way knowledge is organized. We need most to stress the growth of wisdom, so as to bring "the scattered and ineffective mental wealth into effective reaction upon our everyday political, social, economic life".[8]

Increased awareness of ourselves in relation to problem-situations helps us to dream - to make long-range plans - as if we would live forever. Increased wisdom helps us to act as if we would die tomorrow. Both contribute to the quality of life. Both present challenges to information scientists.

Concrete products of increased awareness and wisdom are public policies on matters affecting the quality of life. The subject matter of such policies range widely, from pollution through genetic counseling, transportation, crime and protection of privacy. A key role and function of information science is to provide those who shape public policies with information systems that help to increase their awareness and wisdom.

THE ROLE AND FUNCTION OF INFORMATION SCIENCE

The information sciences can contribute greatly towards increasing the awareness and wisdom of public policy makers. Concerted efforts to make such a contribution can also provide the information sciences with much needed direction, stimulate intellectual leadership and provide better contact with reality. The relationship of information science and technology (IST) with public policy making can benefit both.

Trends

Three important movements in the information sciences that can serve as a springboard for its contribution have recently gathered momentum. They have been called WISE,[9] UNISIST (UNESCO, 1971), and OECD (1971); (we are actually referring to a report by the Organization for Economic Cooperation and Development). WISE is, like UNISIST, a name for: (1) a paradigm, a set of ideas for looking at problems in information science and for assigning highest priorities to problems of organizing knowledge; (2) a socio-political movement to provide wise leadership in the creation of new social organs which are to serve communities as brains serve individuals; (3) an adaptive system for effectively coupling cumulated knowledge, understanding and wisdom with the key social and political problems faced by decision-makers.

While UNISIST assigns top priority to improving world-wide communication among scientists, WISE stresses the awareness and wisdom required by public policy makers and decision-makers. While UNISIST looks forward to a federation or network of voluntarily cooperating libraries and indexing services, WISE aims at rather sweeping educational reforms. Its general aim is to improve man's ability to recognize, select and cope with an increasing variety of problem-situations, so that he can survive - with a higher quality of life - under an increasing variety of conditions. In this respect, WISE resembles OECD, which asserts as its aim that "scientific and technological information must be used more effectively in the future in making public policy, applying to technology, and developing science". OECD recommends that "management information systems for government should be designed to encompass all forms of relevant information and must be timely, structured, selective, and credible".

All three movements are beginning to acquire momentum. The vigorous growth of UNISIST and OECD is well known. A pilot version of WISE on a small scale of helping ordinary people modify their orientation and attitudes to mathematics has been operated experimentally for several years.[9] At a recent meeting, Canadian information scientists have planned to further this movement. Eminent scholars like T. Kuhn, H. Lasswell, K. Boulding, J. Platt, D. de Solla Price, A. Inkeles, C.W. Churchman, and many others have expressed interest in serious discussions to evaluate the desirability and feasibility of WISE. A number of other projects and enterprises are contributing some of the necessary background experiences that may help further the WISE movement[10] and various steps towards information-network indexes on world problems such as those of M. Mesarovic at Case Western Reserve University (Cleveland), J. McHale at SUNY (Buffalo), J.N. Judge's Yearbook on World Problems and Human Development (Brussels), J. Wellesley-Wesley (Rome), 'Q Operation Survival' (London), and G. Feinberg's Prometheus Project at Columbia University (New York City).

These are sources of experience in organization and management of novel enterprises. That has been the greatest lack in information science. At the same time, scholarly programs are also emerging. An outstanding example is Harvard's new program on information technologies and public policy.[11] It is a program of research, teaching, and communication. It aims to develop a coherent understanding of information technologies and their policy implications as well as the illumination of public policy alternatives. It is concerned, for example, with such questions as the extent to which existing institutional structures prevent new technologies from meeting needs. Such questions are central in assessing the desirability and feasibility of a movement such as WISE.

Another important trend in the information sciences which helps qualify it for its roles in public policy is the development of information and referral services. Individuals in every community are beset by complex problems. These range from where to get medical services for themselves or their families to the need for more insight into the issues or candidates in an election. Resources to help people with problems exist to a larger degree in a community than most people realize. Even if a runaway teenager, a rape victim, someone who had been arrested or taken poison senses the existence of agencies responsible to help him if he seeks it, he might not know where to turn. That is the task of information and referral (I & R) centers. Their study is a new chapter in information science[12] likely to change its course towards contributing more to the quality of life.

Information Sciences and Technologies (ISTs) can play two roles in relation to public policies that affect quality of life (PPQLs). These are: (1) the techniques of ISTs can help shape and enforce PPQLs; (2)(a) public policies are needed to regulate the development and use of information systems which are to help improve quality of life; (2)(b) selected concepts of IST can serve as intellectual stimuli for the substance of public policies.

Techniques for Policy-Making

When an I & R Center meets some of the information needs of people in a community, there is an important by-product. The I & R Center can also collect data about the patterns of problems in a community, and how these change. It can serve as a sensitive barometer of a community's quality of life. The techniques by which an I & R center grafts itself into a community so that it can accurately reflect the needs, by which it collects and records reliable data, organizes it, and makes valid inferences from it, all these methods can be used to provide policy-makers with the background to draft appropriate policy statements.

Once a policy is formulated, the same techniques can be used to assess the extent to which the policy is enforced, and the extent to which it is effecting the changes in the community for which the policy was intended. The simplest technique is to continuously monitor certain indicators of the quality of life in a community, much as economists monitor a selected number of leading and lagging economic indicators. A more complex technique is that of intelligence analysis. This involves the early detection of unusual patterns likely to reveal changes in the quality of life. If we restrict our concern with the quality of life to awareness of problem-situations and wisdom to choose among options, then both monitoring of indicators and early-warning intelligence of unanticipated events fall within the ISTs. The considerable experience of ISTs with analogous military and political tasks and to a lesser extent with health care can then be brought to bear.

Need for Policies to Regulate Information Systems

Suppose that it becomes a matter of policy to identify all "accident-prone" drivers. Assume further that erratic control of a car reported by seven or more different drivers within a month has been proved to be a reliable predictor of a significantly increased likelihood of accident involvement. It would be easy to enforce such a policy by providing each driver with a data recording device and reinforcing him to record the license plate number of any car being driven erratically or dangerously. If there were no abuses or misuses, potentially dangerous drivers could be retested, retaught, warned or removed as sources of danger. This would increase the quality of life by reducing accidents. At the same time it brings into relief one of the most serious problems arising in any information system involving people: protection of privacy.

The number of drivers who are erroneously "detected" as being "high-risk" should be kept as small as possible. A public policy to ensure this is as important for the quality of life as one to minimize the number of undetected high-risk drivers. The public policy in question here is one that regulates the information system itself. Information scientists, in collaboration with researchers in law[13] have a very important role in drafting appropriate policy statements.

With regard to how ISTs contribute to the conceptual substance of public policies, we have already discussed the importance of awareness and wisdom. Both are fundamental information scientific concepts. Thus, information science can contribute towards a clearer and deeper conceptualization of the very notion "quality of life". This is perhaps its most important function today.

TOWARDS PUBLIC POLICIES

The foregoing analyses of the two roles IST plays in public policy-making indicated how information sciences have and can contribute towards improving the quality of life. There is still vastly more it could and should contribute. The following are recommendations to various concerned parties for some of the most important contributions information sciences and technologies should make towards helping policy-makers.

(1) To Executives. Policy-making executives at various levels of government play an important role in forming public policies that affect the quality of life. At the lowest, local levels such as that of a ward or school district there is often an effective informal communication net - a "gossip grapevine" - which provides policy-makers with much of the knowledge, awareness, and wisdom they need. Executives at that level should give information scientists an opportunity and funding to study these informal channels, encouraging them to produce an index of that community's unmet need. Information scientists could test the hypothesis that an I & R center's effectiveness is greater if it is intimately coupled into this informal network. They could also determine the optimal size of a community which should be served by one I & R center.[14]

At the levels of city or county, policies relate more intimately to urban/regional planning than to needs of individuals and organizations in communities.[15] Defining and meeting the information needs for urban/regional planning is a rather new and important area of investigation to which information scientists should contribute. Shifts in residential, vocational, and recreational distributions make it increasingly important for planners and executives to adopt a regional perspective. The city is in some important ways an information engine.[16] The big challenge to information scientists is to design information systems that help cope with ill-defined problems. A base from which to build in this direction has been started recently.[17] Executives at this level are well-advised to support experiments - as well as volunteer to be experimental subjects in them - with pilot information systems aimed at helping them.

At the state and national level there exist a variety of information services. Policy-makers in the U.S. Executive branch seldom express needs for better information to help them. If anything, they are overwhelmed by the amount of information they must take note of and digest. This does not mean that they have the awareness and wisdom they need. Very few individuals have it, and we call them statesmen. Their superior performance in policy-making seems to be an individual trait rather than the consequence of information science and technology. This could be tested by comparing the distributions of the outstanding policy-makers in an administration exposed to an advanced IST with the outstanding policy-makers of an earlier age when there was no sophisticated IST. If IST does not make much difference, we need to know why.

If we cannot take steps so that it makes a difference, then we can at least try to maximize the number of potentially outstanding policy-makers who get elected or appointed and minimize the number of elected or appointed policy-makers who are not so highly qualified. This is a task to which IST can contribute. It is a task of even greater concern to policy-making executives in large businesses and industry, where there are clear-cut losses in failing to fill the key jobs with the people best qualified for them.[18]

At the level of blocks of nations (e.g., NATO, OECD, etc.) or internationally (UNESCO) opportunities for setting up pilot versions of WISE should be searched for and exploited. Organizational and financial problems at this level are much more complex, but can be coped with by imaginative and capable leaders. Executives at all levels of public-policy making, but especially at this level, may find helpful to them statements of policies directing their use of information sciences and technologies.

We suggest, towards such policy-statements, the following. Each policy-making executive should provide in his budget for at least one position on his staff for a professional in the ISTs. This position should have responsibility and authority for advising him on existing and needed information systems, for investigating the effectiveness of his policy-making as it relates to information, for utilizing existing ISTs to help provide the policy-maker with needed awareness and wisdom, to coordinate with IST professionals on the staffs of other policy-making executives and to serve as ombudsman for the policy-maker's constituencies.

(2) <u>To Legislators</u>. Policy-making is more usually identified with legislative than with executive branches of government. High-level executives have to shape their own policies or at least interpret general policies implied by laws so that they can use it as a basis for decision-making and action by their organization. Policies drafted by legislative bodies tend to be regulative rather than directive, to apply to any organization for all time rather than to a particular organization now. With increasing scope goes increased responsibility. This makes it increasingly critical that legislative policy-makers get the best that IST has to offer.

The U.S. Congressional Research Service is an example of what information science contributes to legislative policy-makers. It aims to provide Congress with objective, unbiased evaluations of information and policy proposals received from any source.[19] State legislatures might be well advised to set up analogous services. So might legislative bodies of other countries or blocs of countries. If they do, the questions of duplication, coordination, and decentralization arise. On many topics, the Congressional Research Services could probably do a better job than its counterpart for a state legislature if only because of its larger resources. Yet there might be a reasonable division of labor based on the extent to which the problems depend on and affect uniquely local aspects.

Just as in the hierarchy of the judiciary, policy-making at higher legislative levels (e.g., Congress) should deal only with problems that transcend the competence, jurisdiction and willingness of the lowest legislative level (e.g., city, county, state), and which are referred upward.

This principle of policy-making at the lowest level at which it can be adequately done should be framed as a policy-statement. The implications for information systems are a corresponding division of labor. A similar policy on the executive side would have to be originated by the chief executive. While lower levels of executive policy-making report to, and are delegated from the top down, lower levels of legislative policy-making are independent of higher levels and relegate from the bottom up. Thus, information systems at the lower legislative levels should be increasingly specialized and tailored to needs unique to that level and its constituency.

The different information services at various executive and legislative levels may conflict, compete, abet inconsistent or contradictory policies, even if all policies are based on the best available knowledge. An obvious suggestion would be a superordinate information service to adjudicate and resolve such anomalies. This is itself likely to be inconsistent with the policies for decentralizing information services that we have suggested so far. There is a place for top level information-processing, but it is not on the staffs of top-level policy-makers in either executive or legislative (or judicial) roles of government. It belongs into a fourth as yet non-existent branch of government: the "knowledge" branch.

(3) To Information Science Departments in Universities. If we compile a list of all the conferences, university departments, industrial research departments and books which have used the term "information science and technology", we find that the interpretations differ vastly. They include people with backgrounds in and concerned with computer science, communication engineering, library science, the sociology and psychology of communication, data processing, and statistical analysis, systems engineering, control

engineering, and many other fields. Yet there may be an underlying unity. This may exist at both the conceptual and the technological level.

The basis for conceptual unity may be the striving on the part of some information scientists - in common with their colleagues in other sciences - for a coherent conceptualization of how knowledge, understanding and wisdom is generated and used by people to enable them to cope with whatever problem-situations they encounter. Library scientists are concerned with one special aspect of this problem, of how the graphic records of civilization can be organized to maximize effective social utilization. Concerned computer scientists deal with another special aspect of this problem, of how to maximize the computer's usefulness as a problem-solving tool. The various other disciplines are concerned with yet other special aspects of the same fundamental problem. The kind of unifying paradigm that is the goal of information science as a science is not only an interdisciplinary fusion of all these specialties but the emergence of a new discipline. Perhaps it will be a scientific version of epistemology or perhaps it will evolve from the work towards a science of science.

It is such an intellectual discipline that should concern itself with resolving the inconsistencies and anomalies arising both within the growing discipline and in its applications to policy-making. It is essential that practitioners of this discipline - as indeed of all scholarly enterprises - not owe any special allegiance to any policy-makers, executive or legislative. Their first loyalty must be to truth. In this sense, science stands separate but equal with both rulers, judges, and the people. It should participate in a system of checks and balances as a fourth branch of government, not making or enforcing laws and policies, nor interpreting or judging them and the behavior of those they govern, but searching for such truths as would best help men to draft, enforce and interpret laws to help them cope with problems. Generation and communication of irrelevant, insignificant, invalid, unclear or redundant knowledge would be detected and discouraged not only by members of the "knowledge" branch itself, but by the other three branches as well. Policies and judgments made by the other three branches would be constantly scrutinized and checked by the "knowledge branch".

At the technological level, it is becoming increasingly feasible to substitute magnetic tapes for printed hard copy, to trade storage for processing time or long-distance communication and generally to replace one information technology by a variety of others. Practices in journalism, the printing trades, librarianship, computer graphics, etc., will all become increasingly fused as dramatic economies afforded by rapidly developing technologies begin to overcome institutional obstacles (union objections, etc.). Evaluations, guidance and counseling regarding technologies from scholars in the independent "knowledge" branch might be welcome from those in the other branches with vested interests and whose freedom of investigation is subject to greater constraint and bias.

The recommendation is for a fourth branch of government independent of the other three. To whom should this recommendation be made? At present, perhaps the most realistic course is to choose professors of information science as the audience for this message. They should make it their objective to seek out, encourage and develop potential intellectual leaders who are dedicated to creating such a fourth branch, probably at an international level. They should try to convince themselves, their colleagues and the general public of the desirability of such a move so as to develop a climate of public support.

THE FOURTH BRANCH OF GOVERNMENT

(4) To Professional Societies. Such societies as the American Society for Information Science will obviously play a key role in a fourth branch. Indeed, they may already be the organizational seeds insofar as support comes mainly from membership dues. Private universities with international reputations and private foundations or international research institutes funded by philanthropists can play equally vital roles. Their contributions to policy-making have been highly significant. The traditional approach has been to let these institutions select and spawn the best intellects. Thus, they serve as a source of recruits for the various branches of government, who transfer the people they need to serve them. It is a time-honored practice in human organization for a policy-maker to place greater trust in a staff advisor who reports to him than in the impartial judgment or advice of an intellectually and organizationally independent scholar. Scholars often consider it a call of duty or an honor to advise the highest levels of government, and do not often regard academia as an institution capable of exerting influence on policy-making. The question is whether these attitudes and the realities underlying them will change.

Professional societies could perhaps make some useful moves in the direction of such a change. They could try to reach a broader general public with messages from their members that pertain to public policies. Perhaps they can, by proven performance in helping to enlighten the public, obtain its support for a fourth branch to check the other three branches of government.

At this time in history, public appreciation of the value of intellectual and scholarly enterprises is at a low. There is correspondingly greater awareness of complex problems that need urgent attention. Because the problems are complex and urgent, wisdom in framing policies is of critical importance. The organizational roots of the fourth branch - professional societies, etc. - have an unusual opportunity to show what they can contribute, to make their bid for leadership. The move towards attaining greater public support can wait. If the present intellectual community proves its leadership to the public, it will have done a great deal to prepare the ground for restoring the public support in the future which is necessary for a "fourth estate".

References

1. Lasswell, H. and Kaplan, A., *Power & Society: A Framework for Political Inquiry*, Yale University Press, New Haven, Conn., 1950.

   Maslow, A., *Motivation and Personality*, Harper & Row, New York, 1954.

   Foa, U.G., Interpersonal and economic resources, *Science*, 171, 345, January 29, 1971.

   Chapin, F.S., *Urban Land Use Planning*, University of Illinois Press, Urbana, 1965, 25.

2. Lotka, A.J., *Elements of Mathematical Biology*, Dover, New York, 1956.

3. Morison, R.S., Death: process or event? *Science*, 173, 694, August 20, 1971.

4. Kass, L.R., Death as an event: a commentary on Robert Morison, *Science*, 173, 698, August 20, 1971.

5. Ayer, A.J., *The Problem of Knowledge*, St. Martin's Press, Inc., New York, 1956.

6. Weisskopf, V.K., *Physics in the 20th Century: Selected Essays*, MIT Press, Cambridge, 1972.

7. Keynes, J.M., *The Economic Consequences of the Peace*, Harcourt, Brace and Howe, New York, 1920.

8. Wells, H.G., World encyclopaedia, *World Brain*, Doubleday & Doran, Garden City, New York, 1938. Also, reprinted in *The Growth of Knowledge* (Kochen, M., ed.), Wiley, New York, 1967.

9. Kochen, M. and Dreyfuss, G., An experiment in teaching college mathematics, *Intl. Journal of Mathematical Education in Science and Technology*, 3, 315, December 1972.

10. Boyko, H., ed., *Science and the Future of Mankind*, (World Academy of Art and Science [Publication] 1), W. Junk Publishing, Den Haag, 1961.

de Grazia, A., The universal reference system, *The American Behavioral Scientist*, 3, 1965.

Fuller, B., The project for humanity, *Saturday Review*, August 29, 1964.

Stulman, J., *Evolving Mankind's Future*, (The World Institute: A Problem Solving Methodology), Lippincott, Philadelphia, 1967. Also, *Main Currents in Modern Thought* and *Fields Within Fields ... Within Fields*, The World Institute Council, Vol. 1-5.

11. Oettinger, A.G., *Harvard University Program on Information Technologies and Public Policy Status and Plans*, Aiken Computation Laboratory, Cambridge, Mass., February 1973.

12. Kochen, M. and Donohue, J., *Information for the Community*, American Library Association, Chicago, 1973.

13. Miller, A., *The Assault on Privacy: Computers, Data Banks, and Dossiers*, University of Michigan Press, Ann Arbor, 1971.

Westin, A.F. and Baker, M.A., *Data Banks in a Free Society*, Quadrangle Books, New York, 1972.

14. Kochen, M. and Cason, D., A cost-benefit analysis of an information system to help people plan their families, in *Proceedings of Sixth Hawaii International Conference on System Sciences (January 9-11, 1973)*, BB-3, 32, Western Periodicals Co., Los Angeles, 1973.

15. University of California-URISA, *Proceedings of Fourth Annual Conference on Urban Planning Information Systems and Programs*, University of California, Berkeley, August 1966.

Preliminary design of a debugging compiler, in Lew, A., Urban & Regional Systems: Modeling, Analysis and Decision-Making, in *Proceedings of Sixth Hawaii International Conference on System Sciences*, B-1, 90, Western Periodicals Co., Los Angeles, 1973.

16. Meier, R.L., *A Communication Theory of Urban Growth*, MIT Press, Cambridge, Mass., 1962.

17. Badre, A.N., On hypotheses and representational shifting in ill-defined problem-situations, Ph.D. Thesis, University of Michigan, Ann Arbor, August 1973.

Kochen, M., Information systems for urban problem solvers, presented at Conference on Information Systems (December 1972), to appear in *Proceedings*, Tou, J., ed., Gainesville, Florida, 1973.

Kochen, M. and Badre, A.N., Question-asking and shifts of representation in problem-solving, *American Journal of Psychology* (forthcoming).

18. Cohen, M., Ferman, L., Kochen, M. and Peterson, M., *Information Systems for Servicing People*, University of Michigan Press, 1973.

19. Library of Congress, *Annual Report of the Congressional Research Service of the Library of Congress for the Fiscal Year 1971*, U.S. Government Printing Office, Washington, D.C., 1972.

INFORMATION-TRANSFER IN THE EVOLUTION OF UNDERDEVELOPED PEOPLE

Maria Helena F. Lima

Portugal

INFORMATION is of vital importance in any social or political entity. The progress in the field of information shows us how much it could involve people interested in getting in touch with the institutions related with human activities. Man is the main source of all sciences around the world and we can see how much they influence our lives.

During the last century man was close to a new science named Anthropology. Its father, Levy-Strauss, considered all kinds of human relationships as subsidiaries of Anthropology, Ethnology, Ethnography and Human geography. The principal fact in this theory is not the science itself, but the vehicle of understanding - the language. But language is something controversial and we cannot really keep the idea of talking about information forever, as if it were simply a word. One of the purposes of information science should be the study of underdeveloped people who do not know the language of their neighbours but keep their property - intelligence.

Going back to Strauss we see that *Ethnography* is a science devoted to the study of people, races, languages, customs and religions. *Ethnology* studies the origin and formation of the human being, physical character and his distribution on Earth. And finally: Human geography or Anthropo-geography has a real point of contact with other sciences in that they observe very carefully man in his relation with the physical environment like those mentioned, and sociology and history.

Anthropo-geography is the same as human geography. Karl Ritter, the German scientist from the last century (1779-1859), showed in all of his works his preoccupation in relating man to the facts of history. The Ritterian influence appears in the historians Richelet and Elisée Reclus. Although Richelet showed the evolution of people in the past by geographic character only, Reclus through his immortal book "L'Homme et la terre" was the great influence on geographic science and established the tendencies that are followed in our days. Anyway, it was Anthropology and Ethnology together that defined the human being and established the conception of race.

In our modern days we feel able to add a new subsidiary to the Anthropology family - INFORMATION SCIENCE. This subsidiary appears as the most valuable vehicle to construct a bridge among the specialties we follow in order to arrive at a final interpretation of the laws and directions we all keep in mind. Also Information Science must not be developed merely as a discussion of the word only. As I mentioned in the Seven Springs meeting, last year, my point of view is that information must also be accepted as the subject of a science, whenever it is concerned with underdeveloped people, and it should deal with the following matters:

- to protect families but not political groups
- to enrich knowledge of this special group of people but not with propaganda intentions
- to develop education in every place not only in privileged centers
- to condemn uncivilized forms of propaganda or the very cheap way of publicity
- to annihilate public demonstration of pain or charity
- to destroy the terms Savage and Savagery
- to modify general opinion about people especially those who come from underdeveloped areas

In regard to our conception of science and the reasons for this conception that we have been discussing here in this Institute, we must emphasize the fact that "Information" as a science requires a special treatment of "Communication". All common experience points out the fact that when something comes about, each person has a distinct opinion on this commodity. We can assure ourselves at this time that its ultimate meaning is "To inform and to be informed". "COMMUNICATION" in a large sense, results from a process of Information and it will serve not only to communicate something to somebody but also to express things to be developed with the final goal of helping in the construction of a new world. The necessity to explain things for this goal - Construction of a new world - makes information more than just a simple way to offer a lot of specialized news.

As an example, we are aware that those people from underdeveloped countries in Africa, Asia, South America, and some areas of Australia and United States, need so much help and comprehension for their problems and obligations. But they need "Communication" overall. Communication represents in this case personal needs as well as collective ones, where man as a social element is responsible for the bringing up of the family and other social groups. People want to be known not as being different from other human beings, as fools or funny people, limited in time and space, but as an element close to our first human science - Anthropology.

In this field, we must remember that Darwin's theory showed us as beings superior to the "monkeys" because of the volume and formation of the brain, consequently - intelligence. The functional mechanism is similar between man and animal. Differences are found only in the mental faculties but not in the essence. Man was a product, during the consecutive generations, of the environment and we see that the race is not based on a fixed character. We have many examples to explore. The Boers who came from Holland have to adapt themselves to the African warm "planaltos". Not only their psychological type but many physiological details were altered, giving place to a new type disassociated from their origins.

Some alterations could happen by receiving new elements to the "optimum" heredity, but this is not the case with the Boers, who generally don't mix with other races. Consequently, it proves the influence of the "environment" in man's life.

COMMUNICATION is accomplished when a message exists and the information is clear enough between the Sender and the Receiver. The experts point out the fact that the process of Communication could easily change the ideas and actions of the Receiver.

We are arriving at a scientific meaning for expressing this modality of Communication. I would like to take as an example the underdeveloped people in any part of the world, pointing out the necessity of first preparing them to accept appropriate information and disseminate it, having in mind that people who are not well informed and are alien to the problems of Communication. On the other hand we presume that under the influence of Anthropology, Information is fully necessary to enrich man in many ways.

The useful form for applying Information science has been achieved through INFORMATION-TRANSFER. We must not worry about the beauty or ambiguity of the expression because it can provoke emotional impacts.

INFORMATION-TRANSFER is not precisely "Communication". These are distinct phenomena. We are facing now a very crucial point in the exposition. Information Transfer is accomplished when it involves a Sender, a Message and a Receiver. The process of transferring the Message could be named Transmission. We can transmit the contents of a Message without expecting any action as a result.

Important is the Transmission. Whenever the Message is clear to the Receiver opening a way to promote a change in understanding, Information-Transfer is accomplished.

It is now time to emphasize the importance of the MESSAGE. All men in possession of a language, a religion or faith are engaged with some political or social organization. Through some direction, we can accomplish coexistence, changing our cultures, studying new models, living new experiences, introducing appropriate techniques from existing sciences and adopting new developments in modern science. With some effort, all of these commodities could be adopted without destroying personal values, and produce the enrichment of human culture through new methods of Communication (transmission), named INFORMATION-TRANSFER. By increasing the different forms where Message assumes the most important place, we could go on to annihilate preconceptions of inferiority.

We note that Information-Transfer requires a message to bring the ideas to a second person. In order to make Information-Transfer a system indispensable to actual social life, new ideas of Information Science have to be developed. Communication appears when all changes are accomplished through the Message. It is going to produce interesting and unexpected impacts on people around the world, not only the privileged ones but all those interested in communicating to everyone, and listening to everyone else.

I have taken the opportunity to emphasize the fact that Communication influences could enrich since it brings new information to the entire world. This process must be enlarged in order to give new chances for contributing to a better development of this knowledge. Modern Information Transfer must be guided to protect social values and preserve individual freedom against attempts to propagandize and other common attitudes. "Information" as a science must be accepted as offering several ways of interpreting phenomena and problems. Looking around the world we can see people in the most precarious way of living isolated by miles and miles of distance from even middle-sized centers, separated by language and religion and customs more than their incapacity to meet each other, or the political construction of their lives. Millions of human beings partitioned into many geographic zones are in this terrible condition. Information Science should not ignore the fact that Information-Transfer and Communication are necessary to them.

We have taken the opportunity to show you some aspects of different races separated by language more than by religion or customs. The Portuguese language is more understood by some of them than the common dialects of their neighbours. However, in a scientific sense, that is the only way it can interest us, we can see how much the world is alien to their intelligence and how much they are still separated by some old theories of supremacy and ignorance.

How can Information-Transfer be of benefit to improve Communication between members of a group? To establish that principle, a Sender must transmit a Message directly to the father of the group, if necessary through an interpreter, but the people must be in condition to receive Messages and transmit answers, or assimilate the results of Communication.

As we have shown here last year, it is necessary that a Message has a feedback reaction or be usable to more than one specific Receiver. We see in the case of this people that the Transfer process could offer Communication if the Information in the Message is clear enough and usable to the Receiver. And feedback only appears when Communication has been accomplished. A common form for the process with people in this condition would be expressed like this:

TRANSFER - one-to-one- to one- to one - to many

i.e. Government action to a second authority in the area of the people to the professor or the Missioner, to the interpreter and finally to people.

We are facing now a new conception of the world where man takes a place of some importance in the community. The different races are close together. We do not overlook the great error along the centuries which accepts the human being as only distinguished by those qualities - virtues, intelligence, activity and physical appearance - which special conditions have reserved to some groups, some races or some countries.

All human cultures, being different, are called to construct a new community more rich and pleasant, where men can help to live better. This is a primary condition of the evolution of humanity. When a Message occurs between people of different human groups it could better promote the enrichment of cultures and form a splendid block of human relations. This is an example of the evolution of civilization around the world. Here we must emphasize how important Information is, whenever it can change human beings in the universal world. In fact, through a hard and exhaustive search resulting from the knowledge of many races with relations among themselves we could better evaluate the spirit of their cultural differences. The importance is not to live but know how to help people to live.

INFORMATION TRANSFER THROUGH HUMAN SOCIAL-POLITICAL LIFE (HISTORY)

Pablo Lima

Men have been defined as political or economic animals, apparently conflicting descriptions in terms of so-called liberal or socialist theories. We cannot forget, however, that men are moved by different and complex metaphysical beliefs. In matters concerning philosophical speculation or knowledge (ethics, esthetics, psychology and religion as well as codes of morals or behaviour to be put in practice), human beings we know don't care only about the material side of their lives. They fear death; they feel lonely; they want easy explanations for the flood of phenomena surrounding them. They don't just *do* things, but they also think and feel. No definition of man can be complete, but we can consider *doing* and *thinking* and *feeling* of human physical and mental actions - how men act regarding their material and psychic needs, and what they think of the meaning of man and his origins, his life and possible existence hereafter. When men act to modify their environment by perfecting technological means, we see this as the civilized side of men. In *this* case of course they are concerned about their lives on Earth. Then, they call themselves political or economic beings because they are involved in a process of living with or for "progress" in communities which they try to organize according to their political or economic beliefs.

In contrast, is the cultural and spiritual life of mind and heart. Men care about making, keeping, and continuing their total culture. Progress is not only proposed as a way to provide an easy life for everybody, but also to increase justice, liberty, and knowledge everywhere. Despite having a car and a big flat, and a TV set, I cannot feel quite happy if my neighbour is not so fortunate, or if in the suburbs of my town people starve. For we all care about the social environment in which we live. People fear to become slaves of the technological advances they promoted. They protest against pollution, and ask worriedly: What shall we do with nuclear power? Are we wise enough to use it only for a good purpose? This side of man has in it the degree of abnegation, that the material side has of selfishness.

History tells us terrifying things about life in the Roman Empire. The rigid division of classes, the bondage and the extermination of people, the massacre of christians, the mad Caesars, the orgies and cruel scenes in the arenas, where murder was a way of life. But who can fail to recognize the enormous contribution of Roman culture to the development and organization of all Western Europe and the entire world? Can anybody deny that this culture is still alive and dignifying millions of people, giving them the languages of French and Spanish and Portuguese and Romanian with their cultural heritage. How could *that* society leave such a rich heritage?

There can be barbarian cultures, cruel, despotic, like a struggle for life in the jungle, but there are no bad cultures. All of these are "valid" and have something to show and teach.

INFORMATION AND SOCIAL CONTROL SYSTEMS

J.R. Salancik

University of Illinois, Champaign/Urbana, Illinois, U.S.A.

My purpose today is to describe to you what I think are some implications of developments in information technology for social control. Before doing so I would like to describe to you what I understand to be some indications about the nature of social control as they can be derived from experimental research in social psychology, and the role which I think information systems play in producing social control.

Information as a psychological concept is very similar to what Shannon in his outline of the mathematical theory of communication proposed it to be in 1949. Information, quite simply, is any event which reduces uncertainty for individuals. Uncertainty, in turn, is understood to be response uncertainty and exists whenever an individual is faced with more than one response alternative in a situation. In addition, response alternatives are assumed to be evoked by the presence of stimulus events to which responses in the past have previously been paired or to which responses can be inferred from the stimulus. A classic case of response uncertainty can be observed in a rat who is fed food from a dish while at the same time being shocked by a small dose of electricity. A rat who undergoes such a dreadful experience over time will develop an approach-avoidance pattern of responses. The rat will be observed to vacilate between moving toward the food cup, presumably in response to the hunger cues he is experiencing and the prior pairing of those cues with the food dish, and moving away from the food cup, presumably derived from the prior pairing of the food dish with the electric shock. This situation is an obvious case in which alternative responses are evoked by ostensibly the same stimulus event. Anything which results in the resolution of this uncertainty is by definition *information*.

A more commonplace situation in which psychologists have used information concepts is in the areas of cognitive processing and behavior. The contributions of such scholars as George A. Miller, Wendel Garner and Donald Broadbent were enormous in their implications for understanding the human information processing capacity. That capacity can be understood in terms of the entire set of events potentially evoked by a stimulus array and their relative probabilities, much as the communication engineers and physicists such as Denis Gabor and Anthony Weiner had suggested for physical transmission systems. To illustrate this, consider the presentation of the following construction to an English-speaking person, "C-blank-T". If such a set of symbols were presented to a person and he was asked to guess what three-letter word it represented, there is a high probability that the person would say the word "CAT". For those who do not say cat, the remaining responses would be distributed between "CUT" and "COT". Those three responses would represent almost all of the responses which the English-speaking subjects would emit. In contrast, if we were to present the symbols "C-A-blank", the probability of a group of subjects saying CAT would be greatly reduced from the first case. Instead, subjects will report such responses as, CAP, CAN, CAR, CAW, CAM, CAB and CAD.

If one examines this situation carefully, one can note that the stimulus events which we are presenting to individuals and the instructions which we are giving them, in the light of their history and experience with the English language, are contributing much to shape and determine their behavior. First, it is obvious that by requesting the individual to report English words, and then only three-letter English words, we have restricted his selection of potential responses to a very small subset of the verbal utterances which the human is capable of emitting. In addition, by altering the arrangement of the blank and the English-letter symbols, we can evoke eight possible responses from the individual in the case of C-A-blank, and only three possible responses in the case of C-blank-T. And if we rearrange the latter construction into "T-blank-C" we can reduce his responses to only two. From an information theory point of view, each of the different constructions, though similar, vary in the amount of information which they convey. The more the construction reduces a person's responses to fewer alternatives, the more information the construction has.

Although we describe such constructions as varying in the amount of information conveyed, it is equally possible to describe the stimulus arrays as varying in the amount of control which they permit over the responses of the persons taking part in the experiment. In short, information events are events which control human behavior - not all of human behavior, mind you, but probably a good deal of that which we define as cognitive and social behavior.

Beginning with these rather simple ideas of information theory, I would now like to discuss an area with which I am more familiar as a social psychologist, namely, the area of social control. Some of the same concepts apply, as in the case of cognitive behavior; however, they are much more difficult to define and pin down because of the more complex stimulus arrays with which we deal in social reality. There is a clear indication, though, that people experience uncertainty and that such uncertainty can be reduced by careful arrangements of stimulus events which result in influences upon people's behavior.

There are a number of factors which contribute to an individual's feeling of uncertainty. First, there is considerable evidence to indicate that individuals have a need to feel they are correct. Second, there are also indications that individuals have a need to feel that they are worthy. If either of these two feelings are not satisfied at a particular time, individuals will experience what can be called a feeling of uncertainty. When people experience uncertainty they also have a tendency to reduce the uncertainty and thereby reestablish their feeling of worth or correctness.

A major mechanism which exists for the establishment and maintenance of our feeling of worth and correctness is a social communication process. In a good many situations, when people want to find out how they are doing, or whether what they are doing is the correct thing to do, what they do is look around them and see if others are doing the same thing. There is an interesting series of studies in which models are provided in very commonplace situations. One study, for instance, has a well dressed individual standing on the street-corner when the light turns red or a no-walk sign goes on; there is a greater probability that unsuspecting pedestrians will stand and wait for the light to turn green with him than they will if he were not present; if he walks across the street against the light, there is a greater probability that the unsuspecting pedestrians will also cross the street against the light. In a more unique situation, one can increase the probability that cars going at seventy miles an hour will stop and the drivers will get out to aid a woman who is standing on the roadside next to a car with a flat tire. In one condition this experiment was actually run on a Los Angeles freeway and provided a quite striking demonstration of the effect of observing others doing something on our own propensity to do the same thing.[1]

A more discouraging example of the same effect, however, can be found in the response of people to serious situations of danger. Some of you may remember an extraordinary murder in the city of New York in 1964, when a young woman was stabbed some 100 times on two occasions as the assailant went to her, stabbed her, left, and came back and stabbed her some more until she was dead. It was summertime and the windows of a nearby apartment building were open and some thirty-four individuals were standing at one point or another during the ordeal at their windowsills observing the quite obvious events below. This murder and the responses of the tenants across the street were so striking that the newspapers of New York carried the story and its follow-up for months, reporting the interviews of the witnesses, the reactions of the police, etc. etc. The question on everybody's mind was "How could this happen?" How could thirty-four people witness so atrocious an event and do absolutely nothing, not even calling the police at the first sign of trouble. There were many hypotheses: the assailant was huge and too rough for any of them to handle; actually, he was small, slight, and the young female victim knocked him down herself with a kick. There was expressed the possibility that the police, as everyone knew, would not come so why call them; actually it took the police only about two-to-three minutes to arrive on the scene after a call was finally placed. Some put the whole affair down to the possibility that New Yorkers were just cold and/or apathetic, and that such an event would not occur in their city or neighborhood. Actually such events have and will continue to occur in far too many cities.

The answer, as later demonstrated in a number of experimental studies of similar phenomena, seems more likely to lie in the possibility that the people were uncertain about the event and the implications which the event held for them. Although all of them clearly saw the danger facing the girl, they really did not know how dangerous it was nor what was the appropriate action to take in the situation. That is, they did not know how the other tenants, whom they could not see, were reacting to this situation. One man, in an attempt to get clarification actually called up his brother to inquire what he should do about what he saw; the brother, having less information than the caller and therefore seeing no reason to get excited, simply said, "forget it and go to bed". The man did. The reason it is possible that people reacted the way they did is suggested by a series of laboratory events which were conducted by two researchers[2] in which they put subjects in a situation in which a confederate suddenly went into an epileptic seizure and the question was, under what conditions would the unsuspecting subjects come to his aid. By manipulating the conditions to implicate either more or less social information about the seriousness of the victim's plight or the actions of the other subjects, it becomes quite clear that the people who let the young woman die in New York did so because they really didn't know what else to do.

There is an even more telling situation in which it is clear that an individual's desire to reduce uncertainty results in his actually experiencing complex moods. The experiments were conducted by Professors Schacter and Singer[3] of Columbia University and the uncertainty which experimental subjects experienced was induced by injecting epinepharine into their bloodstreams. Epinepharine is a drug that causes general body arousal, and people report such things as stomach flutters, increased heart rate, and flushing of the face. In the experiment in which Schacter injected this drug into subjects, he told them the drug would take a while to take effect; in addition, he told some of the subjects that they would experience certain side effects and described them; others were either not told about the side effects or were told they would be entirely different than in fact they were. The subjects were then told to wait in a room, and while the drug was taking effect they should fill out a personal questionnaire. In the waiting room there was another person, and this person, being a confederate of the experimenter, acted in one of two ways. He either acted in an angry way, cussing and complaining about the questionnaire and generally causing a commotion, or he acted in a very euphoric way by laughing and making jokes and playfully rolling the questionnaire into a basketball for shooting in the wastebasket. The subjects then filled out questionnaires in which they described their moods. Amazingly, the subjects who were aroused by the drug reported moods similar to those observed in their fellows if they had not been told to expect side effects from the drug, or if they were told to expect them but to expect different ones than those which they observed. The interpretation given to this study is that the subjects who did not have an explanation for their arousal (namely the drug) used the information in their environment, provided by the confederate to provide an explanation; thus, in a sense, they concluded, I must feel these flutters in my stomach because I am happy, or because I am angry.

The desire for stabilizing one's own impressions about the world is so prevalent that some have argued the social communication and affiliation process are based upon it. In a series of experiments, subjects are placed in a threatening situation by telling them that they will undergo severe or mild shocks during a subsequent experimental situation. After the subjects are told this, they are asked to wait outside while the experimenter gets the equipment set up. The subjects are given a choice to wait in a room with other subjects about to undergo the same experience or to wait alone; with increasing fear of subsequent shock, there is a substantial increase in the tendency to wait with other subjects. The reason: the subjects anticipate comparing their experiences and fears with those of others in order to determine how badly off they really are in the face of the presumably ensuing shock. This phenomenon is not restricted to subjects undergoing this rather elaborate ruse; if one has ever had the opportunity to observe individuals in any group decision-making or problem-solving situation, one will notice that a prevalent amount of the activity is spent in comparing each other's interpretation of the issue with those around them. Such comparisons obviously serve the important function of allowing each member to determine the correctness of his or her position, and the advocacy system of truth prevalent in the United States assumes that it is through such a process that truth is to be arrived at.

While it is perhaps true that social comparison serves the useful purpose of allowing some sorting out of what is true from what is not, it is also true that conditions can be manipulated to provide comparisons which make an individual's own position quite shaky, in fact so shaky that he will alter it. Experiments designed to demonstrate this are known as conformity experiments, and the classic case is one done by Solomon Asch,[4] in which a person is asked to judge the similarity of two lines. The lines are either equal or not equal, and when a subject is asked to do this task alone, the error rate in identifying the equal lines is virtually zero. However, when the subject is put into the situation in which a number of other individuals make judgments prior to him and make responses which indicate the unequal lines are equal, there is a relatively high probability that the subject will say the same. Moreover, the more equal-appearing the unequal lines are, the greater the tendency of subjects to respond as do the bogus confederates.

The simple conformity situation is made more striking when the responses being manipulated are more astounding. A researcher named Stanley Milgram[5] has conducted a series of experiments which were surprising in their outcome and alarming in their implications. Subjects come into an experimental situation and are met by a researcher who is wearing a white medical coat; soon another subject comes into the room and the experimenter tells the two of them that they are going to participate in a learning experiment in which one of them will be the learner and the other will be the trainer. By a seeming flip of a coin the roles are assigned, and one subject, the learner, is strapped into a chair and electrodes are attached to his wrists and arms, a jaggle of wires spew forth in every direction. The experimenter explains that the purpose of the experiment is to test the effect of shock for incorrect answers on the learner's acquisition of the correct responses. The trainer is seated in front of a control panel with a knob for adjusting the amount of power. A dial face accompanies the knob and has a series of numbers increasing from left to right. Part of the dial face is painted red and across the top of that part there is the word "DANGER". The experimenter allows the trainer to sample the shock at one of the lower levels; it produces a striking but mild sensation of pain. The experiment begins and the learner begins to give his answers to the questions of the trainer. The learner is a confederate of the experimenter and on a prearranged schedule the learner begins to provide the "incorrect" answers to the trainers questions. The experimenter reminds the subject that he is to administer a shock when the learner makes an incorrect response. The trainer selects a dial setting for the shock and presses a button which presumably releases the shock. The learner ejects a mild but loud "Ouch!". The experiment continues, and after a while the learner begins to make more mistakes; the experimenter suggests to the trainer that perhaps the voltage be increased. This time the learner yells. The experiment continues. The learner makes more mistakes. The subject administers more shock. The learner begins to scream. Eventually, the subject moves the dial toward the red painted section of the dial face. The learner begins to scream more loudly, more panicked. The experimenter urges the trainer on. The learner wiggles in his seat, shouts, pleads for mercy, cries out "I have a weak heart; please stop", and then, exasperated, droops over in his chair, slowly revives, and the experimenter suggests the subject continue with the experiment and give him more shock. The subjects went on to the full 450 volts indicated on the dial. In fact an amazing 62% of the subjects tested in this situation went the complete route; the subjects were from all walks of life in southern Connecticut, ranging in age from 20 to 60 years old. When a sample of psychiatrists were asked to predict how many people in the United States would go to the full 450 volts, they predicted one tenth of one percent. Milgram's point was to show that the blind obedience which was attributed to the Germans during World War II in the killing of Jews was not unique to that country; the point was well made.

Explanations of Milgram's experiment are many; the variations in experimental conditions, however, suggest it is primarily because the subject is in a state of uncertainty and conflict about what to do, and the only person available for helping the subject to derive resolution of his conflict is the experimenter - a solid looking and self-assured young man with a white coat and the title of "doctor" before his name.

One of the conditions of variation in the Milgram experiments was the presence of a person other than the experimenter, a person similar to the subject. This person had only to say, "I would not do that", and the subject, with some hesitation, would refuse to comply further with the authority of the experimenter. This too was a simple demonstration, and yet it brings up a second major aspect of a social control mechanism which affects the responses of individuals. That aspect is the extent to which our behaviors are public, and there is considerable indication that public exposure of our own behaviors alters our responses as much as our observation of other people's behaviors. For example, in the classic conformity situation, it is possible to increase or decrease the amount of conformity by either requesting the subject to report his own response either publicly or privately, with significantly less conformity in the latter condition.

Much of my own work is in the area of public exposure and so

I will use it to illustrate what I think are the implications of information systems for social control. One experiment was aimed at simulating the American election system in which a President is elected for two four-year terms, with a reelection between them. One of the arguments against this system is that the actions which the President takes during his first term are influenced by re-election considerations, and where the popularity of an issue is orthogonal to the benefits to be derived from the action, the President will opt for the popular position even when it is not in the best interests of the country to do so. To simulate this, I sampled some 300 possible future actions which the University of Illinois might take and in repeated and independent surveys estimated the extent to which each issue was perceived as popular among undergraduates and the extent to which they were perceived by undergraduates as being beneficial to the University community as a whole. From these three-hundred some twenty were sampled to produce five which were beneficial and popular, five which were neither, and five each which were one but not the other. From these stimuli it is possible to estimate the proportion of decisions associated with each of the two factors - popularity and benefit. The basic experiment then, was to have a subject act as the leader of a group of other students and on their behalf make wise decisions about what the University of Illinois should do; in one case the leader was going to be reelected by his peers for a second session, in another case he would continue as the leader without the bother of a reelection. The results were clear; the reelected subjects, despite the fact that no communication about the feelings of the electorate had been made, made most of their choices on the basis of popularity. The non-elected subjects made their choices on the basis of the issue possessing both benefit and popularity. In another experiment, we manipulated the presence of an observer to the situation; took away the group and added the condition that the administration of the university would be looking at the subject's opinions on the issues. In some conditions, the subject was identified with the opinion; in others not. When the subject was identified with the opinion and exposed to the students he conforms to the popularity dimension; when he is identified with the position to the administration, the subject conforms to the administration (beneficial to the university, unpopular with students) position. In another experiment, this time with regard to subjects' description of their very best friends, we manipulated public exposure to the descriptions. Young male subjects were recruited and an appointment was set up to meet the experimenter, a young woman. At the time of setting up the appointment, the subject was asked the name of his best male friend on campus. When the subject arrived for the experiment a week later, he was told that we had interviewed his friend in order to find out how the friend described him; it was explained that we were studying the accuracy of social perception between friends and nonfriends. The subject was then told how his friend had presumably described him, which was bogus, and the subject was described as a "fairly intelligent person who is friendly but often a bore in conversation and sometimes forgetful of the feeling of others". The description, in short was designed to be somewhat favorable, with a few striking negatives thrown in. We were interested in whether subjects would respond in kind, by themselves being mildly negative, or with kindness. Two independent manipulations were made; on the one hand, the subjects were led to believe that they themselves had provoked the mild negative comment from their friend and on the other hand, they were told nothing about the basis for the comment. In addition, half of the subjects were told that the friend would see the subjects' own description of the friend at a later time; the others were told the friend would not see the subjects' descriptions. It was expected that the public exposure of the subjects' descriptions would produce a powerful constraint upon their behavior, and that any tendency to be more or less kind would be mediated by this constraint. As it turns out, the public exposure had the intended effect, for the two no-exposure conditions were exactly alike and exactly in between the two exposure conditions. The subjects who were not led to believe they provoked their friend into a mild rebuke were much more positive in their own descriptions than those who were led to believe they provoked their friends. Similar experiments have been done in which subjects are asked to relate some personal incidents in their lives in which they have committed relatively mild or severe socially-disapproved actions, in one case failing to give to a charity and in the other hurting someone unnecessarily and unkindly. The subjects are then told that these incidents are going to be given to their friends either to evaluate for their stylistic competence or their indications as to the subject's personality. In some cases, the subjects are not told the descriptions will be given to friends but to strangers, or to no one. Initial analysis of the data suggests that subjects alter their own evaluations of the incidents to enhance their self-esteem in the eyes of their friends, and primarily when they know their friends are going to evaluate them themselves. However, this does not mean the subjects excuse themselves more under conditions; in fact, as expected, it means they more often interpret their severely negative act as their own responsibility, while they interpret the mild negative act as the result of the external circumstances in which they find themselves. We took this to indicate that the subjects in the presence of their friend's scrutinous eye were fellows who could admit their honesty, which is reasonable since their friends would have been able to observe the same. Thus, in essence, they maintain their esteem in the eyes of their friends.

Another well-observed effect of public exposure is not that it affects your decisions or beliefs about issues or things before you have stated them, but that it affects your commitment to them after they become public. The classic experiment for observing this effect is to ask a person to state his position on an issue and then either identify him with the position in presentation to others or not so identify him. Following this, you present the person with strong and convincing counterarguments against his own position. What is measured is the extent to which the subject changes his position in light of the new arguments represented in the attack. The results invariably are that the subject when previously identified with his position will be more resistant to the attack, and in some cases will actually become more in favor of his own side of the issue. This is a common occurrence to anyone who has watched people argue and observed the extent to which, over time, their positions diverge until they are both forced into inescapable corners.

The plain simple fact is that public exposure of our beliefs is a constraint which reduces our freedom for further action. If I have said one thing, and you know that I have said that one thing, then how can I now come and say the opposite. Would you not think I was quite idiotic, always changing my mind, never really certain about anything? Indeed a careful analysis of this situation indicates that the primary contributor to the effect is the extent to which you are uncertain about your position initially. The tendency toward resistance to change your position is partly due to your not wanting to evoke more uncertainty for yourself on an issue which you have already settled. One experiment I did to illustrate this was to manipulate subjects' initial potential uncertainty about the correctness of their decision. The setting was a management decision education testing situation. The subjects were told we were testing out some material to measure people with good management capacities. Their task (they were all students in the business program) was to read a case study and choose one of two alternative actions which the management of the company in the case should take. One choice was overwhelmingly preferred for the case. To manipulate subjects' beliefs about their own capability in making decisions of the type called for, two strategies were tried. One was to let them make some practice decisions and then inform them that they were quite in agreement with all other subjects, without implying whether it was right or wrong. The other strategy was simply to administer a bogus personality test to the subjects and inform half of them that they would be strikingly good managers according to the test while telling the other half they were strikingly poor. This latter manipulation was quite effective for gloom descended upon their faces when informed of this bogus possibility. It was obvious that the commerce students thought that by being in the business program they must have good management capabilities. After the subjects made their decisions on the cases, they were presented with a series of bogus quotations from executives of respectable U.S. companies which indicated the opposition decision should have been made. What we were interested in was the effect which public exposure would have on resistance to change the subjects initial decision in the face of the strong attack from the executives and the prior state of the subject as being more or less certain of his management capabilities. As expected, when publicly exposed, the more uncertain subjects were more resistant to change than the certain subjects; the results were the opposite for the non-exposed subjects. This was expected because for the subjects who were uncertain about their capabilities initially, having taken a stand necessitated that they reduce their uncertainty, and having been publicly identified with the stand places them in the potentially vulnerable position of having to reintroduce themselves to more uncertainty. It is easier in matters of opinion to generate good arguments to support one's own position, and in fact enhancing your own self-esteem by pointing out the flaws of the executives. This seems to happen, for the subjects were asked also to indicate which of the arguments of the executives they think ought to be printed in any educational material which was prepared. There was a tendency to recommend the weaker counterarguments be published rather than the stronger ones. It is far better to show the other side up as a dunce than oneself.

The experiments which I have described in this report to you are intended to illustrate a very simple point: It is, in my opinion, extremely easy to control the responses of individuals by simply controlling the social comparison-communication system within which they operate and by controlling their exposure to a social-observation system. This observation would be trivial were it not for the fact that our lives are enmeshed in a social information-communication system, and often unbeknownst to ourselves we are being affected by those systems. Indeed, there are new developments in information technology which are being made right now, some of which are the subject of conversation in this set of meetings. While there is nothing wrong or right about such developments in and of themselves, I would like to suggest that there are two important dimensions upon which such developments should be evaluated for their contribution to new and powerful social control mechanisms.

The two dimensions which I have been emphasizing in my description of social control are (1) the centralization of the system in terms of the extent to which it provides for or inhibits multiple informational input; and (2) the public exposure potential of the system in terms of the potential it provides for or reduces exposure of ourselves to others.

I have selected these two aspects of information systems as being important for social control because it seems to be that many of the developments in information systems and technology are leading to greater centralization of informational inputs and to greater public exposure of individual behaviors. The epitome of such a system is the recently-discussed use of the computer to conduct retail transactions from point of purchase to transfer of funds from the customer's bank to the bank of the retailer. It is not likely that such a "cashless society" will come about for perhaps twenty years because there may not be sufficient data transmission facilities available or planned to be available to handle the volume this would represent. However, even if it were to come about, one cannot begin to imagine the controlling power such a system could potentially provide. Clever people would get special laws passed to gain access to the system for credit checks, the information would be passed on the marketers of every variety. And while the developers of the system would assure congressmen and consumers alike that extensive security systems would be built into the system, anybody with a little common sense would know that if security systems can be built into the device then counter-security systems could be added on to it.

To take another example, there is a good deal of discussion these days about community information systems of all kinds. The proposals are as varied as the imagination. Some suggest ombudsmen offices and clearinghouses where the community comes to deposit their complaints against the institutions of society - the businessmen who cheat, the social service offices which discriminate or deal unfairly or officiously with community people. Others suggest computer-linked planning systems in which the community is linked directly with the decisionmakers of the community, and experts are linked as resources for both. One can argue of such systems that they represent excellent means for getting input and information to policy-making bodies. One can also argue however that, to the extent they do serve that function and to the extent they are centralized operations with mechanical or human switching devices and filtering devices, they represent important sources of control over information from the community. By simply arranging the sequence of presentation of points of view one could potentially alter the points of view which follow; by structuring the format into computer programs one could possibly restrict or amplify the types of material which are obtained and the kinds of inferences which will be drawn from them.

But one does not have to look into potential future information systems to recognize the implications which the inherent structure of those systems have for controlling events. In the United States, for instance, there is an extraordinary case study developing and unfolding every day - namely, the cluster of events popularly known as the "Watergate Affair". From the reports that dribble out like the teeth of small children as they bump against doorknobs, one can envision two striking things about the organization and the communication system of the White House. The first, which was obvious from President Nixon's announcements of his organization in the fall of 1968, is that the information system is filtered primarily through one or two men. Needless to say, the President is in a position of having only one source of information and consequently nothing to compare it with to attest to its reasonableness or its compatibility with the impressions of others. That this has resulted in the apparent damage it has resulted in is not surprising. Although many have perhaps unkindly suggested the President's image of the chaos in the United States perpetrated by students and anti-war protestors was something which bordered on the paranoid, it may have been quite realistic from the information he was receiving. The second feature, which is perhaps a problem for all echelons, is the extent to which the materials and ideas and beliefs which were developed in the White House were openly disclosed to those who belonged to the "inner circle" of aides and officials while they were carefully guarded and kept from the gaze of outsiders. The results of such an arrangement are obvious and were thoroughly studied in the United States during World War II: a sense of loyalty and cohesiveness develops firmly within the inner circle, with constant reinforcement of the prevalent beliefs, and a tendency to develop beliefs compatible with the exclusion of those outside the circle by suggesting they are perhaps "enemies".

One of the most effective forms of this kind of design of information systems for social control is employed by armies, probably throughout the world. Recruits are brought into isolated camps where they are dehumanized and all their ties with the previous existence are severed along with the cutting of their hair, the assignment of numbers in place of names, and the arrangement of them into small groups of other dehumanized individuals called a platoon. They are taught to depend upon this small group of individuals, eat with them, sleep with them, train with them, etc., etc. And then when the time comes for the young soldiers to be taught how to kill other people the group is effectively used to reinforce an otherwise unusual event; the language system is further devised to reinforce the actions, and people are described as "gooks", and interesting stories are told about how these "enemies" have done terrible things to their friends. And so on and so on, the young men - most of them - learn the difficult business of killing.

The same threads run throughout social communication systems, and while I do not wish to suggest whether these things are right or wrong in general, I would like to emphasize that they are important in that they very definitely influence and control behavior and should be considered seriously in the design of new information systems.

References

1. Bryan, J.H. and Test, M.A., Models and helping: naturalistic studies in aiding behaviour, *Journal of Personality and Social Psychology*, 6, 400, 1967.

2. Darley, J.M. and Latane, B., Group inhibition of bystander intervention in emergencies, *Journal of Personality and Social Psychology*, 10, 2, 15, 1968.

3. Schacter, S. and Singer, J., Cognitive, social and physiological determinants of emotional state, *Psychological Review*, 69, 379, 1962.

4. Asch, S.E., Studies of independence and conformity: I. A minority of one against a unanimous majority, *Psychological Monographs*, 70, 1956.

5. Milgram, S., Some conditions of obedience and disobedience to authority, *Human Relations*, 18, 57, 1965.

INFORMATION TECHNOLOGY AND DEMOCRATIC GOVERNANCE

Irving M. Klempner

State University of New York at Albany, Albany, New York, U.S.A.

Knowledge has always implied power and, unquestionably, modern information technology has made possible the extensive ingathering organization and use of its basic ingredient: information. Clearly, advances in telecommunications technology permitted the rapid collection and transmission of information; mass memory devices, coupled with computer hardware and ancillary programming techniques, have made feasible the storage, manipulation, and recall of vast information resources. Thus, the transistor and printed circuit have revolutionized not only the basic element of information technology, the computer, but also, through radio, television, telephone and satellite transmission devices, made possible for millions of individuals fast and flexible access to sources of information. A number of other electronic and optical devices permitted remote monitoring, scanning, and related information-gathering and recording activities.

What are the implications of these developments for traditional democratic-governance processes? How do these developments affect the individual citizen within our society and, indeed, governments and society at large? Are we to look upon information technology as representing simply an extension of man's physical and intellectual capabilities, or are we to consider it as a force in its own right determining, in effect, the parameters of human decision-making and subsequent human action? How valid is the *laissez innover* philosophy which demands that information technology be given free rein[1] and that it not be subjected to specific developmental checks and controls? Does information technology lead to elitist administrative structures which, in the long run, are inimical to representative governance? Can information technology have the effect of widening the cultural and political gulf which now exists between the upper and lower strata of our citizenry - i.e., between those who govern and the vast and silent majority, the governed? Is it within the realm of possibility that information technology, used unscrupulously and systematically by bureaucratic elites, may be employed for the enslavement rather than the liberation of the human mind? And if one or more of these projected dangers were shown to be a distinct possibility, what should be the obligation and responsibility of the information technologist toward his profession and toward humanity at large?

These are, without a doubt, most difficult questions, and it would be presumptuous on my part to attempt an in-depth exploration of each of them. Moreover, they are questions which confront not only the individual engaged in the field of information technology, but scientists and technologists at large; they are questions which should be considered most acutely by all of those who are in a position to influence long-term trends in applications of science and technology in general, and most assuredly, of information technology in particular. For, as has been noted in the literature, technology, and certainly information technology, touches the lives of individuals and societies at many different points and in a great variety of contexts, and therefore the ways in which it can increase the power of a few over the majority are correspondingly many and various.[2]

While scientists have frequently given considerable thought to the possible negative societal implications of their discoveries, technologists, by and large, have historically shown little concern with the possible consequences of the unscrupulous use of new technology. Thus, belatedly, among the many boons of modern technology are now listed "twelve standard horrors", including electronic invasion of privacy,[3] a product within the field of information technology. Moreover, it can be argued that complex information technology systems - whether these be intelligence-gathering satellites, crime identification center networks, or the more conventional information storage and retrieval systems, require sophisticated and enormously expensive equipment. Of equal significance is the fact that such systems require relatively scarce and expensively-educated supporting staffs which may include electronic engineers, systems designers, programmers, communication specialists, linguistics experts, systems analysts, and technicians. It can be further argued that the very costs, skills, and expertise required to assemble, operate, and coordinate the systems products of information technology would inevitably result in their concentration within an elite group of institutions and would tend to have their use controlled by a group of highly-placed, highly-trained, and, eventually, self-selected bureaucrats.

Thus, while modern information technology may permit the proper collection and organization of information, the increased power through information access may be wielded and controlled by an elite technocracy which, more often than not, and as illustrated by the Watergate hearings, finds it easy, convenient, and desirable to operate outside the traditional checks and balances imposed by a Constitutional system of government. Coupled with a set of clearly stated objectives, systems analysis, and protected by a curtain of secrecy, information technology can be instrumental in the creation and successful operation of a state-within-a-state governing system. Certainly modern information-gathering techniques and communication methods have enhanced this possibility. And if the system were to function properly, only through accident would its internal mechanism ever be exposed to general public scrutiny. The system's objectives, as may already be the case in many a management information system, may represent and may be programmed to attain the goals of the chief executive or a small group of individuals rather than include the more comprehensive input of those affected by the system. Thus, while legitimate power would be established to rest within an overt and formal legal structure, real power would be wielded by a select set of individuals, or by a single executive, operating a logical and highly efficient decision-generating apparatus. The inexorable trend toward real-time systems further illustrates the feasibility and possible effects of modern information technology on governance processes. As has been noted by H. Sackman:[4]

> In real time computing systems ... collection, organization, and storage of information leads directly to action, to integrated surveillance and control over the object environment. This dynamic marriage of information and control in real time systems is a fusion of knowledge and action, and through directed action in real time, information is expressed as power.

Power, then, expressed in real time, may be used in support of objectives which may be unknown to the citizenry and which, in effect, may be inimical to their welfare. While the rights of the individual citizen do not seem as yet to be significantly impaired by privately-generated and operated information technology systems,[5] there seems to be substantial support for the belief that *government-operated systems, operating remotely and in secrecy*, programmed to reflect secret objectives, are, indeed, posing a number of major threats.[6] Let me illustrate a projected White House use of information technology, coupled with a set of official objectives, rules, an administrative mechanism, and a secrecy environment for their implementation.

Consider the following excerpts from a Presidential memorandum relating to a domestic intelligence-gathering plan:[7]

> The President has carefully studied the special report of the Interagency Committee on Intelligence (ad hoc) and made the following decisions:
>
> *Interpretive Restraint on Communication Intelligence:*

> National Security Council Intelligence Directive Number 6 (NSCID-6) is to be interpreted to permit N.S.A. to program for coverage the communications of U.S. citizens using international facilities.
>
> *Electronic Surveillances and Penetrations*: The intelligence community is directed to intensify coverage of individuals and groups in the United States who pose a major threat to the internal security ...
>
> *Mail Coverage*: Restrictions on legal coverage are to be removed, restrictions on covert coverage are to be relaxed to permit use of this technique on selected targets of *internal security interest*.
>
> *Surreptitious Entry*: Restraints on the use of surreptitious entry are to be removed ...
>
> *Development of Campus Sources*: Coverage of violence-prone campus and student-related groups is to be increased. All restraints which limit this coverage are to be removed ...

While information-gathering via mail coverage, surreptitious entry, development of campus sources are certainly not to be considered new developments, what *is* new is that information technology in the form of electronic scanning devices, listening devices, automatic recording devices, radio communication devices, reprography devices, would very likely be used in the attainment of each of the above objectives.

In a separate memorandum, the organizational and implementation group, officially designated as the Interagency Group on Domestic Intelligence and Internal Security (IAG), was given the following instructions:[8]

> Knowledge of the existence and purposes of the group shall be limited on a strict "need to know" basis. Operation of, and papers originating with, the group shall be classified "top secret handle via Comint channels only".

Obviously, a blanket of secrecy was cast over the objectives of the intelligence-gathering plan, its methods of operation, its management, and its overall total existence. Implicit in the above outlined domestic intelligence-gathering plan is the creation of an environment conducive to maximum use of information technology. A fundamental question that needs to be considered here is whether, in the absence of information technology "advances", such a directive would have been issued by the President or whether its basic provisions could have been carried out.

Consider also, another, formerly secret, White House directive which creates the Interagency Group on Declassification and Release of Official Documents, and which calls for recommendations with respect to the following:[9]

> - types of classified information to which public access should be granted, conditions under which, and specific groups to whom, such access should be granted, and criteria and procedures for granting such access;
> - specific steps which can be taken, both administrative and legal, in cases of release or disclosure of classified information or material except in the manner authorized ...
> - changes which ought to be made in that Executive Order to strengthen both the controls over and penalties for the release of classified material except in the manner authorized ...

Obvious from the above excerpt is the attempt to strengthen the rules of secrecy, rules for legal and administrative sanctions for unauthorized release of information, and rules for the selective, i.e., purposeful release (leakage) of information.

Consider, too, the following proposed legislation, introduced in the U.S. Senate, and designed to revise existing criminal laws under the guise of the "Criminal Code Reform Act of 1973":[10]

> 1126 "b" "classified information" means any information regardless of its origin, which is marked or designated pursuant to the provisions of a statute or executive order, or a regulation or rule thereunder, as information requiring a specific degree of protection against unauthorized disclosure for reasons of national security.
>
> 1124 "a" A person is guilty of an offense if, being or having been in authorized possession or control of classified information, or having obtained such information as a result of his being or having been a federal public servant, he knowingly communicates such information to a person not authorized to receive it.
>
> 1124 "d" It is not a defense to a prosecution under this section that the classified information was improperly classified at the time of its classification or at the time of the offense.

Thus, the circle seems to be complete. Objectives have been set by the President. Information technology, whether legally or illegally used, would no doubt be fully utilized in their attainment. The whole operation is covered by a shroud of secrecy. An internal system of controls and penalties is established to protect the secrecy operation. A legitimizing overt legal cover is placed over the state-within-a-state operation by granting the President the right to classify documents secret, penalizing those who disclose their content, regardless as to whether the documents warrant or do not warrant security classification. The individuals involved in the information-gathering system are responsible only to the President.

Appropriate is the following foreboding comment made by Aldous Huxley almost three decades ago:[11]

> In the past, personal and political liberty depended to a considerable extent upon governmental inefficiency. The spirit of tyranny was always more than willing; but its organization and material equipment were generally weak. Progressive science and technology have changed all this completely. Today, if the central executive wishes to act oppressively, it finds an almost miraculously efficient machine of coercion standing ready to set in motion.

Indeed, it is the fear of the "central executive", i.e., government of men rather than laws, that persuaded the framers of the U.S. Constitution to institute a system of checks and balances. It is the Congress that would pass the laws; it is the Executive that would execute them; it is the Judiciary that would adjudicate them - all accomplished under a doctrine of separation of powers and a system of due process, as prescribed in the U.S. Constitution. However, the Huxley observation of several decades ago, the concerns of the framers of the U.S. constitutional system of government, have increasingly come into sharper focus, since current events increasingly bring forth more and more urgent questions with respect to the now fully-emerging influences of information technology as they affect the structure and functioning of our political system.[12]

If we were to include, within the previously-cited information-gathering system agency, access to a government-wide and computerized network of information sources, data banks, military files, social security records, credit ratings, medical records, criminal records, political dissenter lists, one would be compelled to ask whether, in this age of future shocks, if 1974 is almost here, can 1984 be far behind? Certainly the government and not the citizen has access to, and can make use of, these files. Perhaps our only salvation may be the normal, if not purposeful, overload of information.

What, then, is our responsibility toward the positive and beneficial uses of information technology? Like the technologists in the nuclear weapons field, should we seek out legislation, agreements, limits and channels for the proper control and guidance of our technology? Is it within our realm of responsibility to influence trends toward the adoption of interactive machine-based systems for the purpose of decentralizing rather than centralizing bureaucratic decision-making? Can we create information technology systems which would permit the technologically unsophisticated individual to interface with the system's data flow? Can we insist on legislation to assure privacy of selected data banks? Should we influence systems decision-making by illustrating to management, from a technical and ethical point of view, the dangers inherent in quantifying social values? In short, to what extent should we devote our energies toward the development of an information technology which, in the words of Erich Fromm, becomes "a functional part in a life-oriented social system and not a cancer which begins to play havoc and eventually kills the system".[13]

References

1. McDermott, J., Technology: the opiate of the intellectual, *The New York Review of Books*, 13, 25, July 31, 1969.

2. Huxley, A., *Science, Liberty and Peace*, Harper, New York, 1946, 2.

3. Calder, N., *Technopolis: Social Control of the Uses of Science*, Simon and Schuster, New York, 1970, 15.

4. Sackman, H., A public philosophy for real time information systems, in *Information Technology in a Democracy*, Westin, Alan F., ed., Harvard University Press, Cambridge, Mass., 1971, 223.

5. National Academy of Sciences, Washington, D.C., Project on Computer Databanks, *Databanks in a Free Society; Computers, Record-Keeping, and Privacy*; report [by] Alan F. Westin, project director [&] M.A. Baker, Assistant Project Director.

6. Schwartz, H., Review of Westin's *Databanks in a Free Society* in *The New York Times Book Review*, July 8, 1973, 19.

7. Domestic Intelligence, *White House Decision Memorandum*, July 15, 1970 in *The New York Times*, June 7, 1973, 36, col. 5.

8. Organization and Operation of the Interagency Group on Domestic Intelligence and Internal Security (IAG), in *The New York Times*, June 7, 1973, 36, col. 8.

9. U.S. National Security Council, *Procedures for Declassification and Release of Official Documents*, National Security Study Memorandum, 113, January 15, 1971.

10. U.S. Congress, Senate, *S1400, A Bill to Reform, Revise, and Codify the Substantive Criminal Law of the United States* ... 93rd Cong., 1st Sess., March 23, 1973, 43.

11. Huxley, *op cit.*, 3.

12. Ferry, W.H., Must we rewrite the Constitution to control technology?, *Saturday Review*, 51, 50, March 2, 1968.

13. Fromm, E., *The Revolution of Hope: Toward Humanized Technology*, Harper, New York, 1968, 96.

ASSESSING THE IMPACT OF INFORMATION SYSTEMS ON USERS

Charles T. Meadow and Robert S. Wood

U.S. Atomic Energy Commission, Washington, D.C., U.S.A.

INTRODUCTION

One of the characteristics of Western civilization seems to be our faith in technology, including its ability to solve social and behavioral problems. Recently, we have begun to see this notion challenged, particularly the idea of growth as almost an end in itself. (In fact, ecologists now tell us that growth may *be* the end.) Information systems are generally less perilous than out-of-control bio-systems or industrial complexes. Nonetheless, they have their dangers and we are quite sensibly beginning to question them. Some examples include: fully mechanized air traffic control or anti-ballistic missile defense systems; highly centralized, mechanized management information systems; or large, "efficient", but privacy-destroying, government data bases. Information systems do not directly affect their physical environment, but they inevitably change their social environment. We believe that, "The supreme task of managers in the years ahead will be ... not the further development and application of technology and marketing but rather the development of humanistic social and economic mechanisms to cope with the undesirable and generally unexpected consequences of the technological age."[1] It is the nature of this change and what we can do to control or guide it that we wish to address.

THE MODEL OF THE SYSTEM

In designing our information systems, we have a tendency to concentrate on sources and sinks of data and data flows. We almost always begin with the implicit assumption that the problem being studied has an information technology solution. We may question the economic desirability of a technological solution, but we rarely question the feasibility - can it actually be done? We tend to give much less attention to the processes performed on data than we do to the externalities of flow. Worse, we do not look too intently at the people who will use, feed or even be replaced by the "system".

One result is, we believe, that we tend to design information systems to "solve" a *model* of a problem, in the formal sense of the word model - a representation of reality, one in which certain attributes are represented accurately and others either simplified or omitted. Using models for problem solving works well in many disciplines: aircraft design; nuclear weapons effects (how else?); even, we are told, behavior of an electorate. Models of physical performance work well, especially when it is well understood what attributes are being ignored or simplified and what the relationship is of these to the ones represented. One difficulty with information systems is that we construct the model, too often, subconsciously. We do not consciously select those attributes of reality we want to include and exclude and justify our choices. We refer to our model *as* reality and lose sight of the difference.

Another result is that we do not look into the behavioral implications of our simplified assumptions concerning people. For example, when we design a query language or other user-to-computer interface language, we often design primarily for the inexperienced user. We assume, correctly, that the users will not be familiar with the new language beforehand, and that there might be some resistance to learning what many call a "programming language".

Based on this assumption - this model of the user - a simplified language is designed. After a year of operation we may find that users are unenthusiastic about the system but are inarticulate about why. One reason is that regardless of initial reluctance or resistance they learned to use the language and to do so skillfully. Now, the simple language is a deterrent to more effective use of the system because it was designed for novices, but the users' behavior was modified by the system and they are no longer novices. Now we have the wrong language. What caused the design error? - the wrong model of a user, the use of a profile of today's user rather than tomorrow's.

We would like to give two hypothetical case histories which, however, we think you might find familiar. They are intended to illustrate, first, how the role of people can be underestimated and oversimplified and, second, how the definition of data can be oversimplified.

CASE I: UNTHINKING MECHANIZATION

The reports staff (R) of a large corporate headquarters prepares reports of projects and programs carried out at lower levels and maintains figures on budgets and expenditures. They are responsible for a large number of scheduled reports and some unscheduled, or *ad hoc*, ones. Daily, they process dozens or hundreds of input documents establishing or changing budgets, establishing or changing statements of objectives of projects, reporting program contents, production, sales and expenditures. In processing these input forms, the staff scans them for errors, edits them (i.e., changes values or formats, encodes data, etc.), establishes some form of auditing control, schedules computer runs, reviews computer-generated error lists, and so on. On demand, they will determine how *ad hoc* questions can be answered, bargain with the requestors over definition, accuracy desired, timeliness, etc. R is not considered a decision-making organization. They are not shown on corporate organization charts. They are regarded, in fact, as a clerical organization.

Higher management has begun to complain, in ill-defined terms, of a lack of precision and responsiveness by R. The contributors of input (subordinate divisions, field offices, etc.) have begun to complain that: (1) the content of the reports they are required to submit to R is increasingly less descriptive of what they are actually doing and (2) the format and transmission media are expensive to produce.

Why is this? This is staff office-to-staff office reporting, based on regulations established years ago. Corporate accounting and program control procedures have changed drastically since. Many records are automated but not these particular ones. Nothing has happened to change this because managers themselves are not involved in the reporting sequence and pay little attention to anything except final output. Sensing their own growing obsolescence and management's increasing impatience, the staff is growing uncomfortable, but they, too, cannot articulate the problem.

Onto the scene now comes the information systems staff (*ISS*), armed with the conviction that their technology can solve any information processing problem. They study the problem and prepare a performance specification which all concerned are asked to endorse. The uneasiness of R increases because the words do not really mean anything to them (they do not accept the model) and because they are basically unimaginative and conservative people who do not understand what this proposed new system will do to them. It is a mistake to judge this reaction simply as one of fearing for one's job. The same reaction will arise from the most securely tenured people. The fear is of uncertainty of change, not of anything so specific as loss of a job. To speed the story up a bit, the "system" (data, computer programs, data forms, new procedural manuals) is built, tested, declared operational and put into production to produce a voluminous set of year-end financial reports. These are late and inaccurate. Management effectively loses *this form of control* over its organization. It does not lose total control, because, when necessary, a telephone call to a division head establishes all the information the vice-president needs. Once they accept the loss of scheduled, previously sacred, reports, management finds it can still operate.

In R and *ISS* there are accusations and counter-accusations and reorganizations. A year and an additional $200,000 later, the "system" is finally operating at the same level of responsiveness as did the previous manual system. The project is a failure but is called a success. It cost too much and did not reduce reporting delays or clerical manpower; yet it was able, finally, to produce reports.

What happened? Communication failure started it off. Each group had different ideas about what was wanted and none could set its ideas down in full, unambiguous detail - a perfect example of the fable of the seven blind men and the elephant. The staff, R, knew nothing of computers, were frightened of them, did nothing to help, and may even have intentionally hindered the system. The information systems staff started with the assumption the problem could be solved and, therefore, never considered the critical questions, "What really is the problem?" and "Can it be solved?" Management was too busy managing tangible things - people, materials, money - to concern themselves with *information*.

At this point, it seems clear that all our listeners have had this experience or have known of its occurring. Let us examine but one facet in more detail - the impact of the system on the actual user group, Staff R.

First, someone external to the staff declared the existence of a crisis of some sort. The staff was vaguely aware of it, but they were suddenly portrayed as the source of a deficiency without fully understanding it.

Second, although they were consulted during the "systems study" leading to a performance specification, they still did not understand the problem and did not recognize the proposed solution as a solution. They had previously spent their lives collecting and processing data in response to specific demands and they had never been involved in policy decisions about what to collect or how the results were to be interpreted. They did recognize - but to their chagrin no one else did - their own great knowledge of the data, for example their ability to spot an out-of-range number on the basis of 20 years' experience reading such numbers, but without any explicit validation rules.

Third, the staff was asked to operate a new information system with virtually no training. Starting from a position of ignorance about computers, they moved, variously, to positions ranging from a hostile, total disbelief to a bewildered willingness to believe anything.

Finally, when data began to flow through the new system, people no longer knew how to catch and correct errors or how to remove the effects of an erroneous correction. Because patterns were broken, a clerk did not know who to call when a number looked suspiciously like a misprint. Instead, a cascading error *might* have been introduced into the data base.

John Dearden has indicated that part of this problem arises as early as the feasibility study stage of a project. He states: "All feasibility studies assume able systems and computer people, acceptance of middle management and support of top management. Yet, it is the lack of these that most often causes failure."[2]

CASE II: INTEGRATED FILES - A TECHNICAL IDEAL

An integrated file or data base is one in which each element of information need appear only once, interconnections among elements are made explicit, and data elements are defined in a uniform manner.[3] Thus, if a corporation's entire data base were integrated, there would be only one representation of a division's budget, only one sales forecast file, and only one entry of each employee's identification number. Proponents of this method of organizing files argue that it saves storage and eliminates confusion caused by multiple versions of the same data. It seems a technical ideal: all elements well-defined and uniquely represented, storage space minimized, and semantic confusion eliminated. Let us take a hypothetical tour of a typical business firm as it moves into the era of integrated files.

As in any large organization, each division and lower subdivision keeps its own budget. Replacing the various individual budget systems with a single corporate budget file was one of the promised benefits of integrated files. Integration would still recognize the rights of individual divisions and even protect divisional files from access by other divisions or unauthorized corporate staff members. The integrated budget file was implemented, but a survey of computer use, a year later, indicated that divisions had *increased* the amount of computer time devoted to budget programs since pre-integration days. The reason: "bootlegged" systems. Divisions, independently, began to run their own budget systems to feed the corporate system. The latter became, not a working file, but an output file. A unified sales forecasting file was also planned, but never implemented. The information systems department had been unable to get marketing, produce development and the controller to agree on the basis for forecasts - the kinds of products to be assumed available, the nature of the market, the value of goods and currency. Since an integrated file design requires the concurrence of all major users, effort on this one is stymied.

The corporation's Defense Division has recently had a major cost overrun. The MIS failed to provide a warning until it was too late. By then, production was irrecoverably behind schedule. While a debate is still going on about the reasons for the overrun, one point was brought out early in the investigation: cost figures were as much as 12 weeks out of date at the time of their review by management. A file of correspondence was discovered that disclosed a bitter controversy between accounting and several line project managers. The accountants insisted that cost data could not enter the system until a firm commitment was made for it (contract signed, or invoice approved for payment) and could not be released until after the close of the accounting month. The line managers wanted approximate expenditure data entered and reported, as soon as the authorizing manager felt he knew approximately what he was going to spend. Then, as precise data became available, the approximation would be improved. For example, the division rented time on a computer time-sharing system. The managers wanted to estimate expenditures each month, the accountants wanted to record actual expenditures after being billed. Top management ignored what few early warnings were given, possibly out of sheer disbelief. One now-resigned vice-president said that the situation lacked the appearance of an overrun. When that had happened in the past, division headquarters heard from all sides at once. He had gotten only a single warning from his management information system, hence assumed the warning was wrong.

What has gone wrong in this corporation? First, a technical solution was imposed upon a social body - the firm. Second, no one understood what this solution really meant, in mechanical terms; just who was to do what to bring this about. Third, incredibly, management "bought" this idea with little resistance, far less than a proposal for a new building or advertising campaign would have engendered. The promised reduction in storage cost and redundancy seemed to warrant the effort. No one challenged the premise that what is good for the computer is necessarily good for the corporation. This is not, incidentally, a condemnation of integrated files and certainly not of integrated file software. It is a condemnation of the assumption that, without proof in each case, the complete integration of corporate files is desirable or even possible.

Somewhat akin to the problem of integrated files and illustrative of a similar approach to problem-solving, is a proposal for a "total communications system" in the new town of Columbia, Maryland. The system, proposed in 1965, was to be designed eventually to include such offerings as: cable television, remote library reference services, remote shopping and merchandising, legal reference and consultation, tutoring service, on-the-job training, vocational retraining, and time-shared computer service, all electronically delivered to the home. Inhabitants were to be able to subscribe to any mix of services offered.[4]

The "benefits" of such an advanced application of technology were offered as self-descriptive illustrations. Thus, the "benefit" of an on-the-job training link would be an on-the-job training link. Problems which these technology applications would ostensibly solve were only obliquely mentioned and nowhere enumerated. The ideas were fine, but they seemed to promote technology for its own sake rather than to solve extant problems. In 1973, for all practical purposes, no part of the system exists.

THE LESSONS

Taken collectively, many of our information system problems can be categorized simply under the heading of shortsightedness. We do not look at a broad enough segment of the problem; our models, as we said earlier, are faulty. We need to examine the tacit premise of many information systems projects - that an automated information system *is* a feasible solution. We need to increase the scope of information systems projects, but adding depth as well as breadth. In short, we need fewer pious proclamations of need and capability, and more description of how; also, more description of how the social change will be made, not just the mechanical change.

We cannot be satisfied with a statement of user needs as proof that the need can be fulfilled and that, if it is, the original problem would indeed be solved. (In fact, can a need be said to exist if it cannot be fulfilled?) And as we have discussed earlier, oftentimes agreement cannot even be reached on what "user" needs are.

We cannot be satisfied to describe an information system in terms of a file and a set of reports. We must consider: How do the data get into the file? How accurate, how timely are the data? Who will create, transmit, enter, edit and correct the data? What is known about the people who will perform these functions; their attitudes, their state of training and competence now; the same attributes when the system becomes operational?

We must recognize the changes our systems will bring about and propose how to cope with them *as part of systems design*. For example, information tends to generate demand for more information. What new kinds will be wanted? How will the system respond? There will be social changes as individuals adapt differently to a new environment and increase or decrease their contribution to their organization, becoming perhaps more productive or perhaps a burden.

These proposals are not radical, nor even new. There is a literature on justification of management information systems.[5,6] The gist of it is that the setting of an information system should be considered along with technical requirements and cost. In the management context, emphasis is placed on determining what management problems will actually be solved. Here is a quotation from a discussion of how to select computer projects. "No matter what the size of the organization, deciding what to do and what comes first is a perplexing task. In the face of increasing competitive drives and increasing competitive opportunities and threats, the ... manager must come to grips with computer project selection on a measured, analytic basis. The effective screening, selection and evaluation of computer projects require: (1) criteria for making judgments, recommendations, and decisions about authorizing and maintaining projects, and for assigning and shifting resources and (2) a system, a method, a set of procedures. ...We can no longer afford to plunge into the beckoning opportunity without a serious reckoning of costs and returns. We must focus on producing the most valuable output of our available human and machine resources."[7]

TECHNOLOGY ASSESSMENT

Another approach to a broader look at problems is that of technology assessment. The U.S. Congress has paid increasing attention to the need to examine the probable effect of a new technology on all facets of society. Former Representative Emilio Daddario and Senator Edward Kennedy have been primary spokesmen for a formal mechanism to assess new technologies. Their work, and the work of many others in and out of Congress, culminated in October 1972 in the creation of the U.S. Office of Technology Assessment.[8]

According to the expressed views of Congress, technology assessment attempts to "provide early indications of the probable beneficial and adverse impacts of the applications of technology". The assessment process includes the identification of existing or probable impacts of technology; ascertaining cause-and-effect relationships; and identification and comparison of alternative technological solutions. By definition, technology assessment is "the thorough and balanced analysis of all significant primary, secondary, indirect, and delayed consequences or impacts, present and foreseen of a technological innovation on society, the environment or the economy."[9] Technology assessment is also concerned with long range, sometimes irreversible effects, such as changes in educational techniques in primary schools, regulation of mass media, or air pollution. Rarely must we in information science contend with problems of such great significance, although significance is clearly relative. When a company goes out of business due to an information failure, its employees and stockholders may have a different view of the significance of business information than others.

Yet all this may be easier said than done. Although generalized methods for technology assessment have been publicly discussed, they have by no means been completely tested or proven effective. With so great a task and with so great a diversity of technologies to be assessed, a "cookbook" approach does not exist. One author suggests that "the problem is akin to that of how one can eat an elephant. And, considering the magnitude of the task, it is difficult to argue that one place is better than another for the biting to start. And, after a considerable amount of biting has taken place, the elephant remains largely unscathed..."[10]

With a balanced view of technology assessment in mind, we should now ask how this methodology may be transferred from a macro situation to a micro situation. The process is difficult, primarily because individual organizations are much less predictable than the aggregate of organizations in a society. Nevertheless, we think that the general methodology is valid in both instances. The analytic nature of assessing the impact of technology on a particular organization is little different from that of assessing societal impacts. The process of technology assessment of any scope is very similar to any "systems" analysis. The assessor must state the problem as completely as possible, identify alternative solutions, analyze the probable impacts of these alternatives *at as many levels as possible*, and make his decision accordingly. Identifying the whole mix of results which could accrue from a particular action is critical. Perhaps, here, our behavioral and organizational theorist colleagues can tell us more than we know now.

SUMMARY

To summarize, we have used hypothetical examples (which, however, are based on true experiences) to point out some of the greatest failings of information systems. Among them are these: that systems are often technology-inspired rather than need-inspired; that designers ignore the people who will work them, sometimes to the extent of designing unworkable systems; and that their designers do not consider the new environment which their systems will operate in or what will actually happen when the system is implemented.

We have pointed out, on the other hand, that there is beginning to be recognition of the problems and even the development of a methodology to cope with them. What is mainly needed now is a willingness to bear the extra cost and time that these methodologies will add to systems development. They seem, however, investments sure to pay off. Relatively new analytic tools offer one approach to the technology assessment dilemma. Statistical methods, model-building and other techniques markedly increase the power of technological assessment. Yet, paradoxically, these very tools, as we have indicated earlier in the paper, in some senses are themselves technological "problems". What must be done is to strike a workable balance between the truth of reality and the intelligibility of simplicity.

We would like to conclude with a quote from *The Limits to Growth*. "Our intent is certainly not to brand technology as evil or futile or unnecessary. We are technologists ourselves, working in a technological institution. We strongly believe ... that technological development ... will be absolutely vital to the future of human society. We would deplore an unreasoned rejection of the benefits of technology as strongly as we argue here against an unreasoned acceptance of them."[11]

References

1. Schwartz, M.H., Identifying universal principles in MIS design, Keynote Address, *Third Annual Conference of the Society for Management Information Systems*, Denver, Colorado, September, 1971.

2. Dearden, J. and McFarlan, F.W., *Management Information Systems; Texts and Cases*, Richard D. Irwin, Inc., Homewood, Illinois, 1966, 31.

3. Plagman, B.K. and Altshuler, G., A data dictionary/directory design within the context of an integrated corporate data base, *Proceedings of AFIPS Fall Joint Computer Conference - 1972*, AFIPS Press, Montvale, New Jersey, 1972.

4. Stone, C.W., *et al.*, *A Library Program for Columbia*, Pittsburgh, Pa., 1965.

5. Blumenthal, S.C., *Management Information Systems; A Framework for Planning and Development*, Prentice-Hall, Englewood Cliffs, N.J., 1969.

6. Cohen, B.J., *Cost-Effective Information Systems*, American Management Association, New York, 1971.

7. Schwartz, M.H., Computer project selection in the business enterprise, *Journal of Accountancy*, April, 35, 1969.

8. Hahn, W.A. and Chalk, R.A., *The Technology Assessment Act of 1972* (P.L. 92-484), Congressional Research Service, Library of Congress, Washington, D.C., December 1972.

9. *Technology Assessment for the Congress*, Staff Study of the Subcommittee on Computer Services of the Committee on Rules and Administration, United States Senate, U.S. Government Printing Office, Washington, D.C., 1972, 7.

10. *Ibid.*, 64.

11. Meadows, D.H., *et al.*, *The Limits to Growth*, A Potomac Associates Book, Universe Books, New York, 1972, 154.

SOME (MORE OR LESS) PHILOSOPHICAL THOUGHTS ON INFORMATION AND SOCIETY

Eric de Grolier

Institut National des Techniques de la Documentation, Paris, France.

UNISIST'S DIAGNOSIS

At the end of its first chapter, the *Final Report* of UNISIST, three years ago, defined the "ultimate target" of this new international system as follows: "to find new means for reducing the irrationalities and injustice which vitiate the processing of scientific information throughout the world."[1] "Irrational", "injust" - such deprecatory epithets would appear rather severe in the eyes of many experts, whose natural bent of character is an indomitable optimism. Some other people, inclined to consider contemporary events with that dispassionate, cold (their adversaries would say "cynical") outlook which Machiavelli used for describing the beginnings of our modern civilization, would uphold the opinion that irrationality and injustice are essential, intrinsic characteristics of present societies (the most pessimistic, or most lucid, among them would add "of *all* societies"), and that the activities of international organizations like Unesco are no more than an *alibi* with which the Great Powers (or Superpowers) cover their attempt to control social evolution with a view to achieving their own, selfish (or even imperialistic) interests. In the following remarks, I will not try to decide between these conflicting opinions, but to present some arguments in favour of each side, within the limits fixed by the three questions propounded by the "Technical Program Guideline" of the present Institute.

DISCOVERING THE IMPACT OF INFORMATION TECHNOLOGY

What are the best ways to discover the impact of information technology on the individual and on institutions? My personal answer here would be entirely negative: there are *no* best ways leading to such a discovery. One must use a variety of approaches in trying to clear up some of the many problems which are raised by the interaction between information technology and social development. Each one of the four main methods which could be proposed for that purpose has its own limitations, and it is only their combination which would, possibly, produce some useful data on which the penetrating insight of able, gifted futurologists could try to delineate the preferred lines of conduct to be submitted to the decision-makers, to the men of action.

Delphi Method

This is the method recommended by Georges Anderla.[2] However, some serious doubts are to be raised concerning its value, at least in our field. If one looks at the predictions reported by Anderla himself, on the basis of the Swedish Delphi study[3] and the quasi-Delphi Japanese study,[4] one cannot fail to be struck by the confidence with which the "experts" who were consulted have placed in the year of grace 1987 the advent of completely automated, personalized information services.[5] From my point of view, this extreme optimism does not take into account the economic, political and social forces which in fact resist the generalization of innovations.

When, retrospectively, we look at the fate of certain innovations in the past, we note that, as the proverb says, "there's many a slip 'twixt the cup and the lip". More than twenty years ago, the biggest multinational firm in the data processing field triumphantly advertised the full success of an experiment in automatic translation, and if a Delphi study had been conducted at that time, there is little doubt that the majority of "experts" would have predicted the complete replacement of men as translators by "the machine" within less than a decade. Nowadays, probably no really responsible "man of knowledge" in this field would take the risk of advancing a precise date for the day when it will be *economical* to obtain a mechanical translation at least as *reliable* as a good man's translation. Indeed, many linguists would even assert that this is a sheer impossibility, as Bar-Hillel contended in a famous report.[6] Or, let us take on-line information systems. The first successful, operational system of this kind was set up in 1957 - the SAGE system. Sixteen years have elapsed, and on-line systems in civil information networks are still the exception even in the United States; in Europe, they could be counted on the fingers of one hand, and less than two years ago the director of France's military information center discarded bluntly their possibility, on economic grounds. About thirty years ago, Gérard Cordonnier in France announced that the microfiche would soon replace the book as we know it, but three decades were not enough to produce a real boom in microfiche, and it is still far from replacing even the hard copy technical report. The modern school library was born, in the United States, about seventy years ago: it is still an innovation in France, and almost unknown in many developing countries. Indeed, a well-known expert in educational planning writes that it takes about two generations for any important educational innovation to be generalized, and this opinion is probably even too optimistic for any country, except the most developed ones.

Turning to another domain, that of agriculture, it is rather distressing to re-read the predictions of the best experts made in the sixties, which envisaged such an over-abundance of agricultural products that the decision-makers of the European Economic Community were urged to convert into fallow land about 12 million acres (in the then six countries of the EEC). To-day, it is scarcity and not overproduction which we are facing! I would not conclude from all these pitfalls that Delphi exercises are always worthless, but only that experts are, like all men, fallible and that their opinions are to be taken with a grain of salt.

Cross-cultural Comparative Studies

The cross-cultural comparative method as practised, among others, by political scientists and sociologists, has produced recently rather interesting results, especially the various surveys conducted under Professor Stein Rokkan's leadership, by the International Social Science Council Standing Committee on Cross-cultural comparative research. However, its applications to information science are still in their infancy: the so-called "comparative librarianship" is a discipline in its early stages. At the Institut national des techniques de la documentation, we introduced, in 1970, a course of study on information networks, one result of which will be a book edited by Mr. Guilloux, co-ordinator of the course, to be published next year. I am at present writing its Introduction, and I experienced great difficulties trying to synthesize the lessons to be taken from the various experiences described, both on the national level and on the plane of international subject-fields networks. Different socio-economic systems are producing widely divergent national information "systems" (which, rather frequently, do not really deserve such an ambitious label). It does not seem that the experience acquired, by many trial and error processes, in one more advanced country is easily made use of by another nation, even if its own social, economic, political, and cultural characteristics are very similar. The same is true of different "fields" - either disciplines or "mission oriented" fields: the lessons drawn, say, from the organization of scientific and technical information services in chemistry are not so easily transferable to physics or biology - not to speak of egyptology or, say, informatology. However, in my opinion, the nascent discipline of cross-cultural comparative informatology is a promising one, and its development is certainly to be fostered;

the detailed and critical comparison of the various modalities according to which STI is organized regionally (in the widest sense of the word) and by disciplines is, by all means, a most valuable method; let us only remark that the study of failures is at least as interesting as that of successful ventures!

Case Studies

This is a method which is intermediary between cross-cultural comparative surveys and historical studies, so it is appropriately placed here. I would strongly recommend to the persons serving as advisers for decision-making bodies on STI policy the careful study of such first-hand accounts of major discoveries like *The Double Helix*,[7] or the story by Professors Jacob and Monod of how they found the transfer RNA (contained in their acceptance speech for the Nobel prize), or, again, of serious, well-documented analyses such as the book just published by Chertok and De Saussure called *The Birth of the Psychoanalyst*.[8] Case studies of that quality are, unfortunately, all too rare, and it is highly desirable that their production be encouraged. Some community studies conducted during the last two decades or so included one or two chapters on the ways and means by which people are informed (or misinformed); for instance, the one in France on Auxerre by Bettelheim *et al*. But, to the best of my knowledge, practically none up to now has provided valid data for answering the question put, about a hundred years ago, by Ernest Renan: "Peut-on travailler en province?" - which we could translate by "Is it possible to conduct research in provincial towns?".

A recent report by Karl Deutsch *et al* on major innovations in the social sciences has proved that discoveries occurred in very few privileged places, which correspond more or less to the towns where "centers of excellence" exist (first class universities, major research centers). I would be very pleased if some Ph.D. theses could be devoted to the detailed evaluation of STI resources available within "library centers" which Louis Round Wilson detected in his pioneer study called *The Geography of Reading*.[9] I myself made a similar attempt for France in the forties. For example, a very preliminary survey which I undertook concerning psychological periodicals subscribed to by French libraries revealed that only three French provincial cities (Lyon, Montpellier, and Strasbourg) seemed to offer collections strong enough in this field for supporting advanced research, but that even these were still very far from reaching the Parisian level (there is no doubt that this wide discrepancy between the STI resources in Paris - and its suburbs - and in the provinces is one of the main reasons for the deplorable concentration of scientific research in our capital, dramatically described in the Lisle report, just published).

Historical Research

It is a rather curious fact that the history of information techniques and systems is in such a state of underdevelopment as we can note at present. It is true that we have at our disposal some excellent monographs, like Shera's social history of the beginnings of public libraries in the United States,[10] or Mme Dufresne's biography of Ameilhon; one could cite a few dozen scattered studies of this quality on as many individual newspapers, propaganda enterprises (among others, the Creel Committee, or the Nazi propaganda studied in Tchakhotine's *Le Viol des foules*) or "invisible colleges" (like the one in France known as the "Cercle d'Arcueil"). But syntheses are sorely lacking, and very few authors have the required thorough knowledge of cultural and social history, and/or of the history of science, to enable them to place the evolution of means of communication and information in its larger context, so that one could find more enlightenment reading a book like François Jacob's *La Logique du vivant*, which is a general history of biological theories, than a hundred specialized articles on one or another aspect of STI during the last three centuries since the "take-off" which occurred in 1665.

The methods of the new school of quantitative history have so far been almost never applied to the history of information systems, and we can deplore the fact that after the pioneer study of Wyndham Hulme, half a century ago, bibliometrics remains in its "pre-disciplinary" stage (except, maybe, in USSR); that is a sideline for a few isolated individuals. Even in France, where the development of the "Ecole des Annales" had some favorable consequential effects (François Furet's team studying the book and society in the 18th century, Estival's thesis on "La Statistique bibliographique en France sous l'ancien regime"), we cannot so far perceive the signs of concerted efforts in this field. Reading such authoritative reports as that of J.-C. Gardin for UNISIST, or G. Anderla for OECD, one is astonished to find so many inaccuracies, omissions, and distortions in their historical accounts part. The resulting lack of valid historical perspective is extremely detrimental to futurological predictions. For one sound, scientifically-based bibliometric study, like that of Professor Boutry on the evolution of the physics literature since the First World War, there are dozens of superficial, ill-documented, methodologically unreliable essays, which quite naturally mislead well-intentioned but too self-confident model-makers. There are, however, some indications that this lamentable situation is beginning to change. New periodicals (like the *Journal of Library History*) are appearing which pay more attention to the history of information systems; studies stimulated by the interest recently shown in the so-called "Bradford's Law" and in Goffman's "epidemiological" theory have attempted to collect historical data verifying (or invalidating) these models.

Integration of Methods

There is a great need for integrating various methods - those sketchily evaluated above, and maybe also others - within the framework of interdisciplinary research programs. One of the main conclusions of the second part of Anderla's report for OECD was for recommending the creation of an Institute of Information Sciences and Technologies, "to put an end to the lack of coordination and the present confusion in the matter of training and advanced research".[11] It is rather disappointing to remark that something quite similar was already advocated by Douglas Waples almost forty years ago,[12] and that in his book *Investigating Library Problems* - unfortunately out of print for a long time, and still not replaced - this same author pioneered in providing a remarkable outline of the whole array of research methods to be employed in such a multidisciplinary institute. I personally made various proposals with the same purpose (back in 1933, 1939, 1942, and again in 1956, 1971 ...) - without any success. Apparently, neither in the United States nor in France did any academic or governmental consensus exist for supporting a large scale attack on the whole front of informatological research. Circumstances are now perhaps different and public feeling more mature - at least let us hope so.

INTERACTION BETWEEN INFORMATION TECHNOLOGY AND SOCIETY

In this section I would like to bring forward some arguments and hypotheses which could serve as a point of departure for more thorough and extensive studies - which would go far beyond the limits of a report for an Institute - on what I prefer to describe (instead of using the terminology of the "Technical Program Guideline"), the *impact* of information technology, as the *interaction* between this technology and societal development.

Let us begin with a very simple, positive, factual statement: information is, with matter and energy (and, perhaps, movement), one of the most elementary, indefinable actions. It is intimately connected with structure and order; indeed, absence of information or disorder are quasi-synonyms. Differences in quantity (and possibly in quality) of information do make essential differences in nature among "things": as Monod aptly remarked, the only difference between crystals and living organisms lies in the different amount of information (by several orders of magnitude) which is present in their respective structures. According to a perfect Hegelian law, quantity becomes quality, and quantitative differences involve qualitative "jumps" when nature's evolution is passing from procaryotes to eucaryotes, or from *préhominiens* to Man: at each cleavage, at each new level of "emergence", one can notice a sudden increase in the quantity of information, which is correlated with a more complicated, refined structure. The deepest, most fundamental characteristic of the various stages of human evolution, from australopithecines to Homo Sapiens, lies without doubt in man's capacity for processing information, and human societies, from clans to empires, are best gauged according to the amount and complexity of information which is necessary for the functioning of their social structures. As André Lebeau recently wrote, "what characterises most fundamentally the degree of development of a population is the quality of the information which is at the disposal of its members when they are confronted, individually or collectively, with their problems".

Now, it would be a grievous mistake to consider that this evolution is linear, and that progress has been uniform, at least from Cro Magnon to the 20th century. In the same article by Lebeau, from which I just cited a passage, there is a very curious argument put forward for countering the thesis of those who challenge "the ways of the present technical society, and more particularly those of [economic] growth". Lebeau writes that this attitude "seems to run counter to a natural law, which is that evolution does not regress." This assertion - which has the characteristics of an article of faith - is at variance with many facts, even in the field of non-human evolution (let us cite only the phenomenon of parasitism and remind the reader that, some decades ago, a book appeared in Paris under the title of *L'Evolution Régressive*, which the majority of biologists considered as grossly exaggerating the regressive movements, but did not treat as a simple fake). But, coming to Man, Lebeau's creed is simply untenable. We could give him counsel to read such books as Levi-Strauss' *Tristes tropiques*, or Georges Friedmann's *La Crise du Progrès* for convincing himself that the belief in indefinite and continuous progress is to be left to the old curiosities shop, and in fact has been abandoned by western social philosophy since Rousseau. Paul Valéry once wrote that "Nous autres, civilisations, savons maintenant que nous sommes mortelles" (We civilizations, do know now that we are mortal). Now, the death of a civilization can best be interpreted as a break of continuity in its information mechanisms - as the death of a man is established by EECs proving the irreversible disruption of his central nervous system. The civilization of Ancient Egypt, as well as that of the Mayas, were definitively dead when nobody could any longer read their writings; the burning of Alexandria's Library and the closing of the last (Hypatia's) chair in Athens rang the knell of our own Ancient World. And it seems rather frequent that such a break occurs just after the moment when information structures are at the acme of perfection. Between Trajan's death (117) and the end of Septimus Severus' reign (211) the Roman system of information was at its summit: the wonderful network of Roman roads was completed, the Acta Diurna ensured the smooth transmission of orders, decrees, etc., and the work of the great jurisconsults (Ulpianus *et al*) guaranteed the uniform Rule of Law throughout the Empire. But the next hundred years saw the great persecutions of the Christians and the construction of Aurelian's

Wall, which showed that the civilization of the Eternal City was now on the defensive; and another century had not yet passed when, in Bordeaux, Ausonius lamented the end of "the" (i.e., his) civilized world. Communications were everywhere (in the Western Empire) endangered or broken, and before long, instead of great libraries with hundreds of thousands of rolls, monasteries would be proud of collections containing some dozens of codices.

A majority of our contemporaries are doubtless convinced that such things could no longer happen. Are we sure? Really sure? The more complex, the more sophisticated from a technological point of view our systems of communication and information become, the more vulnerable they are. If the whole "memory" of contemporary civilization was to be stored in a Giant Computer, with hundreds of thousands (or even millions) of terminals connecting it to the remotest places on Earth, through satellites for direct tele-distribution - as some of our experts envisage - I would be very much afraid of the possibility that one day Barbaric people might burn it, as they burnt the Great Alexandrian Library, and that 99% or even 100% of our "cultural heritage" would be so lost. But, a civilization's death probably begins from the interior, before it is finished off from the exterior. From Cumont (*Les Religions Orientales dans l'Empire Romain*) up to now, many authors have pointed out how the growth of mysticism, of "counter-cultures" and contestation movements undermined the Roman Empire much before it was seriously imperilled by Barbaric invasions. The late Professor Sorokin, to explain these (and other) facts, propounded a cyclical theory, according to which all civilizations are passing through periods of rationalism, scientific and technical "progress", then of criticism, doubt, and finally reach a wonderful era of solid faith (the last of these enviable states having been, for our Western World, the times of Albertus Magnus and Thomas Aquinas) and, according to Sorokin, a new period of that kind is now approaching.

The International Social Science Council is now proposing a major research project on what its chairman calls "World Dissatisfaction"; the representatives of disciplinary international associations which are members of the Council have approved this idea. Dissatisfaction, apparently, is present everywhere - but is it for the same reasons? The Brazilian Junta is buoyantly conducting a Pacification War against the last unknown Indian tribes which could oppose the construction of the Transamazonian (and at the same time, maybe, hinder the destruction of the last great oxygen-producing forest of our planet); according to a television show recently transmitted by the ORTF, the poor "last savages" so pacified are dissatisfied by being deprived of the gadgets of their peacemakers, and especially those which serve to transmit their "superior culture". At the UNISIST Intergovernmental Conference, in October 1971, many delegates from Unesco's Member States representing the Third World nations were deeply dissatisfied by a "system" which they considered to be designed for the sole benefit of the developed countries and which, in their opinion, did not take into account the urgent, specific needs of their people. In NATO countries, one has only to read the underground press, or such non-conformist magazines, widely bought by the younger generation, as (in France) *La Gueule Ouverte* or *Charlie-Hebdo*, to be convinced that large strata of our populations, and the most "activist" ones, are refusing *en bloc* the dominant culture - not only the atomic bomb or nuclear plants, pesticides, and chemical fertilizers, but also, and perhaps above all, advertising, the educational system, television, the Press tied to the Establishment, in short the whole network of communication and information media which is at the very basis of our present civilization. I saw recently in one of these publications, a violent, and somewhat naive, denunciation of all components of our STI system, from patents to the "invisible colleges", which at first made me smile, then irritated me, but in the end provoked in my mind some mournful thoughts, and some doubts about the moral justification of what I am proposing as progressive models to my students.

Do we know how much people are "satisfied" or "dissatisfied" in the USSR, in Popular Democracies or Mainland China? From such facts as the existence of "Samizdat" and internments of non-conformists in psychiatric hospitals in Russia, the rather widespread distrust of the official press in Eastern European countries, and the sudden disappearance of practically all conventional STI media in China during the Cultural Revolution, one might be tempted to conclude that there also, "tout n'est *pas* pour le mieux dans le meilleur des mondes" (all is *not* for the best in the best of all possible worlds).

Turning now to science fiction writers, we could be struck by the fact that so many of them envisage the emergence of some kind of World Empire (before the end of our century), thus sharing the predictions of more sober observers of the current evolution of multinational firms, such as Robert Lattès. But they see it almost immediately followed by a sudden collapse - involving, of course, the total disruption of communication and information networks - after which would prevail a state of anarchy, of complete anomy (see, for instance, René Barjavel's *Ravage*, published about thirty years ago, as well as Michel Jeury's *Le Temps Incertain*, 1973).

Let us return to less futuristic aspects of our problems. One rather important remark which can be made, studying the history of that fashionable subject, the "diffusion of innovations", is that up to now a crucial role has been played by personal contacts and by a kind of "group effect" in information transfer. S. Dejider has most convincingly shown that the phenomenon of "brain drain", so much lamented by newspaper men and, occasionally by government representatives, is not at all new, and that, on the contrary, it can be detected as having been a major factor in scientific development in Athens, Alexandria, as well as more recently during the late Middle Ages and early Renaissance. Many other studies - among them an excellent one on the diffusion of innovations in England from prehistoric times to the 19th century - pointed out the part taken by "zones of contact", places where the mutual influence of different cultures can most easily occur. Examples are here too numerous to deserve more than a mere mention: from the Ionian coast to Venice and Padua, from Alexandria to Bagdad or Sicily (for these last two, especially as centers of planned translation efforts), from the Rhine region (Gutenberg!) to the Paris of the first part of the 19th century. Travel and emigration exercise the same kind of influence: let us just remind ourselves how his Paris and Nancy sojourns contributed for Freud to his elaboration of psychoanalysis, or how American scientific psychology profited from so many young professors of U.S. universities going to their German colleagues as students or junior researchers, or, again, what vigorous impulse was given to the still developing U.S. science by migrant Jews and liberals, socialists, etc., coming from Germany and Italy during the late thirties. What we call the "miracle grec" (Greek miracle) is possibly nothing more than such a "group effect" in a place which was at the crossroads of many cultural influences. And the Renaissance was probably caused mainly by the cumulative effect of information transfers from various cultural centers (the Arabs, Central Europe, Byzantine refugees during and after the collapse of the Eastern Empire). It is very interesting to note that, during the Cultural Revolution, the flow of printed information to and from Mainland China was almost completely interrupted, as well as the internal circulation of scientific journals, etc., but that China's science and technology were not at all hindered, owing to the fact that personal contacts (symposia, attendance at congresses and fairs) were, on the contrary, increased.

These facts must serve as a warning against any STI policy which would put too much emphasis on the use of scientific and technical literature, and of automated systems capable only of giving long (or not so long) lists of references more or less relevant and dropping, in virtue of their poor recall performance, 40% or so of what could be of use (and, maybe, the best and most advanced things being in this 40%). Albeit so many experts now are predicting the end of person-to-person meetings, as video-phones and other devices of that kind become in general use (which, at least in France, is not for tomorrow), I would, if I were a "decision-maker" in this field, still very much rely on sending as many of my people abroad for getting information directly from "the man who knows". And I would probably decide on the immediate suppression of the *Bulletin signalétique* of the C.N.R.S., allocating the millions of dollars which it costs instead to some "Fund for Sending Young Researchers to Foreign Centers of Excellence" (FSYRFCE). I'm pretty sure that this simple measure would provoke an almost immediate boom in French R & D results (provided, of course, that when they return to the Mother Country, these youngsters are not left unemployed).

No one could deny the effect of improved means of communication on STI development, and, by way of consequence, on scientific and technical development itself. Let us only remark that international scientific and technical congresses, conferences, symposia, etc., began where, during the second half of the 19th century, railways lines were first interconnected, and received a vigorous impulse from the progress of air travel after World War II. Stein Rokkan repeatedly proposed to Unesco's General Conference (up to now, unfortunately, without success) the creation of a Fund for helping social scientists from the Third World to participate in international scientific colloquia, seminars, etc. Such a measure would, to be sure, be most beneficial, instead of Unesco's squandering so many hundreds of thousands of dollars a year for subsidizing bibliographies which nobody reads. However, a word of caution must be given. Not all cultural contacts are beneficial. The Conquistadores destroyed pre-Columbian cultures without any benefit, either for themselves or for the subdued Peruvians, Mexicans, etc. And, recently, an Indian author has shown, in a special issue of the *International Social Journal* of Unesco, what powerful defences the traditional Brahmanist culture of India built up against the diffusion there of Western science and technology.

We know very little, after all, about the role of information in creativity. Too much information is perhaps as detrimental as not enough information. The role of "marginal men", not wearing the blinkers of prejudices or of inherited obsolete theories, has perhaps not sufficiently been emphasized. It is not by mere chance that Sigmund Freud - a Jew coming from a multinational state - and not Pierre Janet - a member of the Establishment of the self-confident "fin de siècle" France - was the founder of psychoanalysis. Neither was it by accident that Karl Marx, a stateless émigré, was the person who unified French utopian socialism, English economics and German philosophy into such an explosive theory. Why do certain small countries, placed at the confluence of several cultures, like the Netherlands or Israel, show better performance in scientific and technical research, in relation to their population and to their GNP if not because there it is easier to absorb various influences, to be freed from the "esprit de clocher" (parochialism)?

About one and a half centuries ago, in a deplorably now almost forgotten little book, the young Guizot set forth a theory according to which the spectacular development of European civilization was due to the fact that here, and nowhere else (at least not to the same degree), has there been what he called the "co-existence de principes divers" (coexistence of diverse principles). It does seem to me that this hypothesis is worthwhile reviving, and that it has something to do with the development of a theory concerning "the impact of information technology on the individual and on institutions". I will return to this point when dealing with deontological problems.

Now let us discuss a little the language question. Peter Marler, in a stimulating study on animal communication, has shown what a strong correlation exists between the evolution of language and the structure of animal societies. Zoosemiotics is now, apparently, more developed than sociolinguistics, and, apart from a few exceptions, sociolinguists have sadly neglected the subject of scientific language (in fact, so far as France is concerned, the half a dozen books written in that field, for instance by Pius Servien and Dagognet, were not authored by professional linguists). The most important remark which is to be made is that, from the beginnings of scientific thought, there has always been a *dominant* language, which served as a kind of *lingua franca* for the large majority of living scientists. The first one was probably Egyptian, then it was Greek - for a most important period. Then the supremacy was divided between Chinese and Arabic. Medieval and Modern Latin was the language of science, from about the time of Roger Bacon and Oresme up to Newton, with, perhaps, a short interlude for Italian just before the judgment of Galileo. The 18th century, and the beginning of the 19th was the glorious, but rather short, era of French, with the great abstracting and translating enterprises of the Académie des Sciences. From 1830 to World War I, German was dominant. Now, it is English (or Anglo-American). What next? It seems that the phenomenon of "accélération de l'histoire" (acceleration of history) described by Daniel Halévy is noticeable here too: the flourishing period of each new dominant scientific language is apparently becoming shorter. Another fact which we can note is that the eternal reign of one ISL (international scientific language) happens to be celebrated just at the moment when its predominance is coming to an end: Rivarol's *Discours de l'Universalité de la Langue Française* was published a few decades before French ceased to be really universal. If I would risk a prophecy, I should predict that before the end of this century, English will be replaced by some other language for STI - but which one I would not dare to say - maybe an artificial language.

Against this possibility, however, objections crowd upon one's mind. No artificial pasigraphy or pasilaly, since the Real Character of Bishop Wilkins, has ever met with any appreciable measure of success. *A posteriori* systems, from Esperanto to Lancelot Hogben's Interglossa, took for granted the superiority of indo-european roots (in one or the other of their historical transformations), disregarding the plausible claims of, let's say Chinese, for providing the lexical basis of a new, neutral language. *A priori* systems - which were practically the only ones imagined for two centuries - have been condemned almost unanimously since Couturat, and

Gardin's report for UNISIST recently brushed them aside with a few disdainful remarks. Besides, up to now, no language dominated the STI scene except by virtue of being that of a dominant power - empire, nation, or, for Medieval Latin, church. The advent of a new candidate would depend on the strength of its supporting forces, and the present lameness of international organizations does not allow serious hopes for their capacity for imposing on the scientific and technical community the use of any other language than that of the current Super-Super-Power.

From another point of view, it is not quite certain that STI really requires the use of a *spoken* lingua franca. On the contrary, the historical example of Chinese ideograms, and the more recent experience with mathematical signs (one can recall the fact, widely reported by newspapers, that at a recent international congress of mathematicians, the audience followed for two hours a demonstration at the blackboard, entirely in silence, and at the end burst into applause), logical operators, structural chemical formulas, electrotechnical symbols, etc., does give some credibility to the thesis of Ch. K. Bliss, according to which "semantography" with only graphemes (or could we say: semantographemes?) would provide the solution. However that may be, we must admit that the present situation is a most uncomfortable one for the large majority of the earth's population, whose mother language is not American. If no other factor was present, this one would alone be quite sufficient for ensuring a privileged position for Anglo-Saxon scientists and engineers - and for being responsible for one of the main injustices deplored by the UNISIST report, but against the perpetuation of which it offered remedies almost entirely worthless.

The limits of a report for an Institute such as this do not allow us to delve into details of the impact of information technologies. Let us nevertheless examine, albeit superficially, some techniques, or "tools" (adopting the extensive definition of "tool" proposed by Illich.[13] The auditory signs which Homo Sapiens inherited from his ape and monkey cousins had, contrary to the chemical signs preferred by so many other animal species, a major disadvantage: the lack of durability. Their partial substitution by writing was therefore such a progress that major frontiers between our disciplines - history/prehistory, sociology/ethnology - are based on the presence or absence of this criterion: possession of a writing system. Marcel Cohen was certainly right when he entitled his great book on the history of writing *The Great Invention of Mankind*. Concerning writing, at least three remarks are worthwhile introducing. First, the rate of diffusion of this innovation has been, after all, extremely low, if we consider not only some privileged populations, but Earth as a whole. Dozens of Central Asian languages were first provided with writing by the Soviet Union in the 1920's. When I was in Senegal, three years ago, discussions were still active on the best writing system to be given to serere, wolof, and other vernaculars which are spoken there. Secondly, even among literate peoples, the rate of illiteracy (or quasi-illiteracy) is still high, except in the most developed countries. Thirdly, the textbooks and grammars used for teaching people to become literate, and among them, interestingly, those used by the two languages most important internationally, French and English, are hopelessly inadequate, leading to an enormous waste of time and energy for the poor pupils having to master (more and more inefficiently, if we judge by our examination papers) the intricacies, subtleties and incoherences of the official "orthography". A fourth remark, of a different character, would be that "writing" - either as handwriting or, more frequently now, typewriting - has *not* been made obsolete by other, technically more advanced tools - typography, computer listings, videotapes, etc. - and that, especially in the forefront sectors of research, it has retained all its value, on the same plane as the personal "man to man" contacts to which I alluded above. In a field like transplantation, in which no less than 13 new international periodicals began publication during the last five years, our best French specialist, Jean Hamburger has just written[14] that

> it is difficult ... to recommend lists of books or articles to be read, as progress is so fast that written [he means "printed"] information is almost always obsolete when it is published: direct and constant contacts from researcher to researcher are the only way for following the advanced front. Here is a domain where telephone, airplanes, *letters* and unceasing small meetings on one or the other side of the Atlantic have become major research instruments (my italics).

Here we encounter the first manifestation of a phenomenon which is probably quite general in informatology - namely that new techniques up to now never totally *replace* the old ones, which still remain valid in their own field of application. If and when the letter, today a useful "tool of research" as it was in Huygens' and Father Mersenne's time, is definitively supplanted by, say, telefacsimile devices, is a matter of conjecture.

Printing was known in China and Korea about six centuries before Gutenberg. The reason why it was reinvented in Mainz was that it was socially and economically indispensable, at that stage of development of European universities, because the "pecia" system no longer provided the required study materials. As Aesop's tongue, it could be depicted as the best or the worst thing. Without printing, modern science, modern nations - in short, modern society - are unthinkable. Freedom of expression is, to be sure, inseparable from the printing press, which made possible Milton's *Areopagitica* as well as Voltaire's or Thomas Paine's pamphlets. However, printing also permitted the diffusion of *Voelkischer Beobachter* and the mass of tabloids which are just now intoxicating hundreds of millions of NATO countries' citizens. The rapid strides made, since the turn of last century, by what we now call reprography are perhaps even more important than the development of conventional printing: I would not hesitate to putting in the place of honour, in an imaginary Gallery of Great Men of Informatology, the portraits of the inventors of Roneo, Gestetner, Rotaprint - as it is they who made possible the technical reports boom which has had such a tremendous impact on STI organization since 1940 or so. The same is true for xerography and other electrography devices. All these tools, comparatively cheap and easily operated by non-specialists, brought to STI a hitherto unknown flexibility, and to society at large better potentialities for small groups, counter-culture movements, and non-conformists generally speaking, to voice dissenting opinions. The case of micrographics is somewhat different. Its impact up to now, after all, has not been dramatic - to use a rather fashionable word in these Institutes. As Klaus Otten aptly noticed in a recent article, technology is here in advance of the social, economic, and even cultural or political conditions which could make possible its more efficient use. And, profiling on the horizon, there is now holography, the expansion of which might have rather spectacular effects. Non-book materials - to use a rather inadequate, but widespread terminology - are still in their incunabula era, and their Gesner is not yet born. The Council of Europe recently decided to set up, probably in Florence, a Center for insuring their bibliographic (extending the meaning of this word!) control and fostering their utilization in higher education. Some recent documents (among which two special issues of *Library Trends* are most noticeable) are pointing to the quasi-revolution which they could provoke in educational technology. What is perhaps less known are the benefits which ethnologists can derive from their application for preserving the cultural heritage of pre-literate societies, which otherwise could easily be lost forever. In this respect, let us mention the exemplary work done by the Archives Culturelles du Sénégal during the last decade, with the help of the French ORSTOM (Office de la Recherche Scientifique et Technique Outre-Mer).

Here again, as for the printed "book", the *Janus bifrons* nature of information technology is to be emphasized. The same "mass media" which do contribute so much to the disintegration of ancient, venerable but highly vulnerable cultures different from the Western one, and the apparently irresistible levelling of cultural traits the world over, to the homogenization of different civilizations which they are dissolving (or melting) into a uniform magna ("Les enfants de Marx et du Coca Cola" - the children of Marx and Coca Cola - as it is said in a movie by Jean-Luc Godard) - these *could* be transformed into powerful instruments for cultural revivals. But the preconditions for such a transformation are now quite rarely satisfied. CATV, videocassettes, satellites for direct teledistribution - these novelties will not be exceptions to this general rule. It is extremely interesting to follow the vicissitudes which characterize CATV developments in European countries. Such a "liberal" (in the 19th century meaning of the word) country as Belgium accepted rather early this new medium, but did not allow the right to exploit its best potentialities by authorizing private groups of citizens to insert their own programs into the circuit. Italy, after a period of more or less clandestine tolerance, bluntly forbade, a few weeks ago, all derogations to the RAI state monopoly, convenient pretext for avoiding the horrifying perspective of being obliged to grant some freedom of expression to possibly anti-establishment movements. Our own French governments followed a somewhat tortuous policy. Arguing, like the Italians for the state monopoly of the ORTF (and of the PTT), they refused to authorize more than extremely limited experiments. They created a semi-public, semi-private firm, "Vidéogrammes de France", associating ORTF with the biggest oligopolistic group in the press, printing, and book publishing industry, Hachette; whereupon the chief negotiator of this agreement from the governmental side became General Manager of that group. They promised that, in the future, larger scale "experiments" (in the "villes nouvelles" - satellite towns - the local press would be duly involved). Then, the Hachette group hastened to buy two small weeklies, in the two regions which would be first concerned by that "liberal" decision. An ill-intentioned observer could add that the Hachette group constantly - and intelligently - supported the Gaullist part, and that it was De Gaulle himself who, in 1945-46, vetoed its nationalization. In all these cases, we are rather far from the true liberalism of Canada, with its CATV and its video groups promoting the self-expression of many small circles, with divergent - and not always pro-governmental or pro-establishment - opinions.

Three years ago a symposium, convened under the auspices of the Encyclopedia Britannica and the University of Chicago (the proceedings of which have been edited by Sackman and Nie), gave an opportunity for examining the social and political consequences of what will be, almost without doubt, the great Revolution in information technology of our time: the fusion of two hitherto separated techniques, that of computers and that of telecommunications with some other components, among which are holography, micrographics and videographics. It would already be necessary to bring the symposium results up to date, if only for taking into account the far-reaching effects of satellites on direct teledistribution. Let us remark, in this connection, that U.S. technology is here about 15 years in advance of the European one: the experimental American satellite ATS-F will operate its first retransmissions next year in India and Brazil; on the other hand, on our continent, the German Federal Republic is the only country preparing something similar, but for the horizon 80.

Present-day television is, we must admit whether we like it or not, the best instrument for the alienation of the whole mankind, and the most demagogic, by nature, of all mass media. Illich explains why he refuses to speak on television:[15]

> "Because the television discourse is unavoidably demagogic. A man speaks on a small screen, millions of men and women are listening. In the most favorable case, the maximal reaction of the public could only be 'bip, bip, I agree' or 'bip, bip, I disagree'. No real interchange is possible."

(The case of the newspaper is a bit different: at least, there is the "Letters to the editor" possibility.) It is precisely this interchange which the new technologies will make possible - and, maybe for the first time, they could abolish, or at least make more permeable, the barriers which up to now existed, and were even gradually reinforced, between the domain of interpersonal communications, where constant feedback was allowed (STI), on the one hand, and that of "mass" communications, where this feedback was practically non-existent, on the other hand. There are, however, some prerequisites for such a happy ending. One of them would be that statesmen must be wise enough to ensure that these new tools do not become segregative instruments, as has been the fate of the book (according to Illich). Another one is that scientists themselves must make more extensive and coherent efforts to speak what the head of the department of economics and sociology at the French National Institute of Agricultural Research recently called "un discours intelligible" (an understandable discourse). This would entail much more than one annual Kalinga Prize.

What about libraries, documentation centers, information analyses centers, and the like? No doubt, according to the official creed of our time, here are truly democratic servants of Civilization (with a capital letter). It is true that libraries, the world over, are used much more (and in certain countries, like Italy and even France, almost exclusively) by the middle and upper middle classes, professionals, students and housewives. And the Third World (undeveloped countries), and even more the Fourth World (that is, those lower classes of people in developed countries which are more or less "marginal": Negroes and Chicanos in the States, or Portuguese, Arabs, Turks, etc., in Western Europe, and also the "poor whites") are almost deprived of these cultural institutions. Library resources are as unevenly distributed as any other resource, or even more so, among countries and within countries (underdeveloped areas in developed countries are, generally speaking, not better served than the so-called "developing countries). One of the main reasons for the brain drain to the United States might be that their great libraries offer unequalled facilities to scholars. In fact, even in Paris, I myself not infrequently dream of Pittsburghian or Chicagoan collections! Very, very few governments really organize their library networks with a view to allowing free access to all their resources for any "man in the street" in the remotest village: Denmark, England, Switzerland - the list is, with these three names, probably complete. Besides, nonconformist literature is almost everywhere either completely banished from public (and even university) libraries, or its use is restricted. Moreover, there is a curious distortion in the library structure of underdeveloped countries (and, from that point of view, even France is underdeveloped). The "lower strata" - school and children's libraries, public libraries for the "common people" (bibliothèques "de lecture publique" in France, Büchereien in Germany) - are much weaker than the "upper strata" - university and, above all, special libraries. I remember vividly the situation in Dakar: the University library, set up and managed by the French, contained beautiful series of French, English, and even German scholarly bound books and periodicals (I noticed, among others, the *Handbuch der Physik*) jealously preserved, in "restricted" areas open only to professors, and practically never read. When I was shown, by contrast, the small "public libraries", organized by Senegalians with private funds and manned by unpaid volunteers, I found a few hundred worn out paperbacks and old magazines.

On December 7, 1972, at the National Center for Scientific Research in Paris (CNRS), a heated, and sometimes stormy discussion occurred during a meeting held by the three major professional associations concerned with STI (ABF, ADBS, ANRT) on the theme of means of access to ST documents. It showed, very concretely, how in a so-called highly civilized and developed country like France, it is in fact difficult to have access to the sources of STI, what barriers - economic, institutional (without speaking of those due to military and industrial secrecy and to the Great Professors' mandarinate) - are erected across the passes, and finally how any poor chap, who does not have the good fortune to be a member of a (preferably Parisian, if not at least Lyons, Marseilles or Strasbourg) team connected with an international invisible college, is in fact at a loss when he or she tries to get in time, the information which would be most essential for his or her work. It is, after careful consideration, extremely surprising that such an evidently most efficient institution as MLSST has not yet been imitated by any other country, or that NTIS (with its predecessors) was copied in USSR and FGR with a lag of about two decades, and that French official circles have been speaking about the advantages of that kind of service for a decade without having been able to decide on its creation. A little more than one year ago, on May 4, 1972, the then officially-appointed decision-maker in the field of French STI, Mr. J. Delorme, presented at a meeting of AFCET a paper on STI policy, in which one can find this extraordinary statement:

> "It is out of the question, taking into account the volume of the information [coming from foreign countries] and the cost of its processing, that we could have at our disposal in France at least one copy of all world scientific and technical publications. We must, on the

contrary, make a selection, essentially based on the importance of the documents and the modality of procurement abroad".

(What was meant by this "modality of procurement" - "modalité d'obtention" - was left in the dark; neither was it explained who would judge the "importance" of such and such a document, and according to which criteria the judgement would be made.) What a wonderful acknowledgment of impotence, for the representative of a government so proud of its power! To the credit of this eminent polytechnician, it can be said that nobody mentioned to him the existence in the United States of a certain Farmington Plan, or the fact that Lenin, in 1921, ordered for a country ruined by war, revolution and foreign intervention, to undertake precisely what he considers impossible for the prosperous French Fifth Republic. Under these conditions, it is not surprising that many French researchers or engineers nowadays prefer to order direct from NLIST the Xerox or microfilm copies of the documents which they need. After all, it could be reasonably maintained that, Europe being such a small continent, a policy for exhaustive acquisition of world STI would have to be determined at the level of European institutions - either the EEC or the Council of Europe.

If we turn our attention again to the Third World, we cannot be astonished if we see that, at the UNISIST Intergovernmental Conference, one delegation made an interesting observation, during the discussion of a certain Recommendation No. 17 concerning the study of specifications for modern information transfer networks: it pointed out a number of practical obstacles barring the way to the development of such networks, notably the shortage of means for reprography. It would be wise to aid the developing countries in solving such practical problems before beginning to set up networks of more sophisticated services.

Let us now have a look at indexing and abstracting services. One thing must be noticed at once: only two countries in the world have been able to sustain an almost complete and self-sufficient system of such secondary publications, the United States and the USSR. And, even for them, there are certain gaps in these networks: the U.S. depends on Great Britain for physics, and many Soviet chemists apparently prefer CAS to the *Referativny Zhurnal* for their current work. Germany, valiantly fighting to maintain the famous *Chemisches Zentralblatt*, was finally constrained to cease its publication, after one hundred and forty years during which it had been the symbol of German supremacy in this field. The French experience is no less interesting. In 1940-42, the CNRS, turning to the best account the circumstances created by World War II and the Nazi occupation, scuttled the various abstracting services maintained, for better or for worse, by a few professional scientific societies and replaced them by a sort of French *Zentralblatt*: the *Bulletin analytique*, later on rechristened *Bulletin signalétique*, as its "abstracts", never longer than five or six lines, did not deserve a better label than that of a mere descriptive notice. Nobody knows how much it cost the successive governments of the Etat Français, the IVth and Vth Republics, to continue this prestige undertaking, which is still supposed to be indispensable in avoiding the dismantling of the already all too much threatened position of French as a scientific and technical language, and the rallying of the scattered members of Francophonie to the clientèle of Anglo-Saxon or even (who knows?) Russian STI enterprises. The result is however, from a practical point of view, derisory: French researchers, when they can, do use the American or English abstracting services. I have the deepest respect for the competent and devoted persons who are the editors of various parts of the *Bulletin signalétique*, but it could not prevent me from thinking that it is an enormous waste of time for the young researchers who are obliged to write the pseudo-abstracts, and for the documentalists who compose the bibliographic descriptions or make the indexes, of paper for their printing, and finally of money, both for their publishing and for the libraries subscribing to them. The sooner it is stopped the better.

My country also has the privilege of lodging, and partly of subsidizing, the editing staffs of a number of international annual indexing publications in the social sciences and humanities (history, geography, classical philology, etc.). At the risk of being looked upon, here again, as an iconoclast, I would advance that this formula is now badly antiquated, and that these services just survive as a manifestation of the general tendency for all beings to persevere in their being. Generalizing the debate a bit, I would say that there is nothing to be gained for any country which is *not* one of the two superpowers in spending more than the smallest possible amount of money in these secondary services. What it must ensure is that its national ST literature be adequately covered by the Anglo-American (and Russian) abstracting and indexing journals, and that indexing be properly done for its own primary publications (and, possibly, the primary publications in its language printed outside its borders). A small slice of the budget squandered for the B.S. would be sufficient to provide Francophones with the equivalent of the H.W. Wilson indexes, which are now sorely lacking. On the other hand, these smaller countries would also have to develop information analysis centers for those quite specialized fields for which they possess the required competent manpower. This is, in fact, the path which is being followed in France for textile technology, cancerology or economy of energy. Annual reviews and state of the art reports are another species of the large family of STI productions, sometimes labelled as "tertiary publications". It was once a German specialty, and is now an American one for something like 80% or more; however, in certain technical fields, for example foundry, the German effort is still quite remarkable. The Boutry report recommended in vain that French authors be encouraged to devote more energy to that kind of exercise; it's a pity that the inertia of our public powers prevented any follow up of this good idea. Data banks are fashionable and any comprehensive survey of STI institutional framework would no doubt have to examine very carefully their distribution and development. In France, the Delorme paper already cited above devoted seven lines to this question, concluding that a "common policy" would have to be defined by the BNIST (National Bureau for STI, the creation of which was delayed until February 1973) and the Délégation à l'Informatique (Delegation for Computer Science). As far as we know, this policy has not yet been made public. Unesco's Department of Social Science has officiously proposed that the International Social Science Council take the lead in scheduling a project for the establishment of a European Data Bank or Data Archive within the framework of the Center for Coordination of Research and Documentation in the Social Sciences, which ISSC set up, with Unesco's support, ten years ago in Vienna. This could be a most interesting venture, as data banks and data archives for the social sciences are rather undeveloped still in Europe, and could set a lot of rather crucial problems (of secrecy, protection of privacy, information retrieval systems for facilitating secondary analysis, etc.) which are far from being solved.

To conclude these scanty, but already too tedious remarks, I could submit for thinking over by this audience some generalizations which could serve at least as working hypotheses for a more elaborate study:

(1) Interpersonal, direct communication, with immediate feedback, which was the rule in small, "primitive" societies - Hopis, or Greek city-states, or Swiss Cantons six centuries ago - but was driven back by the expansion of mass media, can be revived (on a much wider scale!) by new technologies. However, this will not be an automatic process; on the contrary, it would require concerted efforts from social groups which might profit from this new situation, and would probably encounter much resistance from various components of "the Establishment" as, in the long run, it could signify the end of the State as we know it.

(2) Up to now, *all* developments in information technologies were instrumental in fostering the polarization of richness, power and "culture", in increasing inequalities between poor and rich nations as well as social differentiation inside developed countries, in dismantling "primitive" and "popular" cultures to the benefit of a world, homogenized "consumption society". The coming revolution in communication and information technologies will give to mankind the possibility of "reversing the engine". It is not at all sure that this opportunity will be seized.

(3) There is a kind of "Matthew effect" in information, para-

llel to that which Robert K. Merton described for science, or another which concerns the distribution of income and the influence on it of "social transfers" and which motivated last month a controversy in the Paris newspaper *Le Monde*. The countries where there exists a good network of information systems of all kinds, from telephone lines to libraries, are those where the new technologies receive their fastest and most extensive applications. The social groups already best served by good bookshops, fine libraries, high-standing newspapers, will be the first beneficiaries of the new "information utility".

(4) "Information is power", and information is the basis of decision-making. André Lebeau describes this as follows: "the professional activity of that class of the population which analyses, judges, orients, decides, is exclusively reduced to information transactions". Paul Nizan was more severe; describing the European colony of Aden, he writes:

> "hidden in their hygienic dens, they work under fans' wings in offices where ... typewriters inscribe without respite a few black signs. The life of folk from our countries consist in combining them, undoing them, recombining them. It's a crazy game."[16]

Anyway, people for whom confidential, secret, private information is the main source of their privileged position do not easily share it with others; thousands of examples could be cited here, and some exceptions only confirm the rule.

(5) From Plato (at least) down to our time, the progress of information technology has been correlated with the development - and maybe the hyperdevelopment - of specialization. Knowledge has been crumbled, fractionalized, split up into watertight compartments. One's specialist jargon is impervious to another's, even if their scholarship provinces are contiguous, Leonardo da Vinci was the last universal genius, Poincaré the last universal mathematician: when the limit of this process is reached, the situation will be the one frequently described, where the perfect scientist "knows everything about nothing". Efforts at restoring a minimum of encyclopedism have been, on the whole, deceptive. At a time when everybody is chattering about pluridisciplinarity, multidisciplinarity, etc., communication among different disciplines has never been so hazardous. And it is symptomatic that, studying the various proposals for a typology of the main services to be ensured by future automated information systems, Anderla could mention only one author who had taken into consideration the need for "initiation to a field of activity new for the user". Professor Jean Wyart, devising the *Bulletin analytique* of CNRS as an instrument for interdisciplinary research, had imagined that bringing together summaries of articles, in the shortest form, from a variety of specialized periodicals, would satisfy the needs of researchers exploring the literature outside the kernel of their familiar journals. We

do see now that it was a fallacy. Other less mechanical devices will have to be invented (by the way, I doubt very much if this kind of service will ever be, as Anderla predicts, entrusted to a computer - at least not before "artificial intelligence" research bestows upon the machine the capacity for synthesis which, up to now, it does not possess at all).

(6) There is at the same time too much information and not enough information: scarcity of valuable, high quality information, and superabundance of redundant, worthless, or even noxious, spoilt information. Modern civilization produces mental pollution as well as physical pollution, spiritual nuisances at the same rate as material ones. Fighting this degradation of our moral, social environment is surely as urgent as preserving or restoring our natural environment - but certainly even far more difficult.

DEONTOLOGICAL QUESTIONS: THE "CODE OF ETHICS" OF INFORMATOLOGISTS

Deontology is now a subject in vogue. Psychologists are studying an international code of ethics for their profession; the APA is revising that which it adopted ten years ago or so, and in France a hot discussion has been provoked by a severe pamphlet of M. de Montmollin, "Les Psychopitres" (The Psychobuffoons). The International Social Science Council has proposed to its members - disciplinary associations - to undertake a project for comparing the various rules of conduct followed in the different fields of social research. In domains which are not very far from informatology, public opinion research and advertising, controversies are actively carried on for determining what is permissible or not; the deontology of journalism is periodically debated. But I must confess that I do not understand very well the question put on the agenda of our Institute by the "Technical Program Guideline": "In what areas of human endeavour is it necessary that a science of information establish standards and a code of ethics?" - and that for two reasons. First, it seems to me that information is pervading *all* areas of "human endeavour"; it is not appropriate to make a choice among them. Research, management, home and foreign policy, education - all these fields make constant use of information, and its use or misuse could be governed by some "ethical" rules which are not so very different one from the other. Secondly, it is quite apparent that deontological rules are formulated and endorsed by professional associations, not by men of science. As informatology is as yet not covered by a professional association embracing all its branches, I do not see how a practical code of ethics could be established.

The Boutry report, in France, contained a proposal for what it called a "code de bonnes manières" (Code of good breeding), which covered such trivialities as the desirability of authors' abstracts, avoidance of double (or multiple) publication of the same results in various periodicals, etc. On the other hand, *The Double Helix* and numerous publications by R.K. Merton have shown what kind of unwritten "code of ethics" did exist among scientists concerning their treatment of information: the war to the knife for priority leaves very little place indeed for moral considerations. And so far as mass media are concerned, I remember a book by Stuart Chase which described all attempts at decontaminating their domain as futile "dykes of straw" - that was about forty years ago, but apparently if things have changed, it's not for the better.

Of course, it would be rather easy to outline the elements of a generalized informatological ethics - for a future Utopia! For the sake of diverting us a bit from more serious tasks, I propose that it could follow the motto which, in my country, is inscribed on the façade of town halls, schools, churches, and prisons: "Liberté - Egalité - Fraternité" (Freedom, Equality, Fraternity). After all, our Fathers of the Revolution knew how to stir public opinion in favour of worthy ideals. Freedom first. It is quite appropriate. Freedom of expression and freedom of access to what has been expressed. Against any attempt for concealing the Truth, restricting the availability of useful information, or monopolizing it for egoistic (individual or group) purposes. Authorizing minority, non-conformist groups to use freely every available means of communication. Then Equality. Ensuring that any man on Earth, as destitute as he might be, can have access to the Treasure of Information of the whole of Mankind, irrespective of his social status, race, geographical location. Abolishing all privileges coming from birth, wealth, political and managerial power, so far as information is concerned. Finally, Fraternity. With, maybe, part of the huge world armament budget, arrange an international system permitting the Third World nations to close the "information gap" which is one of the most tangible obstacles to their development.

As soon as we come down from the ethereal realm of Dream to the hard ground of Reality, we see at once what a chasm opens up between the ideals of fancy and the facts of our earthly world. We could argue that freedom of expression is more restricted in Pompidolesque France than under the benevolent laxity of Mr. de Malesherbes, and freedom of communication more impeded in today's Europe than in the cosmopolitan society of the Prince de Ligue and Goethe. Passports and visas were practically unknown during the Belle Epoque, before 1914 (except in Russia) - what a difference now! During the Napoleonic wars, British scientists were allowed free entrance into the enemy's citadel, Paris. Would this ever again be possible? Research conducted under the direction of Pierre Bourdieu at the Centre de Sociologie Européenne has most clearly shown the illusory character of the proclaimed Equality of access to knowledge, even in countries like France. What about India or

Brazil or South Africa? And one can discern precursory signs of a resurgence of racist theories - the negation of fraternity - in the States (look at the controversy about IQ racial differences) as well as in France (liberty left to the Ordre Nouveau movement to propagate its neo-Nazi doctrine).

Turning to the less poisoned domain of STI, strictly speaking, A. De Grazia's studies have dismounted the mechanisms of recent examples of scientific intolerance. Huge increases in the cost of indexing and abstracting services - especially with the advent of magnetic tapes - make it impossible for the average researcher to keep abreast of ST progress, unless special measures are taken (as in Canada) for helping him. All this is not particularly comforting. I might be taxed with undue pessimism, but what I see in this last quarter of our century is pointing more to a return of the Dark Ages than to the entrance into a new era of Enlightenment. In any case, one cannot but subscribe to the recommendation of Mr. Anderla's report, asking that we begin preparation of what he calls an "International Information Charter".[17] The danger would be in building up unrealistic, idealistic standards, which real facts would immediately invalidate. What we need at present is not so much a "code of ethics" as research into the "science des moeurs" (scientific study of manners and customs) applied to information transfer.

INFORMATOLOGY AND PUBLIC POLICY

We are now arriving at the last question on the agenda of our Work Group: "What is the role and function of information science in public policy?" Let us first observe that this formula seems to imply that there is a one-way causal effect, from informatology to public policy. I wonder if this is true. Until now, statesmen did exert much more influence on the orientation of information systems than informatologists carried weight on decision-makers in this matter. This is not a situation peculiar to information science, by the way. Informatology belongs more to the social sciences than to natural sciences, and social scientists generally speaking have been of very small account in Public Affairs as has been demonstrated by several recent American books.

So far, no information scientist has been a member of the Brain Trust of any head of State - if one refuses to give this qualification to the late Dr. Goebbels (who, however, was certainly one of the best experts ever known in the art of the "Rape of Crowds", which in fact is one of the most important parts of applied informatology). It might be sustained that Gabriel Naudé's role as Mazarin's counselor was a historical precedent. Let us notice that Naudé was not only the author of the *Avis pour dresser une bibliothèque*, but also of a very interesting book on the technique of the Coup d'état, in which he revealed himself as a worthy disciple of Machiavelli. It is true that, so we are told, both President Nixon and Mao Tse Tung were in their prime youth employed as clerks by libraries, but it did not apparently very much influence their subsequent attitudes as politicians. One can also sustain that, albeit indirectly, the British Museum and the German-Swiss Leihverkehr were together responsible for the Russian Revolution (as there would have been no *Kapital* without the documentary resources put at the disposal of Karl Marx by the excellent work of Panizzi, and the writing of both *State and Revolution* and *Imperialism, Last Stage of Capitalism* required the smooth functioning, in Geneva, of what was then the best interlibrary loan system.

Nevertheless, it does appear that the effects of public policy on information science - and on information techniques - have been far more considerable. Studying them adequately would require the use of at least the last three of the four methods we considered at the beginning of this paper. One would have to begin, probably, with Ptolemy's policy in Alexandria; then one would consider the remarkable plans submitted by Bishop de Talleyrand in 1791, and which the Convention tried, vigorously but with mitigated success, to put into practice. One century and a quarter later, one could take into account England's creation of the National Central Library and the Research Associations, and then evoke Lenin's famous decrees for the organization of a comprehensive information network covering the whole Soviet Union. The poet Archibald McLeish's role under Franklin D. Roosevelt would be evaluated. And now, after the Sputnik shock in 1957, it would describe the measures taken first by the U.S. government, and then, with a time lag varying between ten to sixteen years, by other NATO Member States for reorganizing their STI systems. On the international plane, beginning with the Académie des Sciences' remarkable enterprises under the Ancien Régime, followed by the Royal Society's leading role during the 19th century, one would have to evaluate the results of efforts sustained by national and international professional associations and the - as yet mostly unsuccessful - action by Inter-Governmental Organizations for regulating the world's information flow. Yes - here is a vast field for future research, and NATO as well as other international sponsors (OECD in particular) would be well inspired to devote to it a reasonable amount of money.

CONCLUSION

I am very conscious of having raised more questions than provided solutions or given definite answers. My possible justification is that the role of such an introductory paper is to provoke discussion, even at the risk of appearing too controversial and rather unorthodox.

Many interesting problems have been left aside in my cursory remarks. Among them is that of *objectivity* - this ideal aim of STI system proponents, which is so difficult, and even impossible to achieve. That of the technocratic misuse of information, and of the dehumanizing role which one can attribute to too many modern information technologies. That of the social consequences of the rise of information utility industries, and in particular centralization vs. decentralization. That of the real significance of the rather numerous signs which appear, here and there, of a coming crisis in the traditional STI institutions: from the severe restrictions in the budgets of great public and university libraries in the U.S. and the catastrophic situation which has confronted French universities for the last two years, (many of which were forced to cancel *all* book acquisitions to maintain at least their periodical subscriptions), to the CNRS's incapacity for coping with the increase of STI production in the coverage of the *Bulletin Signalétique*. And, finally, the long range effects which will unavoidably derive from the worldwide recognition of Man's right to information - on a par with the right to freedom, to work, to security. But let us leave all this for some future, permanent, International Institute of Informatology!

References

1. UNESCO, *UNISIST. Study Report on the Feasibility of a World Science Information System*, 1971.

2. Anderla, G., *Information in 1985. A Forecasting Study of Information Needs and Resources*, OECD Document, 1972, 66.

3. Anderla, G., *op. cit.*, 67.

4. Anderla, G., *op. cit.*, 68.

5. Anderla, G., *op. cit.*, 71.

6. Bar-Hillel, Y., *Report on the State of Machine Translation in the United States and Great Britain*, Office of Technical Services, U.S. Department of Commerce, 1959.

7. Watson, J.D., *The Double Helix*, Atheneum, New York, 1968.

8. Chertok, L. and De Saussure, R., *Naissance de psychanalyste de Mesmer à Freud*, Payot, Paris, 1973.

9. Wilson, L.R., *The Geography of Reading*, American Library Association and University of Chicago Press, Chicago, [1938].

10. Shera, J., *Foundations of the Public Library*, Shoe-String Press [1949, reprinted] 1965.

11. Anderla, G., *op. cit.*, 128.

12. Waples, D., *Investigating Library Problems*, University of Chicago Press, Chicago, 1937, 201.

13. Illich, I.D., *Tools for Conviviality*, Harper and Row, New York, [1973], 23.

14. Hamburger, J., *The Power and the Frailty*, Macmillan, New York, [1973].

15. Illich, I.D., *op. cit.*, 22.

16. Nizan, P., *Aden, Arabie*, Rieder, Paris, 1931, 90.

17. Anderla, G., *op. cit.*, 130.

Chapter 6

PROFESSIONAL ASPECTS OF INFORMATION SCIENCE

Professional Aspects of Information Science

Editorial Summary

"To be or not to be", the Shakespearean phrase, has applicability to the treatment offered by Atherton, Belzer and Welt in approaching the educational and research activities of the Information Scientist. Perhaps the training of future information scientists should be towards specialization in a specific field of information such as physics or chemistry, coupled with ways of dealing with acquiring, processing and disseminating technologies. On the other hand, perhaps we should not train the specialist but rather the generalist. Perhaps we should have competencies, but should these competencies be directed at science or service, or both? Are the competencies for science and service the same? It would seem from a historical perspective that information scientists relish the debate, and are not in the mood for commitment at this time.

It is clear, however, that Atherton suggests an indictment that cannot easily be dismissed if we take our identity seriously. If we are scientists, then the data collected by Atherton and her colleagues concerning research activities are quite provocative and alarming. Of particular note is her claim that information scientists have not been measuring up to the requirements of science. Atherton quotes Einstein in identifying the requirement: Science "is the endeavor to formulate invariate relationships of observational data, that is, invariate relationships between specific measurements. Science seeks to show the relationship between various entities of the world as these entities are measured by the scientist". Scientific activity normally involves referring to previous work done by others, and great care in attending to the assumptions underlying analytic or sampling techniques. Care is also taken to state hypotheses clearly and identify independent and dependent variables. Apparently, Atherton's report shows that these practices do not characterize the majority of the scientific reports from the field. The very fact that Atherton believes that there should be a commitment by information scientists to at least these requirements of science should be suggestive of the necessary and sufficient standards for those whose careers are committed to a particular version of the information-scientific paradigm.

Welt emphasizes a fundamental point concerning educational philosophy in Information Science, namely the responsiveness of all educational programs in meeting the needs of the community. Welt believes that his program at American University is oriented directly toward training individuals who have responsibility to the "real" world to approach and solve problems. This laudable objective cannot and perhaps should not be challenged. One might question however, how long such training will remain relevant when related to the rate of change of the technology and the technical structure which constitutes the problems of the real world. Considering the vast growth of technology, and changing managerial and operational problems whose solution depends on information specialists, it is not clear to what extent that Welt's philosophy of education in this field can sustain itself. It appears dependent on derivation and implementation of a highly efficient continuing education program.

Whatever the philosophy underlying an educational program in Information Science, the dominant questions seem to be: what should the subject content be? and what should the overall goals be?

Belzer believes that Information Science has three concerns, namely, theory, engineering, and service. It is within this context that the basic competencies which an educational program should attempt to develop in the individual and its course content should find definition. What are these competencies? Some examples are skill in analysis (e.g., use of operational analyses), and skill in the knowledge and use of computers and related technologies (e.g., displays), ability to design and evaluate systems. Moreover, ability to instruct, to counsel, and to manage are fundamental skills that an individual trained as an information scientist should possess. Educational programs in information science should make explicit their competence in their area of major concern, namely, the use of media to communicate, and an appreciation of the role of the human dimension in information dissemination and use.

The presentations implicitly demonstrate the interdisciplinary character of information as an area of concern to man. The spectrum of courses, objectives, and philosophy all indicate that a competence to deal with information requires an interdisciplinary perspective. Unfortunately, our speakers make little or no suggestion as to the nature of interdisciplinary pedagogy.

The relevant competencies reflect a broad spectrum. There are specific content areas, but the relatedness of one area to another is not clear. There are differing goals (theory, design, and service), but how these goals are tied to each other in an educational program is not clear. What is clear is that whatever information science programs are producing, they are not reflecting science as scientists conceive it. This may be a symptom of a young science which inherits talents and interests that are quite heterogenous and idiosyncratic. This is to be expected from a profession whose leadership has come from individuals in many different fields. But the conclusion from this state of affairs presents a paradox, for one would expect that such a constituency would facilitate the articulation and development of an interdisciplinary science. Apparently this is not so, and from what evidence Atherton has been able to produce, the interdisciplinary constituency remains disciplinary. It is in the next generation of information scientists now being trained that we place the hope for interdisciplinary education.

RESEARCH IN INFORMATION SCIENCE: AN ASSESSMENT

Pauline Atherton

Syracuse University, Syracuse, New York, U.S.A.

The most promising methodologies and techniques for future productivity in our field should be those which contribute most to the development of information science as a *science*. A critical appraisal at this time[1] might contribute something to our growth and maturity as scientists. Albert Einstein[2] defined science: "in a technical sense, it is the endeavor to formulate invariant relationships of observational data, that is, invariant relationships between specific measurements. Science seeks to show, with the greatest possible exactitude, the relationship between various entities of the world as these entities are measured by the scientist. ... Of course, conceptions and formulations of relationships are always subject to revision."

Can we say in 1973 that research in our field is scientific in this technical sense? Does research in our field, specific measurements, contribute to our understanding of the invariant relationships of observational data? Is research in information science a systematic and rigorous gathering and analysis of data with which to test or generate hypotheses, with the long-term goal of developing a set of logically inter-related propositions which describe, predict and explain the occurrence of particular phenomena?

If the answers to these questions are negative, how can we continue to call our field "information *science*"? If most of our "research" is done to seek answers to immediate problems facing libraries and information systems, why don't we change the name of the field to library and information technology or library and information engineering? To probe a true answer to the above questions a few of us at Syracuse University analyzed the characteristics of research published in the leading journals in the field. The major portion of this paper is a report of our findings. This could give us empirical data to verify or refute our statement that information science is a scientific field of research.

If Einstein's definition of science is too exact, perhaps you would prefer William Bevan's definition of *Science*:[3] "Science is, in its most fundamental sense, an approach to solving problems. Its goals may be both specific and general, concrete and abstract, practical and theoretical, and immediate and long-range, and several centuries of experience have demonstrated a significant reciprocity between conceptual and practical advance."

This is a broader definition, but we still must determine if the research reported in our field contributes solid data upon which to base decisions and move in new directions, or if the measurements are the result of a most haphazard and fortuitous process of collecting data. Can we now make any predictions or place any faith in the conclusions and recommendations of research done in our field? Has the researcher asked the right questions? Has he designed his study carefully? Has he described his methodology carefully? Has he taken the greatest care in obtaining the data? Were his goals to test, evaluate, develop, or test a prediction?

To answer these and related questions, we undertook a critical analysis of research methods in the published literature in the field of information science. We first counted citations in all the chapters of the last four volumes of the *Annual Review of Information Science and Technology* to arrive at the four most cited journals. Under the supervision of Dr. Jeffrey Katzer we developed and tested a "Research Methods Checklist", reproduced in Appendix I of this paper. Several trial runs and versions of the checklist assured maximum consistency among the analysts by the time we each began our chore of analyzing the literature.

The work reported here could not have been done without the cooperation of two doctoral students, Pat Moell and James Sodt, and two other faculty members, Dr. Jeffrey Katzer and Dr. Ken Cook. In all, we worked over 200 man-hours to complete this task. To my knowledge it is the only "hard data" available about the research methods used in the field of information science.

STATE OF THE ART OF RESEARCH IN INFORMATION SCIENCE: AN ASSESSMENT OF 430 PUBLISHED PAPERS

In an attempt to review the state of the art of research in the field of information science, the *research methods checklist* was designed and used on all the papers published in four journals during 1969-71 (430 papers). The journals selected were *Journal of The American Society for Information Science* (107 papers), *Journal of Library Automation* (41 papers), *Information Storage and Retrieval* (68 papers) and the *ASIS Proceedings* (214 papers). These journals were chosen because they were the most cited in the last four volumes of the *Annual Review of Information Science and Technology*. After several tests of the instrument for consistency among the judges (analysts), each person proceeded to analyze all the articles in a single journal, published in the 3 year period.

Most of the survey results will sum totals across the four journal titles. If any journal contributed a disproportionate amount to any figure, that will be noted. To summarize the data and findings a series of questions will be answered. The data are also given on the checklist appended to this paper in the order in which it was collected.

General Questions

Was the paper a report of scientific research?

211 of the 430 papers were considered a detailed report of a study defined as "scientific".

55 of the 430 were termed "nonoriginal report of research, more a status check of a project".

164 of the 430 were not considered scientific research (as defined earlier).

The percentage of papers which were *not* scientific research in the four journals was: JASIS - 13%, ASIS Proceedings - 43%, IS & R - 53%, and J. Lib. Auto. - 54%.

What was stated overall major goal of paper? (Some had multiple goals)

(a) 26% papers had as their goal *testing or evaluation.*
(b) 38% had *descriptive* goal.
(c) 16% had a *developmental* goal (tracing history of project).
(d) 13% had a *model building* goal.
(e) 12% had a *prescriptive* goal.
(f) Less than 5% had either a *predictive or exploratory* goal.

What was basis for findings, conclusions, or recommendations of study being reported?

158 of the 430 papers based findings on *quantified measures.*

137 of the 430 papers based findings on *author's introspections.*

Fewer than 40 papers based findings on any of the following: *nonquantified measures, literature reviews,* or *expert opinion.*

Identification of variables. For the 158 scientific research papers based on quantified measures, one might assume that the dependent and control variables would not be difficult to distinguish.

Based on our analysis this assumption is correct only in a qualified way. Granted, some 94 papers did specify a dependent *single* variable, but when it was described it was not operationalized. Such terms as system performance, precision, co-occurrence of keywords, judge's rating of indexing, etc. were the names assigned to this variable, often with no explanation of how this variable could be measured in an unambiguous manner.

Some 78 papers mentioned a single control variable, but here again, there was no operational definition, merely a naming of the variable, e.g., search method or search strategy, index language, temperature, etc.

> Comment. This portion of the checklist points out a very serious weakness in research methodology in information science, namely, the lack of an operational definition of the variables being studied. Inkeles, in his review of national character studies in the *Handbook of Social Psychology* makes some comments which could also be pertinent criticism of our field: "What we have is a clinical-idiosyncratic mode of analysis. There is much to commend such an analysis, particularly in the initial development of an area of investigation, when both theory and sheer information are so limited. Nevertheless, the lack of a standard analytic scheme creates a number of problems. For one thing, it means there is no rigorous test for the occurrence of omissions or distortion of analysis. If certain characteristics are not mentioned in the analysis, it is not clear whether they are absent in the system being studied, or are present but have been neglected by the analyst. This hinders inter-system comparison or cross-system generalization."

What ways were used to measure each major independent variable? Of the 158 scientific research papers based on quantified measures, 105 of them used only one method of measuring the major independent variable. I don't know how this compares with other fields of science, but I think it is safe to assume that we would have more reliance on research in information science if variables were measured in more than one way.

Questions Related to Evaluation of Research Reporting

Were decision-rules and hypotheses clearly stated? Fewer than 25 of the 211 scientific research papers specifically stated decision-rules to draw conclusions or recommendations from the findings or evidence considered. (A decision-rule would allow for more than one outcome or possibility.)

Only 20 of the 211 scientific research papers were rated *high* regarding their specific statement of one or more hypotheses. Another 45 were rated *average*.

Instrumentation - Is there reliability and/or validity of measurement? 32 of the research papers rated *high* on the degree to which their instruments were pretested and their reliability and/or validity assessed. 37 others were rated *average*.

Was there recognition of weaknesses in study and attempt to compensate for them? Only 13 papers were rated *high* on this matter and another 69 were rated *average*.

Is research a replication of a previous study? 21 papers were rated *high* and 125 were rated *average* on this matter. This number includes research papers which were not based on quantified measures. It would include descriptive studies, surveys, etc.

Could study be replicated? 71 papers were rated *high*, i.e., the reader knows enough from the paper to get the same answer if he used the exact same sample of observations. Another 137 papers were rated *average*. Here too, more than the research papers based on quantified measures are included.

Are findings applicable to similar situations? The analysts rated most of the papers in the study as either *high* (119) or *average* (230) regarding the degree to which the findings, recommendations, results, or conclusions apply to other similar situations. (Only a few papers from *Information Storage and Retrieval* did meet this test.) However, only 33 papers rated *high* regarding the degree to which the author attempted to integrate his findings, results, conclusions, or recommendations into the current state-of-the-art. Another 172 papers were rated *average*. In other words, fewer than half of the 430 papers integrated their report with the current state-of-the-art.

> Comment. This is another weakness of reports in the field of information science. More often than not, the references appended to most papers refer to the author's previous work and not to related work done by others. Only 91 papers of the 430 were rated *high* on the degree to which the author attempted to generate his objectives, problems, questions, or hypotheses out of a review of the current state-of-the-art. 199 papers did either a modicum of a literature review or presented some authoritative rationale by the author. Most fields of science have found that long term growth depends upon the integration of known information with current findings.

Were the objectives stated and achieved? As might be expected almost all of the papers did a fair job of stating objectives (205 were rated high and another 192 rated average). The analysts felt that three-fourths of the papers (276) achieved their objectives to a *high* degree, and another 94 did an *average* job.

Questions Related to Subject and Type of Research

The most frequently used phrases in the open-ended question on the checklist regarding type of research method were the following:

System description
Experiment
Mathematical deduction
Progress report
Analysis from experience
Opinion survey/think piece
Prescriptive essay
Simple comparison

Some of the statistical techniques and other analytical techniques mentioned were: analysis of variance, Chi-square, multiple correlation, interval testing, mathematical model building, algorithm deduction, Delphi method, cost analysis, failure analysis, operation-research model, simulation model and sociogram.

What was being observed; what was unit of analysis, environment, data source, and method of analysis? In almost all of the scientific research papers based on quantified measures, 158 papers to be exact, some observation or unit of analysis which provided one score or value of measure on the dependent variable was reported. The unit of analysis varied greatly, but the ones most often cited were: citations (or abstracts or documents, or catalog records); search requests (or queries); keywords; groups, people, indexers, students, profile, etc.; "system".

The number of observations reported varied from one to thousands, e.g., one system, 5 abstracts, 1 profile, 76 persons, 4 indexers, 3 systems runs, 250 messages, 150 searches, 11,005 citations, etc. When the *environment* was reported, more often than not, it was a *realistic* environment (109 papers). Only 75 papers reported data being collected in a highly controlled environment.

The *data source* was almost unanimously collected from *observations* or events or incidents (152 papers).

The *method of data analysis* was primarily *quantitative* (105 papers), using only the simplest tabulations or percentiles.

35 papers reported a nonquantitative analysis, 27 an inferential analysis (using analyses to make statistical inferences), and 28 put real or simulated data into formulae to get output. Remembering that 65 papers had as their goal model building, this

last figure is somewhat of a surprise.

120 papers attempted to *sample* from the available observations, but only 48 of these 120 clearly and carefully stated how they selected them. Another 32 papers used all available observations.

Only three papers used random numbers and another seven used computer simulation.

> Comment. It would appear that most of the research reported is in the real environment, but few researchers reported analytic or sampling techniques being used to help make inferences or generalizations. Perhaps this is characteristic of a pre-science state.

Summary of Findings

The "Research Methods Checklist" used in this study proved to be a useful instrument for discovering the following weaknesses in research reports in information science:

(1) The lack of operational definitions of variables being studied.
(2) The lack of predictive studies.
(3) The failure to use more than one way to measure the major independent variables in the study.
(4) The failure to make a specific statement of hypotheses and/or decision-rules to draw conclusions or recommendations from the evidence considered.
(5) The failure to pretest instruments of measurement.
(6) The failure of most papers to do an exceptional job in integrating the findings of the study into the current state-of-the-art.
(7) The over-concentration on system descriptions and progress reports in the published literature.
(8) The lack of emphasis on the human information processing problem and the use of relevant techniques to explore this subject scientifically.
(9) The failure to collect data in a controlled environment, and the failure to perform inferential analyses, or careful sampling of the data.

Personal observations. Have we failed to make a distinction between knowledge and skill, between knowing how and knowing that (this phrase is borrowed from a paper by David Hays, to be published in *Daedalus*)? Are the goals of our research reports to describe our skills, to test them, to prescribe action based on that experience, more than to answer certain fundamental questions which will contribute to our knowledge about those skills or the new skills we may need? The significance of each measurement we make is restricted by the manner in which we have designated each entity. If our findings in this survey are valid, we need more mathematical models, conceptual frameworks, laws, world pictures, and theories. We need researchers who can appraise correctly the significance of the measurable factors and weight and judge the non-measurable factors. The subjects of our research have been system variables more than human variables. I would like to suggest a corrective measure for research in the field. We need to ask and try to answer, scientifically, some of the following questions:

(1) How do people seek information?
(2) Why do people seek information?
(3) Why do people fail to seek information?
(4) Do different types of people seek information in different ways, and if so, how can information systems be designed to suit different types of clientele?
(5) How do people process information?

In trying to handle these complex questions, we may have to discard much of the methodology most prevalent in the field today. Inadvertently, perhaps, we have ignored the nearly invisible social processes which affect our work the most, human information processing. Our reliance on systems analysis and related techniques has inhibited our study of the more substantive and more difficult aspects of our world of information science, namely the *human being who is processing information*. Maybe there is a growing awareness that interpersonal relations (between staff and user of information systems, between staff and staff, between manager and workers, etc.) will contribute a great deal to the optimal efficiency of systems. Our research efforts should reflect this awareness and point us toward the right decisions to make in system design of on-line, computer-based information systems, etc.

Our concentration over the past ten years on the citation, the abstract, the query, the system run, the index language may not have been the proper unit of analysis. It may be that it is the source of difficulty, but our symbol system should be expressive of the most important concepts and processes we must deal with. The cognitive style and other personality variables of information seekers (such as degree of motivation, anxiety, frustration, need for reassurance and reward, need for feedback, work orientation and situation contexts such as time pressure, stimulus for obtaining information, intended use of information, etc.) are all factors which have important implications for our continuous effort to understand and improve information systems.

There is a need for researchers in this field to become more sophisticated in their survey and experimental research methodologies. Cognizant of the disciplines of cognitive psychology, organizational psychology, and sociology of science, we may uncover some fundamental laws of human information processing.

As researchers in a field we will always be interested in the surrounding technology. Nevertheless, in my opinion, the most promising approach for our field would be to engage in new avenues of research that directly relate to *the human information processing* problem. We can leave much of the systems design and evaluation to more competent researchers who are only on the periphery of our field. We need to invite into our circle and re-train ourselves and future researchers in information science in the methods and techniques that contribute to our understanding of *that* central problem.

APPENDIX I

FINDINGS CONCERNING 430 PAPERS IN 4 IS JOURNALS USING "RESEARCH METHODS CHECKLIST"
for Papers in IS Journals

(Note: checklist devised by Jeffrey Katzer; numbers represent data collected by Syracuse University Group.)

Citation: ______________________________
____

CHECKLIST - check as many as applicable in each case, but only if each item forms an integral part of the study.

An "NR: stands for "not relevant". Blanks are treated as NR.

I. Analysis

A. This study is a
[N=430]
- 55 Report of Research (nonoriginal report, less detailed, including a status check of a project)
- 211 Research Report (the first, detailed, report of a study)
- 164 Not research

B. Overall major goal is
[N=430]
- 111 Testing or Evaluative
- 65 Model building (Major goal is contruction of a "New" model or the application of an existing model to a "New" situation)
- 68 Developmental (traces history of project, etc. over time)
- 30 Predictive (to test a prediction)
- 30 Exploratory or Speculative
- 53 Prescriptive or Suggestive (e.g., "To improve services you should ...")
- 164 Descriptive (not over time - see developmental)

C. The findings, conclusions or recommendations of this study are mainly *based* upon
- 158 quantified measures
- >40 nonquantified measures
- 137 author's introspections and thoughts
- >40 expert opinion (other than author or literature review)
- ____ other:

IF ANY OF THESE ARE CHECKED, THE REMAINING ITEMS IN THIS PART MAY NOT BE APPLICABLE. Continue, responding to those questions appropriate to this study.

D. [N=158] Independent and dependent variables are difficult to distinguish. In this study there are ____ (how many) variables.

> 94 - *single* variable
>
> EXAMPLES:
>
> System performance; precision; co-occurence of keywords; judge's rating of indexing, etc.

Brief Description of the Major Ones:

SKIP to Item G if this item is completed. Be sure to describe use of variables in boxed area after item M.

E. Dependent Variables: those which are analyzed, added, measured, etc. Usually the contents or body of a table, or the vertical axis of a graph represents the dependent variable.

How many dependent variables are there?

> 94 papers -
>
> no operational definition of dependent variable

F. Independent or Control Variables: those which are used to group or categorize observations (e.g., male, female). Usually the column headings of a table or the horizontal axis of a graph represent the independent variable.

How many independent variables are there?

78 papers -
no operational definition of single control variable
EXAMPLES: Search method; Search strategy; index language; temperature, etc.

Give OPERATIONAL DEFINITION of the major ones.
Be sure these are *variables*, not levels of the variable.

G. Observations or Unit of Analysis: That which provides one score or value or measure on the dependent variable. Usually each item in the body of a table came from one observation. It may help to answer part K first.

Range: one to several thousand

[N=158] ___ How many observations are there?
__________ What is one observation or unit or analysis? (e.g., person, book, card, etc.)

Examples from study: citation *or* abstract *or* document; search request *or* query; keywords; system; indexes

H. If data is collected it comes from
75 A highly controlled data collection environment
109 A very realistic data collection environment

I. If data is collected it comes from
152 Observations or Events or Incidents (looking at people, systems, etc. Essentially noninteractive)
____ Discussions (informal oral interaction)
____ Interviews (formal oral interaction using interview guide or a consistent set of questions)
____ Questionnaire (written response to a structured set of questions)

J. If data are collected they are analyzed
35 Nonquantitatively
105 Quantitatively (using the simplest tabulations or percentiles)
27 Inferentially (using analysis to make statistical inferences)
28 Analytically (putting real or simulated data into formulae to get output. In this case the formulae

are of the most important - see B-2 "Model Building"

____ other:

K. Sampling

32 All available observations were used in this study (entire population)

72 Some of available observations used in this study, but there was no apparent care taken in terms of choosing them.

48 Some of available observations used in this study, and *care* seems to have been taken selecting them Such care must be specifically stated.

L. Use of Random Numbers

3 Are data or occurrences artificially generated randomly?

____ Is Queueing Theory mentioned as being the major procedure employed?

____ Are Markov processes mentioned as being the major procedure used?

M. Simulation and Gaming

7 Is computer simulation involved?

____ Does role-playing or acting-out the components of a system produce much of the data which is later analyzed?

EXAMPLES:

System description; experiment; mathematical deduction; progress report; opinion survey; simple comparison; analysis from experience; prescriptive essay, etc.

Brief title or common name for overall RESEARCH METHOD used. If above checklist inadequately describes the methods used, briefly describe them.

II. Evaluation of Method - unless otherwise noted use a 5-point scale for each of the following (5=high degree; 3=average; 1=no degree). NR is also a legal response.

a: N = 430 published articles
b: N = 211 scientific research papers
c: N = 150 research papers with quantified measures

N.[b] ____ {32-*high* 37-*ave.*} Instrumentation of Measurement: To what degree were the *instruments pretested*; was there any attempt to *assess* the *reliability and/or validity of the measurement*? Instruments may be humans, paper-and-pencil tests, scales, etc.

O.[a] ____ {71-*high* 137-*ave.*} Publicness, Objectivity or *Replicability*: To what degree could you replicate this study (within the limits of sampling error) using mainly the information presented? Do you know enough to get the same answer if you use the exact same sample of observations?

P.[a] ____ {205-*high* 192-*ave.*} How well were the objectives of the study *stated*?

Q.[a] ____ {276-*high* 94-*ave.*} If the objectives were stated, how well were they achieved?

R.[c] ____ {105-used 1 way.} How many different ways, instruments, or techniques did the author attempt to measure *each* major dependent variable? (DON'T use the 5-point scale.)

S.[a] ____ {91-*high* 199-*ave.*} To what degree did the author attempt to generate his objectives, problems, questions, or hypotheses out of a review of the *current state-of-the-art*? (GUIDELINES: 1=none, 2=either a modicum of literature review *or* some authoritative rational by author, 3=some of both, 4=more of same, 5=high quality)

T.[a] ____ {33-*high* 172 *ave.*} To what degree did the author *attempt to integrate his findings*, results, conclusions or recommendations into the current state-of-the-art (this refers to what is written after the results, recommendations, or conclusions section)?

U.[a] ____ {21-*high* 125 *ave.*} To what degree is this study a replication of a previous one?

V.[b] ____ {13-*high* 69-*ave.*} To what degree did the author *recognize* the *weaknesses* in his study and attempt to *compensate for them*, or deal with them honestly? (Author must specifically state weaknesses or treat them so clearly that it is unmistakeable.)

W.[a] ____ {119-*high*} To what degree do findings, recommendations, results, or conclusions *apply to other similar situations*?

{230-*ave*. (Would you trust them if you had to apply them?)

X.[b] ____ To what degree did the author specifically state one or more hypotheses?
{20-*high*
{45-*ave*.

Y.[b] ____ Were decision-rules specifically stated to draw conclusions or recommendations from the findings or evidence considered? (Decision-rules must allow for more than one outcome or possibility.) DON'T use 5-point scale.
{>25)

APPENDIX II

SURVEY OF PARTICIPANTS IN NATO INSTITUTE

This survey of publication/communication channels used by participants was made to test the adequacy of the sample in the paper.

30/80 respondents (10 speakers)

U.S.A.	- 13
Canada	- 5
U.K.	- 4
France	- 1
Sweden	- 1
South Africa	- 1
Greece	- 1
Australia	- 1
Anon	- 3

Survey of Information Channels

30 respondents (13/30 in U.S.A.)

	Named	Frequency (>2)
Publishers	31	7
Report Literature Distributors	29	2
Journals	90	14
Conference Proceedings	30	3

Most Cited Channels

Publishers	Report Literature Distributors	Journals	Conference Proceedings
Wiley	US/NTIS	JASIS	ASIS Conference
Academic Press	US/ERIC	Journal of Documentation	AFIPS Proceedings
Marcel Dekker		Information Storage and Retrieval	Polytechnic Institute of Brooklyn
		American Journal of Psychology	
		Aslib Proceedings	
		Information Scientist	
		Journal of Librarianship	
		Library Quarterly	
		Library Trends	
		Nature	
		Science	
		Special Libraries	

Book Publishers (31 named; 7 more than twice)

Wiley	-	4
Academic Press	-	3
Marcel Dekker	-	3
AFIPS Press	-	2
Bowker	-	2
Scarecrow	-	2
Greenwood	-	2

McGraw Hill
ALA Press
EB/ASIS
Syracuse University Press
Heath
Methven
Yale University Press
Harcourt
Longmans
University of Toronto Press
University of Chicago Press
etc.

Report Literature Distributors (29 named; only 2 more than once)

US/NTIS	-	10
US/ERIC	-	3

US/ONR
U.S. Army Electronic Command Technical Reports
University of Western Ontario/SLIS
U.S. Office of Telecommunication Policy Tech. Report
NASA Special Publications
Army Research Institute Reports
GSLIS/University of Pittsburgh
INID (France)
Information Industry Association Report
Human Development Corporation Report
FID Report Series
ISSC Reports
etc.

Journals (90 named; 14 named more than twice)

JASIS	-	8
Journal of Documentation	-	4
Information Storage & Retrieval	-	3
American Journal of Psychology	-	2
American Psychologist	-	2

Aslib Proceedings	-	2
Information Scientist	-	2
Journal of Librarianship	-	2
Library Quarterly	-	2
Library Trends	-	2
Nature	-	2
Science	-	2
Science Studies	-	2
Special Libraries	-	2

Acto Oto - Laryngolica
Australian Library Journal
Business Quarterly
Canadian Journal of Physics
Drexel Library Quarterly
Human Factors Bulletin
IEEE Transactions on Antennas and Propagation
International Journal of Social Science
International Journal of Control
JACM
Journal of Computer and System Science
Journal of Information and Control
Journal of Library Automation
Management Science
Management Information
Operations Research
17th C. Newsletter
etc.

Conference Proceedings (30 named; 3 more than twice)

ASIS Conference Proceedings	-	4
AFIPS Proceedings	-	3
Polytechnic Institute of Brooklyn, Symposium on Computers and Automata	-	2

ASEE
American Mathematical Society
ACM
IEEE
MORS (ONR)
NATO Advanced Study Institutes
EDUCOM
FID/VINITI
Annual Information Retrieval Colloquium
International Congress on High Speed Photography
International Conference on Low Temperature Physics
Conference on Instructional Technology (NRC - Canada)
Ontario Universities Computing Conference

Conference of Library Association of Australia
Université de Tours Symposium
=
National Telemeting Conference
etc.

References

1. Based on presentation made at Conference on Information Science and Biomedical Communications Research held on October 5, 1972. National Library of Medicine, National Institute of Health.

2. Einstein, A., *Out of My Later Years*, [Essays], Philosophical Library, New York, 1950, p. 41.

3. Bevan, W., editorial in *Science*, 26 May 1972.

PROFESSIONAL ASPECTS IN EDUCATION

Jack Belzer

University of Pittsburgh, Pittsburgh, Pennsylvania, U.S.A.

INTRODUCTION

Under a grant from the Bureau of Libraries and Learning Resources of the U.S. Office of Education, the Center for the Study of Information and Education established three task forces to explore the relationship of information systems and educational systems. The study reported here, includes partial analysis of the questionnaire data obtained in the Development and Evaluation of Curricula in Information Science under a grant from N.S.F. GZ-2525 and is based on the work of one of the three task forces: the task force on Curriculum Reform of Library and Information Science. (The task force consisted of C. Walter Stone - MARC, Inc., Pacific Grove, California, James Brown - San Jose University, and Jack Belzer - University of Pittsburgh.) One of the other two task forces worked on Information and Information Needs, and the other on Alternative Information Systems for Emerging Institutions. All of the task forces were established by the Center for the purpose of exploring the relationship of information systems and educational systems and to consider alternative uses of information.

Students in professional schools are trained not only for jobs which exist now or tomorrow, but also to fill jobs which are now non-existent, to deal with problems which will arise 5 or 10 years from now, to utilize technology which will be developed to solve such problems. In order to establish some context for the conclusions which will be drawn in an effort to develop an appropriate educational program in information science, it seems appropriate to extrapolate from what is known to future needs.

The predictions which follow will thus represent an educated guess only.

TENTATIVE FORECASTS

The structure of our society is becoming more information-dependent, and the cost of the efforts which are being made to meet these societal needs is soaring. Bigger and more expensive buildings required to hold materials for possible future needs are being built and getting access to these materials is becoming more difficult. People want information and not the media on which it is stored. Moving large stones on which information is recorded is primitive and moving books and papers is similarly archaic. Information will be stored on media with densities of many orders of magnitude greater than present day, and access to any of it would be almost instantaneous. Electronically-controlled reservoirs of information would be meeting all of the existing needs of our society, including personal, educational, vocational, and avocational information demands. Information storage, retrieval, transmission, and display methodologies would be utilized to reach those who need information and bring it to them rather than moving people to libraries and information centers. Information scientists would also have the responsibility to assure more effective utilization of the information provided to users.

Recent literature appearing in professional journals with regard to future development of library and information service programs and its related projection of manpower needs do not preclude the possibility that new types of local and regional information service centers would become the primary means of getting access to the world's store of recorded knowledge, as well as meeting community and societal information needs. These local centers would be mere tributaries linked directly to regional, national, and international centers.

TYPES OF CURRICULA REQUIRED TO MEET FUTURE NEEDS

(1) The total number of personnel for establishing and maintaining such centers would necessarily be large. It would demand a wide range of expertise covering a broad spectrum of competencies, the most essential of which would include: (a) human communication and learning processes; (b) familiarity and demonstrated ability to use the most modern technology for storing and getting access to the world's store of knowledge; (c) to possess subject expertise in one or more fields of basic human interest and endeavor, to provide sound interpretation of user needs and employ information resources, on behalf of the user, effectively; (d) to permit the user access directly to the information source, for him to interact with and utilize it as he wishes.

(2) Taking into account these forecasts, the content of the

educational programs must reflect the needs of the expected professional practices. These would require: (a) increased knowledge of the information and information needs and related understanding of personal and inter-group communication and learning processes; (b) abilities to design, develop and utilize improved information handling systems; (c) the capability to utilize effectively the most advanced technology, e.g., computers, communication networks, and displays, for storage, processing, transfer, and distribution of information; to develop leadership to be "active" rather than merely passive in terms of social effects of information distribution and be willing to accept personal responsibility for identifying needs and disseminating information in terms of these needs.

(3) Educational needs are identified with career opportunities as well as with the individual interests and talents or paths which individuals wish to pursue. Three major areas or paths which information scientists can pursue have been identified as: (a) the nature of information: Research information scientist and academician whose careers would be identified with universities or government and industrial research organizations. These professionals would be teaching and conducting research at the frontiers of present knowledge and extending the state-of-the-art in the field; (b) systems-oriented careers in the information field where personnel would be responsible for planning, designing, installing, and evaluating information systems of many varieties; (c) service-oriented careers provide opportunities for professional information scientists to come in direct contact with system users, to administer information centers and, in some instances, perform such tasks as indexing, abstracting, etc.; (d) one aspect which is frequently overlooked is the need for improvement in instructional methodology. Not only is cost of education climbing and something must be done to slow it down, but effectiveness of teaching can be reflected in the product as well as save valuable time to the student. The need is to be moving away from traditional lecture courses and rote memory associated with textbook-centered approaches. Emphasis should be on individualized and small group seminar learning techniques. Introduction of case studies, real-life situation environments, or field conditions, student self-pacing and self-evaluation activities, and possible internship arrangements should be pursued.

IMPROVEMENT IN EDUCATION IN INFORMATION SCIENCE

The following are areas identified as needing improvement in the educational process, especially as related to information service-oriented careers:

(1) Goals of preparation should be defined as: (a) to help make it clear that prime concern is with information needs and with

provision of access to information as such and *not* simply with the location and delivery of library materials; (b) to highlight the importance of learning basic functions as distinguished from typical institutional routines; (c) to prepare information specialists to reach people where they are, faster, utilizing modern media; (d) to meet societal needs, i.e., to better meet the needs of special groups in the population.

(2) Organizational thinking should emphasize: (a) development of the field as a broad spectrum ranging from paraprofessional to professional, or consideration should be given for the establishment of more cooperative programs, sharing of scarce resources, awarding of joint degrees, ready transfer of credits (and development of close ties with State agencies, professional associations, etc.); (b) three main areas of study at the graduate level: Information Science; Technology; Service; (c) specialization among the schools in accord with special interests, locations, student body needs, etc.

(3) Content changes should emphasize: (a) the giving of more attention to human relationships as distinguished from management and technology; (b) establishment and enforcement of demands for adherence to appropriate standards for admission to graduate study; (c) provision of more options for specialization in given fields of career interest (defined in terms of functions and constituencies); (d) abandonment of obsolete "cores"; (e) greater subject specialization; (f) use of new tools, i.e., statistical techniques, personnel management, etc.

(4) Instructional methods employed should: (a) be geared to competency-based curricula - provide more field experience components; (b) make more and better use of new teaching techniques and technologies; (c) stress self-learning approaches.

(5) Federal (or other) support should be sought to mount a nation-wide program for the regular inventory, production, distribution and appraisal of new learning "packages", etc., relevant to information science education on the graduate level and for use in continuing education.

EDUCATION FOR COMPETENCY

The vital question is, "What specific competencies should professional information scientists possess in order to satisfy future demands?"

Below are lists of types of competency which are recommended to be acquired by students at some point in the sequence of their education for professional work in information science. Professional work is defined arbitrarily to include performance of functions such as the following: (a) management (organization and personnel), communication, persuasion; (b) counselling and guidance, liaison; (c) information acquisition, organization, analysis and description; (d) instruction and instructional program development; (e) media production and utilization; (f) research activity; theoretical aspects; (g) subject knowledge interpretation and evaluation, application; (h) systems analysis and design; (i) technical systems development, supervision, and maintenance.

In listing job functions and in suggesting competencies, it is not presupposed that any one information specialist could or should be expected to acquire all of the professional knowledges and skills believed desirable or that any one educational institution can or should provide the full range of learning experiences needed to operate all types of information service programs in the future. Both people and institutions necessarily will specialize. However, the list is believed to be indicative of the range of competencies prerequisite. And, although improvements come gradually (except in newer institutions) the direction in which such changes should lead are fairly clear.

PROFESSIONAL COMPETENCIES

In the interest of establishing some priorities, the statements which follow present in some detail more formal definitions of nine areas of competency which should receive increased emphasis in developing educational programs to serve the future. The areas bear chiefly on program planning, management, and on utilization of resources.

Established as prerequisite for students interested in pursuing graduate study in the field are basic understanding of logic, statistics, linguistics, computers, artificial languages, and some level of advanced mathematics. To this list of prerequisites there is frequently added an "information science" core which (whether completed on undergraduate or graduate levels) can provide a common foundation for information science graduates. Studies in the following areas are said by specialists to meet core needs: (a) Introduction to Information Science; (b) Systems, Theory and Application; (c) Mathematical Methods in Information Science; (d) Computer Organization and Programming Systems; (e) Abstracting/Indexing/Cataloging, Theory & Application; (f) Information and Communication Theory; (g) Research Methods in Information Science.

NINE AREAS OF COMPETENCY

(1) Ability to use appropriate methodologies and environments to develop *resource-sharing systems*.

(2) Ability to develop *mathematical models and/or simulations*

of information science systems (utilizing these techniques for evaluating systems and design of alternative approaches).

(3) Ability to use *operations research methods* in the analysis or design of information systems for purposes of optimization (including use of microphotographics, networks, union catalogs and other systems).

(4) Ability to develop and/or use appropriate criteria and methods of evaluating information systems and their performance (by developing *statistical data and analyses* from which significant conclusions may be drawn)

(5) Ability to use *computers** for performance of many tasks in the information science environment such as:

(a) Identifying and selecting information content in documents by *automatic indexing and abstracting, linguistic analysis*, and producing *thesauri* and Methods for Regeneration of Information.

(b) Maintaining "bibliographic control" by conceptually designing, implementing and utilizing *information storage and retrieval systems* such as *Medlars, NASA information centers, Chemical Digests*, etc.

(c) Managing *control* of library and information functions such as: (i) acquisitions; (ii) cataloging; (iii) circulation control; (iv) serials control; (v) budget control.

(*Note: One major change in professional education curricula wrought by demands for increasing information science content is represented in the fact that, during the past four years, the number of computer science courses has increased from twenty to thirty-nine percent of the total curriculum almost exclusively at the expense of library science courses as such. The recognition that the sheer bulk, both intellectual and physical, of contemporary information requires new means for dealing with this problem is reflected in the change in the curricula during the last four years.)

(6) Ability to develop *question-answering systems* and methods for regenerating information.

(7) Ability to develop systems for *selective dissemination* of information.

(8) Ability to design and/or use appropriate methods and criteria for determining the effects or *influence of information on users* and the extent to which information materials actually satisfy or fill needs.

(9) Ability to choose many various ways of *displaying information* which are most appropriate to the purposes for which the information is being developed.

CONCLUSION

The competencies in demand for professional practice in the field of information as derived in this paper are based on tenable predictions regarding future of information requirements. The goals of such professional education may need redefinition in terms of these broader perspectives responsibilities. Competency-based approaches to curriculum development should be considered with the prospect of restructuring presently on-going programs to accommodate future needs.

A PRAGMATIC APPROACH TO THE EDUCATION AND TRAINING OF INFORMATION PROFESSIONALS

Isaac D. Welt

The American University, Washington, D.C., U.S.A.

Discussions concerning the education and training of information specialists have been going on for at least a dozen years. Meanwhile, external conditions and circumstances have been changing, particularly those involving the market, for the services of this group of professionals. As an attendant at most of the meetings and conferences at which these discussions have taken place and as an avid reader of the proceedings which they inevitably generated, I have never felt that I could come away with suggestions and guidelines which could be easily adapted to the solution of my own problems as an educator in a graduate school program. I have also been somewhat reluctant to engage in polemics concerning questions of basic educational philosophy on the one hand and specific curricula and course topics on the other, since I could not circumvent the problems resulting from differing educational backgrounds.

Some twenty years ago, I left the laboratory bench where I conducted research in biochemistry to enter the then-emerging area of science information handling. Like many of my contemporaries, I felt that there was a need for the scientist and engineer to do something about the so-called information explosion. To put it in the more popular jargon of the ecologist, it was necessary to engage in cleaning up the products of one's own intellectual metabolism, for no one else was going to do it properly. This is still my basic approach, especially after nine years as a full-time professor and curriculum director of an active graduate program. Accordingly, my contribution to this Advanced Study Institute is heavily biased by my own experience and somewhat peculiar background.

Hence, I am strongly convinced that the education and training of information professionals must be sufficiently flexible to allow us to cater to the needs of a large variety of individuals preparing for a diversity of positions, many of which do not presently exist. My location in the city of Washington, D.C., which probably contains the largest number of information enterprises in the entire world, reinforces this attitude rather strongly. To standardize our training and educational programs now or in the near future would result in a loss of "credibility" insofar as the workaday world is concerned. It might even lead to a deflated graduate degree designed along the lines of the "least common denominator".

The major problem which seems to recur during all of the discussions in the past is that we are culturally inbred, we are talking to ourselves. I would like to see increasing input from those who employ our graduates, from the managers of information centers, from the research and development ranks; in short, from those individuals who provide the material basis for a profession, namely, employment opportunities with a high degree of status and expectations for continued professional growth. We must have a goal in sight if we are to compete for well-qualified students in what has only recently become a buyer's market, at least in the United States. Let me cite the recent paper by Jean Tague of the School of Library and Information Science and John Carroll of the Department of Computer Science at The University of Western Ontario,[1] entitled "Information in an Informationless World or the making of an Information Scientist" published in the January-February issue of the *Journal of the American Society for Information Science*. I assume that it is representative of the thinking of our younger professionals:

> But, he must have learned, at some point, the technique of making connections between users and sources, devising the transformations between receiver sets and generator sets. How can he learn his skill?
>
> A one-word answer, again, is appropriate. Experience. Two kinds of experience - experience with the language of the users and experience with the operation of present-day information systems. That, and an inquring mind which, faced with an information-seeking set, will not be restricted to present solutions.

After describing, in passing, the seminar-based, graduate program at Western Ontario, they conclude that:

> The kind of training programme we have outlined could provide a body of specialists that would constitute a major social resource. We believe that the graduate portion of the programme is worthy of the same kind of public support now afforded the training of librarians, if not to supersede it.
>
> The work-study format permits development of the program to advance in concert with the development of information centres.

At this point, a discussion of our experience at the American University can logically begin. First of all, our program, for want of a better name, is called "Scientific and Technical Information Systems", with these terms to be interpreted in their broadest scope. Thus, we have recently had an excellent student paper on "Information Handling in the Field of Archeology". Our definition depends to a large extent on the kind of student body we are able to recruit. It is housed, administratively, in the Center for Technology and Administration, which, as of this year, has become part of the university's College of Public Affairs.

Our basic educational philosophy and training goals can best be exemplified by the following quotation from a recent brochure describing the Center for Technology and Administration:[2]

> The discipline, Technology of Management, exists apart from Business Administration, Government and Public Administration, and the classical fields of Mathematics and Science. It evolved from the acceptance of the electronic digital computer with its associated communications capabilities, as well as from the emergence of a broad and substantial quantitative theory of the properties, structure and flow of information. It approaches problems in all fields involving information transfer, transformation, and storage. The discipline should be included in the education of those who will be involved in the management, or design, or development of large, complex systems involving people, resources and objectives. It includes but is not limited to the use of computers in systems analysis, systems dynamics, statistics, computer-assisted instruction, command-and-control systems, data systems, information storage and retrieval, simulation and modeling and quantitative policy development.

During the nine years which I have spent in developing the program, I have never had to face the agonizing problem of defining information. We state simply that whatever it is, it facilitates the decision-making process. Without adequate and relevant information, the manager in government or industry cannot make the best or optimum decisions. The old intuitive approaches no longer suffice in a complex world, particularly where decisions of a technical nature are involved. We carefully avoid confusion with data collection and data processing by bringing the human being into the system; by stressing that scientific and technical information (STINFO) ensues as a result of the evaluation of data by experts.

We are not even sure that we are educating information scientists. Some twelve years ago, after our Center had been functioning for several years, it was decided to develop courses in what was then called "Documentation". Such courses were taught by Lea

Bohnert and Si Newman. In fact, there are still some of us who are not fully reconciled to the name change. We are not sure that a true science exists. Perhaps then, we are educating and training information engineers or technologists, to use the terminology first introduced by Robert Taylor.[3] It really does not matter to us as long as we know that we are fulfilling vocational needs in a changing society and so long as we can grant master's and doctorate degrees based at least partially on our course offerings. It is somewhat strange that those of us with rigorous training in the so-called "hard sciences" tend to be quite conservative in this respect, while others with a humanistic background, are determined to label us for all time.

Still, in our overall outlook, we tend to agree, at least partly, with Borko, when he states what information science is, in his opinion:[4]

> It is concerned with that body of knowledge relating to the origination, collection, organization, storage, retrieval, interpretation, transmission, transformation, and utilization of information.

It is only proper that we try to relate this to the now-classic paper of Otten and Debons, "Towards a Metascience of Information: Informatology":[5]

> First, we want to distinguish between information and operations on information.

Then, further on:

> Operations on information, on the other side can be compared to the various forms by which energy can be manipulated, e.g., the conversion of heat into electricity, chemical energy into heat, etc. These manipulations on energy obey certain fundamental laws. In the same way, one can postulate that operations on information will obey its own set of fundamental laws.

This may well be the case. It has taken the science of physics several centuries to arrive at this point and chemistry is just now developing a theoretical underpinning. In the meantime, our own program which, I must strongly emphasize, has never received a cent or a penny of outside financial support, must remain relevant in its teaching program, to our own clientele, both actual and potential. In the Washington, D.C. area this constitutes the Federal Government and its numerous contractors and grantees.

The success of the center and of our information program is very largely dependent upon the caliber of our students, who are among the best in the country. We are dealing mainly with a group of mature, highly-motivated men and women, recruited mainly from the ranks of science, engineering, and business administration majors who are desirous of continuing and up-dating their education. Most of them attend classes on a part-time basis, being employed full-time in various positions, from computer programmers to at least one Rear-Admiral in the U.S. Navy. Thus there is usually no question of motivation. With a median age of over 30, we have at least chronological maturity. Prominent among the student body are retired officers and enlisted men from the nation's armed forces, retired civil servants embarking on a second career, and last but certainly not least, the housewife, emerging from the wilds of suburbia. They help us to maintain a decent sex-ratio, which is about four-to-one in favor of the men. We strongly stress the concept of life-long continuing education, and to this end, we welcome all graduates of approved four-year collegiate institutions as non-degree students. We urge them to apply for graduate standing only if their undergraduate records indicate that this is virtually a certainty. Otherwise we urge them to take several courses with us so that their undergraduate record can be forgotten and forgiven. We have been aware of the well-known phenomenon of the "slow-starter" ever since the inception of our program. Classes are held in the evenings and on Saturday mornings. Evening sessions last for two-and-one-half hours with a short break. Some of our students, after an eight-hour workday and the well-known commuting hassles, have been known to sign up for two periods per evening. Such devotion to duty helps bring out the best in our faculty. It must be emphasized that very high standards of academic performance can be maintained with this type of student body. There is a strong distinction made between graduate and undergraduate courses in our program with the student expected to learn to conduct himself accordingly as he proceeds along with his studies. There are surprisingly few failures as such, although the drop-out rate presents a problem. Some thirty-three semester hours (or eleven courses) of graduate credit are required for the MSTM degree (Master of Science in the Technology of Management). To complete the program, two fields of concentration must be selected from among six. These, in addition to STINFO are Management Information Systems, Computer Applications Systems, Research and Development Administration, Operations Research, and Environmental Management Systems. Four or five courses are required in each field together with certain general courses which will be mentioned later on. Most of the students who elect STINFO also take MIS (Management Information Systems) so that these two areas of concentration exemplify our approach to information science. Each program is capped with a final seminar course with active student participation and a comprehensive examination covering the entire field including subjects not covered in the required courses. It must be emphasized that we do not train librarians or computerniks. The STINFO program has had the constant problem of maintaining its identity between the Scylla of computer science and the Charybdis of library science.

During the first five years or so, many of our students were employed in information facilities throughout the metropolitan area. Like many in our field, they had more-or-less drifted into information work without formal academic credentials. Some of these individuals had published scholarly papers in our area and held responsible positions. They were also active in professional societies. Their major reason for returning to school was the increasing demand for advanced degrees on the part of their employers in order to insure their continued success in their careers. Essentially, they wanted to have their many years of work-related, on-the-job experience "legitimized" by the award of an academic graduate degree. Such a clientele presented a formidable challenge to our faculty, for frequently some of our students had had more experience in certain areas of information handling than did their mentors. The problem was easily resolved by the encouragement of a democratic, cooperative, give-and-take classroom atmosphere. I, for example, have yet to attend a graduate seminar during the course of which I did not learn some concept or some new approach to a problem. We soon became accustomed to meeting with our students at gatherings of our professional societies. During the course of the years, an "invisible college" developed which still exists today and is responsible, to a certain extent, for the successful placement of our graduates in desirable positions both in government and in industry. These characteristics of our student body make it difficult for other information science programs to take advantage of our decade of experience. We got used to the idea that students were simply colleagues who had not undergone certain academic experiences of a broadening nature. They tended to have most of their work-experiences limited to a narrow aspect of information science, such as library automation, natural language systems, etc. In recent years, when we began our course and faculty evaluation programs, it was gratifying to learn that even the "old-timers" among our students had gained some new insights from our course offerings. They had learned, for example, that for some information storage and retrieval systems, such as those in the field of chemistry, a controlled vocabulary as exemplified by the thesaurus, was not necessary. They had become more aware of other approaches than those of the agency in which they were employed. This decrease in parochialism, we feel, was a major attribute of our educational and training approach for the type of student body which we were obliged to service.

Before proceeding to develop a full-blown STINFO curriculum, we had to assure ourselves that we were indeed meeting real needs for professional manpower, for many of us had simply moved into the field without formal educational backgrounds. Why develop courses on a formal basis when the rudiments of information handling could easily be picked up on the job? Furthermore, unlike the situation in librarianship, there were no accrediting agencies. To this very day, it is still possible for a computer programmer or a librarian, for that matter, to call himself an information scientist without credentials of any kind, be they academic or the more valuable evidence of scholarly publication and job experience. The use of the designation, library/information science, has distressed many of us and has made our recruitment problems more difficult. In fact, it is only now that the U.S. civil service is considering the development of a series of position descriptions for information science. In this respect, I am sure, the British and the Canadians are well ahead of us.

Fortunately, 1963 was the year of our "Magna Carta", a report issued by the President's Science Advisory Committee under the title "Science, Government, and Information" and subtitled "The Responsibilities of the Technical Community and the Government in the Transfer of Information". It is more commonly known as the "Weinberg Report"[6] after its chief instigator, the physicist, Dr. Alvin M. Weinberg, the Director of the Oak Ridge National Laboratory. Here we had a clear indication from the "user group" that there was need for the kind of educational program which we had cautiously initiated several years before at our Center for Technology and Administration.

This document, though more than ten years old now, is still required reading for all of our students. I quote from page 28, section G:

> *The Technique of Handling Information must be Widely Taught.*
>
> But our schools and colleges will have to do more than insist on proficiency in handling the language. They will also have to insist on some proficiency in the techniques of information retrieval. The technical man, as an author, contributes to the information explosion; as a user of information, he is overwhelmed by the explosion. He must therefore be able not only to express himself clearly and succinctly and with proper regard for subsequent retrieval of what he writes; he must be acquainted with the new tools and techniques of information handling. Imparting such skills to our new generation of technical people is the job of our colleges, universities, and technical schools. They will have to teach, much more aggressively than they have in the past, the techniques of technical communication.
>
> Schools of science and technology have offered some training in use of the literature and in the techniques of communication, but their efforts have been sporadic. Only in those fields, notably chemistry, where the information crisis has been clearly discerned, has much formal training in literature retrieval been given. Some chemistry departments have for years required their students to take short courses in the use of the literature, and many have adopted the more effective procedure of making active use of the literature a necessary part of the work in such courses as qualitative organic analysis.

This explains why so many of us, myself included, started out as chemists.

To be more specific, the report recommends that:

> Our colleges and universities must provide ... instruction in these techniques as part of the regular scientific curriculum. They must also educate in the art of handling information more professionals who can lighten the burden of the technical man and can invent new techniques of information retrieval. In short, knowledgeable scientific middlemen who themselves contribute to science are the backbone of the information center; they make an information center a technical institute rather than a technical library. The essence of a good technical information center is that it be operated by highly competent working scientists and engineers - people who see in the operation of the center an opportunity to advance and deepen their own personal contact with their science and technology. Proliferation of the specialized information centers will therefore require many such 'information scientists', dedicated and knowledgeable technical men who help interpret and assimilate the literature for others working in the field.

The involvement of subject-matter specialists in the training of information personnel was emphasized in a short paper published as part of the Proceedings of the 1963 meetings of what was then the American Documentation Institute (now the American Society for Information Science) in which I stated that:[7]

> Close cooperation between many segments of the academic community is necessary in order to develop an adequate program for the education and training of information specialists in biomedicine. The medical school and the basic science departments of the liberal arts school can provide subject matter courses as well as the laboratory for the conduct of research in documentation. The foreign language departments can easily provide elementary reading courses in the major languages of importance to biomedicine, while the library school can handle the course offerings pertinent to information science. In addition, courses in data processing, mechanization of storage and retrieval, etc., can be provided by the business or government administration departments of the university.

I do not know whether such a program now exists in the U.S. or elsewhere although Darling,[8] a leading medical librarian, indicated its importance ten years ago. However, some five years ago, the American University, with the whole-hearted cooperation of the

Chairman of our Chemistry Department, inaugurated such a pilot program which thus far, has been limited to the training of chemical information specialists. Suitably-prepared students are admitted to the graduate program of the Chemistry Department but are required to take about half of their courses in the field of scientific and technical information systems. A special course in chemical notation systems has been developed by Alan Gelberg for such students and offered during the fall semester each year. In addition, chemists are also advised to take our regular course in publication techniques, which emphasizes computer applications in the areas of photocomposition as well as the economic and management aspects of journal publication. Until this year, this course was taught by Joseph Kuney, who is responsible for its development. This fits in very well with our university's long-term concern with this area, which represents the output phase of scholarly communication. Dr. Lowell H. Hattery of our faculty has long been a leader in this hitherto neglected subject. It should also be emphasized that the Washington, D.C. area is a major publishing center as a result of governmental activities as well as activities of the many scientific societies which have headquarters there. Foremost among them is the American Chemical Society which has been a pioneer in the application of computer-assisted photocomposition to the publication of both primary and secondary journals. We feel strongly that this area is a neglected part of the information science continuum and we have accordingly sponsored institutes for the printing and publishing industry[9] in order to acquaint them with the latest technological developments. After all, this involves the final step in the system of communication between the producers and the users of information and can easily become a bottleneck.

At any rate, we have graduated a number of students in recent years with a Master of Science Degree in Chemical Information Systems, conferred by the Chemistry Department. Within the coming school year, we hope to confer our first Doctor of Philosophy degree in this chemical specialty. The American Chemical Society has had its Division of Chemical Literature for many years and publishes the highly-regarded *Journal of Chemical Documentation*. There is no reason why a chemist who intends to make his career in the field of chemical information handling should be forced to take laboratory courses as part of his graduate work or to conduct laboratory research for his dissertation requirement. There are certainly many problems in chemical information handling which are worthy of graduate research.

It is our ambition to develop similar cooperative graduate programs, at least at the Masters level, with such academic departments as Physics, Biology, etc., for in the long run, what is most needed in our profession is the "hyphenated" specialist, the physics-information scientist, the engineering-information scientist, etc. A cursory examination of the employment opportunities which are advertised should quickly convince the sceptics that this is the type of individual who is much sought-after by government and industry. Nor should the subject expertise be limited to the "hard sciences"; there is need in the social sciences as well. Some of our former students came to us with degrees in psychology and experience in working for the American Psychological Association. Although this may invariably lead to a fragmentation of our nascent profession, there is little that can be done about it in the face of the increasing fragmentation of knowledge itself. In librarianship, for example, there is fragmentation and specialization, if not in the library school curricula, certainly on the job. It seems to me that the only thing which the elementary school librarian and the medical librarian, to cite an extreme example, have in common is that they both deal with books. The medium of communication is, at least to my mind, an insufficient reason for a commonly based educational philosophy.

Our meager resources do not permit us the luxury of presenting special courses in such basic areas as statistics, linguistics, psychology, logic, the history and philosophy of science, library science and computer science. Instead we refer our students to courses offered by other units within our university or elsewhere in the metropolitan area. Since we are dealing largely with mature individuals who, in many cases, know what they are interested in, it is possible for us to "customize" course offerings for each student, if necessary with a minimum of "required" courses. We are thus able to waive our requirements in such areas as statistics, management mathematics, etc. for many students who have been exposed to their subject matter in the course of their undergraduate program or even on the job. It would be counter-productive and downright silly to expect a student who has had many years of programming experience to take our courses in FORTRAN, COBOL and PL/1.

Despite our emphasis on the "science" of management, we still adhere to the old-fashioned idea that education consists primarily of the interaction between student and faculty. We are, then, indeed fortunate to have available to us, a distinguished faculty, both full-time and part-time. It has been our basic philosophy that in the emerging discipline of information science, an "ivory tower" approach is intolerable. All of our instructors must have had practical experience in the workaday world if they are to communicate their know-how and expertise to our students. Otherwise, the many excellent textbooks and self-study courses would be adequate. Although we do not have an active research program, all of our faculty have had and continue to have responsibilities for the development and the day-by-day administration of information facilities. This is particularly true of our part-time faculty who constitute the bulk of our instructional staff. Putting it succinctly, we expect our professors and lecturer to teach in the evenings what they are doing during the workday. Thus, our students are constantly exposed to the real-life problems which face members of our profession and to participate, even vicariously, in their solution. Of course, there are obvious disadvantages. Part-time instructors may not always be available when students wish to confer with them. Here, the telephone is a valuable asset. My students, for example, are urged to call me at the office or at home whenever they have questions which can be answered relatively briefly There is a certain "turnover" in faculty but then, this also makes it possible for us to replace instructors who don't work out much more easily and promptly. Each teacher may approach his course with a different point-of-view, despite common syllabi. This is an asset rather than a liability and exposes our students to diverse approaches to the same subject matter. At a graduate level, we prefer to teach rather than to indoctrinate or to "spoon-feed". Fortunately, the Washington metropolitan area is replete with scientists and engineers who are frustrated teachers and who would, in many instances, prefer the academic life if they could afford to and if suitable positions were available. As a result, we have many more applications from qualified individuals than we can handle. Thus, modern management is taught by modern managers and not out of someone else's obsolescent textbook. Feedback from our highly-critical student body both in the form of formal evaluations and word-of-mouth testimonials has convinced us that this system works. Furthermore, it is cost-effective. It would be remiss of me not to express our sincere appreciation to these underpaid men and women who bring to our program their lifetimes of experience. Some of them, such as Joseph Kuney are "household" names in the profession.

We make every effort to expose our students to professional activities at an early stage. They are actively recruited into the ranks of the American Society for Information Science and usually form a significant minority of the attendees at the local meetings of the Society's Potomac Valley Chapter where some of them have held elective office. We prefer and they prefer not to be segregated within student chapters. They are also in evidence at the national meetings and have contributed to the doctoral forums during the 1971 and 1972 conventions. They have submitted brief communications and letters to the editor to the *Journal of the American Society for Information Science* in recent months. Being made aware of the importance of the "invisible college" as an important vehicle of professional communication, they have, in many cases, developed their own "in-groups" and inform each other of job vacancies and employment conditions as well as counsel the new students concerning course offerings. They also perform the valuable service of evaluating our instructors on an informal basis, since we do not wish to have a "silent majority" of dissatisfied students who simply "drop-out". To make sure that they are made aware of the subject-matter content of the courses for which they register, they are provided with detailed syllabi, broken down as to the topics to be covered during each class session, before registration. In addition, they are urged to visit the campus bookstore and browse among the texts which are shelved on the basis of course requirements. Such measures have proven their worth by reducing the number of failures. Our full-time faculty prefers to counsel each student individually despite the presence of competent professional counsellors at the university.

Despite our unique situation, we keep scanning the literature on education in the information sciences in the hope that we could find similar programs, if not in the United States, then perhaps abroad. We have attended several meetings where educational philosophy was discussed. The first of these were the two conferences held at the Georgia Institute of Technology in 1961[10] to be followed by yet another one convened in 1964 at the Western Reserve School of Library Science[11] where I presented a short discussion of our then-emerging program. The 1965 meeting at Airlie House[12] was quite disappointing to us in its outcome. In 1966, Marilyn Bracken (one of our students and our first Ph.D. to write a dissertation in the information science field) and Charles Shilling[13] published a survey of educational and training programs in the United States, including our own. We first became aware of developments in the United Kingdom and in Canada as the result of papers delivered by Saunders and Schur of the Postgraduate School of Librarianship at the University of Sheffield[14] and Pitternick and Brearley of British Columbia.[15] We were gratified to find a close philosophical relationship between these programs and our own. The Swedish approach, as described by Bjorn Tell[16] stressing as it does, systems analysis, was also in the same vein. We have found the British publication by Schur and Saunders, entitled *Education and Training for Scientific and Technological Library and Information Work*[17] most helpful in learning what our colleagues overseas were doing.

To summarize, we find that our basic educational approach and the goals of our program closely resemble those of our British colleagues. This impression was strengthened as a result of my contact with the Israeli information scientists during the brief course which I conducted in 1971 under the auspices of the Centre for Scientific and Technological Information, and again in 1972 when I spent three months in that country as a UNESCO expert. According to Farradane,[18] the content of the syllabi, as described by Schur[19] "follows mainly the British model". It seems, therefore, that we are not alone in our emphasis upon the pragmatic even at the expense of the theoretical. Now it is evident to us why we use mostly British texts for our courses. Before coming to this meeting, I discovered a new one, the excellent book by Ronald Stamper of the London School of Economics entitled *Information in Business and Administrative Systems*. It covers almost precisely the areas with which our Center for Technology and Administration is concerned,[20] since he states in his preface that:

Only recently has it become necessary for many people

> to discuss information with rigour in a context that does not fall conveniently into the framework of a single discipline such as psychology or communications engineering. How to use information technology to solve organisational problems is a question which cannot be answered within the confines of any single established discipline, hence the need for a new framework such as the one offered in this book.

NATO should be made aware of the necessity for a "New framework" since the organization, as I recently learned, also conducts Advanced Study Institutes in technology transfer. Yet, during the course of its sessions at Novotel-Evry, France, held two months ago, there were no information scientists in attendance and the obligatory role which information science must play in the process of technology transfer was not included in the abstracts which I had an opportunity of seeing.

Perhaps the differences between the U.K. approach and the American attitude toward education and training stems from a fundamental difference in what information science is all about. Farradane in his review in the sixth volume of the *Annual Review of Information Science and Technology*[18] states it most succinctly:

> In Great Britain the term 'information science' has been used to cover both the theoretical and practical aspects of information handling, from searching the literature, answering enquiries, abstracting and writing reports, to classification, indexing, computer-based information retrieval systems, and the testing of such systems. It has not, in Great Britain, been envisaged as only a discipline of investigative science, as defined by Taylor or by Borko, and as it now appears to be widely accepted in the United States. Nor has it been quite the counterpart of 'information technology' (techniques of information handling).

Perhaps the differences are at least partly due to the fact that many leading British information scientists come from the ranks of engineering and have had many years of practical experience in providing information services in an industrial or governmental environment. Perhaps I may conclude this discussion by quoting from yet another British source. Brittain, in his recent book, *Information and its Users*,[21] presents what is essentially our own educational philosophy when he states that:

> The advent of computer-based information storage and retrieval systems, the application of systems theory to the management and planning of information centres and libraries, and the sheer cost (and perhaps novelty) of these systems have played a part in drawing attention to the design (including user requirements) of systems. These systems involve a feedback from users that could not be incorporated in a non-computer-based-system.

> There is, in short, a movement towards a receiver-controlled system and a movement away from a source-controlled system.
>
> The attention to user-oriented information systems is only one example of a growing phenomenon. Attention to, and a demand for, user-orientated services is currently to be seen in a wide range of human activities. Examples of user enquiries, other than into information systems, include investigations of the users of public transport (with the object of reorganising transport systems to provide services that are required and used frequently), consumer research (although this pre-dates user studies of information requirements), and the re-design and creation of courses of instruction that are student, rather than teacher, orientated.

In other words, we teach what our students and the job market consider to be information science at least 'at this point in time'.

References

1. Tague, J. and Carroll, J., Information in an informationless world or the making of an information scientist, *Journal of the American Society of Information Science*, 24, 40, 1973.

2. *Master of Science in Technology of Management*, Center for Technology and Administration, College of Public Affairs, The American University, Washington, D.C., 1973.

3. Taylor, R.S., Professional aspects of information science and technology, in *Annual Review of Information Science and Technology*, 1, Cuadra, C.A., ed., 15, 1966.

4. Borko, H., *The Conceptual Foundations of Information Systems*, System Development Corporation, Santa Monica, California, 1965. (Report No. SP-2507) (AD-615 715).

5. Otten, K. and Debons, A., Towards a metascience of information: informatology, *Journal of the American Society for Information Science*, 21, 89, 1970.

6. U.S. President's Science Advisory Committee, *Science, Government, and Information: The Responsibilities of the Technical Community and the Government in the Transfer of Information; A Report*, U.S. Government Printing Office, Washington, D.C., 1963.

7. Welt, I.D., The education and training of information scientists in biomedicine, *Automation and Scientific Communication*, Part 2, 167, 1963.

8. Darling, L., Development of training programs in American medical libraries, *Bulletin of the Medical Library Association*, 51, 339, 1963.

9. Hattery, L.H. and Bush, G.P., eds., *Automation and Electronics in Publishing, (The American University Technology of Management Series)*, 3, Spartan Books, Washington, D.C., 1965.

10. George Institute of Technology, *Proceedings of the Conferences on Training Science Information Specialists*, Atlanta, Georgia, 1963.

11. Goldwyn, A.J. and Rees, A.M., *The Education of Science Information Personnel - 1964*, Center for Documentation and Communication Research, School of Library Science, Western Reserve University, 1965.

12. Heilprin, L.B., Markuson, B.E. and Goodman, F.L., *Education for Information Science, Warrenton, Va., September 7-10 (1965)*, Spartan Books, Washington, D.C., 1965.

13. Bracken, M.C. and Shilling, C.W., *Education and Training of Information Specialists in the U.S.A., Biological Sciences Communication Project,* The George Washington University, Washington, D.C., 1966, mimeographed.

14. Saunders, W.L. and Schur, H., Special library and information services for science and technology departments in British universities and the implication for professional education, in *International Conference on Education for Scientific Information Work*, 38, 1967, FID 422.

15. Piternick, G. and Brearley, A., Canadian training of scientific information workers, in *International Conference on Education for Scientific Information Work*, 73, 1967, FID 422.

16. Tell, B.V., Inclusion of systems analysis and programming in curricula for documentalists, in *International Conference on Education for Scientific Information Work*, 200, 1967, FID 422.

17. Schur, H. and Saunders, W.L., *Education and Training for Scientific and Technological Library and Information Work*, Her Majesty's Stationery Office, London, 1968.

18. Farradane, J., Professional aspects of information science and technology, in *Annual Review of Information Science and Technology*, 6, 399, 1971.

19. Schur, H., Postgraduate science information course in Israel, *UNESCO Bulletin for Libraries*, 24, 251, 1970.

20. Stamper, R., *Information in Business and Administrative Systems*, A Halsted Press Book, John Wiley and Sons, New York, 1973; B.T. Batsford Lt. in Great Britain, 1973.

21. Brittain, J.M., *Information and its Users. A Review with Special Reference to the Social Sciences*, Wiley-Interscience, New York, 1970.

WORKING GROUP REPORTS

REPORT OF WORKING GROUP 1 ON "THE NATURE OF INFORMATION"

Robert F. Barnes and Klaus W. Otten

Group Leaders

I. Objectives

The objectives of the group were seen by the participants to have an essentially threefold nature. For our deliberations, therefore, we attempted to develop the following three components:

(1) a general conceptual framework broad enough to portray a wide variety of information phenomena in a way acceptable to information scientists in general;

(2) within such a general framework, a number of sufficiently detailed features to allow a precise identification of whatever particular aspects of information phenomena are of interest to an individual information scientist and to facilitate the systematic relation and comparison of such interests;

(3) from such a framework, to suggest areas of study which seem especially critical to our understanding or promising in results.

II. Approach

The composition of the working group provided us with a nice microcosm of the generally heterogeneous character of Information Science and its practitioners. Not only were there differences in professional training and viewpoint (our group contained engineers, mathematicians, psychologists, philosophers, ...), but even differences in conceptions of the purpose and nature of science itself were noted. Such heterogeneity has challenging positive aspects, of course, but poses difficulties as well. One is thus faced with a difficult choice - whether to "legislate" a single conceptual framework and the concepts and terminology it embodies, or to "tolerate" as many frameworks as are introduced, with the attendant multiplicity of terminologies that this involves. In an extreme form, the first approach is authoritarian and poses a very real danger of stultifying progress. The other extreme, however, is undesirably anarchical and produces conceptual chaos.

It was therefore felt that the best approach to deal with this problem would be to try to pose a very general schematic sort of framework - ideally one broad enough that almost anyone might find within it the resources to specify clearly and precisely the meaning of any of his terms. In such an approach to definition, the emphasis is more on the role of *definientia* in *expressing* key concepts than on the role of *definienda* in *naming* key concepts. It is only after clarity of expression is attained that we can ultimately reach consensus on appropriate terms and their definitions.

The group members also expressed a number of distinct views as to the directions and concerns that this year's working group ought to assume in its deliberations. Some were concerned with the development and understanding of a very general concept of information, applicable to a broad range of phenomena - ranging from those dealt with in genetics and physiology to those encountered in the operation of document retrieval, management information, and other such systems. A majority of the participants, however, wished to concentrate on various aspects of human information processing. Of these, many tended to focus on activities of the individual person as an information processor, while others were concerned with information processing as a socially-oriented phenomenon. Finally, among those concerned with information phenomena on the individual level, there were still major differences in interest - with some concerned chiefly with the syntactic and semantic foundations of information processing, and others attending primarily to the utilization of such information (in decision-making activities, for example). It must be emphasized, however, that this diversity had an unmistakably "centrifugal" effect as well as the obvious "centripetal" one. Several times, after a decision to divide into smaller, more homogeneous subgroups had been made, the interplay between the different interest groups was stimulating enough that we postponed the subgrouping; in fact, it never did occur. This shows, we believe, that information scientists can develop interests that unite them which can become more important than the differences that divide them - a fact of considerable promise for the prospect for Information Science as a discipline.

III. "Prolegomena to any Satisfactory Future Theory of Information Phenomena"

Whatever may be one's perception of "information" and the

phenomena in which it is involved, certain features stand out. These, then, pose certain requirements for any theory which is to account satisfactorily for such phenomena. Not everyone, of course, will agree on *all* these features, but the working group was able to establish a consensus on a number of them:

(1) the *multi-level structure* of information phenomena, reflecting different types and complexities of information processing;

(2) the *system-specific character* of information, depending upon the internal state of the system, which includes past experience (and, in man, cultural bias and conditioning) as well as teleological factors;

(3) the *environmental relativity* of information at each level of the multi-level structure;

(4) in simplest form, the *sequential nature* of the processing, with each level's processed output serving as the next level's "raw data"; in most complicated form, the *heterarchical nature*, with processing at any level drawing upon resources from all levels;

(5) the *interactive character* of many information systems, in which the information-processing activity also serves to restructure the system for processing subsequent inputs;

(6) ultimately, but not necessarily, at each stage of the processing, the *goal-oriented nature* of information processing;

(7) the often *bi-directedness* of the information process, involving interaction between two or more systems or between system and environment.

These, we believe represent key aspects of information phenomena, which must be taken account of by any satisfactory theory of information phenomena.

IV. The Conceptual Model

The general conceptual framework discussed and adopted by the group is essentially that presented by Klaus Otten in the paper he gave in one of the plenary sessions. To avoid the danger of misrepresenting in a necessarily oversimplified summary, we refer the reader to that paper. We can say at this point, however, that the model was seen to be of the desired general character, applicable to all the information phenomena we considered, with features of various sorts which account for all the previously mentioned key aspects of information phenomena.

V. Recommendations for the Future

As our perceptions of the nature of information became clearer, feelings began to crystallize as to certain appropriate guidelines for the future development of Information Science. They are of various sorts - some being more substantive in character and some more of a methodological or procedural nature, though no clear-cut boundary between the two types seems to exist.

(1) In the present "pre-paradigm" state of Information Science, each researcher has a responsibility to be aware of *and to clearly specify* his own biases, preconceptions, presuppositions, working assumptions, etc.

(2) To avoid terminological confusion (of which the working group experienced more than enough!) each person should clearly specify the meanings of his technical terms, by reference to some standard model, if possible. Only then can we hope that certain usages will become widely enough accepted so as to eventually constitute a standard vocabulary, which is a highly desirable goal.

(3) A useful stimulus for research would be the identification of those presently-established theories (including a large body of *specific* laws, hypotheses, and results) which relate directly to the study of information and information systems. (Some examples of this sort of effort appear in the report of the 1972 working group, but a greater, more systematic effort is needed.)

(4) Within the conceptual framework presented here, we need to develop at all points specific operational procedures and predictions which can be tested against empirical data. Within the body of presently-established empirical research, we need to see what can and what cannot be interpreted in terms of the framework.

(5) In the development of alternative models, it is crucial to make clear those points of similarity to, difference from, and improvement upon existing models.

(6) For any future meeting, a careful identification of well-defined professional specialties (e.g., neurophysiology) *and of specific persons* within those specialties whose participation in discussion would be highly desirable.

(7) At all times, as "information scientists", we should more explicitly direct our attention to the information-transfer aspects of our own research (e.g., the existence of informal information networks, special interest groups such as ASIS-SIGFIS, etc.)

Working group leaders: Robert F. Barnes
Klaus W. Otten

Working group administrative assistant: Inez Fitzgerald

Working group participants:

- Serafino M. Amoroso
- Lynn E. Baker
- Yehoshua Bar-Hillel
- B.C. Brookes
- Paula C.E. Burghraeve
- Fernando Campos
- Spyros Diamessis
- Robert A. Fairthorne
- Mario Gnamignaui
- Glynn Harmon
- Esther E. Horne
- Joseph C.R. Licklider
- Maria Lima
- Frank Maher
- Thomas Minder
- Ian Mitroff
- Xavier O. Monasterio
- John S. Nicolis
- Joseph R. Royce
- Tefko Saracevic
- Jimmie R. Suttle
- Thomas J. Triggs
- Michael S. Watanabe

REPORT OF WORKING GROUP 2 ON "INFORMATION TECHNOLOGY AND HUMAN NEEDS"

Orrin Taulbee and Robert Landau

Group Leaders

The breadth and complexity of the aspects of human needs and technology as well as the number of participants in the panel suggested a grouping in terms of common interest if such grouping could be found. Toward this direction a proposal was made to identify groups and permit the panelists to select a particular group. The proposal was for the three groups:

(1) human needs
(2) information systems that contained a data base
(3) information support systems, i.e., containing no management-provided data base.

Among the participants there did not appear to be much interest in the latter group, so only two groups were formed.

These groups met separately, the first to identify the dimensions of human needs in relation to technology, the second to identify the dimensions of technology to satisfy human need. There was an evident disparity in the first group: those concerned with human information needs irrespective of technology and those who felt that in many instances technology as exemplified by the computer was not desirable. The second group addressed the question of technology under the assumption that well-defined needs had been expressed. Each of these groups developed a set of points which were presented and discussed by the panel. These will be put in evidence in various forms in succeeding portions of the panel report.

In order to bring the diverse viewpoints of the panel into closer harmony it was felt that the title should be changed to

Human Information Needs and Supporting Technology. Two principal concerns appear to come to the fore with regard to information needs. The first is that of the concern of the user who, in order to use the system, must use the protocol and procedures established, procedures that are inflexible in nature. Another way of stating this is that the individual is controlled or directed by the system rather than the reverse being true. Obviously there is room for trade-off here if economics dictate that certain things would be prohibitively expensive which the user desires.

The second major concern of information needs recognizes that needs are very close to rights. It was proposed that there be put forth a "declaration of human rights with respect to information". This declaration recognizes the right of the individual to have access to information which will be advantageous to his welfare. It also recognizes the right of the individual to withhold information about himself so that it cannot be put into a system which makes information accessible to others. These human rights should be recognized and dealt with in the design of information systems. A sample declaration is specified below:

(1) Information necessary to maintain awareness of and control over information systems, i.e., continuous public information about the input, content, and product (including use) of *all* information systems in which "data" on humans are included.

(2) Information necessary to optimize the welfare of humans, e.g., current information pertaining to health and education.

(3) Information necessary to maintain human (individual) dignity, i.e., every individual has the need of information to avoid the possibility of being viewed ignorant by others having privileged access to information.

(4) Information necessary to maintain "progress", to advance the frontiers of knowledge, e.g., scientific information for scientists, technologists, society, etc.

(5) Information necessary to maintain a level of intellectual awareness compatible with an individual's intellectual curiosity and capacity.

(6) Information necessary to pursue day-to-day activities with a minimum of accidents (wrong decisions, unjust decisions, stupid or unadaptable acts, etc.) which could have been avoided with available information; e.g., weather forecasts, road conditions, etc.

(7) Information necessary to avoid being a victim of need (perhaps artificial) created by information systems; e.g., mass

media, state information services, etc. Part of being human is *knowing* when these things are being done to you.

(8) Information about the intended and possible uses of the data being collected about you.

After establishing a social framework under which all "human information needs" should be examined, the panel undertook an examination of more specific needs. The discussion quickly demonstrated that the concept of information need is at best elusive. Who is to say what is the information need of an individual or a group of individuals? Is it the individual or group itself or is it an external source which specifies the need? When is a perceived need real? An individual or a group may not perceive a need but an external source may in fact identify strongly-felt needs and try to convince the individual or the group that this is a real need. One may ask the individual or the group for the information needed; of course, it is well known that the response is usually expressed in terms of generalities. This results from inability to project future needs and inability to articulate precisely. Thus if a designer undertakes a task to develop an information system to satisfy the needs, then of necessity he must speculate as to what the precise needs are and design accordingly. Through an interactive process the system is modified and adjusted with the aim of satisfying the initial and ongoing needs. Toward this end of determining human needs for information it should be recognized that there are several classes of needs associated with different individuals depending on the functions that they perform:

(1) User.
(2) Management of system to provide information.
(3) Technical specialists, such as programmers and analysts.
(4) General public.

To indicate the diversity of information needs we give here several examples along with the technology which should be considered in order to fulfill those needs:

(1) Career planning. Direct access to a major data base which provides for search and browsing capability to permit an individual to assess the desirability of pursuing a particular field of future endeavour. This system would be capable of providing probability estimates of student success based on similar past performances and psychological profile compared to others that have entered the field, ability to interact with other professionals or students in the field through the information system, etc.

Technology:
Computer-generated verbal output
Interactive capability

Computer acceptance of verbal input
TV-like output (computer-generated)

(2) Medical. Immediate access to a health analysis system with a sensor/communication link to a regional information system. Such a system would be able to provide an immediate determination of significance of a variation in physical well being.

Technology:
Implanted sensors for a variety of measurable functions (human)
Communications interconnect
Personal data base against which variance can be determined
General medical analytical algorithms

(3) Social interaction. Access to interactive data base is desirable to permit discussion on any particular subject matter with individuals with similar interests and/or social inclinations. System must provide for free interchange of opinion and ideas.

Technology:
Interactive terminals which require little or no special skills to operate

(4) Education. A capability to select from a wide variety of subjects for study, CAI programs which can be engaged in from the home. Study could result in obtaining degrees or not. The system should have capability to suggest future courses of instruction leading to specific degrees and should accumulate instructional results for certification of specific successful accomplishment.

Technology:
Interactive terminals which require little or no special skills to operate

(5) Recreational. Access to a data base through which vacation planning can be done, immediate recreation alternatives can be evaluated and selected. This might include access to a data base which provides capability to proceed on an automobile trip from any point to any other point without benefit of maps. Directions would be provided automatically as alternative possibilities occur.

Technology:
Simple interactive terminal with system capability to make reservations, payments, etc.
Computerized roadways
Automobiles with auto data bank interaction

Since much of society's information needs are met by interaction with one or more information systems it seems then appropriate to restrict our attention to the information needs of users or potential users who interact with an information system containing a data base. Thus needs and technology will be explored on the basis of this restricted view. Information needs may be approached from the point of view of the total system or of the functions to be performed within the system. From the total system point of view we identify four needs:

(1) Reliability
(2) Quick response
(3) Adaptability
(4) Economy in delivery

Specific needs as identified by the panel in relation to functional aspects of the information system are:

(1) Input
(a) Compatability of language in communicating with system
(b) File creation - input and transformation of data to a form acceptable by the system
(c) Broad commonality of input language (this is in partial contradiction with adaptability aspect of overall system)
(d) Training. Tailored to the experience level of the user of information systems.

(2) Processing
(a) Access - including portability of access devices
(b) Browsing
(c) Updating
(d) Awareness of content of information system

(3) Output
(a) Compatability of language in communicating with system
(b) Output selectivity in quantity and type
(c) Ability to manipulate output and re-enter it as input into system

Information technology can similarly be grouped (in terms of hardware or software) by input, processing, and output. Other areas of importance include system support and economics.

Major input hardware includes key punching (KP), Optical Character Readers (OCR), Magnetic Tape Data Entry (MTDE), Direct Terminal Entry (DTE). The great majority of information being put into machine readable (MR) form is by KP; but that is only a very small percentage of all information being created. For economical reasons, OCR equipment requires a high concentration of data entry at one place. OCR is primarily useful for information in print restricted to a limited number of fonts. OCR will continue to be applied in limited situations as required and justified. Since the late 1960's, there has been a rapid increase in the use of MTDE systems using both small tape cassettes as well as standard computer tapes. This method is replacing the KP method. This trend will continue and increase. In the early 1970's DTE systems appeared and will increasingly take over from both KP and MTDE.

The importance of input and output software is dependent upon the degree of decentralized computation needed in the particular system, and the degree to which full graphic display is required. The rapid increase in "intelligent" terminals and mini-computers indicates a thrust to provide decentralized computer power. The continued lowering of information transmission cost will further compete with low-cost decentralized computer power. Hardware processing has been a major thrust in the last two decades. Systems readily available now can process millions of records (and their many sub-elements) in parts of a second and provide the results of queries from hundreds of terminals back to those terminals simultaneously within seconds. Software processing has been developing in combination with the hardware improvements; there are now more software languages than natural languages in common use. The major development in the early 1970's has been the trend from batch processing to random access, direct terminal systems. This reverses the process of removing the user from his data base caused by batch systems, to once again allow a close symbiosis between the user and the information store house at a new higher level of interrelationships. Access to very large information bases is now instantly available; electronic browsing among these and updating is possible from a remote terminal.

Major output hardware includes printers, computer output microfilm (COM) and terminal display cathode ray tube (CRT) for alphanumeric or graphic information. Printers can be used in high speed form (200-1600 characters per second) at the computer site or in low speed form (10-200 characters per second) at the user's desk using standard dial up telephone lines. CRT's can be used also at the user's desk with or without the printer at the same speed range. COM allows the great amount of material being produced in machine readable form to be stored in a compact and inexpensive form. To date there has not been widespread acceptance of microfilm because of the difficulty of updating, reading and portability of the reading equipment as compared with printed output.

There are numerous support activities that augment the hardware and software developments. System techniques, decision theory, communication facilities and training techniques have all moved forward along the scale to augment the hardware and software developments.

The cost of information technology indicates a clear trend

downward thus assuring an increasing use of computers, communication facilities, and micrographics. Storage, computation, communications, and terminal costs are decreasing dramatically. There is every reason to believe that the total effect will be a continuing increase in speed and a decrease in cost of information delivery. Economical considerations relative to satisfying information needs are an over riding factor. Needs may be agreed upon but if meeting these needs cannot be justified in terms of cost then the determination of those needs represents only an academic exercise. This means that it is necessary to establish means of determining precisely the costs of production of various user services. Information system managements have had little or no experience in developing and collecting those costs.

Up to the present time most information systems have been operated on a "research and development" basis, or an overhead basis in which there was little concern with the cost of individual services. But today with the high cost of information systems a new era is evolving. More and more information systems managers are being required to justify their existence on the basis of income from sales of service, and there appears to be no change in this direction. This results from administrative concerns with accountability and demonstration of the value of an information system.

CONCLUSIONS

(1) Technology will drag individuals with information needs along whether they want to be or not. Consequently, they should take an active role in shaping the future direction of technology.

(2) Need is sometimes created because of presence of technology and service; witness the example of xerox - the need for reproduction capability was grossly under-estimated.

(3) Rights of the individual with respect to information should be preserved in the design, development and operation of information systems.

(4) Flexibility should be preserved and expanded as much as possible in the technology of information systems so that the human is unfettered in his information seeking.

(5) Since different individuals use different information and usually use identical information in different ways, the development of an information system is an evolutionary process responding to the aggregate need of its body of users.

(6) Technologists and those with the need for information should join together in the development of information systems; in this, appropriate checks and balances with trade-offs between what

is desired and what is possible is effected.

(7) The information needs of an individual or a group which are real or perceived may or may not be satisfied depending on the availability of technology to deliver, and the economic justification of the process.

RECOMMENDATIONS

(1) We recommend that studies on information technology and decision-making be made for other fields, such as health care, welfare services, education, etc. As envisaged, these studies would be similar to the business study reported in the 1972 Conference Board Report on "Information Technology: some Critical Implications for Decision Makers". This will be directed at the placement of information production and delivery services in these selected areas on a business-like basis.

(2) We recommend that studies be made of the various ways in which technology has stifled or restricted man's search for information with the view of either removing the technology or making it more adaptable.

REPORT OF WORKING GROUP 3 on "IMPACT OF INFORMATION ON SOCIETY"

Manfred Kochen and Eugene Rathswohl

Group Leaders

MAJOR ISSUES OF THE WORKING GROUP

The working group agreed that major issues could be subsumed under two questions: (1) what are the *effects* of information systems, and (2) what are the policy implications of these effects. The effects may be (a) physical, psychic, social, or (b) personal, ethic, moral. These effects might be studied in the following circumstances:

(a) on individuals and special groups;
(b) in developing vs developed countries;
(c) on the development of Third World countries;
(d) on system operators and programmers;
(e) on participatory democracy;
(f) on management;
(g) on other systems.

The determination of policy implications might take into account the following studies:

(a) to identify and critique underlying ideology of information systems;
(b) to differentiate long vs short range social planning;
(c) to investigate implications for the control of information;
(d) to study the role of information professionals;
(e) to determine interrelationships between problems of information (acquisition, processing, dissemination) and political problems and climate;
(f) to determine the payoff between designed and randomly-developed information systems.

APPROACH OF THE WORKING GROUP

The group decided to adopt a case-study approach. Two types of information systems were identified that have structurally different characteristics:

1. The National Crime Information Centre (NCIC) which represents a relatively well-structured system dealing with specific kinds of information problems.

2. PLATO and POLIETARIA as examples of relatively ill-structured systems that serve a variety of community services, such as social service systems and community planning.

The rationale of this approach was to discuss the characteristics of each type of system and identify the characteristic effects (impacts) of each type.

Findings

1. The characteristics of NCIC were identified mainly from the outline in Weston, *Databanks in a Free Society*. The file is maintained by the Federal Bureau of Investigation (FBI). The records include under "persons" only "fugitives from justice". The records identify items with unique serial numbers, e.g., the hot car list. However, personal property without unique identification that has been stolen would not be included in the file. Thus, the file consists of narrowly-defined records. Contributors to the file report to NCIC on a voluntary basis. Information is released only on specific request, so there is restricted automatic dissemination. NCIC consists of an on-line computer system with 40,000 teletype terminals each located in a police station with 100 remote control terminals. Input to NCIC is a RAP sheet, but possibly in the future, probationary reports will also be input. This input is decentralized. It comes from many levels and locations of police forces.

Issues, impacts, effects of NCIC

The name NCIC has strong psychological effects because of the power of the name. NCIC is strongly related to the RAP sheets which are not wiped out or destroyed. The RAP sheets create a large amount of paperwork for the officer filling out the report. There is room for errors and some errors may not be detected or deleted when the information is reported to NCIC. The FBI is assumed not to make errors. Arresting officers quite frequently do. The system design involves at least one major issue, that of the problem of standardization of the reporting system. The RAP sheet must define the crime that the person was arrested for. Different jurisdictions have different definitions of "crime". Although system design favors a unified, homogenous crime-reporting system, it is a difficult political problem to achieve it, even if it were desirable. "Unification" or "standardization" is inappropriate in many ways. The tendency of data processing, however, is to centralize. NCIC may be misused. Access is relatively easy for legitimate organizations, such as banks, credit services, etc., and is also relatively easy for those who know the procedures. However, NCIC files could easily be used for other purposes, both legitimate and non-legitimate. For instance, the file data could be used in a non-legitimate way to construct a "profile" of criminals. Files also have implications for surveillance, arrest, and suspicion. Those who argue for their use in this way claim that presumption of innocence does not prevent surveillance. Those who argue against it point to the effect of surveillance on the surveyor - one who is surveyed most is the most suspected. They also point out problems of balancing errors of commission with errors of omission. Not only do we not know what effects the system has on an officer reporting his data on a RAP sheet, but we do not know the effects on the user of the file, especially if data in the system are never deleted from it.

2. The characteristics of PLATO were determined mainly from Valarie Lamont's paper presented in the plenary session. The general concern is to involve citizens in community planning. PLATO is a teaching computer system using the CDC 6700. It is capable of simultaneous access by 4,000 remote terminals and consists of a plasma display unit which permits computer generation of slides, graphs, texts, and other printout capabilities, for instance xerox. The input console is a keyboard with the possibility of audio input and the programming language is TUTOR. Possible uses of PLATO were determined to be:

(a) referral;
(b) inter- and intra-group communication;
(c) education and training.

The programs already available in PLATO are:

(a) information storage and retrieval;
(b) issues in mental health conferencing; inter- and intra-state communication.

Thus, PLATO is an example of a mass information technology. How can it be used to involve citizens at all levels of education and income and all persuasions to participate in community planning? PLATO makes it possible for citizens to use the system via terminals to be better informed on community issues.

Issues, impacts, effects of PLATO

These were put mainly in the form of questions. How can the community safeguard input and output? Who controls the input? Should the input be controlled? In what ways? Why? Who develops the software for the system? What assumptions do the programmers work with? What controls are put on the programmers? Who edits the input and the output? How should editing be done to ensure fruitful, representative citizen input? How is the database updated? In what ways is editing productive and counter-productive? What is the grammar of the information media?

There are also questions of citizen participation and system design. Who has access to the technology itself? Specialists only or the general public? Is there any user control over the hardware or software of the system?

Questions were also raised concerning the relationship between technology and community participation. What are the psychological effects of PLATO technology on users? What are the differential effects of information technology on children versus adults? What are the effects on how the issues are discussed? In what ways does it facilitate discussion of policy issues? In what ways does it militate against it? Can participatory democracy be maintained or even facilitated by contemporary media of information?

The characteristics of POLIETARIA were derived mainly from a presentation by Dr. M.R. Alvum-Costa. It is an example of an information system in a developing country. POLIETARIA is conceptualized as a scheme of information dissemination concerning the general problem of the protection of childhood:

(a) biological protection
(b) mental protection
(c) legal protection

and is conceived as an educational system to promote generation and dissemination of useful information for improving the quality of life. It is an example of an information dissemination system that is not necessarily technologically based, but based rather on man-to-man communication. The general concern here is the relationship between generators of information and the receivers and users of that information, which is seen as a cooperation of further education concerned with (1) studying the relationship of social norms, quality of life, science and technology, and human ambition, and (2) studying and implementing ways of designing better conditions of social adaptation for people who require it.

Issues, impacts, effects of POLIETARIA

What is the cultural bias built into western information technology? What are the effects of various methodologies and observation technologies on information concerned with children? How artificial is information when generated in abstraction away from real-world situations? How can we determine the psychic effect of information technology on young/old, pre-literate/literate/post-literate people?

(1) Questions of information generation.

(2) Questions of dissemination and utilization.

Are the two concepts interrelated, nonseparable? What factors affect the utilization of information? How do we solve problems of the acceptance of information? If the channel of information is unwelcome, will the information packages be oversophisticated? How will the information be used? How can the disseminator relate to people within their own context? How will he determine the level of demand? How can we use our various media in a way that makes sense in a different culture context? How can a user learn to use a disseminator as a resource? How can a disseminator learn to use the user as a cultural resource?

In sum, the group found itself talking about the cultural anthropology of information systems. The issue of cultural bias led the group to question what effects (both productive and counter-productive) would be created by applying western information technology to underdeveloped countries. This issue revolves around the fact that the means of dissemination are people, and we are concerned with the adaptability of people - both disseminators and receivers. A new learning would be required by both the disseminator and the user. An investigation into the factors or dimensions of man which relate to his capacity to accept information is needed: his capacity for stimulation, level of knowledge, level of learning, state of health - biological, mental, and social. We also need to study factors of the dimensions of man which relate to his capacity for transferring or disseminating information.

GENERAL QUESTIONS BASED ON THESE FINDINGS

(1) How can we design information systems that will be sufficiently sensitive to the problems we have identified?

(2) What sort of society can emerge, given certain ideologies, certain assumptions of information? How do certain assumptions or ideologies affect society?

(3) What does it mean to have self-corrective (information) systems that are not demeaning of man? Indeed, are information systems demeaning of man? In what ways? How can we ensure that information systems do not become demeaning of man?

(4) How can we ensure the provision of a system that accomplishes what we intend with a minimum of unintended, deleterious consequences?

(5) How can we decide for/against extension of a system? How can we replace an existing system?

(6) Two major kinds of questions, perhaps, should always be asked:

(a) What are the effects of information technologies and organizations? What are the services and disservices?

(b) What are the effects of information on particular people? (We need analyses of the kinds of people affected.)

(7) How can we re-present information, i.e., how do we design an interface to achieve the intended effect in the target population? Why *should* there be a system? What is the rationale of the information system? Is the rationale "legitimate"?

(8) In what ways is the information system productive? Counter-productive? Does the system actually facilitate the effect opposite to the intended? e.g. does NCIC actually facilitate crime? In what ways does NCIC improve education processes? The effort to challenge the assumption that a system is 'correctly' designed will be an attempt to understand before we act.

(9) A general issue for all community information systems: in what ways will a system increase or decrease information dissemination in the community? What are the quantitative and qualitative dimensions of dissemination? How are the processes of dissemination changed via new technologies? What are the effects of media (a) on individual identity, (b) on tribal community identity?

In developing countries, what are alternative ways of making all information available? In other words, how do we make information programs relevant, taking into account the political dimensions of the problem?

SUMMARY

The group reduced the results of their discussion to three classes of question:

(1) Who are the major constituencies or groups of people affected by the system? What are the characteristics of the people affected? Some examples of characteristics were as follows:

(a) Users - people who know the system exists and actually make use of it:

(i) managers;

(ii) operators and maintenance people;

(iii) generators of information (police, special interest groups, educators, etc.) who input data to the system;

(iv) users themselves.

(b) Non-users - includes two types of groups, (i) those who know the system exists and do not use it, and (ii) those who do not know the system exists but are affected anyway.

(2) What are the effects? What are the characteristics of the effects - physical, psychic, social? Are they productive or counter-productive? How do we take into account the 'cultural anthropological' dimensions of information generation, dissemination (transfer) and utilization?

(N.B. This question of effects was the major thrust of the working group.)

(3) What are appropriate assessment procedures for studying effects? What should be our methodology? Some suggestions were:

(a) Technological assessment (assessment can be done prior to, during, and after an information innovation).

(b) Studies similar to concept of user-need studies (much emphasis on technical, experimental procedures of user needs).

(c) Ways of overcoming the general problem of assessment (i.e. that assessment data is not effectively utilized in the system design or in a system redesign).

Working group leaders: Manfred Kochen
Eugene Rathswohl

Working group participants:

Y. Bar-Hillel	C. Meadow
D. Conway	B. Nevitt
J. Finnerty	E. Rathswohl
M. Kochen	B. Rayward
V. Lamont	V. Rosenberg
P. Lima	J. Salancik

REPORT OF WORKING GROUP 4 ON "PROFESSIONAL ASPECTS OF INFORMATION SCIENCE"

Jack Belzer and Frank Gibbons

Group Leaders

The group considered the proposed agenda as indicated in the tutorial presented by Frank Gibbons and it highlighted the following aspects as being of particular relevance:

(1) Proposed International Institute for Advanced Study and Research in Information Science.

(2) Identification of areas of research relevant to information science.

(3) Manpower requirements.

(4) Status of the Information Science profession.

(5) International co-operation.

In response to these five items, the Working Group has made the following recommendations:

1. With regard to the International Institute for Advanced Study and Research in Information Science a proposal to NATO for a Research Evaluation Conference in 1974 has been drafted and appears in Appendix A. The following steps were taken toward that end:

(1) It was proposed that an International Institute for Advanced Study and Research in Information Science be established. In order to establish such an Institute a NATO Research Evaluation Conference in Information Science should be organized:

(a) To select a governing body for the Institute and ensure its continuity as a functioning body.

(b) To identify sources of financial support.
(c) To set goals and objectives.
(d) To identify possible locations.

(2) To assist in selecting the governing body as well as the participants in the above conference a "straw vote" was conducted at the second NATO Advanced Study Institute in Information Science held at Aberystwyth in August, 1973. The results are given in Appendix B.

2. The following recommendations were made with regard to identification of areas of research relevant to information science.

(1) Research in information science is well identified and it should be recognized that the records are maintained by F.I.D., UNESCO, ASIS, Aslib, etc.

(2) Areas of research in other disciplines relevant to information science should be identified.

(3) A centre for creating and maintaining records of such ongoing research should be established and maintained. Appendix C is a memorandum indicating a methodology for obtaining such information. NATO could be an appropriate organization for supporting this proposed research program.

3. Manpower requirements are based on career opportunities as well as attempts to meet those future needs which are identifiable.

(1) Career opportunities.

(a) The following tasks were identified which are being performed by information personnel: abstracting, indexing, thesaurus construction, computer file searching, literature searching, interface with system users, report writing, systems analysis, classification, programming, etc.

(b) Personnel on the Ph.D. or equivalent level are essential for advancing the state of the art of information science and technology although the percentage of the total manpower requirements is small.

(c) Career opportunities in various channels of the information science profession exist in the following three major areas:

(i) Service-oriented and administrative professional activities in information science, e.g., personnel responsible for interface with systems user, administration of centers for such interface, and/or carrying out tasks such as abstracting, indexing, etc.

(ii) Systems-oriented careers in the information field - personnel responsible for the planning, installation, and evaluation of information systems.

(iii) The research information scientist, and the academician, who would be engaged at universities, or government and industrial research establishments, teaching and conducting research into the theory and practice of information science.

(2) Manpower needs.

(a) New developments in information technology will create needs for increased manpower.

(b) Opportunities for education and research in information science will have to be provided for scholars and students from other disciplines.

(c) Continuing assessment of manpower needs should be maintained. A unit to maintain such assessments should be included within the structure of the proposed new Institute in Information Science.

4. Status of the Information Science profession.

The need for improving the status of the profession of information science and technology is recognized by this group. It recommends that the following steps be taken to improve this image.

(1) The mass communication media must be used in order to bring the importance of this field to the attention of the public.

(2) Action directed towards policy makers in government, industry, international organizations and also academia should be taken, in order to alert them to the contributions that information science can make towards the functioning of these institutions. They must be made aware of the fact that the decision-making processes depend on the support they receive from information science personnel.

(3) Continued efforts must be made to extend the interaction of information scientists with colleagues in other disciplines to prevent isolation from the real world.

5. International co-operation.

(1) Steps should be taken for the creation of an international federation of educational institutions engaged in the education and training of information science personnel. We believe that the nucleus of such an organization exists in these halls.

(2) Measures are to be taken to facilitate the exchange of students, faculty, and research personnel on the international level.

(3) The feasibility of ensuring better international coverage of information science and technology activities for inclusion in the Annual Review of Information Science and Technology should be

explored to give it international quality.

(4) We view with favour the UNISIST approach to international co-operation in ensuring compatability amongst systems and development of standards for recording, transmitting and utilization of information data bases.

6. Implementation of recommendations.

A mechanism should be established, here and now, for ensuring that these recommendations are implemented.

Working group leaders: Jack Belzer
Frank Gibbons

Working group participants:

- R. Adams
- P. Atherton
- J. Belzer
- J. Brittain
- E. DeGrolier
- F. Gibbons
- A.B. Gupta
- J. Kesting
- I. Welt

Appendix A. Proposal to NATO for a Research Evaluation Conference on Information Science in 1974

Following on the two successful NATO Advanced Study Institutes in Information Science held in 1972 and 1973, we propose a Research Evaluation Conference of five days duration which would help identify the direction that research and development in the field of information science should take. A group of experts not to exceed 50 would be invited to consider the past, present, and future of information science.

The objective of this Information Science Research Evaluation Conference (ISREC) would be the following:

(1) Preparation of critical state-of-the-art reports on existing areas in Information Science and the identification of priorities for the next decade.

(2) Preparation of a set of recommendations which could guide the scholars in a proposed Center for the Advanced Study of Information Science. This might possibly set a specification of the most important research questions in the field or guidelines and priority areas for exploration.

(3) Presentation of proposals for the allocation of human and educational resources which could further the programs of the NATO Science Committee.

Attached to this proposal is a list of names which is typical of the group we intend to convene.

Planning Procedures for ISREC

(1) Establish a three- to five-person organizing committee to draft the invitational letter and statement of responsibility for the critical state-of-the-art reports by invited experts.

(2) No later than two months before the conference circulate each of the reports to two or three other invited experts who would be the Rapporteurs, writing critiques and presenting a more balanced picture to the Conference.

(3) At the conference the state-of-the-art reports and the critiques will be reviewed by all participants. Based on this, recommendations will be made for further action.

Appendix B. "Straw Vote"

Working Group 4 - Call for Assistance

HELP US GO TOWARD AN INTERNATIONAL INSTITUTE FOR INFORMATION SCIENCE (IIIS)

Introduction

1. In Working Group No. 4 (on Professional Aspects of Information Science) we have been considering one of the topics Frank Gibbons raised in his Tutorial, *viz:*

> the need for a permanent international institute, the major objectives of which would be to support research and teaching that would further the development of information science as a discipline.

2. As an intermediate step it was suggested that we should consider the convening of a meeting (perhaps of two or three days' duration) in order to further this cause. Dr. Kester suggested that NATO would perhaps support such a meeting if presented with an appropriate proposal within the next month.

3. The objectives of the meeting should be:

(1) agree upon a set of objectives of the Institute;
(2) establish criteria for location of an Institute;
(3) propose an administrative structure;
(4) evolve a set of criteria for selection of scholars.

4. One of the first practical problems would be to decide who should be invited to attend such a meeting for they could conceivably be considering "the founding fathers of the discipline of Information Science".

5. In our Working Group we tried listing the names of three persons whom we judged would be suitable. We wanted to find out if there was any agreement amongst the members of our small group. We then tallied the results from eight persons and found that there was considerable agreement about three names. We should not disclose these names until later in case it influences you in the small task we would like you to do.

6. We would now like to find out if this Institute's participants would also come up with a fairly small number of names chosen frequently and a large number of names chosen only *once* or *twice*. It would indeed be interesting if the names at the top of the list were the same as we produced in our Working Group.

PLEASE WRITE DOWN THE NAMES OF FIVE PERSONS, FROM ANY FIELD, WHOM

YOU CONSIDER, BY NATURE OF THEIR EXPERIENCE, LOCATION, STANDING, PERSONALITY (OR OTHER CHARACTERISTICS), SHOULD BE INVITED TO SUCH A MEETING WHICH HAS AS ITS OBJECTIVES THOSE LISTED IN 3.

In making your choices please bear in mind the following:

(a) As the Institute would be International you can consider persons from any country.

(b) It is expected that one of the purposes of such a meeting would be to select a much smaller group of persons who would then be charged with the initial states of the development of the institute, including the negotiation with appropriated bodies for funds to ensure the founding and subsequent existence.

In response to this call, the following are the results of the "straw vote" the selection of a committee to organize the NATO RESEARCH EVALUATION CONFERENCE IN INFORMATION SCIENCE.

This vote is presented in toto with the idea that a selection can be made for those who will organize this conference as well as suggest possible participants.

Name	Number of Votes
Debons	23
Licklider	16
DeGrolier	14
Cameron	9
Brookes	8
Watanabe	8
Atherton	6
Bar Hillel	5
Barnes	4
Belzer	4
Gibbons	4
Friedman	3
Mitroff	3
De Solla Price	3
Samuelson	2
Fairthorne	2
Landau	2
Monasterio	2
Nevitt	2
Otten	2
Saunders	2
Smith, H.	2
Welt	2
White, H.	2

The following received one vote: Alvum-Costa, Bhattacharyya, Brittain, Ceccato, Burks, Churchman, Cleverdon, Day, M. Dejider, Foskett, Gardin, Garfield, Gorn, Griffiths, Haberman, Hemming, Kabrisky, Kochen, Lowrey, Harmon, Herner, Martell, Martyn, Marzocco, Neelameghan, Perrot, Saracevic, Shera, Svenonius, Tobler, Tarski, Unger, Vickery.

Appendix C. Proposed Memorandum aimed at identifying research projects of interest to information science

Two NATO Advanced Study Institutes, the first at Seven Springs, in Pennsylvania, USA, in 1972, and the second at Aberystyth, Wales, in August, 1973, have now been held in order to bring together representatives of many countries and disciplines with a view to establishing the foundations of a study of the science of information.

Information is as necessary as oxygen to the survival of the human organism, but, until recently, no systematic attempt has been made to study the phenomenon called information in all its aspects which involve the generation, transmission, dissemination and utilization of information.

A number of disciplines and professions are making important contributions to a study of this subject such as psychology, philosophy, mathematics, computer science, linguistics, cybernetics, systems analysis, operations research, sociology, library science, and management sciences, areas with which we have been concerned in the NATO Advanced Study Institutes in Information Science.

We believe that a great deal of relevant research in this area is probably already in progress and that a most useful purpose would be served by identifying institutions and individuals who are involved in such activities. To accomplish this aim we are circulating this memorandum to learned societies, professional organizations, academic institutions, and similar research groups throughout the world. We hope to obtain details of all research work which is connected in any way with the themes mentioned above.

CONCLUSIONS

E. DeGrolier, S. Watanabe, Y. Bar-Hillel, and J.C.R. Licklider

E. DeGROLIER

Well, only one observation on the Report of Group 2. I think that in the list of the various functions given in this report, it might be useful to add one, which is, that if you wish a subdivision of what is called social interaction in the report, it would be good to specify that the system has to consider the human being as a citizen. That is, as in his communication with the political parallel, the political system, he has to be informed, better informed, on what is going on in the state, government and international affairs, and so on, and I suspect that the reporter of Group 2 more or less agrees with me on this. Of course, it is included, if you want, in the social interaction but I believe it could be useful to specify these needs of human beings as political citizens, political man. You know that President DeGaulle said once that man is a political animal, well Plato had already said that and it was taken over by DeGaulle! It is true, man is a political animal, and he has special needs as a political animal and it is possible that this side of his needs has been neglected up until now. So it could be interesting to ask if you have recreational needs, educational needs, health needs, and so on, but I think that political needs is something special, specific, in that it can be good to indicate how human we are. Apart from that I have no further comments to make.

Cameron: Would you not be prepared to comment on the Institute as a whole as you were there last year?

DeGrolier: Oh yes! I was. Well, anyway, it was a useful one. Maybe otherwise I would have left before this. I think that compared to the first, it has been a bit more scientific as we had at least four or five presentations which were based on a real scientific approach. I will cite only one - Brookes' - for this side of the Atlantic, and one for the other side of the Atlantic, well Garvey-Griffith's as they acted more or less as a team. I think that this approach is a quantitative approach. It is probably the best one and the one to which in the future even more weight can be given. Then I think that our group 4 has at least one good idea. The others were also good but the idea of this International Institute is something which to me is very important because I don't think that Institutes of the present kind which exist only for a fortnight or two and then disappear up in the mist after the end of the year, though they are useful of course, are a substitute for real Institutes, that is something permanent. The second comment I would give is that, in my opinion, it would be extremely important to set up, apart from the Institute which is recommended in the fourth group some links on the professional level: scientific professional. We speak of military - industrial complexes, understand. I would speak of scientific - professional complexes, uniting research people and professional people engaged in management of the systems. You have this in America, you have ASIS, which functions more or less as the link between the research people and the professional people. In Europe, on the contrary, we have only professional groups, scattered on the national level only, not on the European level and I believe strongly it is time now to organize something like the ASIS for European countries. And this is my second comment. This Institute and the preceding one has put in contact a number of people from different countries of Europe with a number of Americans. There are two ways of doing that. There is one way which Dr. White suggested, that we establish chapter of ASIS in Europe. This I consider as very difficult to achieve because of the sensitivities of European people who probably would not be happy with a solution in which they would be like a subsidiary of the American organization. The second way is to set up a mechanism for establishing a framework for a European society for Information Science which would be very closely connected with the American one, but as brothers are connected, not son and father. I believe that this has been done very successfully for experimental social psychology. The Social Science Council of the United States has been instrumental in provoking a joint meeting of a selected number of American experimental social psychologists and their European counterparts and this led to the establishment of a European society for Experimental Social Psychology, acting in very close contact with its American counterpart, but not a member of it. A partner on the stage of equality. I think that this could be done in Europe and that would be a very useful consequence of this Institute if we could do this, not on the professional level, but on the scientific level. If such a society were really acting to establish the foundation of Information Science as a science, with professional people involved, but with a strong, firm, experimental, and scientific objective, in the sense which Pauline Atherton mentioned one evening, that would be an extremely important result of this Institute.

S. WATANABE

First of all I would like to congratulate Dr. Debons and Dr. Cameron as well as other organizers who have brought this conference to a successful conclusion.

Among the things that happened here, I think that the talks, the lectures, and tutorials were very good, particularly because you gave enough time for each speaker and adequate time for discussion, because in most conferences you are given very little time. It was good particularly because the discussions were also very vivid, including the caustic remarks of Dr. Bar-Hillel, which were both very destructive and constructive, and very stimulating.

There is some discrepancy, some big gap between information scientists and those that came from the librarian side and those gaps should be more or less figured out somehow in the future. Some kind of common ground must be found.

Now as far as the working groups are concerned, I appreciate the efforts of the discussion leaders who wound up with some kind of report and I think they were very well written but as they took account of too many opinions the results turned out to be too banal, which is probably inevitable in a discussion of this sort. However, the procedure or the discussions which led to these conclusions were perhaps even more important than the final write-up for each group, because many problems were mentioned and many remarks, which tended to be important, were excellent stimuli for each of us and I think this is the unwritten results of this discussion or working groups. As far as I am concerned, I am satisfied with one thing and not satisfied with another thing. But I am satisfied that that popular word which dominated last week's conference ("paradigm" in the sense that Kuhn has introduced) has practically disappeared from the conference. Another thing that I am not satisfied with in the reports is that they have not expressed one of my beliefs - that a major target of Information Science is to get rid of unnecessary or useless information. That aspect, to make information more succinct and important, was not mentioned as best as I could hear.

Finally, I wish success to the future of this institute. Personally I wish to see an international Institute of Information Science be formed and work, as DeGrolier has said, on an international basis - not America or Europe, but also Japan, Formosa, or China probably - and that would be a worthwhile enterprise.

Y. BAR-HILLEL

I came to this institute to find out if the seven years that have passed since I last dealt actively with the topic of information science or information sciences, or informatology, whether something substantially new has happened, or even to see if the old misunderstandings that have bewildered the field since the late 40's have been reduced to a considerable degree. It will by now surprise very few of you to hear that my impressions in this respect are rather unequivocal. The old misunderstandings and misconceptions, for instance, in regard to the relations between physical, selectional, and semantic entropy (where the same label is used for "different bottles" to quote Fairthorne), which I discussed about 18 years ago in a lengthy paper I find are still very much alive, though often covered up by so many partly new labels, which pile up confusion on confusion. Just to have labels of physical, syntactic, semantic, and pragmatic information, I'm afraid, as such, does nothing to clarify the situation, since there still exists no serious concept of physical information. There exists a very serious concept of physical entropy and some people believe that since physical entropy is a serious concept, and since entropy and information are somehow closely connected, there must therefore be something that should be called physical information. Maybe, but I don't know of it! I also don't know of the existence of the serious concept of biological information though I have attended two or three conferences during the last year dealing with biological information. I still don't know what those people were talking about! I know of no attempt to clarify what could possibly be meant by syntactic information. Selectional information has nothing to do with information in any interesting sense but only has to do with signal transmission. Now semantic information, which hopefully has something definitely to do with information in the ordinary sense, has nothing to do with transmission or communication. Pragmatic information finally, perhaps, could be explicated as relative semantic information in some interesting sense of relative, which however, has to be clarified. Semantic information is not something that can be generated. Selectional information is not something that can be used. So long as these very foundational issues are not clarified it is difficult to see how serious progress can be made on substantial matters.

I did not attend last year's institute and can only infer from the report I received on its proceedings, but I understand that the same lack of clarity of basic issues must have bedeviled the last institute as I'm afraid it did this one. Unfortunately, many participants in this, and probably also in the former institute, seem to believe that all one has to do to get rid of the well known equivocations is to provide definitions, not realizing that this process of providing definitions, if one can get to it at all (and not every process of application will wind up with the process of definition), is the last step of the process of clarification or explication; the prior step being the so-called clarification of the explicandum, to use Carnap's terminology. Since this step has by no means been accomplished either last year or this year, and some I'm afraid don't realize the vital importance of these considerations, I believe that we are still pretty well stuck in the foundational mud.

Library scientists, due to their professional preoccupation with classification, always had a tendency (from an inadequate sampling of my own) to develop some encompassing schemes of systematizing human knowledge, starting with the very great librarian Leibnitz. Let me submit that the library scientists are not really qualified to do this particular job and they would do better to spend their valuable time otherwise. This schema of an all-encompassing classification of human knowledge in the 1970's seems to me to have a Don Quixotic aura anyway! So the so-called experts I'm afraid will be able to make little headway and I think library scientists should deal with what they are qualified for.

Now as the last negative aspect - I will come to the positive ones soon: The two working groups in which I have participated, have performed much worse than is reflected in the very able summaries. I think that the two summarizers have been able, out of their own understanding, to add a great amount of importance to those so-called summaries. Yet I think they should be treated as what they did rather than what the working groups did. What this means for the future is of some importance and I submit that Dr. Debons and Dr. Cameron should think about it; that this type of working group is really not fruitful and should be changed in some way. I won't make any suggestions now.

Now to the positive aspects. There are a number of them. First the extremely pleasant atmosphere both from a meteorological and from a human point of view. Secondly, the limitless devotion of the organizers and I shall not expand upon this because it has been so obvious. There have been a number of non-speculative, well-organized, well-substantiated, and well-documented presentations that enlarged my horizons and hopefully some of the others and which contributed important information for further research. Let me mention almost at random DeGrolier and Brookes. I think Professor Brookes has been able to set the stage for a new sub-branch of informatology, namely statistical bibliography. Statistical bibliography I think is going to stay with us and I think its presentation showed us that some important insights can be gained in this particular field.

Now fourthly and not lastly but perhaps most importantly - this conference or the participants in this conference have shown particular awareness of the humanity and social responsibility of the scientist, a realization that the well-informed man is not only more powerful but also has an obligation to take into consideration the potential misuses of this knowledge he creates or helps to disseminate; that we should not become slaves of a technology and so on. All of this terminology has been used many times on other occasions so it is well known to every one of you. What impressed me was that a good number of the participants are thinking in those terms and are trying to impress us that we should rethink our issues in those terms. I think they are right in principle, though it can easily be exaggerated however, and lead easily to a deterioration of the professional aspects, and therefore lead to a compromise between the human tendencies (not necessarily humanistic) of any field of science and its professional knowledge-expanding aspect. The latter has to be found but not necessarily by us. I'm not sure if we are prepared or trained to do this.

Now to the recommendations. I think that we are still as much in need of a clarification of the foundational, basic aspect of information sciences which I assume was an objective of last year's institute. I am not sure that we made any serious headway in this respect, the major question still being whether, in the various things that we dealt with under the heading of information science, there is sufficient coherence, or to use Wittgenstein's term, whether there is a sufficient family resemblance between them to warrant a departmentalization under a common roof. So I am not sure that information science is a going concern. The term exists but I'm not at all sure if it covers something substantial. I could easily identify dozens of important needs, dozens of important occupations, a good number of largely independent concepts that are connected with the topic of information science but I'm still far from convinced that this particular departmentalization is opportune. I suggest that either in the next institute or on some other occasion, the particular issue should be dealt with, i.e., whether there is a sufficient reason for trying to bind these things together into such a new department and whether it can be justified without prejudging the outcome.

Finally, as a citizen of a non-NATO country let me therefore take my recommendation with a large grain of salt. I would think that there is no good reason for the field of information science to be treated by NATO or some other specialized organization. I am convinced that this field has to be treated on a truly international basis with financial support coming from NATO, Warsaw pact, Japan, SEATO ... I don't care, but the field could only suffer from being reduced from a really international basis to some subpart of it. So I propose that NATO continue to support these investigations but will not restrict them to NATO as such. But as you see from my participation as well as by a number of others that it was not restricted to NATO citizens. Therefore, all together, the NATO label is really not appropriate and should be replaced by something less particularistic as some of my sociological friends would put it.

J.C.R. LICKLIDER

After these three masterful statements the only clear observation that I can make is that I see only one thing left that I think needs to be said. I don't believe I am best equipped to say it, but I would like to try to express some thoughts that have organized themselves in my mind these two weeks.

The keynote, I believe, just evolved during the first two days. It bore within it a little opposition, a little discord. The keynote had to do with the diversity of the world views of the participants in this meeting and of people in the field of information science. It had to do with the concept that was most frequently voiced under the term, "paradigm" and "ideology" and it seemed that, that would cause trouble as a thing opposed to Tony Debons' strong desire to have an integration, to have a forward movement in the organized formulation and understanding of the field. I think that after two weeks, I now see that there wasn't a fundamental discord there, but actually, the whole thing is understandable in terms of what is going on in the large world outside and that bears on the future of this movement. It seems to me that it is an emphatic fact that we are diverse as to a world view, more diverse here than we are back home in the United States where these last four years we have seen that we have a rather strong, if not diverse polarization. I think that the very important, basic, business of any international cooperative organization is to make some progress in understanding and then hopefully, to some extent, in reconciling those world views. I think that this is probably necessary and I believe that in this meeting it was just a fact that the forward motion of organizing and integrating has to wait for what one might say from an unsympathetic point of view is a lot of milling about and talking about what we basically believe. But again from a sympathetic point of view one has to say that this is the basic business of international cooperation, and accept that NATO is in some sense evolving from a kind of military treaty into an organization that is seeking to promote international cooperation on a broader level.

I haven't said much yet, but I'm trying to express the feeling that it is important to understand our diverse world views and I would like to relate that feeling to the other sentiment that I have had (and I guess I can say this without any presumption of arrogance) that during the last 25 years, a lot of the energy and drive and funding for things internationally scientific have come from the United States. In the present mood in the United States, I think there is a kind of insistence that a meeting like this be tangibly productive. That it make some forward progress in terms of the established paradigm which is challenged, but is still strong. I think in the United States there's a tendency to say, "Well if a group like this doesn't really make such a forward motion it hasn't done a very good job". But I see in Europe (inclusive of course of the area in which we're having this meeting) and in Japan and in other parts of the world, a surge of energy and of enthusiasm for moving in science and technology and information that presents an opportunity that is much needed - an opportunity to start with a due deliberation on these matters of paradigm and to take a longer term view and be more patient about productive, tangible progress in a given direction that is assumed to be good. This is related (and I've heard some interesting words from Dr. Gibbons about this and perhaps he might want to talk about it) to the energizing and funding of the proposed institute. I think it is important that a lot of it come from other sources than the U.S. If it does, it will make it easier for the U.S. to get back into the game, so to speak. So I see that as the serious part of what we've done and I'm truthful in saying that I didn't realize that was what we were doing until very near the end of the meeting. As seen from some established points of view, it wasn't a very good meeting. I would agree with Bar-Hillel's statement that there were some papers that were, as individual papers, very good, but they were kind of diverse and didn't really fit. I humorously suggested in working group 1 that the co-chairmen should not write a representative report, they should write a good report and Bar-Hillel suggested that that happened in others. But my point is that it's not important to evaluate this thing from that pre-established, not-often-enough-critized or considered, point of view. We're really dealing with something more basic. Having said all that, I think I can skip specifics (I had several) and come to my appreciation of the persistent strength of the drive exhibited by Tony, exhibited by Barry Nevitt, exhibited by the co-chairmen of working group 1 (and I believe by others too that I haven't seen as closely) to produce some kind of organizing synthesis, despite the kind of keynote diversity, and make something that I think is going to turn out to be, as judged by even the most critical bias, a pretty good report. Tony, I think you're in better shape this time, to have a good report than you were the last time. You slaved, you worked very hard the last time to get a good report out of the last one; this one you won't have to work quite so hard, but I hope you will work very hard, because it's important to have success even on that measure. I think there is no question about success in terms of this other measure that I've been trying to describe here.

SUMMARY

William J. Cameron and Anthony Debons

WILLIAM J. CAMERON

I suspect my comment is going to be impressionistic, really. I got the impression in comparing the institute of last year with this year, that to a certain extent, that business of "being lost in the clouds of unknowing" was partly a result of people going back over old ground, to see where we were at. The image that occurred to me was that of a runner or jumper who wants to make a great leap forward. Obviously, when Tony introduced what he wanted out of this institute, it was a great leap forward, if we could manage it. But it seemed to me that the leap was not possible until one would go back and measure the ground over again and look at where we are.

I was very impressed by the fact that Kuhn's paradigms never came up last year - not once. Yet it has dominated a great deal of the present discussion this time. It seems to me that in going back over this ground, it is a scramble in the fundamental mud, if you like, but it provides a real possibility of moving forward - at least recognizing how muddy the track is, before we take that great leap, whether it's toward unification, or to a family resemblance, between different portions of the discipline of information science. I at least get a very optimistic picture from the discussions. To coin a phrase, I think we are concerned much more with the ground than the figure. But the various figures that have come out very clearly, Mr. Brookes' presentation and others, I think only take significance from the attempt to survey that ground both internationally and interdisciplinary. So that my fundamental impression is that Tony will have to do some work, yes, but that he has some very good ground to work from.

ANTHONY DEBONS

Well, I believe my comments will be colored by my background and interests in this institute. First of all, I think the major thing that I would like to express is that I don't wish to suggest any pretentiousness on my part to understand the depth of this problem beyond what my colleagues and the working groups have brought forth in this meeting. The level of effort, their concentration, their conscientiousness, all of it are so formidable, that anything that I can add to it at this point would be pretentious. But I do want to comment on a technical detail or two briefly enough to permit an 11:30 closing time so as to get you to your respective ways of travel. First of all, I think Dr. Licklider was right in saying that there was expressed a drive for integration, for developing something that has meaning and perhaps I should explain what that motivation is all about. I believe the motivation is ingrained in the recognition of the existence of real problems of information for which people are seeking solutions. They are confronted with day-to-day problems with information and they're seeking information scientists to lead them to solutions. Someone has to formulate their problems and their respective possible approaches to their ultimate solution. It's true that from a scientific view we have to be extremely critical of ourselves. We are obliged to be very careful in our definitions. We need to be conservative in what we can contribute, but in essence, these constraints need to be realized in terms of constraints faced by people who are really confronted with information problems, day in and day out. They need solutions.

The second basic motivation underlying this institute relates to education. All of us have a direct responsibility for educating those who have made a commitment to the study of information. Students of information are seeking, not muddled, unclear, hypothetical assessments, but rather they are seeking scholarly leadership for definition and clarity in concepts. Individuals going into this field should not be subjected to ambiguity without reason. Consequently I have felt a sense of responsibility to do something in this area, to provide some sort of framework upon which these motivations could be in fact satisfied.

Have these motivations been satisfied in any measurable way by this institute? I think that the answer to that is probably implied in the ability of the students of information in the future to see meaning and definition in what we all have said. Hopefully it is these individuals who will be able to express our position with a positive perspective 20, 30, or 40 years from now. The last point I wish to make is that the concept of diversity is a measure of reinforcement to the attempts of this institute. I consider diversity in our points of view is not a negative view, rather a positive. I would not like to disturb this in any way. I want the proceedings of this institute to represent a challenge to our young people to provide the stability for (and I want them to become a member of the community, as was suggested by Dr. DeGrolier, and others) the community of international people, where they can, in effect, perceive themselves, not in the national cast but in an international cast where the cultural dimensions of a problem become real to them. I would like the proceedings of this institute to represent a challenge to those whose interest is the study of information. The proceedings being a reflection, not of national interest, but rather of international spirit should provide the future scholars of information with a sense of the immensity of information problems. In the study of information they are in the grasp of a powerful phenomenon - a phenomenon encompassing all of the human dimensions, as well as those dimensions that relate to his quest for understanding himself - his technology and his value systems.

I think this has been a superb experience for me. As a result of the first and second institutes, I can say that I owe much of my understanding of information to my distinguished colleagues. All of you have given me a new dimension on how to proceed and I am obliged to do everything in my power to continue our quest for understanding in the future.

AFTERTHOUGHT

William J. Cameron and Anthony Debons

The institute has provided some distinct impressions that the editors would like to share.

It is interesting that some of our speakers talked about information as like energy. Indeed information can be likened to energy for the matter of this Institute. Like water, information is all about us and man requires information as much as water for survival. Yet, like water, information needs to be marshalled and used discriminately. It would seem that the concerns of the speakers and participants were in that direction. Like all forms of energy, information can be dissipated and wasted. It would appear that man possesses the marshalling mechanisms, yet the distributing and using channels are not clear. With significant advances in technology, it seems apparent that greater efficiency can be realized in proper marshalling and distribution; but whether or not these abilities really represent potentially significant advances to man in dealing with the environment still needs to be pondered systematically.

Man's great temptation (sometimes born of necessity) is to perceive energy as something to be controlled. It is through science that the hope for the proper use and direction of energy is to be achieved, for science and technology are deemed rational mechanisms for this purpose. If nothing else has been established, the proceedings made emphatically clear the importance of the human aspect of information. Whether the debate was toward clarity in perceptions about information, or whether it was in consideration of tools that enable us to talk to and understand each other - the foremost thought was the importance of man. Over and over again, a convincing plea was made for a balance between our craving and quest for objectivity in understanding information in one hand, and on the other, a recognition of the power as well as role of human intuition in the shaping of a science intended to deal with information.

The hope that information systems can provide a way for man's ability to deal with vast increases in knowledge is clear from the proceedings. What is also clear is that information systems will influence the human condition in some way. The significant aspect of the proceedings is the overwhelming concern as to what these effects could be. Perhaps a science of information could tell us. Indeed it may, but a caveat is offered. Let us be skeptical of the approach that we can find the science by the application of the customary measuring instruments and looking glasses. Then there is also the problem with attitude. For those of us who seek order in science, the possibility should be considered that a science of information may require less order and more flexibility than we have been accustomed to in our thinking about science. If the requirement for order should still persist, it should be an order impelled by clarity rather than conformity.

Both in the Seven Springs institute and in the present one, the theme or keynote was a striving, a search for identity and, perhaps, unity. Both institutes revealed the complexity of issues, the morass of belief systems. The wide range of problems, both theoretical as well as practical, the uncertainties in approaches for resolving them, and in general the heterogeneous character of the information experience was well portrayed by the speakers and participants. Yet, this very complexity provides a source of identity for the science we are attempting to discover. Perhaps it is not a singular science but rather a confederation, an aggregate, of sciences, comprising individuals of varying scientific, social, and philosophical interests and predispositions who apply such proclivities to the understanding of a human problem. As such, the search is not for a new science. Rather, the quest is for a formula of a meta-experience, interdisciplinary in scope, that provides a base for emerging new identities - identities that are directed to complex human problems.

The heterogeneity of subjects and approaches, the variety of attitudes so explicit in the deliberations of the institute should not lead to a perception of hopelessness - and the conclusion that nothing has indeed been achieved. In this contemplation, an analogy may be advanced. The analogy is that of wine: its production and human consumption. As we all know, the pleasure of wine derives from aging, given the proper environmental conditions which in combination produce a liquid rich in texture and pleasant in taste. Similarly, in science the ingredients of knowledge are a mixture of rationality, objectivity, and intuition; essentially human observation and experience. Left to ripen, these ingredients under the appropriate conditions (as history has taught us), provide the texture for an emerging wisdom. Our suspicion obtained from our review of the collective assembly of thoughts, observations, debate, and speculation is that the ingredients for a science of information may be indeed embedded in the two institutes. Hopefully, the fermenting process has been initiated. Whether or not anything has been accomplished must be left - as with wine - to time. Only time and those who will taste the wine will tell. The fineness of the wine we should add is in the making and not in the color of the bottle, the cork, or for that matter in the label.

INDEX

* The following contractions are used:

Def - Definition
irt - in relation to
qirt - Quoted *irt*

PARTICIPANTS IN THE NATO INSTITUTE (1973)

Adam, Ralph
Alvun-Costa, Maria Rosa
Amoroso, Serafino
Atherton, Pauline
Avivi, E.
Baker, Lynn E.
Bar-Hillel, Y.
Barnes, Robert
Belzer, Jack
Bernier, Roger
Bouthillette, Jean
Brittain, Michael
Brookes, B.C.
Burghraeve, Paula
Cameron, William J.
Campos, F.L. Borges
Cetincelik, Muammer
Conway, Derek
Costa, Fernando Vasco
Debons, Anthony
DeGrolier, Eric
DeMey, Marc
Enslow, Philip H.
Fairthorne, R.A.
Finnerty, James
Fitzgerald, Inez
Friedman, Lee
Garvey, William
Gibbons, Frank
Gramignani, Mario
Griffith, Belver C.
Gupta, Anand B.
Harmon, Glynn
Hogberg, Lars G.
Horne, Esther E.
Irvine, James
Jabbour, Fouad
Janes, Donna Jean
Keen, E.M.
Kester, Tilo
Kesting, Jean G.
Klempner, Irving M.
Kochen, Manfred
Lamont, Valarie
Landau, Robert
Leighton, Jeremy
Licklider, C.R. (Joseph)
Lima, Maria
Lima, Pablo
Maher, Frank
Martell, John
Marzocco, Frank
Meadow, Charles T.
Meincke, Peter
Minder, Thomas
Mitroff, Ian
Monasterio, Xavier O.
Mullin, James
Nevitt, Barrington
Nicolis, John
Otten, Klaus
Rathswohl, Eugene
Rayward, Boyd
Rosenberg, Victor
Royce, Joseph R.
Sackman, Harold
Salancik, Gerald R.
Samuelson, Kjell
Saracevic, Tefko
Simpson, Ian Stuart
Smith, Edward
Smith, Hugh
Steinacker, Ivo
Strasser, Alexander
Suttle, Jimmie
Taulbee, Orrin E.
Triggs, Thomas J.
Unger, George
Watanabe, Michael Satosi
Wellisch, Hans
Welt, Isaac D.
White, Herbert S.
Wooster, Harold

GPSR Compliance

The European Union's (EU) General Product Safety Regulation (GPSR) is a set of rules that requires consumer products to be safe and our obligations to ensure this.

If you have any concerns about our products, you can contact us on ProductSafety@springernature.com

In case Publisher is established outside the EU, the EU authorized representative is:

Springer Nature Customer Service Center GmbH
Europaplatz 3
69115 Heidelberg, Germany

Batch number: 10824463

Printed by Printforce, the Netherlands